Metal Organic Frameworks for Wastewater Contaminant Removal

Metal Organic Frameworks for Wastewater Contaminant Removal

Edited by

Arun Lal Srivastav
Chitkara University

Lata Rani
Chitkara University

Jyotsna Kaushal
Chitkara University

Tien Duc Pham
Vietnam National University

WILEY VCH

The Editors

Arun Lal Srivastav
Chitkara Universiy
Himachal Pradesh
India

Lata Rani
Chitkara Universiy
Himachal Pradesh
India

Jyotsna Kaushal
Chitkara Universiy
Himachal Pradesh
India

Tien Duc Pham
Vietnam National University
Hanoi
Vietnam

Library of Congress Card No.:

British Library Cataloguing-in-Publication Data
A catalogue record for this book is available from the British Library.

Bibliographic information published by the Deutsche Nationalbibliothek
The Deutsche Nationalbibliothek lists this publication in the Deutsche Nationalbibliografie; detailed bibliographic data are available on the Internet at http://dnb.d-nb.de.

Print ISBN: 9783527351923
ePDF ISBN: 9783527841547
ePub ISBN: 9783527841530
oBook ISBN: 9783527841523

Cover Design: Schulz Grafik-Design, Fußgönheim, Germany

Typesetting: Set in 9.5/12.5pt STIXTwoText by Integra Software Services Pvt. Ltd, Pondicherry, India

Printing and Binding: Bell & Bain

Printed on acid paper

Contents

Preface

Water pollution is a major problem throughout the world. Water pollution is caused by organic inorganic, biological, and radioactive contaminants. Various anthropogenic activities such as agricultural, industrial, and urbanization, attributed to an increase in the level of contaminants in water bodies. Among various techniques, adsorption is having additional benefits like low cost, environment friendly and low chance of by-products generation.

Metal-organic frameworks (MOFs) adsorbents have brought a revolution in the area of wastewater treatment as they have possessed unique characteristics like high surface area, large pore size, high selectivity, and high contaminants removal efficiency. These properties are unique over the traditional adsorbents. Present book discuses about the application of MOFs for the removal of emerging water contaminants such as antibiotics, endocrine disruptors, pharmaceuticals, microplastics, pesticides etc. Moreover, preparation methods of MOFs are also described in details along with their physico-chemical properties. Some case studies are also included in the book so that it can motivate the researchers for better learning experiences.

Hence, this book may be a mile stone in the field of water treatment as it will provide better insights to the research community.

July 2023

Arun Lal Srivastav, India
Lata Rani, India
Jyotsna Kaushal, India
Tien Duc Pham, Vietnam

1

Application of MOFs on Removal of Emerging Water Contaminants

Nguyen Minh Viet[1], Tran Thi Viet Ha[2], and Nguyen Le Minh Tri[3,4]

[1] *VNU Key Laboratory of Advanced Material for Green Growth, Faculty of Chemistry, VNU University of Science, 334 Nguyen Trai Street, Thanh Xuan, Hanoi, Viet Nam*
[2] *Faculty of Advanced Technologies and Engineering, VNU Vietnam – Japan University, Luu Huu Phuoc Road, Cau Dien Residential Area, Nam Tu Liem District, Ha Noi, Viet Nam*
[3] *Laboratory of Advanced Materials Chemistry, Advanced Institute of Materials Science, Ton Duc Thang University, 758307, Ho Chi Minh City, Vietnam*
[4] *Faculty of Applied Sciences, Ton Duc Thang University, 758307, Ho Chi Minh City, Vietnam*

Abbreviated list

AC	**Activated carbon**
AOPs	**Advanced oxidation processes**
FO	**Forward osmosis**
MF	**Microfiltration**
MOFs	**Metal–organic frameworks**
NF	**Nanofiltration**
PPCPs	**Pharmaceuticals and personal care products**
RO	**Reverse osmosis**
UF	**Ultrafiltration**
WWTPs	**Wastewater treatment plants**

1.1 Introduction

1.1.1 Sources of Emerging Water Contaminants

Water is an essential factor that can define the properties of all living species, and there is no doubt that our Earth cannot exist without water. With the rapid development of human society and the economic sector, the types of contaminants are

Metal Organic Frameworks for Wastewater Contaminant Removal, First Edition. Edited by Arun Lal Srivastav, Lata Rani, Jyotsna Kaushal, and Tien Duc Pham.

becoming increasingly hard to control. One of the particular concerns is called "emerging contaminants".

Emerging contaminants can be defined as contaminants that have new sources and new pathways to the human body, with novel adverse effects on human health, the environment, and the ecological system. Emerging contaminants are common chemical substances, and modern analytical techniques can detect them. They can be classified depending on their nature, origin, potential effects, or the possible fate of the contaminants. For example, the NORMAN database classified the emerging contaminant into 21 groups with over 700 counted compounds: PPCPs, organic dyes, herbicides and pesticides, industrial compounds/by-products, and nanomaterials [1–5]. These contaminants have been dispersed in wastewater and sewage water via different routes, and lots of them can even be found where they have never been used. Emerging contaminants enter the water environment from different sources, and the origin of the emerging contaminants in the water is often difficult to detect.

One of the primary sources of emerging water contaminants is wastewater treatment plants (WWTPs). Humans use PPCPs, nanomaterials, and food additives in their daily life, and produce a tremendous amount of waste or wastewater, then wastewater moves to WWTPs. It should be mentioned that most emerging contaminants are outside the scope of regulation, or they are not regulated. Besides, current WWTPs are not designed to treat emerging contaminants due to their low concentration. The biosolids obtained after WWTPs are another source of these contaminants.

Other remarkable sources of emerging water contaminants are agriculture and livestock. In modern agriculture, pesticides, herbicides, antibiotics, hormones, etc., are used frequently, introducing the residue chemicals to the environment, including air, soil, and water. The contaminants can be transported between different phases of the environment by spreading, leaching, and runoff.

The landfill, which is the final destination of waste from everywhere, is another source of emerging contaminants. The components of waste are very complex. The leachate from landfill sites consists of emerging contaminants that can come into the groundwater, streams, or WWTPs, leading to adverse environmental and human health effects.

1.1.2 Emerging Water Contaminants Treatment Methods

Identifying the treatment methods for removing emerging contaminants in the water is a priority that must be concentrated on the best practices for ensuring the use of safe water for the community. Recently, many studies have investigated new processes to eliminate emerging contaminants from water, including membrane technology, adsorption, biodegradation, advanced oxidation processes (AOPs), and so on (Table 1.1).

i) Membrane technology

The membrane process is a phase-changing process that can be applied for emerging water contaminant treatment. The membrane can be made from many different materials (e.g., carbon nanotubes, polyurethane, polysulfone, and polyvinylidene), and the contaminant that can be retained or passed through can be decided based

Table 1.1 Summary of advanced treatment technologies for organic pollutant removal.

Technology	Target emerging contaminant	Removal efficiency	Note	Ref.
1) Membrane				
a) IF				
Polysulfone- and polyvinylidene UF membranes	Bisphenol A (BPA) and 17β-estradiol (E2)	75–98%	Polysulfone from Koch Membrane Systems	[6]
b) NF				
NF-200 and NF-90	Acetaminophen	18–81%		[7]
	Phenacetin	70–78%		
	Caffeine	62–93%		
c) FO				
FO: Hydration Innovations (HTI, Albany)	1,4-dioxane	55–68%		[8]
	Acetaminophen	45–89%		
	Metronidazole	70–99%		
	Phenazone	85–99%		
	Caffeine	80–99%		
	Bisphenol A	40–99%		
	Carbamazepine	65–99%		
	17α-ethinylestradiol	85–99%		
d) RO				
RO: Aromatic polyamide membrane (Midland, MI)	Ibuprofen	90–99%		[8]
	Naproxen	95–99%		
	Fenoprofen	95–99%		
	Gemfibrozil	95–99%		
	Ketoprofen	95–99%		
	Acetaminophen	100%		
2) Adsorption				
a) Activated carbon				
Commercial AC (PB 170, PB 170–400, PC 1000, WP 235, Carbsorb 28, Cyclecarb 305, W 35, SA Super, LP 39, MP 25 and Hydro XP 17)	Atrazine, Diuron (pesticides)	40–75%	Wastewater with an average DOC of 5.6 ± 0.9 mgC/l, C_0 of 6 and 57 ng/l, AC doses of 10 mg/l	[9]
Polymer-based spherical activated carbon particles	Estradiol (hormone)	96–99%	In combination with UF	[10]

(Continued)

Table 1.1 (Continued)

Technology	Target emerging contaminant	Removal efficiency	Note	Ref.
Granular activated carbon	NO_3^--N Metoprolol and diclofenac (PPCPs)	95% for NO_3^--N 80% for PPCPs	Integrate with solid-phase denitrification (SPD) with biodegradable polymer poly-3-hydroxybutyrate-co-hydroxyvalerate (PHBV)	[11]
b) Biochar				
Tea-waste biochar	Caffeine in aqueous media (PPCPs)	15.4 mg/g^{-1}	Pyrolyzed at 700°C using steam activation (TWBC-SA)	
Magnetic porous biochar obtained from a biomass garlic skin	Paraben, metronidazole, and oxytetracycline (PPCPs)	415, 287, and 822 mg/g^{-1} for paraben, metronidazole, and oxytetracycline, respectively		[12]
Biochar (BC) and biomass carbon quantum dots (CQDs) from reed straw	Carbamazepine (PPCPs)	96.43%	BC and CQDs for the modification $MgIn_2S_4$/BiOCl (MB) heterojunction photocatalyst with Z-scheme structure	[13]
c) CNTs				
Multiwall carbon nanotube (MWCNT)	Ciprofloxacin hydrochloride (PPCPs)	1.7446 mg·g^{-1} or >88%	Initial ciprofloxacin hydrochloride concentration of 4 ppm	[14]
Commercial multi-walled carbon nanotubes	Chloramphenicol, thiamphenicol, florfenicol, sulfadiazine, sulfapyridine, sulfamethoxazole, sulfathiazole, sulfamerazine, sulfamethzine, sulfaquinoxaline, ibuprofen, carbamazepine and diclofenac	Freundlich constant (KF) values were 353–2814 $mmol^{1-n}$ Ln/kg, 571–618 $mmol^{1-n}$ Ln/kg, and 317–1522 $mmol^{1-n}$ Ln/kg for sulfonamides, chloramphenicol, and non-antibiotic pharmaceuticals, respectively		[15]

Table 1.1 (Continued)

Technology	Target emerging contaminant	Removal efficiency	Note	Ref.
Multi-walled carbon nanotubes	Tetracycline antibiotics	253.38 mg/g^{-1}	CNTs were synthesized in the CH_4/H_2 mixture at 700 ^{0}C by a chemical vapor deposition method using Co-Mo particles as catalysts	[16]
3) Biodegradation				
Activated Sludge	Estrone 17β-Estradiol Estriol Estrone-3-Sulfate 17β-Ethinyl estradiol 4-Nonylphenol mono- and ethoxylated nonylphenol polyethoxylated nonylphenols	79% 0% 45% 36% 34% 0% 88% 66%	Aerobic and anaerobic WWTP in the UK	[17]
Soil Filtration	Estrogens 17β-Estradiol 17β-Ethinyl estradiol Triclosan Ibuprofen	26% 99% 27% 90% 18%	Aerobic	[18]
4) AOPs				
Photocatalyst ZnO rod in coupling with activated carbon fiber	Tetracycline	Removal capacity of over 99%	The initial concentration of 40 mg/L	[19]
N-Cu co-doped TiO_2@CNTs	Sulfamethoxazole	93% of COD and 89% of TOC	Sono-photocatalytic degradation eliminated within 180 min	[20]
UV/H_2O_2 treatment	Meprobamate Carbamazepine Dilantin Atenolol Primidone Trimethoprim	> 90%		[21]

(Continued)

on the properties of pore size, surface charge, hydrophobicity, oleophobicity, etc., of the membrane. Membrane technology can be classified into microfiltration (MF), ultrafiltration (UF), nanofiltration (NF), forward osmosis (FO), and reverse osmosis (RO). MF cannot remove contaminants with a size lower than 1μm; therefore, it is not capable of removing emerging contaminants. UF, NF, FO, and RO have smaller pore sizes than MF, and they have been used to remove emerging contaminants in water. When the pore size decreases, emerging contaminant removal efficiency improves significantly.

Due to pore blocking and pore restricting, membrane fouling may be the main obstacle to the application of the filtration process, which generally causes high costs for the operation and maintenance of the system.

ii) Adsorption

Similar to membrane technology, the adsorption process can transfer the contaminants from one phase (i.e., in water) to another phase (i.e., in solid), and it has been widely reported on for the removal of different emerging contaminants. Activated carbon (AC) is the most popular material for the conventional adsorption treatment of contaminants in water due to its high porosity and high surface area. AC can be made from different raw materials, such as wood and coconut shells AC's adsorption can be integrated with other treatment processes, such as UF, coagulation, and AOPs in order to improve the total removal capacity. Biochar is also an adsorption material that can be applied to emerging contaminants in water. Biochar is a charcoal-based material that is made by heating biomass via a pyrolysis process at high temperatures in conditions without oxygen. The feedstock for biochar synthesis may come from agriculture by-products such as straw, rice husk, and corn husk. Carbon nanotubes (CNTs) with very high surface area are also popular materials that can be applied to emerging water contaminants.

The structure of the adsorbent material plays a significant role in the process's efficiency, and it is required to count on the treatment parameter, such as pH and temperature, to achieve the best performance of the materials. Besides, the key disadvantage of this kind of material is related to the sustainability of the production process; scaling up can lead to high energy consumption, and the carbon footprint is too large to consider the biochar or AC as an effective material to treat environmental problems.

iii) Biodegradation

Depending on the type of pollutants, some biological processes can be applied in different conditions: aerobic and anaerobic. Activated sludge seems to be the most common, and other systems using biological filtration and membrane bioreactors have also been investigated. Removal levels can range from low to nearly complete removal, and biological processes may not eliminate some common biodegradable contaminants under any conditions. Another research direction developed recently is the combining of the biological process with the adsorption or electrical process.

It should be noted that the biggest disadvantage of this method is managing and treating the biosolid produced during and after application. Another area for improvement is the need for analytical methods to identify and quantify the compounds in a very complex matrix. However, it can also be an excellent opportunity for further research related to biodegradation for emerging contaminants in the water environment.

iv) AOPs

AOPs gained much attention from researchers because of their high capacity in the removal of pollutants in water as compared with other processes. Emerging contaminants in water can be treated using photocatalysis, UV photolysis, Fenton, sono-chemical oxidation, or ozone in combination with hydrogen peroxide. By the activities of free hydroxyl radicals (•OH), AOPs can help deal with complex and durable emerging pollutants when other processes seem not to work. AOPs are considered a highly effective, novel, and green process for environmental treatment in general and emerging contaminants in particular.

However, a gap exists when discussing AOPs, and it is about how to scale up the process and apply it to the industrial system.

1.1.3 MOFs as Exceptional Materials for Water Remediation

MOFs are porous structures prepared from metal ions and organic linkers or bridging ligands via the formation of coordination bonds. Recently, MOFs have been considered a promising material that can effectively deal with the issues related to water pollution due to (i) large surface area, (ii) stability in water, (iii) ease of functionalization, and (iv) capacity to produce on a large scale. Interestingly, MOFs can work as selective adsorbents, the catalytic material, and a combination of them. MOFs and the composite of MOFs with other components have been proven by several studies to successfully remove a wide range of contaminations, such as dye, PPCPs, pesticides, and herbicides [22–25].

The first research on MOF was published in 1995, opening a new direction for adsorption materials with a much better surface area, high porosity, and high accessibility to contact the adsorption site. MOFs also have advantageous properties compared to conventional porous materials because the porosity of MOFs can be turned through a reticular approach through inorganic nodes and organic linkers. MOFs are new for removing emerging contaminants in water, and there is an increased interest in it. The following sections of this chapter will discuss detailed strategies and examples for applying MOFs in water remediation.

1.2 MOFs Strategies in Water Remediation

Three important strategies related to MOFs have recently been applied to water remediation: (i) adsorption of emerging water contaminants by MOFs, (ii) catalytic oxidation of emerging water contaminants by MOFs, and (iii) a combination of adsorption and catalysis for the mitigation of emerging water contaminants by MOFs.

1.2.1 Adsorption

Compared to other water remediation technologies, which have many drawbacks such as low efficiency, high sludge production, and expensive disposal, adsorption is a popular, attractive, and versatile treatment method due to its effectiveness, economic benefits, simple and easy operation, and potential reversibility [26, 27]. Adsorption refers to the process in which molecular or ionic compounds in the liquid or gaseous phase are kept on the rigid surface of solids. Adsorption has been commonly used in environment-related fields such as sewage treatment and removal of toxic gases due to its low price and easy operation [28–31]. Various factors can strongly impact the efficiency of the adsorption process, including the fate and concentration of contaminants, the types and physicochemical characteristics of adsorbents, and operation parameters such as temperature, pH, and adsorption time. Some current adsorbents such as zeolite, silica, etc., face problems with slow adsorption rate and low adsorption capacity [32, 33]. Unlike these, MOFs possess a large surface area, high porosity, and varied pore properties, which makes it possible for them to become competitive adsorbents for the mitigation of different species such as organic dyes, heavy metals, PPCP, herbicides, pesticides, etc. [34–38]. For instance, Ji et al. developed a MOF-based carboxyl adsorbent (MIL-121) for the selective removal of Cu^{2+}, Pb^{2+}, and Ni^{2+} ions from aqueous solutions [39]. More than 99% of the heavy metals were removed from the water in the presence of co-existing 10 000 mg/L of Na^+ ions. When applied to a real electroplating wastewater sample, the adsorbent could produce 3000 mL of clean water per gram material. Similarly, selective adsorption of Cr^{3+} in a solution containing other co-existing ions such as Pb^{2+}, Cd^{2+}, Cu^{2+}, Ni^{2+}, and Cr^{3+} was reported on formic acid and amino-modified MOFs Form-UiO-66-NH_2 [40]. The modified MOFs were fabricated via formic acid modification and were used as adsorbents for selective adsorption of Cr^{6+}. After modification, the material exhibited a higher specific surface area (919 m^2/g), total pore volume (0.49 cm^3/g), and pore diameter (2.14 nm) than the unmodified material, whose specific surface area, total pore volume, and pore diameter were 879 m^2/g, 0.34 cm^3/g, and 1.86 nm, respectively. The good texture of Form-UiO-66-NH_2 enabled its outstanding Cr^{6+} adsorption capacity (338.98 mg/g), which was ten times higher than that of the unmodified MOFs [40].

The adsorption of Cr (VI) over MOFs was also reported by Valadi et al. [41]. In their study, they synthesized a variety of zirconium-based MOF and chitosan composites, namely MOF 808/chitosan, and used them as efficient adsorbents for homogenous adsorption of Cr (VI). The MOF 808/chitosan had the maximum adsorption capacity of 320 mg/g and could be reused after six successive adsorption cycles. For the remediation of Cu (II), Hg (II), As (III), and Pb (II), a magnetic Fe-BTC MOF composite was synthesized as an adsorbent by a solvothermal method [42]. The maximum adsorption capacity was found to be 55 mg/g, 57 mg/g, 147 mg/g, and 155 mg/g for Cu (II), As (III), Pb (II), and Hg (II), respectively. The Langmuir isotherm model was best fitted to the adsorption of Cu (II), Pb (II), and As (III), while the Temkin isotherm model was best fitted to that of Hg (II). The MOF composite also showed the selective adsorption of those heavy metals in their mixture.

In addition, many authors have reported using MOF-based materials as adsorbents for the removal of organic dyes. Ahmadijokani et al. used the zirconium-based MOF, namely UiO-66, to selectively adsorb different organic dyes, including cationic dyes (malachite green and methylene blue) and anionic dyes (methyl red and methyl orange) [43]. The adsorption capacities of UiO-66 for anionic dyes were higher than those for cationic dyes. The excellent adsorption capacities of UiO-66 were the consequences of its large specific surface area (1276 m^2/g) recorded on a gas adsorption system. In an attempt to control the specific surface area and charge of the MOF framework, Li et al. introduced a mixture of ligands containing 1,4-benzenedicarboxylic acid (BDC) and 1,2,4,5- benzenetetracarboxylic acid (BTC) with different stoichiometry in order to construct multivariate MOFs denoted as UiO-66-n$(COOH)_2$ MOFs ('n' represents the molar fraction of BTC, n = 0.25, 0.5, 0.75, 1.0) [44]. It can be deduced from their study that the decrease of specific surface area (from 1620 to 179 cm^3/g) and the zeta potential (from 36.11 to -29.78 mV) indicated that the surface charge of the zirconium-based MOFs changed from positively charged to negatively charged. Thereby, their adsorption behaviors toward interested dye molecules can be easily regulated.

While the pristine UiO-66 only showed its ability to adsorb anionic dyes, the multivariate MOFs could adsorb cationic dyes selectively. The adsorption selectivity of the MOFs was attributed to their high surface area and the electrostatic interaction between their charged surface and the dyes. With the introduction of Fe_3O_4 and covalent organic frameworks (COFs) to the structure of MOFs (M = Ti, Fe, Zr) via a facile solvothermal method, Wang et al. were able to fabricate magnetic and porous Fe_3O_4@MOF (M = Ti, Fe, Zr)@COFs hybrid composites with good recyclability for effective azo dye adsorption [45].

For the adsorptive removal of sewage containing a pharmaceutical-like sulfanilamide, Jia et al. developed a novel hierarchical porous starch-chitosan-UiO-66-COOH composite in which the UiO-66-COOH-type carboxylic zirconium metal-organic framework was nearly immobilized within porous starch using a one-pot chitosan-adhesive route [46]. Due to the high affinity between the Zr-O bond within UiO-66-COOH and the sulfonic atoms, sulfanilamide could adsorb on the composite. The adsorption obeyed the pseudo-second-order kinetic and Langmuir models and was considered homogeneous monolayer chemisorption.

By grafting UiO-66 into polymerized chitosan with the aid of epichlorohydrin as a linking agent, Chen et al. successfully prepared porous chitosan / UiO-66 composite foams [47]. The materials were successfully applied to the effective adsorption of ketoprofen. The maximum adsorption capacity of the porous chitosan / UiO-66 composite foams was 209.7 mg/g, which outweighed most reported adsorbents. Also, working on zirconium metal-organic frameworks, Zhao et al. could achieve a magnetic Zr-MOF structure by loading Fe_3O_4 nanoparticles on MOF-525 via a secondary-growth approach [48]. The material improved the removal of pharmaceuticals such as tetracycline and diclofenac sodium with maximum adsorption capacities of 745 and 277 mg/g for diclofenac sodium and tetracycline, respectively. Different kinds of interaction, such as hydrogen bonding, electrostatic interaction, π-π interaction and anion -π interaction, were found to be involved in the adsorption process. Similarly, electrostatic interaction, hydrogen bonding and π-π interaction also played major roles in the adsorption of two popular pharmaceuticals such as clofibric acid and carbamazepine on the rigid MIL-101(Cr) and flexible MIL-53(Cr) reported by Gao et al. [49].

When it comes to the removal of herbicides and pesticides by MOFs, several studies have reported them. For example, a recent study discussed the ability of a new hollow NiO/Co@C magnetic MOF to be used as a solid-phase extraction adsorbent for the effective adsorption of organic nitrogen pesticides [50]. The results showed that the hollow NiO/Co@C magnetic MOF is a promising candidate for water purification and removal of hazardous organic nitrogen pesticides. Moreover, the role of MIL-53 MOFs in the effective adsorption of metolachlor herbicide was emphasized in the study conducted by Hamza et al. [51].

In general, the physicochemical properties of MOFs determine their pollutant adsorption behavior. High porosity, surface area, selectivity, mechanical stability, and ease of functionalization are the critical features of MOFs [34]. The higher the surface area that MOFs have, the more adsorption sites they have. Their active sites and surface charge are tunable for the accumulation of a specific pollutant and can be functionalized by surface modification. The accumulation of various functional groups on the surface of MOFs and their unique structures make it possible for them to adsorb emerging pollutants effectively. The adsorption efficiency of MOFs can be enhanced by using large organic linkers, creating defects in MOF structures, introducing functionalities in metal nodes and the organic linkers, and fabricating various MOF-based composites [52]. Understanding the adsorption mechanism is pivotal because it enables us to design different MOF materials for future applications. Plausible adsorption mechanisms of emerging contaminants on MOFs are acid-based, hydrophobic, electrostatic, hydrogen bonding, or π-π interactions or a combination of these interactions.

1.2.2 Catalyst

Unlike adsorption, which only transfers pollutants from a solution to the surface of an adsorbent, requires costly posttreatment for the elimination of recalcitrant pollutants, and may cause toxic secondary pollutants, catalytic oxidation using various catalysts can completely degrade or convert water contaminants into biodegradable, less toxic, and benign substances. Recently, AOPs have received much interest in wastewater treatment due to their excellent reproducibility and simplicity [53].

In AOPs, the oxidative degradation is driven by free radicals such as •OH, $O_2{\bullet}^-$, and/or $SO_4{\bullet}^-$, which have high oxidation power (e.g., 1.8–2.7 V for ·OH, 2.5–3.1 V for $SO_4{\bullet}^-$) and can oxidize organic compounds to harmless and benign substances, such as CO_2 and H_2O, under appropriate conditions [53, 54]. AOPs include photocatalysis, Fenton reactions, Fenton-like reactions (electro-Fenton, photo-Fenton, and photo-electro-Fenton), and sulfate radical-based oxidations. Photocatalysis is the AOP in which photo-active materials or semiconductors such as ZnO, TiO_2, CdS, WO_3, etc., are used as homo- or heterogeneous photocatalysts and are capable of generating photo-induced charge carriers (electrons and holes) under irradiation of

UV or visible light [55]. Those charge carriers can then react with adsorbed O_2 and H_2O molecules on the catalyst's surface to generate $O_2\bullet^-$ and $\bullet OH$ radicals, respectively. Meanwhile, Fenton and Fenton-like processes refer to the AOPs in which ferrous ions catalyze the decomposition of H_2O_2 (Fe II) to generate $\bullet OH$ and peroxide radicals capable of oxidizing organic molecules [56].

In contrast, sulfate radical-based oxidations are alternative approaches to Fenton reactions and are driven by highly reactive sulfate radicals ($SO_4\bullet^-$) produced by successful activation of certain oxidants, such as persulfate (PS) and peroxymonosulfate (PMS), by UV, heat, alkaline pH, sonication, and transition metals. In the last decade, many studies have emphasized the role of MOFs as heterogeneous catalysts in different AOP reactions to remove emerging water contaminants. MOFs can act as semiconductors under light irradiation because their organic ligands can absorb light to activate their metal sites through a ligand-to-metal charge transfer mechanism [57]. For example, Khosroshahi et al. introduced the nanocomposites of MOF-808 and $NiFe_2O_4$ as photocatalysts for effective photocatalytic degradation of meropenem and photocatalytic reduction of Cr (VI). The synthesized materials have low bandgap energies and were active under visible light irradiation [58]. In another study, Khodkar et al. explored the photocatalytic degradation of paraquat, a popular herbicide for killing weeds in agriculture, using a magnetic MOF composite α-Fe_2O_3@MIL-101(Cr)@TiO_2 [59]. The effects of experimental parameters such as the dosage of catalyst, pH, the initial concentration of paraquat, and reaction time on the paraquat degradation efficiency were also investigated in their study. The results showed that the maximum photocatalytic degradation efficiency could reach 87.46% after 45 min under the optimal conditions: catalyst dosage = 0.2 g/L, pH = 7, and initial paraquat concentration = 20 mg/L [59].

Moreover, the role of MOFs as heterogeneous catalysts for the removal of emerging water contaminants was also found in studies of Fenton and Fenton-like processes. For example, in the study of an ultrasound-assisted heterogeneous Fenton system, Geng et al. provided new insights into the catalytic performance of environmentally friendly Fe-based MOFs (MIL-53, MIL-88B, and MIL-101) in the degradation of tetracycline hydrochloride, a universal therapeutic medicine [60]. The active coordination sites of metals of MOFs were responsible for their catalytic performance. MIL-88B showed the best catalytic performance owing to its great Lewis acid sites, representing the most coordinated active sites in the structure of MIL-88B [60]. Recently, in Ye et al.'s work, they have reported the role of Fe-bpydc MOF (bpydc is short for 2,2′-bipyridine-5,5′–dicarboxylate) as a highly active catalyst showing excellent catalytic performance in the heterogeneous photo-electro-Fenton degradation of the lipid regulator bezafibrate [61]. Its excellent catalytic performance can be explained by the enhanced mass transport and charge transfer, the reduction of H_2O_2 to $\bullet OH$ and of Fe (III) to Fe (II), and the excitation of Fe-O clusters. In addition, the use of MOFs as catalysts in sulfate radical-based oxidations has been known in the last decade. A recent study reported using B,N-doped carbocatalyst (Fe@BPC-XBN) as an effective peroxymonosulfate activator for bisphenol A degradation [62]. The catalyst was synthesized from the iron-based MOF, boric acid, and boron nitride, providing it with more catalytic centers and helping it improve its stability.

1.2.3 Synergistic Effect of Adsorption and Photocatalyst

Developing MOFs with desirable adsorption and catalytic degradation performance is another strategy for removing emerging water contaminants. This strategy is based on the fact that adsorption and catalysis are not independent. Obviously, the immobilization of the contaminants on the surface of MOFs is one of the requirements for their excellent catalytic performance. Therefore, combining adsorption and catalytic degradation of pollutants by MOFs will open the door to efficiently removing emerging water contaminants. For example, by using titanium metal-organic frameworks (Ti-MOFs) and COFs in combination with triazine frameworks through covalent bonding, Yang et al. were able to construct a composite material that has high adsorption capacity and good photocatalytic activity for the removal of bisphenol A (BPA) [63]. The efficient synergistic adsorption and degradation of BPA (100 ppm) by the covalent-integrated Ti-MOFs@COFs could be achieved within ten minutes, which is more preeminent than other materials.

Similarly, in another study, Zhao et al. introduced a series of nanosheet-structured manganese-carbon (MnO_x@C) composites obtained from the calcination of 2D manganese-1,4 benzenedicarboxylic acid-based MOFs under different temperatures [64]. The composites could adsorb and degrade 4-aminobenzoic acid ethyl ester (ABEE) through the PMS activation. The MnO_x@C calcined at 900 $^{\circ}$C exhibited outstanding ABEE adsorption (27.0%) and degradation performance (91.3%).

1.3 Emerging Water Contaminants by MOFs

Recently, due to (i) large surface area, (ii) stability in water, (iii) ease of functionalization, and (iv) capacity to produce on a large scale, MOFs are widely utilized for the removal of pollutants in aqueous solutions.

1.3.1 Organic Dyes

In recent years, synthetic dyes have been widely utilized in medicine, printing and dyeing, and cosmetics industries due to the development of the dye industry. Therefore, a massive amount of dye wastewater was released into the environment. The dye wastewater composition is complex; most are refractory organic matter, high chroma, and continuous solid pollution. Nowadays, there is much-advanced research on the removal of organic dyes based on MOF materials. MOFs removed organic dyes through adsorption and the photocatalytic process.

1.3.2 Adsorption

PEI-modified Cu-BTC was used to adsorb congo red and acid blue with high concentrations (1200 and 100 mg/L^{-1}) [65]. The adsorption capacity for the two dyes mentioned above of PEI-modified Cu-BTC is 2578 and 132 mg/g, respectively. Not only does it have excellent processing efficiency, but the reusability of PEI-modified Cu-BTC is also highly effective, with the ability to be reused six times.

Li et al. successfully synthesized MIL-53 (Al) with the adding amine functional group and applied this material to adsorb methylene blue and malachite green [66]. The results showed that the removal capacity of MIL-53 (Al)-NH_2 was superior to that of MIL-53 (Al), while the adsorption capacity of MIL-53 (Al)-NH_2 for methylene blue and malachite green only reached 45.2 and 37.8 mg/g, respectively. In contrast, the adsorption capacity of MIL-53 (Al) is only 3.6 and 2.9 mg/g.

Hamedi et al. [67] successfully synthesized composite material MIL-101(Fe)@PDopa@Fe_3O_4 by solvothermal method and used it for the adsorption of two dyes: methyl red and malachite green. The results show that the processing ability of MIL-101(Fe)@PDopa@Fe_3O_4 is excellent, with high processing efficiency. Through experiments, the adsorption capacity of MIL-101(Fe)@PDopa@Fe_3O_4 for methyl red and malachite green was determined to be 833 and 1250 mg/g.

Activated carbon is a normal adsorbent used to treat common industrial colorants. The activated carbon-enhanced HKUST-1-MOF derived from copper salts is established by Azad et al. [68] with the treatment of 10 mg/L^{-1} of three dyes including crystal violet, disulfide blue, and quinoline yellow. At optimum values of pH 4, 0.02 g adsorbent was very effective with the adsorption capacity of Ac-HKUST-1 for respectively 133, 130, and 65 mg/g.

1.3.3 Photocatalytic and Electrostatic Activities

Wang et al. proposed a new carbon material—CQD—which was used to modify NH_2-MIL-125(Ti) [69]. The results show that the modified material has a smaller specific surface area than NH_2-MIL-125(Ti) (198 $m^2/g < 478$ m^2/g) but the visible light treatment efficiency for Rhodamine B of CQD/ NH_2-MIL-125(Ti) is better than that of NH_2-MIL-125(Ti). The reusability of the two materials is also remarkable, as the two materials can be reused seven times.

After adding the NH_2 group to enhance the electrostatic interaction in the framework, MOF-199-NH_2 was denatured with $BaWO_4$ [70]. UV-Vis measurement results show that the band gap of the modified material is smaller than that of MOF-199-NH_2 (3eV < 3.2eV). The band gap is reduced, so the photocatalytic ability of the modified material is also better than that of the original material (98% > 38%). Although the photocatalytic treatment results are positive, the light of the photocatalyst must be in the ultraviolet region.

1.3.4 PPCPs

PPCPs have attracted much attention from scientists as newly emerging water pollutants. These products mainly include analgesics, antibiotics, fungicides, etc. The amount of PPCPs has increased significantly every year due to human demand. PPCPs are persistent in bioaccumulation and toxicity and will release into the water many ways after use, causing severe environmental issues. Therefore, MOFs are considered safe and effective materials to degrade/remove PPCPs.

1.3.5 Adsorption

Cheng et al. [71] enhanced the sulfamethoxazole (SMX) adsorption rate of only ten minutes with a maximum adsorption capacity of up to 284.7 mg/g^{-1} by imprinting the molecular polymers (MIPs) on the surface of UIO-66. The received MIP-IL@UiO-66 exposed superior identification toward SMX based on both π-π and/or electrostatic interactions, hydrogen bonds, and molecular structure. The material could be applied as a commercial solid-phase extraction (SPE) adsorbent when it can be cycled five times.

The metal-organic frameworks had high adsorption for tetracycline or/and norfloxacin based on the ferric fabricated benzenetricarboxylic (Fe-BTC) material [72]. The pollutant adsorption capacity in the isolation system for each antibiotic tetracycline and norfloxacin was 714.3 and 334.6 mg/g^{-1} and the effect of norfloxacin is significant in a binary system. The dominant adsorption mechanisms are coordination bond, π-π interaction most attributed to norfloxacin, and pore diffusion, an active site for tetracycline.

Fe_3O_4@Cd-MOF@CS microspheres were prepared through facile coating and applied to amoxicillin removal. The adsorbed microspheres showed the best removal efficiency at a pH of 8.1. The hydrogen bonding, electrostatic interaction, and π-π stacking were the three main factors for the adsorption of amoxicillin when the highest adsorption capacity was found at 103.09 mg/g^{-1}, respectively. After five reused cycles where the material was washed with 40% methanol solution, the Fe_3O_4@Cd-MOF@CS microspheres showed good amoxicillin removal efficiency with a slight change from 73.92% to 62.07% [73].

A composite of MIL101(Cr) combined with magnetite nanoparticles was successfully synthesized for the selective extraction of parabens (preservative) through an adsorption process reached from hydrogen bonding, hydrophobic, and π-π interactions. The synthesized composite showed good stability and accuracy and achieved high extraction recoveries [80–96%] [74].

Copper mixed-two multivariate triazole frameworks (CuMtz) [75] were fabricated employing a solid solution approach to use for the adsorption of triclosan (disinfectant). The material contains 22% doping ligand, giving the best adsorption capacity as 743.2 mg/g^{-1} even even after four reuses. The motivating force is the hydrogen bonding accompanied by the anion exchange under the basic environment between anionic and BF_4^- of copper precursor.

1.3.6 Photocatalytic Activities

In the modulation of Fe(II)-MOF using different organic acids, Chen et al. [76] indicated that citric acid can regulate the release rate of Fe through chelation and improve the degree of crystallinity and porosity. In comparison with adding other modulators, Fe(II)-MOF-CA showed ideal catalytic activity: the SMX degradation increased from 40.4% to achieve the highest rate of 91.7% within 120 min.

The extracted lecithin from the yolk was used as an auxiliary for $LiY(MoO_4)_2$ quantum dots bond to 1,3,5-Benzenetricarboxylic acid throughout the hydrothermal process. The formed $LiY(MoO_4)_2QD$/BioMOFs is an eco-friendly photocatalyst that has ibuprofen removal higher than 99% after one hour under UV irradiation. The morphology of $LiY(MoO_4)_2QD$/BioMOFs is coral-like with radius <10 nm, and it shows super-radiant photoluminescence (SRP) properties that can be utilized for drug delivery systems in anti-cancer drugs [77].

A novel photocatalyst material was synthesized by the hybridization of graphitic carbon nitride (g-C_3N_4) and titanium-MOF (NH_2-MIL-125) [78]. The synthesized material has a near band gap with the pristine MOF. At the same time, the reduced electron-holes recombination rate led to a higher photocatalytic performance that completed the removal of 10 mg/L^{-1} diclofenac in 2h under LED irradiation. The detected by-products include short-chain carboxylic acids, such as oxalic and formic acid, and chloride and nitrite give the primary avenue of the diclofenac degradation pathway.

1.3.7 Herbicides and Pesticides

Herbicides and pesticides are toxic chemicals widely used in agricultural crop production. However, due to its difficult biodegradability, the remaining herbicides and pesticides in the soil after use will move to the water environment, causing environmental problems [79]. Therefore, removing herbicides and pesticides by MOFs is a better choice for scientists. Herbicides and pesticides can be removed from the water environment via adsorption or photocatalysis.

1.3.8 Adsorption

HKUST-1(Cu) was a metal-organic framework efficiently synthesized at room temperature, similar to ZrO_2@HKUST-1, an easily composited material from HKUST-1. These two materials had large pore size and surface area, which makes both have better adsorption capacity for cyhalothrin. The results showed the adsorption capacity of HKUST-1(Cu) and ZrO_2@HKUST-1 for cyhalithrin was 140 and 138 mg/g after 120 min, respectively [80].

UiO-67(Zr) metal-organic framework was a popular metal-organic framework synthesized by many methods such as hydrothermal and solvothermal. With a large surface area, UiO-67(Zr) was a suitable material for the adsorption of herbicides. In 2018, UiO-67(Zr) was utilized for the removal of the herbicide atrazine and showed entirely satisfactory results with UiO-67(Zr), although the adsorption capacity of UiO-67(Zr) for atrazine was only 12 mg/g [81].

MIL-53(Cr) was evaluated to remove 2,4-dichlorophenoxyacetic acid (2,4-D) in the aqueous solution [82]. MIL-53(Cr) showed fast adsorption (within 1 h) with high adsorption capacity (556 mg/g), followed by its simple reusability by washing the used MOF with water/ethanol.

MIL-101(Cr) is a fairly common MOF material synthesized by the hydrothermal method. The MIL-101(Cr) was applied to adsorb diazinon with a treatment time of about 45 minutes. The adsorption capacity of the MIL-101(Cr) carbon metal frame is 158 mg/g [83]. In the other study, the modification of Cr-MIL-101 by 3,5-Bis (trifluoromethyl) phenyl isocyanate (BTP) metal-organic framework was fabricated to adsorb acetochlor with an adsorption capacity of 322 mg/g. Cr-MIL-101-BTP can treat acetochlor with high concentration and the regeneration ability is extremely promising, with the ability to be used six times [84].

An ionic liquid modified chitosan used to dominate with UiO-66 is also a promising composite material with an adsorption capacity up to 893 mg/g when applied to treat high concentrations of 2,4-dichlorophenoxyacetic acid (up to 500 mg/L^{-1}) for one hour [85]. H-bond and electrostatic interaction mechanisms explain why this has such a good adsorption capacity.

1.3.9 Photocatalytic Activities

Khodkar et al. used α-Fe_2O_3@ MIL-101(Cr) magnetic MOF to reduce bipyridine concentration in an aqueous medium [59]. Under strict control conditions on pH and dosage, the α-Fe_2O_3@ MIL-101(Cr) material had reduced only 45% of contaminants. The combination of pristine material with TiO_2 raised the efficiency of bipyridine removal up to 90% in the first solution with a concentration of 20 mg/L^{-1} [59].

MIL-100(Fe)@ZnO was obtained by Ahmad et al. [86] and fabricated for application as a photocatalyst to treat the herbicide atrazine. MIL-100(Fe)@ZnO synthesized by the solvothermal method had a specific surface area of 654 m^2/g and a band gap energy of 2.63 eV. Under visible light for 120 minutes, 79% of 20 mg/L^{-1} atrazine solution was reduced by the MIL-100(Fe)@ZnO material. The reusability of the material is also entirely satisfactory with five reuses, showing the material's potential.

Fe_3O_4@MOF-2 is a modified magnetic material developed by Sajjadi et al. for diazinon insecticide treatment application [87]. Under visible light, the photocatalytic ability of Fe_3O_4@MOF-2 was demonstrated through a treatment efficiency of up to 99% for 30 mg/L^{-1} of diazinon insecticide within 30 minutes. Not only that, the regenerative ability of this magnetic organic metal frame material is also surprising, with 15 reuse times.

Metal-organic frameworks can be synthesized relatively easily by conventional methods. In their study, Oladipo et al. [88] fabricated AgIO3/MIL-53(Fe) material under room temperature conditions. The AgIO3/MIL-53(Fe) material has a small band gap energy of 2.4 eV and is suitable for photocatalysts to treat malathion and chlorpyrifos. The results showed that, under sunlight, its ability to handle these two pesticides was excellent, reaching more than 90% efficiency with 120 minutes of lighting and a concentration of pesticide solution of 20 mg/L^{-1}.

Zr-MOF@WO_3/graphene oxide is also a metal-organic framework composite material with good photocatalytic ability [89]. Fakhri et al. [89] used this photocatalyst material to reduce the concentration of malathion in water. Within 70 minutes, under visible light, 20 mg/L^{-1} malathion solution was treated quite effectively with an efficiency of up to 83%. The Zr-MOF@WO_3/graphene oxide can be reused with four cycles.

1.3.10 Industrial Compounds/By-products

Various organic pollutants have also been removed from water by MOFs and MOF-based adsorbents or photocatalysts to form by-products or unharmful products [79, 90, 91].

1.3.11 Adsorption

A synthesized MOF-177 had a BET surface area of 2970 m^2/g applied to the adsorption of VOCs compounds. The main results showed the adsorbed amount of benzene was 800 mg/g^{-1} and toluene was 585 mg/g^{-1}. Even at high humidity, the water became the competitor, but the adsorption capacity was still relatively high. However, MOF-177 could be pre-drying to avoid the decomposition of the MOF-177 framework structure [79].

The industrial surfactant perfluorooctane sulfonate (PFOS) has been removed with an adsorption capacity reaching 939 mg/g^{-1} at acidic pH using a novel MOF-808 metal-organic framework. This material has a high specific surface area of 1610 m^2/g with good pore size and abundant active adsorption sites that offer stability in solution under different pH conditions for 7 days. The PFOS adsorption recorded was an endothermic and spontaneous process that related to the electrostatic interaction between the cationic central cluster ($(Zr_6O_4(OH)_4)^{12+}$) of MOF-808 and PFOS anion [90].

1.3.12 Photocatalytic Activities

The bi-functional composite $Bi_{12}O_{17}Cl_{12}$/MIL-100(Fe) (BM200) was synthesized to explore the degradation of bisphenol A (BPA). Under white-light irradiation, the photodegradation of BPA reached 100% after 60 min with the addition of persulfate. The BPA was degraded by $\cdot SO_4^-$ and $\cdot OH$ radicals to form intermediate products such as short-chain aliphatic acids and mineralized products of CO_2 and H_2O [91].

1.4 Challenges and Perspective in Using MOFs for the Removal of Emerging Water Contaminants

Facing contaminants has become one of the main issues of the last few decades, and MOFs, as a new group of porous materials with several advantages over conventional adsorbents, are the outstanding candidate to help humans deal with these related issues. The potential applications using MOFs have been broadened recently for all types of environments, which prominently include emerging water contaminants [92–94]. However, stability under different conditions is a crucial challenge when being discussed in relation to MOFs. Most MOFs are water-labile or unstable at different levels due to water molecules' gradual replacement of metal-coordinated linkers. Thus, many efforts have been made to improve this issue. For example, selecting MOF constituents is a feasible approach to fabricating water-stable MOFs via de novo synthesis [95, 96]. Besides de novo synthesis, many postsynthetic methods have been researched to improve the water stability of MOFs: functional group modification, ion-exchange, hydrophobic surface treatment, and material compositization[97, 98]. Another issue is that many MOFs are built up from toxic metals such as cobalt, cadmium, chromium, and/or noxious organic moieties – even molecules that are themselves considered emerging pollutants that can be released to the environment. Choosing safe components to make greener materials is also a priority that needs to be considered to deal with issues such as toxicity, life cycle assessment studies, cost-effectiveness. Further, the behavior of MOFs as adsorbents, catalysts, or synergetics is still under discussion.

Overall, there is no doubt that MOFs are a great porous material that can effectively be applied to remove emerging water pollutants. The ability to control pore size and nature, access mesoporosity, utilize reticular chemistry, and incorporate active sites in porous compounds as per demand has resulted in MOFs commanding research interest across other areas as well.

1.5 Conclusion

Recently, a group of emerging contaminants has been recognized as significant water pollutants that adversely affect human and wildlife endocrine systems. Natural attenuation and conventional treatment processes cannot remove these micropollutants, which are reported to bioaccumulate in macroinvertebrates, other organisms in the aquatic food web, and humans. With high surface area, stability, and modular composition, MOFs are a potential candidate for environmental treatment in general and emerging water contaminants (organic dyes, PPCPs, herbicides, and pesticides). For the long-term outlook, MOFs have recently been one of the most attractive research directions. The challenges and perspectives in this field illustrate this material's increased interest and vast future.

References

1 Chinnaiyan, P., Thampi, S.G., Kumar, M. et al. (2018). Pharmaceutical products as emerging contaminant in water: relevance for developing nations and identification of critical compounds for Indian environment. *Environ Monit Assess* 190: 288.

2 Robles-Molina, J., Gilbert-López, B., García-Reyes, J.F., and Molina-Díaz, A. (2014). Monitoring of selected priority and emerging contaminants in the Guadalquivir River and other related surface waters in the province of Jaén, South East Spain. *Science of the Total Environment* 479–480: 247–257.

3 Zhu, L., Jiang, C., Panthi, S. et al. (2021). Impact of high precipitation and temperature events on the distribution of emerging contaminants in surface water in the mid-Atlantic, United States. *Science of the Total Environment* 755 (2): 142552.

4 Hidayati, N.V., Syakti, A.D., Asia, L. et al. (2021). Emerging contaminants detected in aquaculture sites in Java, Indonesia. *Science of the Total Environment* 773: 145057.

5 Vasilachi, I.C., Asiminicesei, D.M., Fertu, D.I., and Gavrilescu, M. (2021). Occurrence and fate of emerging pollutants in water environment and options for their removal. *Water* 13 (2): 181.

6 Heo, J., Flora, J.R.V., Her, N. et al. (2012). Removal of bisphenol A and 17-estradiol in single walled carbon nanotubes-ultrafiltration [SWNTs-UF] membrane systems. *Separation and Purification Technology* 90: 39–52.

7 Yangali-Quintanilla, V., Sadmani, A., McConville, M. et al. (2010). A QSAR model for predicting rejection of emerging contaminants [pharmaceuticals, endocrine disruptors] by nanofiltration membranes. *Water Research* 44 (2): 373–384.

8 Linares, R.V., Yangali-Quintanilla, V., Li, Z., and Amy, G. (2011). Rejection of micropollutants by clean and fouled forward osmosis membrane. *Water Research* 45: 6737–6744.

9 Mailler, R., Gasperi, J., Coquet, Y. et al. (2016). Removal of emerging micropollutants from wastewater by activated carbon adsorption: experimental study of different activated carbons and factors influencing the adsorption of micropollutants in wastewater. *Journal of Environmental Engineering* 4: 1102–1109.

10 Tagliavini, M., Weidler, P.G., Njel, C. et al. (2020). Polymer-based spherical activated carbon – ultrafiltration [UF-PBSAC] for the adsorption of steroid hormones from water: material characteristics and process configuration. *Water Research* 185: 116249.

11 Yi, C., Qin, W., and Wen, X. (2020). Renovated filter filled with poly-3-hydroxybutyrateco-hydroxyvalerate and granular activated carbon for simultaneous removal of nitrate and PPCPs from the secondary effluent. *Science of the Total Environment* 749: 141494.

12 Feng, Z., Zhai, X., and Sun, T. (2022). Sustainable and efficient removal of paraben, oxytetracycline and metronidazole using magnetic porous biochar composite prepared by one step pyrolysis. *Separation and Purification Technology* 293: 121120.

13 Qi, K., Song, M., Xie, X. et al. (2022). CQDs/biochar from reed straw modified Z-scheme MgIn2S4/BiOCl with enhanced visible-light photocatalytic performance for carbamazepine degradation in water. *Chemosphere* 287: 132192.

14 Ayça Avcı, İ.İ. and Baylan, N. (2020). Adsorption of ciprofloxacin hydrochloride on multiwall carbon nanotube. *Journal of Molecular Structure* 1206: 127711.

15 Zhao, H., Liu, X., Cao, Z. et al. (2016). Adsorption behavior and mechanism of chloramphenicol, sulfonamides, and non-antibiotic pharmaceuticals on multi-walled carbon nanotubes. *Journal of Hazardous Materials* 310: 235–245.

16 Ali, A.B., Eder, C.L., Afshin, T., Nadali, A., Gholamreza, G., Mehdi, V., Ghasem, H., Mohammad, S. (2016). Removal of tetracycline antibiotic from contaminated water media by multi-walled carbon nanotubes: operational variables, kinetics, and equilibrium studies. *Water Science & Technology* 74 (5): 1202–1216. https://doi:10.2166/wst.2016.301

17 Paterakis, N., Chiu, T.Y., Koh, Y.K.K. et al. (2012). The effectiveness of anaerobic digestion in removing estrogens and nonylphenol ethoxylates. *Journal of Hazardous Materials* 199: 88–95.

18 Carr, D.L., Morse, A.N., Zak, J.C., and Anderson, T.A. (2011). Biological degradation of common pharmaceuticals and personal care products in soils with high water content. *Water, Air, & Soil Pollution.* 217: 127–134.

19 Thi, V.H. and Lee, B.-K. (2017). Great improvement on tetracycline removal using ZnO rod-activated carbon fiber composite prepared with a facile microwave method. *Journal of Hazardous Materials* 324: 329–339.

20 Isari, A.A., Hayati, F., Kakavandic, B. et al. (2020). N, Cu co-doped TiO2@ functionalized SWCNT photocatalyst coupled with ultrasound and visible-light: an effective sono-photocatalysis process for pharmaceutical wastewaters treatment. *Chemical Engineering Journal* 392: 123685.

21 Fernando, L.R.O., Eric, C.W., Shane, A.S. (2010). Evaluation of UV/H2O2 treatment for the oxidation of pharmaceuticals in wastewater. *Water Research* 44 (5): 1440–1448. https://doi.org/10.1016/j.watres.2009.10.031

22 Lijin Huang, M.H., Chen, B., and Bin, H. (2018). Magnetic Zr-MOFs nanocomposites for rapid removal of heavy metal ions and dyes from water. *Chemosphere* 199: 435–444.

23 Du, C., Zhang, Z., Guanlong, Y. et al. (2021). A review of metal-organic framework [MOFs]-based materials for antibiotics removal via adsorption and photocatalysis. *Chemosphere* 272: 129501.

24 Hooriabad Saboor, F., Nasirpour, N., Shahsavari, S., and Kazemian, H. (2022). The effectiveness of MOFs for the removal of pharmaceuticals from aquatic environments: a review focused on antibiotics removal. *Chemistry an Asian Journal* 17 (4): e202101105.

25 Mondol, M.M.H. and Jhung, S.H. (2021). Adsorptive removal of pesticides from water with metal-organic framework-based materials. *Chemical Engineering Journal* 421: 129688.

26 Abdollahi, N., Moussavi, G., and Giannakis, S. (2022). A review of heavy metals removal from aqueous matrices by metal-organic frameworks [MOFs]: state-of-the-art and recent advances. *Journal of Environmental Chemical Engineering* 10 (3): 107394.

27 Singh, R., Singh, S., Parihar, P. et al. (2015). Arsenic contamination, consequences and remediation techniques: a review. *Ecotoxicology and Environmental Safety* 112: 247–270.

28 Tian, S., Zhang, C., Huang, D. et al. (2020). Recent progress in sustainable technologies for adsorptive and reactive removal of sulfonamides. *Chemical Engineering Journal* 389: 123423.

29 Vikrant, K., Kim, K.-H., Kumar, S., and Boukhvalov, D.W. (2020). Metal-organic frameworks for the adsorptive removal of gaseous aliphatic ketones. *ACS Applied Materials & Interfaces* 12 (9): 10317–10331.

30 Li, H., Li, Y., Li, B. et al. (2020). Highly selective anchoring silver nanoclusters on MOF/SOF heterostructured framework for efficient adsorption of radioactive iodine from aqueous solution. *Chemosphere* 252: 126448.

31 Tang, Y., Wang, X., Wen, Y. et al. (2020). Oxygen-selective adsorption property of ultramicroporous MOF Cu[Qc]2 for air separation. *Industrial & Engineering Chemistry Research* 59 (13): 6219–6225.

32 Barczak, M., Dobrowolski, R., Borowski, P., and Giannakoudakis, D.A. (2020). Pyridine-, thiol- and amine-functionalized mesoporous silicas for adsorptive removal of pharmaceuticals. *Microporous and Mesoporous Materials* 299: 110132.

33 Huang, S., Deng, W., Zhang, L. et al. (2020). Adsorptive properties in toluene removal over hierarchical zeolites. *Microporous and Mesoporous Materials* 302: 110204.

34 Ihsanullah, I. (2022). Applications of MOFs as adsorbents in water purification: progress, challenges and outlook. *Current Opinion in Environmental Science & Health* 26: 100335.

35 Kumar Maka, V., Tamuly, P., Jindal, S., and Narasimha Moorthy, J. (2020). Control of In-MOF topologies and tuning of porosity through ligand structure, functionality and interpenetration: selective cationic dye exchange. *Applied Materials Today* 19: 100613.

36 Ghosh, R., Sunny, B., Babu, M., and Thirumurugan, A. (2022). Role of crystal nucleation and growth conditions on the soft-templated preparation of HKUST-1 MOFs with hierarchical porosity involving Gemini surfactants. *Microporous and Mesoporous Materials* 334: 111758.

37 Spokoyny, A.M., Farha, O.K., Mulfort, K.L. et al. (2010). Porosity tuning of carborane-based metal-organic frameworks [MOFs] via coordination chemistry and ligand design. *Inorganica Chimica Acta* 364 (1): 266–271.

38 Chen, L., Zhang, X., Cheng, X. et al. (2020). The function of metal-organic frameworks in the application of MOF-based composites. *Nanoscale Advances* 2 (7): 2628–2647.

39 Ji, C., Xu, M., Yu, H. et al. (2022). Mechanistic insight into selective adsorption and easy regeneration of carboxyl-functionalized MOFs towards heavy metals. *Journal of Hazardous Materials* 424: 127684.

40 Chen, P., Wang, Y., Zhuang, X. et al. (2023). Selective removal of heavy metals by Zr-based MOFs in wastewater: new acid and amino functionalization strategy. *Journal of Environmental Sciences* 124: 268–280.

41 Valadi, F.M., Shahsavari, S., Akbarzadeh, E., and Gholami, M.R. (2022). Preparation of new MOF-808/chitosan composite for Cr[VI] adsorption from aqueous solution: experimental and DFT study. *Carbohydrate Polymers* 288: 119383.

42 Castañeda-Ramírez, A.A., Rojas-García, E., López-Medina, R. et al. (2022). Magnetite nanoparticles into Fe-BTC MOF as adsorbent material for the remediation of metal Cu[II], Pb[II, As[III] and Hg[II]] ions-contaminated water. *Catalysis Today* 394–396: 94–102.

43 Ahmadijokani, F., Mohammadkhani, R., Ahmadipouya, S. et al. (2020). Superior chemical stability of UiO-66 metal-organic frameworks [MOFs] for selective dye adsorption. *Chemical Engineering Journal* 399: 125346.

44 Li, -T.-T., Liu, Y.-M., Wang, T. et al. (2018). regulation of the surface area and surface charge property of MOFs by multivariate strategy: synthesis, characterization, selective dye adsorption and separation. *Microporous and Mesoporous Materials* 272: 101–108.

45 Wang, Q., Zhao, Y., Shi, Z. et al. (2021). Magnetic amino-functionalized-MOF[M = Fe, Ti, Zr]@COFs with superior biocompatibility: performance and mechanism on adsorption of azo dyes in soft drinks. *Chemical Engineering Journal* 420: 129955.

46 Jia, X., Zhang, B., Chen, C. et al. (2021). Immobilization of chitosan grafted carboxylic Zr-MOF to porous starch for sulfanilamide adsorption. *Carbohydrate Polymers* 253: 117305.

47 Chen, J., Ouyang, J., Chen, W. et al. (2022). Fabrication and adsorption mechanism of chitosan/Zr-MOF [UiO-66] composite foams for efficient removal of ketoprofen from aqueous solution. *Chemical Engineering Journal* 431: 134045.

48 Zhao, F., Fang, S., Gao, Y., and Bi, J. (2022). Removal of aqueous pharmaceuticals by magnetically functionalized Zr-MOFs: adsorption kinetics, isotherms, and regeneration. *Journal of Colloid and Interface Science* 615: 876–886.

49 Gao, Y., Liu, K., Kang, R. et al. (2018). A comparative study of rigid and flexible MOFs for the adsorption of pharmaceuticals: kinetics, isotherms and mechanisms. *Journal of Hazardous Materials* 359: 248–257.

50 Zhao, J., Huang, P., Wang, X. et al. (2022). Efficient adsorption removal of organic nitrogen pesticides: insight into a new hollow NiO/Co@C magnetic nanocomposites derived from metal-organic framework. *Separation and Purification Technology* 287: 120608.

51 Ahmad Isiyaka, H., Jumbri, K., Soraya Sambudi, N. et al. (2022). Effective adsorption of metolachlor herbicide by MIL-53[Al] metal-organic framework: optimization, validation and molecular docking simulation studies. *Environmental Nanotechnology, Monitoring & Management* 18: 100663.

52 Li, J., Wang, X., Zhao, G. et al. (2018). Metal-organic framework-based materials: superior adsorbents for the capture of toxic and radioactive metal ions. *Chemical Society Reviews* 47 (7): 2322–2356.

53 Sharma, V.K. and Feng, M. (2019). Water depollution using metal-organic frameworks-catalyzed advanced oxidation processes: a review. *Journal of Hazardous Materials* 372: 3–16.

54 Han, W., Li, D., Zhang, M. et al. (2020). Photocatalytic activation of peroxymonosulfate by surface-tailored carbon quantum dots. *Journal of Hazardous Materials* 395: 122695.

55 Nguyen-Le, M.-T., Lee, B.-K., and Tran, D.-M. (2017). EDTA-Na2-assisted synthesis of rod-like titanate-TiO2 composite architectures with enhanced visible-light-driven properties. *Journal of Industrial and Engineering Chemistry* 56: 225–233.

56 Lu, S., Liu, L., Demissie, H. et al. (2021). Design and application of metal-organic frameworks and derivatives as heterogeneous Fenton-like catalysts for organic wastewater treatment: a review. *Environment International* 146: 106273.

57 Zhang, X., Wang, J., Dong, -X.-X., and Lv, Y.-K. (2020). Functionalized metal-organic frameworks for photocatalytic degradation of organic pollutants in environment. *Chemosphere* 242: 125144.

58 Khosroshahi, N., Bakhtian, M., and Safarifard, V. (2022). Mechanochemical synthesis of ferrite/MOF nanocomposite: efficient photocatalyst for the removal of meropenem and hexavalent chromium from water. *Journal of Photochemistry and Photobiology A: Chemistry* 431: 114033.

59 Khodkar, A., Khezri, S.M., Pendashteh, A.R. et al. (2019). A designed experimental approach for photocatalytic degradation of paraquat using α-Fe2O3@MIL-101[Cr]@TiO2 based on metal-organic framework. *International Journal of Environmental Science and Technology* 16 (10): 5741–5756.

60 Geng, N., Chen, W., Xu, H. et al. (2021). Insights into the novel application of Fe-MOFs in ultrasound-assisted heterogeneous Fenton system: efficiency, kinetics and mechanism. *Ultrasonics Sonochemistry* 72: 105411.

61 Ye, Z., Schukraft, G.E.M., L'Hermitte, A. et al. (2020). Mechanism and stability of an Fe-based 2D MOF during the photoelectro-Fenton treatment of organic micropollutants under UVA and visible light irradiation. *Water Research* 184: 115986.

62 Wan, Y., Zhang, W., Han, X. et al. (2022). B,N-decorated carbocatalyst based on Fe-MOF/BN as an efficient peroxymonosulfate activator for bisphenol A degradation. *Journal of Hazardous Materials* 430: 127832.

63 Yang, L., Wang, Y., Yuan, J. et al. (2022). Construction of covalent-integrated MOFs@COFs composite material for efficient synergistic adsorption and degradation of pollutants. *Chemical Engineering Journal* 446: 137095.

64 Zhao, Y., Zhan, X., Wang, H. et al. (2022). MOFs-derived MnOx@C nanosheets for peroxymonosulfate activation: synergistic effect and mechanism. *Chemical Engineering Journal* 433: 133806.

65 Quan, X., Sun, Z., Meng, H. et al. (2019). Polyethyleneimine [PEI] incorporated Cu-BTC composites: extended applications in ultra-high efficient removal of congo red. *Journal of Solid State Chemistry* 270: 231–241.

66 Li, C., Xiong, Z., Zhang, J., and Wu, C. (2015). The strengthening role of the amino group in metal-organic framework MIL-53 [Al] for methylene blue and malachite green dye adsorption. *Journal of Chemical & Engineering Data* 60 (11): 3414–3422.

67 Hamedi, A., Zarandi, M.B., and Nateghi, M.R. (2019). Highly efficient removal of dye pollutants by MIL-101 [Fe] metal-organic framework loaded magnetic particles mediated by Poly L-Dopa. *Journal of Environmental Chemical Engineering* 7 (1): 102882.

68 Azad, F.N., Ghaedi, M., Dashtian, K. et al. (2016). Ultrasonically assisted hydrothermal synthesis of activated carbon–HKUST-1-MOF hybrid for efficient simultaneous ultrasound-assisted removal of ternary organic dyes and antibacterial investigation: taguchi optimization. *Ultrasonics Sonochemistry* 31: 383–393.

69 Wang, Q., Wang, G., Liang, X. et al. (2019). Supporting carbon quantum dots on NH2-MIL-125 for enhanced photocatalytic degradation of organic pollutants under a broad spectrum irradiation. *Applied Surface Science* 467: 320–327.

70 Ramezanalizadeh, H., Zakeri, F., and Manteghi, F. (2018). Immobilization of BaWO4 nanostructures on a MOF-199-NH2: an efficient separable photocatalyst for the degradation of organic dyes. *Optik* 174: 776–786.

71 Cheng, G., Li, X., Li, X. et al. (2022). surface imprinted polymer on a metal-organic framework for rapid and highly selective adsorption of sulfamethoxazole in environmental samples. *Journal of Hazardous Materials* 423: 127087.

72 Zhou, Y., Fang, F., Lv, Q. et al. (2022). Simultaneous removal of tetracycline and norfloxacin from water by iron-trimesic metal-organic frameworks. *Journal of Environmental Chemical Engineering* 10 (3): 107403.

73 Pooresmaeil, M. and Namazi, H. (2021). Chitosan coated Fe3O4@ Cd-MOF microspheres as an effective adsorbent for the removal of the amoxicillin from aqueous solution. *International Journal of Biological Macromolecules* 191: 108–117.

74 Klongklaew, P. and Bunkoed, O. (2021). The enrichment and extraction of parabens with polydopamine-coated microporous carrageenan hydrogel beads incorporating a hierarchical composite of metal-organic frameworks and magnetite nanoparticles. *Microchemical Journal* 165: 106103.

75 Yang, M., Guo, P., Feng, X. et al. (2021). Solid solution approach to the design of copper mixed-triazolate multivariate-MOFs for the efficient adsorption of triclosan. *Microporous and Mesoporous Materials* 324: 111297.

76 Chen, H., Wan, J., Yan, Z. et al. (2022). Construction of ultra-high defective iron-based metal-organic frameworks with small molecule acid regulator for enhanced degradation of sulfamethoxazole. *Journal of Cleaner Production* 348: 131367.

77 Asadi, F., Forootanfar, H., Ranjbar, M., and Asadipour, A. (2020). Eco friendly synthesis of the LiY [MoO4] 2 coral-like quantum dots in biotemplate MOF [QD/BioMOF] for in vivo imaging and ibuprofen removal from an aqueous media study. *Arabian Journal of Chemistry* 13 (11): 7820–7828.

78 Muelas-Ramos, V., Sampaio, M.J., Silva, C.G. et al. (2021). Degradation of diclofenac in water under LED irradiation using combined g-C3N4/NH2-MIL-125 photocatalysts. *Journal of Hazardous Materials* 416: 126199.

79 Huang, P., Wang, X., Zhao, J. et al. (2022). Hollow Co-MOF-74 incorporated electrospun nanofiber membranes with hierarchical structures for enhanced removal of polycyclic aromatic hydrocarbons by drain-type adsorption. *Chemical Engineering Journal* 449: 137759. https://doi.org/10.1016/j.cej.2022.137759

80 Fan, C., Dong, H., Liang, Y. et al. (2019). Sustainable synthesis of HKUST-1 and its composite by biocompatible ionic liquid for enhancing visible-light photocatalytic performance. *Journal of Cleaner Production* 208: 353–362.

81 Akpinar, I. and Yazaydin, A.O. (2018). Adsorption of atrazine from water in metal-organic framework materials. *Journal of Chemical & Engineering Data* 63 (7): 2368–2375.

82 Jung, B.K., Hasan, Z., and Jhung, S.H. (2013). Adsorptive removal of 2, 4-dichlorophenoxyacetic acid [2, 4-D] from water with a metal-organic framework. *Chemical Engineering Journal* 234: 99–105.

83 Mirsoleimani-azizi, S.M., Setoodeh, P., Samimi, F. et al. (2018). Diazinon removal from aqueous media by mesoporous MIL-101 [Cr] in a continuous fixed-bed system. *Journal of Environmental Chemical Engineering* 6 (4): 4653–4664.

84 Wang, B., Yang, Y., Lu, Y. et al. (2020). Rapid and efficient removal of acetochlor from environmental water using Cr-MIL-101 sorbent modified with 3, 5-Bis [trifluoromethyl] phenyl isocyanate. *Science of the Total Environment* 710: 135512.

85 Huang, X., Feng, S., Zhu, G. et al. (2020). Removal of organic herbicides from aqueous solution by ionic liquid modified chitosan/metal-organic framework composite. *International Journal of Biological Macromolecules* 149: 882–892.

86 Ahmad, M., Chen, S., Ye, F. et al. (2019). Efficient photo-Fenton activity in mesoporous MIL-100 [Fe] decorated with ZnO nanosphere for pollutants degradation. *Applied Catalysis B: Environmental* 245: 428–438.

87 Sajjadi, S., Khatami, A., Bagheri, N. et al. (2019). degradation of diazinon pesticide using catalyzed persulfate with Fe3O4@ MOF-2 nanocomposite under ultrasound irradiation. *Journal of Industrial and Engineering Chemistry* 77: 280–290.

88 Oladipo, A.A., Vaziri, R., and Abureesh, M.A. (2018). Highly robust AgIO3/MIL-53 [Fe] nanohybrid composites for degradation of organophosphorus pesticides in single and binary systems: application of artificial neural networks modelling. *Journal of the Taiwan Institute of Chemical Engineers* 83: 133–142.

89 Fakhri, H. and Bagheri, H. (2020). Highly efficient Zr-MOF@ WO3/graphene oxide photocatalyst: synthesis, characterization and photodegradation of tetracycline and malathion. *Materials Science in Semiconductor Processing* 107: 104815.

90 Chang, P.-H., Chen, C.-Y., Mukhopadhyay, R. et al. (2022). Novel MOF-808 metal-organic framework as highly efficient adsorbent of perfluorooctane sulfonate in water. *Journal of Colloid and Interface Science* 623. https://doi.org/10.1016/j.jcis.2022.05.050

91 Zhao, C., Wang, J., Chen, X. et al. (2021). Bifunctional Bi12O17Cl2/MIL-100 [Fe] composites toward photocatalytic Cr [VI] sequestration and activation of persulfate for bisphenol a degradation. *Science of the Total Environment* 752: 141901.

92 Riley, B.J., Vienna, J.D., Strachan, D.M. et al. (2016). Materials and processes for the effective capture and immobilization of radioiodine: a review. *Journal of Nuclear Materials* 470: 307–326.

93 Rowsell, J.L.C. and Yaghi, O.M. (2004). Metal-organic frameworks: a new class of porous materials. *Microporous and Mesoporous Materials* 73: 3–14.

94 Furukawa, H., Cordova, K.E., O'Keeffe, M., and Yaghi, O.M. (2013). The chemistry and applications of metal-organic frameworks. *Science* 341: 1230444.

95 Yuan, S., Feng, L., Wang, K., Pang, J., Bosch, M., Lollar, C., Sun, Y., Qin, J., Yang, X., Zhang, P., Wang, Q., Zou, L., Zhang, Y., Zhang, L., Fang, Y., Li, J., Zhou, H.C. (2018). Stable Metal-Organic Frameworks: design, Synthesis, and Applications. *Advanced Materials* 30: 1704303. https://doi.org/10.1002/adma.201704303

96 Jiang, H.-L., Ding, M., and Cai, X. (2019). Improving MOF stability: approaches and applications. *Chemical Science* 10: 10209–10230.

97 Gamage, N.D.H., McDonald, K.A., Matzger, A.J. (2016). MOF-5-polystyrene: direct production from monomer, improved hydrolytic stability, and unique guest adsorption. *Angewandte Chemie International Edition* 55: 12099–12103. https://doi:10.1002/anie.201606926

98 Xizhen, L., Dawei, F., Chen, Y.P., Liu, T. F., Wang, X., Zhou, H.C. (2015). The preparation of an ultrastable mesoporous Cr[III]-MOF via reductive labilization. *Chemical Science* 6: 7044–7048. https://doi.org/10.1039/C5SC02587G

2

Metal-Organic Frameworks and Their Stepwise Preparatory Methods (Synthesis) for Water Treatment

Debarati Chakraborty and Prof. Siddhartha S. Dhar

Department of Chemistry, National Institute of Technology, Silchar, Cachar, 788010, Assam

2.1 Introduction

Water is a valuable resource and the foundation of all life [1]. Despite the fact that water covers 72% of the earth's surface, just 5% of that water is freshwater. Due to the exponential expansion of climate change, water resources are becoming increasingly scarce. The need to address the issue of water scarcity is urgent [2]. One of the primary strategies for addressing water shortages is meant to be the reduction of water pollution [3–8]. The newly developed porous MOF has seen extensive use recently [9–15], particularly in separation applications [16–18]. They are composed of multipurpose inorganic metal ions organic linkers [19–22]. They are the primary potential resources for the formation of adsorbents and separation membranes due to their distinctive pore architectures, adjustable pore sizes, and abundance of sites for adsorption, large surface area, and highly porous surfaces [23, 24].

Separation technique was established to be crucial and efficient in the field of wastewater treatment. Due to the fact that it uses little energy, membrane separation has attracted a lot of attention. This is because molecules can be allowed to pass through membrane pores selectively without any phase shift or chemical intervention. In this way, membrane separation has advanced with significant growth over the course of the past few years [25]. It has been considered a method with great potential for resolving energy and ecological problems. There is always a balance of selectivity and permeability when using membranes to separate, purify, and decrease diverse feed liquid elements through selective separation [26, 27]. The combined effect of membrane separation and X lessens the selectivity-permeability balance by merging the best features of the components. Though, due to the filler's lack of compatibility with the polymer, the expansion and strategy of fresh fillers is essential. Materials like zeolite, ceramic, carbonyl, etc., that can be used as membranes are crucial to the process of membrane separation. Selectivity, permeability, and antifouling characteristics in a membrane material are also crucial. To that end, it is

Metal Organic Frameworks for Wastewater Contaminant Removal, First Edition. Edited by Arun Lal Srivastav, Lata Rani, Jyotsna Kaushal, and Tien Duc Pham.

necessary to locate a material that not only offers a solid basis for the progress of high-quality membranes but also makes for exceptional fillers [28–30].

The class of MOFs is widely regarded as one with considerable potential for porous crystalline materials because of their unique functional and structural characteristics and they generally consist of metal centers and organic linkers [31–36]. The organic polymers, which act as linkers, connect to the metal ion, resulting in the basic primary units for the formation of porous three dimension structure MOFs. Hence, the ions of the metal and organic linker used in the synthesis of MOFs are generally called primary building units (PBUs). In general, the first row transition series metal ions such as Fe^{3+}, Cr^{3+}, Zn^{2+}, and Co^{2+} are used as connectors for the formation of MOFs [1, 37, 38]. The metal ions or nodes are connected through organic linkers containing functional groups forming coordination bonds such as carboxylate, phosphate, sulfonate, amine, nitrile, etc. [39–41]. Instead of using just metal ions, organic linkers in MOFs are joined by metal-oxygen-carbon clusters. Two components – secondary building units (SBUs), which contain cluster or metal ion nodes, and organic linkers – can be combined to create MOFs [42].

2.2 Classification of Metal-Organic Frameworks

The following categories of MOFs have been created based on structural characteristics: [43]

a) **Rigid frameworks**: These porous frameworks are strong and robust, and they maintain their porous structure upon the guest molecule adsorption or desorption [44].
b) **Flexible/dynamic frameworks**: Flexible frameworks exhibit a significant change in shape upon the addition or removal of foreign particles and are also influenced by environmental variables like pressure and temperature. Some MOFs display a breathing effect during adsorption and desorption, including MIL-5, MIL-8, and SNU-M10 [45–47].
c) **Open metal site**: The presence of an open metal site can escalate MOF performance. The capability of MOFs to accumulate CO_2 is improved by the presence of water in open metal site MOFs [48].
d) **Surface functionalized framework**: Grafting functional groups onto the surface of MOFs can improve their capacity for adsorption. When arylamine, alkylamine, and hydroxyl functional groups are embedded on the exterior of the porous framework through ligand coordination to an unsaturated metal center, it has been reported that this increases the selectivity and capacity of MOFs for CO_2 adsorption [49–52].

Traditionally, organic linkers and metal ions are combined under mild circumstances to create MOFs, which have a crystalline and porous network. "Modular synthesis" is the term used for this. Different synthesis techniques have been created and used to synthesize these materials over the past 20 years. These can generally be divided into two categories: traditional solvothermal methods and novel methods [53].

2.3 Synthesis of MOFs

2.3.1 Conventional Solvothermal/Hydrothermal and Non-Solvothermal Method

The majority of MOFs seen nowadays are typically created using solvothermal or hydrothermal methods with traditional electric heating. The standard electric heating is a method that includes heating the initial precursor within a sealed vessel with an autogenous temperature and pressure below 250°C [54]. The discovery of several crystal-like materials with simple to complex structures has so far been made using the solvothermal and hydrothermal methods that are also frequently used during the study of phase transitions, crystal development and morphology, and the detection of newly developed porous materials. The word "solvothermal" refers to a reaction where the solvent is an organic solvent, while "hydrothermal" refers to a reaction where the solvent is water [55, 56]. In traditional solvent-based thermosynthesis, a combination of solvent-based metal ions and organic linkers is heated in glass vials for temperatures below 400K and in autoclaves lined with Teflon for temperatures above 400K. The temperature, solvent type, reagent concentration, pressure, etc., are synthesis parameters that can be controlled for a particular structure. An essential variable in MOF synthesis is the temperature of the mixture of reaction. When a reaction occurs at a certain degree greater than the boiling point of the solvent, it is referred to as a solvothermal reaction, and when it occurs below the boiling point, it is referred to as a nonisothermal reaction. Another name for this technique is direct precipitation reaction. The crystal morphology that is generated is strongly influenced by the reaction temperature.

When a reaction occurs in an organic-water mixture of solution at a certain temperature higher than 100°C and also under a pressure greater than atmospheric pressure, the process is referred to as a solvothermal synthesis [57, 58]. The beginning reaction mixture is heated in a secured container with a tight seal, like an autoclave made of stainless steel and Teflon to produce pressure. The autogenous pressure inside the closed vessel rises and frequently reaches 15 bar when the temperature rises above atmospheric pressure. This improves the reaction between the reactants and the solvent during the MOF formation process. Depending on the precursors utilized, the time of reaction regarding this kind of synthesis typically ranges between a few hours and a few days. In a sealed vessel with autogenous pressure, starting ingredients are heated in a process similar to solvothermal synthesis. Water serves as the solvent in a heterogeneous reaction known as hydrothermal synthesis, which takes place at pressures more than 1 bar and at temperatures between 0 and 100°C [59]. The physicochemical characteristics of the reactants and products may change due to the water solvent's ability to speed up the reaction. For instance, Yamauchi and colleagues have shown how to synthesize bimetallic zeolite imidazolate frameworks (ZIFs) by using the solvothermal approach to combine cobalt and zinc metal ions into a single crystal. The precursor solutions for ZIF-8 and ZIF-67 are combined, then receive 12 hours of 100°C heating to create the bimetallic ZIFs. Similar to ZIF-8 and ZIF-67, the bimetallic ZIF crystal has a rhombic dodecahedral form. In this crystal, Zn^{2+} and Co^{2+} coexist and are equally dispersed.

The solvothermal and hydrothermal procedures, on the other hand, have some drawbacks that are harmful to the crystallization process. The former technique makes it challenging for the inorganic precursors to dissolve throughout the synthesis process since it uses a non-polar organic solvent. Conversely, the latter technique uses water which, while useful for dissolving inorganic precursors, produces hydrogen bonds that impede connection between framework species and templates and ultimately obstruct the nucleation process. By combining ionothermal synthesis with ordinary heating, these restrictions can be removed. Ionic liquid is used in ionothermal synthesis, an enhanced form of solvothermal synthesis, as a medium for structure formation and dissolution for the breakdown of initial precursors. Salts known as ionic liquids melt at temperatures lower than those required for synthesis (150–220°C) [60]. Additionally, because ionic liquids have a small vapor pressure, it is possible to carry out synthesis at room pressure without having to worry about safety issues caused by the high autogenous pressure produced [61]

2.3.2 Room-Temperature Synthesis

The quick production of MOF-5 materials by combining the starting materials at ambient temperature was first demonstrated by Yan and colleagues in 2002. An appealing way to make MOFs is by combining metal salts and organic linkers at room temperature [62]. Due to the lack of a source of energy needed to initiate nucleation and development, this approach provides a straightforward, quick, and high yield course for the synthesis of MOFs. In fact, this technique has addressed some issues with conventional electric heating, including the need for thermally insensitive starting materials to create MOFs and sluggish reaction times [63]. Furthermore, scaling up MOF manufacture to suit industrial demand is made easier by room-temperature synthesis. The quality of the MOFs generated is a major concern and has been shown to be poor in some instances, despite the fact that this approach provides a superior pathway for their synthesis. The prevalence of MOFs that have been synthesized at ambient temperature has been widely reported [64]. One of the most researched MOFs that can be produced at room temperature is ZIF-8, for example [65]. It was possible to obtain the ZIF-8 product, which had a 403 nm average diameter and a narrow size distribution. To create smaller ZIF-8 nanocrystals, more MeIm is added compared to the zinc supply. MOF-5, HKUST-1, ZIF-67, MIL-53(Al), and MOF-177 are other MOFs that can be synthesized at ambient temperature [66]

2.3.3 Unconventional Methods

In a mortar and pestle, a combination of metal salt and an organic linker is ground without the use of a solvent, followed by heating it gently to get rid of any water or other volatile molecules that were made as a result of the reaction. Mechanochemical method is the name given to this technique [67]. In this process, breaking intramolecular bonds requires mechanical force, after which there is a change in the chemical reaction. Due to the absence of a solvent in this method, it is known to be safe for the environment and to produce a high quantity of goods. Short reaction periods, typically in the range of 10 to 60 minutes, can be used to generate quantitative yields of tiny MOF particles. Metal oxides were frequently chosen as the starting material instead of metal salts, leaving water as the only waste [68].

The first mechanochemical synthesis of MOFs was reported by James and colleagues in 2006. The term "liquid aided grinding" (LAG), first used in 2006 by Jones and colleagues, refers to a grinding method that uses a small amount of solvent. As a solvent improves the molecular mobility of the reactants, the LAG method may stimulate, accelerate, and direct the structure of mechanochemical processes. Later, the ionic liquid-assisted griding (ILAG) method to enable the mechanochemical production of MOFs was presented by Frii and colleagues [69]. Due to the presence of ionic components in the MOF, a tiny quantity of salt was introduced to speed up the reaction and act as a structure-directing catalyst in the pillared MOF production process. The ILAG method was expanded by the same group to include the synthesis of ZIFs. The ineffective mechanosynthesis of ZIFs from ZnO has been aided by the minor addition of ammonium ions. For instance, a novel non-porous particle with qtz structure was created as a result of the reaction between ZnO and EtIm with varied additions of the salts of ammonia [70].

Stepwise ZIF transformations with topological ANA, RHO, and qtz interconversion were seen when the grinding time was changed. Recent research by Tanaka and colleagues has shown that nanosized ZnO powder may be used to mechanically synthesize trimodal pore structure ZIF-8 with hierarchical organization. To enable the conversion of ZnO to ZIF-8 in a solvent-free medium, zinc acetate dehydrate was used as a Zn supply and catalyst. Surface area, shape, and pore distribution of the resulting hierarchical ZIF-8 were compared to those formed from the standard approach. Their research also supported the hierarchical ZIF-8's superior adsorption capacities [71].

2.4 Alternative Synthesis Methods

In addition to these approaches, alternate paths have also been tried. Different morphologies and crystallization rates can affect the material's characteristics, as well as its particle size, size distribution, and adsorption capabilities. For instance, the capacity of molecules for adsorption, separation, and catalysis can all be directly impacted by guest molecule diffusion in porous materials with different particle sizes. Important detours fall within one of several categories:

2.4.1 Microwave-Assisted Synthesis

Microwave (MW) radiation is used in this method to supply reaction energy. Although it has numerous applications in the quick production of nanoporous materials, it is typically used in organic synthesis [72]. The advantages of this approach are quick crystallization, phase selectivity, limited particle size distribution, and morphological controls. With the help of MW technology, a number of MOFs comprising Fe^{3+}, Al^{3+}, Cr^{3+}, V^{3+}, and Ce^{3+} have been created. The first MOF to be created with the MW technique was Cr-MIL-100.

Dipole polarization and ionic conduction are two significant fundamental processes that enable microwaves to heat the molecules they are in contact with. The earlier process is used on polar substances like water molecules that will arrange themselves with the microwave's quickly varying electric field, losing energy as heat due to the result of molecular rotation, friction, and collision [73]. In the latter process, heat is produced due to friction of molecules and dielectric loss when conducting ions in solution move through the solution in constantly changing orientations. In addition, the electrical resistance of semiconducting and conducting materials causes heating to occur when ions or electrons within them form an electric current. The reaction time for microwave-assisted synthesis typically spans from one minute to one hour, and the temperature must be above 100°C. A few benefits of this approach are the insignificant crystallization time, selectivity of phase, limited distribution of particle sizes, and morphological definition. Contrarily, traditional electric heating takes a long time and is ineffective because it heats the walls of the reactor first and then the reaction by conduction [74]. The identical starting materials were also used to synthesize MOF-5 using traditional electric heating for comparison purposes. The findings demonstrate that crystals produced by microwave synthesis were smaller than those produced by traditional electric heating [75]. Subsequently, 30 minutes later, at 180°C and using solvent ethanol, quantitatively pure HKUST-1 crystals were produced. Interestingly, using DMF as a solvent resulted in the almost quantitative synthesis of HKUST-1, taking only 20 minutes [57].

Figure 2.1 shows MOF-5 crystals obtained using microwave synthesis and conventional electric heating:

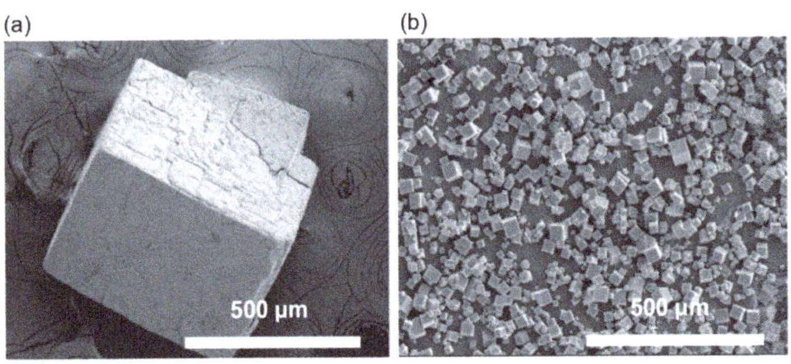

Figure 2.1 Scanning electron micrographs of MOF-5 crystals obtained using (a) microwave synthesis and (b) conventional electric heating [76]. Copyright 2008, Microporous Mesoporous Mater.

2.4.2 Electrochemical Synthesis

Researchers from BASF published the first patent electrochemical generation of MOFs in 2005. Their research emphasis was on the anodic oxidation of the appropriate metals to create Cu, Zn, Mg, and Co based MOFs. The manufacture of MOFs using this synthetic procedure is a metal-free procedure that delivers through the anode's metal ions suspension to a mixture of reaction including a linker, which is organic, and salt conductivity in electrochemical cells. Protic solvents are employed to stop metal from depositing on the cathode, which results in the production of H_2. This strategy is advantageous for industrial applications since MOFs are continually manufactured without salts of metal. Hartmann and colleagues demonstrated the synthesis using two electrodes of copper with a voltage of 15 V and 0.05 mA current [77]. However, in comparison to crystals synthesized using the solvothermal process and reflux under atmospheric pressure, these ones have a reduced pore volume and surface area. According to chemical and thermogravimetric analysis, this is caused by the blockage of MTBS and H_3BTC in the MOF pores.

It is interesting to note that electrochemical synthesis is expanded to include the formation of thin films of MOF on the electrode and subsequently producing MOF in powder form. Electrochemical synthesis has been used by De Vos and colleagues to generate HKUST-1 thin films. By changing the synthetic circumstances, such as the voltage applied, the ratio of water:ethanol and the reaction time, HKUST-1 crystals with sizes ranging from 2–50 m is produced. For instance, increasing the voltage from 2.5 to 25 V results in a larger metal ion concentration near the surface, which leads to the formation of smaller crystal coatings. Water is added to the electrochemical synthesis, which delays the crystallization process and causes larger crystals to form. Additionally, tracks on the printed circuit board were used to demonstrate the electrochemical development of the HKUST-1 layer on patterned copper. The achievement of patterned MOF film growth may make it easier to incorporate MOF material into practical devices [77, 78].

Figure 2.2 shows HKUST-1 crystals synthesized under different synthetic conditions using electrochemical synthesis:

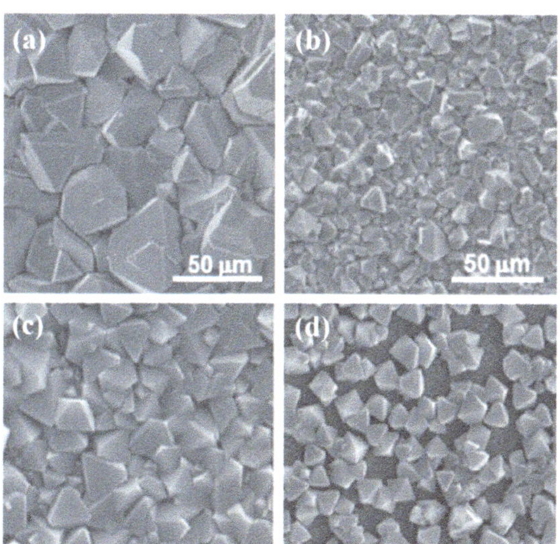

Figure 2.2 Scanning electron micrographs of HKUST-1 crystals synthesized under different synthetic conditions using electrochemical synthesis. (a) 50:50 water:ethanol, 2.5 V, 30min; (b) 25:75 water:ethanol, 2.5 V, 10min; (c) 25:75 water:ethanol, 25 V intermittently (block wave; 0.5 Hz), 45 s; (d) 25:75 water:ethanol, 25 V intermittently (block wave; 0.5 Hz), 15 s.

2.4.3 Sonochemical Synthesis

For many years, sonochemical synthesis has been used in the production of nanostructured materials, analytical chemistry, and organic compounds. This method typically involves the contact of the mixture of reaction with high-intensity ultrasound around 20 kHz–15 MHz [79]. The primary cause of the chemical outcome of ultrasound is audio cavitation, which is the emergence, expansion, and implosion of bubbles in liquids. The sound field will cause the created bubbles to vibrate. They will then develop to a size of typically tens of micrometers while also efficiently storing ultrasonic energy. The foams will then quickly outgrow their boundaries and burst, releasing the energy they had been holding. Hot spots are created as a result of the energy released during deflation of the bubbles, which have cooling and heating rates of >1010 Ks^{-1} and at a temperature of 5000 K. They also have 1000 bar pressure. A laboratory ultrasonic horn can deliver ultrasonication at a 100-fold higher intensity than an ultrasonic cleaning bath, making it an alternative piece of equipment for sonochemical reactions [80]. Additionally, whilst the reaction container is submerged in the ultrasonic cleaning bath, the sample container is immediately immersed in an ultrasonic horn, and this is the major characteristic used to differentiate between ultrasonic horns and ultrasonic cleaning baths. For the synthesis of MOFs, sonochemical synthesis is favored over other techniques because of its ease of use, product selectivity, and speed of reaction [1]. For instance, Ahn and colleagues validated the sonochemical production of MOF-5 in a Pyrex horn-type reactor utilizing 1-methyl-2-pyrrolidinone as the solvent. The superior crystals of S-MOF-5 were contrasted with MOF-5 crystals that were produced under typical circumstances (C-MOF-5). The findings indicated that high-quality S-MOF-5 crystals can be created in as little as 30 minutes, whereas high-quality C-MOF-5 crystals are produced after 24 hours. The S-MOF-5 crystals formed were 60 times smaller than C-MOF-5 (900 m), and they are single-phase high-quality cubic crystals with a size of 5 to 25 m, according to SEM pictures. Higher than that of C-MOF-5 (100°C), the temperature of the reaction employed for the production of high-quality S-MOF-5 crystals was maintained at 155°C. However, it was shown that only poor-quality MOF-5 crystals are produced when C-MOF-5 is synthesized at 155°C. This research demonstrated that sonochemical synthesis, as opposed to conventional synthesis, may produce high-quality crystals in a shorter amount of time.

In the kinetic study, Jhung and colleagues effectively created MIL-53(Fe) crystals utilizing an ultrasonic horn and sonochemical conditions at a comparatively low temperature (50–80°C). We compared these superior MIL-53(Fe) crystals to those made using traditional electric heating and microwave synthesis. The crystals produced by sonochemical and microwave synthesis were much smaller and more homogenous than those produced by traditional synthesis. This might be because MIL-53(Fe) simultaneously crystallized and nucleated under normal circumstances [81, 82].

Figure 2.3 shows images of different forms of synthesis of UMCM-15, MOF-5, and MIL-101:

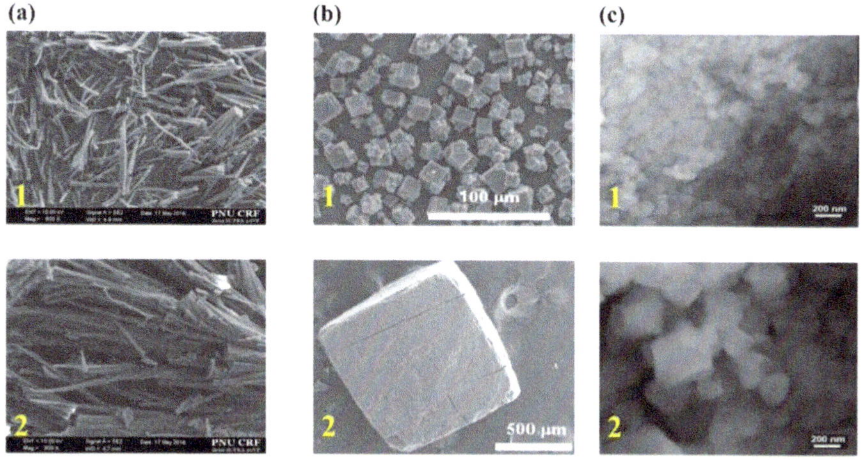

Figure 2.3 SEM images (a) 1. Microwave-assisted synthesis of UMCM-15 (15 min) 2. Solvothermal synthesis of UMCM-15 (32 h) (b) 1. Ultrasonic-assisted synthesis of MOF-5 (30 min) 2. Solvothermal synthesis of MOF-5 (24 h) (c) 1. Mechanochemical synthesis of MIL-101 (Cr) (4 h) 2. Solvothermal synthesis of MIL-101 (Cr) (20 h) [83]. Copyright 2022, International Journal of Hydrogen Energy.

2.4.4 Surfactant-Assisted Synthesis

A method of soft templation known as surfactant-assisted synthesis has recently become popular for the synthesis of MOFs and mesoporous silicas and oxides of metal. Typically, in this method, surfactant is used as a template for controlling the production of MOFs. The choice of surfactant is important because it will have a noteworthy impact on the structure, surface area, and pore size of the MOF materials that are produced at the end of the reaction. Anionic, cationic, and nonionic surfactants are the three main categories of surfactants. Surfactant molecules group together to create micelles during the process, which are followed by rodlike micelles [84]. These micelles serve for the synthesis of the framework. The final output must then be produced by removing the template [85]. However, the quantity of Pluronic F127 utilized has no effect on the size of the nanocrystals. Although, with the addition of concentrated acetic acid results in the synthesis of bigger nanocrystals, because of the carboxylate linkers' deprotonation, which slows down crystal growth, is reduced. Recently, the Li and Zheng groups reported the ZIF-8 nanocrystals synthesis with the assistance of surfactant with particle sizes of >100 nm and high surface areas of 1360 $m^2\ g^{-1}$. The structure-directing agent is a 1:1 molar mixture of nonionic surfactants Span 80 and Tween 80.2. Their research showed that altering the quantity of the combined nonionic surfactant in the reaction mixture had no effect on the size of the nanocrystals [86].

It is interesting to note that the mixed nonionic surfactant is crucial for the creation of the ZIF-8 phase. The size of ZIF-8 nanocrystals will alter as the temperature of the reaction changes. Owing to their slower crystal development, relatively tiny, irregular crystals can be produced at lower reaction temperatures. Mesoporous MOFs can be synthesized using the surfactant-assisted synthesis method (mesoMOFs). Metal-organic frameworks known as mesoMOFs have pores that range in size from 2 to

50 nm. Mesoporous MOFs are a possible candidate for a number of uses, including storage of gas, heterogeneous catalysis, and adsorption, due to the presence of mesopores. The processes used to create mesoMOFs are the same as those used to create microporous MOFs. For instance, using a cooperative template system, Zhou and colleagues have effectively synthesized mesoMOFs [87]. Citric acid, a chelating agent, is in charge of establishing a bond via coulombic attraction and interaction between the surface active molecules and the metal ions. The citric acid serves as the co-template during the synthesis of the mesoMOFs, while the surfactant CTAB directs the structure. After the surfactant/chelating agent is employed separately, mesoMOFs cannot be produced. This is brought on by the minimal interaction between the molecules of the surfactant and the MOF precursors [88].

2.4.5 Layer-by-Layer Synthesis

The approach of layer-by-layer preparation is employed for thin-film MOFs. The procedure is built on surface chemistry and involves immersing a functionalized organic interface successively into metal ion and organic linker solutions. It has been found that the order in which the reactants are added affects the thin film's orientation [89].

Surface plasmon resonance (SPR) spectroscopy has been used to examine the kinetics for stepwise creation in this process. The two main elements that influence the pace of growth of MOFs are the source of metal and termination of surface. For the substrates functionalized with several functional groups, including COOH and OH, highly oriented growth was seen [90]. In addition to these synthesis pathways, other pathways have been used to create MOFs, including chemical solution deposition, postsynthesis modification, and the ionothermal method [91]. In a paper, Banerjee and colleagues describe the preparation of Fe-based MOF using gel-degrading Fe-metallogels. Other MOFs, however, have not yet been reported using this method [92].

Figure 2.4 shows the various pathways to synthesize MOFs:

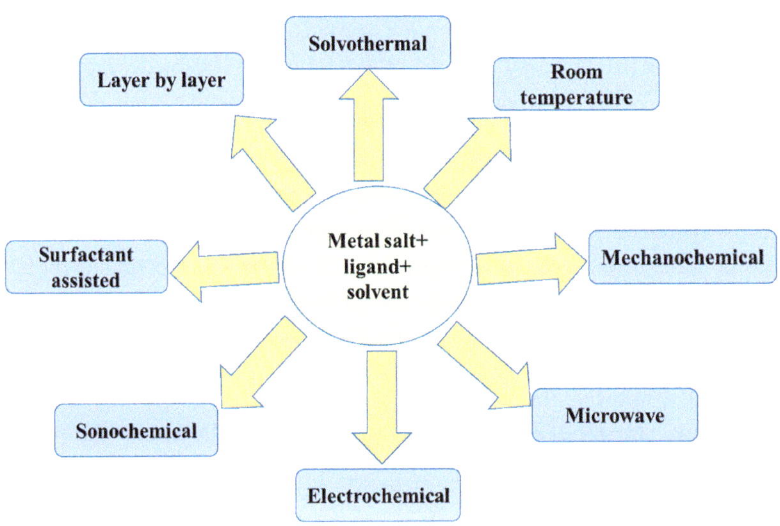

Figure 2.4 A schematic diagram for various pathways to synthesize MOFs.

2.5 Factors Affecting the Synthesis of MOFs

2.5.1 Solvents

The solvent system is vital for both the MOF's synthesis and determining its morphology. Solvents can act as space-filling molecules or they can coordinate with metal ions. They instead serve as a structure-directing agent. Solvents utilized in MOF synthesis should be polar and have a high boiling point (Figure 2.5(a)) [93].

The reaction medium has an impact on the MOF synthesis process because of the solvent's polarity in addition to the solubility and protolysis characteristics of the organic linker. Additionally, it has been noted that the same reaction conditions using various solvent systems can produce MOFs with various morphologies (Figure 2.5(b)) [94]. This might occur as a result of the variable rates at which organic linkers are deprotonated in various solvent systems. According to Banerjee et al., various crystal structures of MOFs containing magnesium and PDC (3, 5-pyridine dicarboxylic acid) were created under the same circumstances utilizing various solvent systems. They discovered that the MOF network's dimensionality is

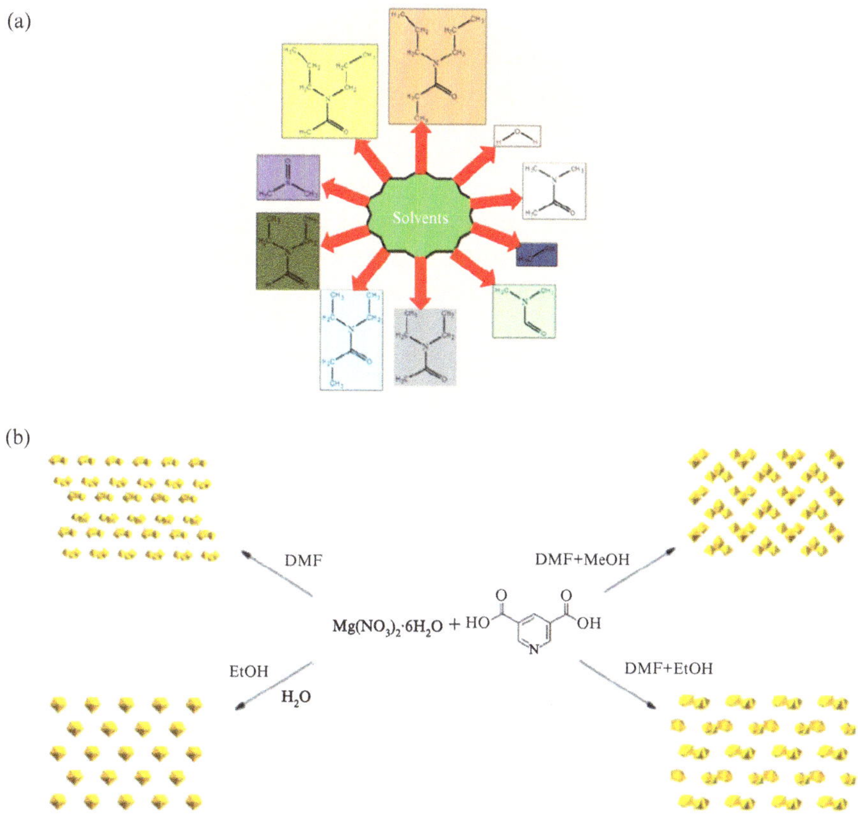

Figure 2.5 (a) Solvents used in the synthesis of MOFs (b) Effect of solvent system on the morphology of MOFs [97]. Copyright 2011, Journal of Coordination Chemistry.

determined by the solvent's capacity to coordinate with metal. When DMF/MeOH and EtOH/H_2O are utilized as solvents, H_2O has the highest affinity toward Mg while EtOH and MeOH have no propensity to coordinate with metal centers [95]. Different solvent systems cause the synthesis of MOFs with varying pore sizes in addition to structure. The size of the solvent molecule has been found to be associated with the change in pore size caused by various solvents. The order of the molecules' sizes in the three solvents employed is DMP > DMA > DMF. The reduction in pore size occurs in the same pattern [96].

2.6 Temperature and pH Effects on the Synthesis of MOFs

MOF synthesis is significantly influenced by temperature and reaction media pH. At various pH levels, linkers can adopt various coordination modes. Additionally, as the pH value rises, the linker deprotonation intensity rises. It has been noted that an interconnected network forms at higher pH levels and an unconnected network does not [98]. The pH of the reaction media also affects the color of MOF compounds. Additionally, it has been confirmed that higher-dimensional molecules develop at higher pH levels. Table 2.1 shows the impact of pH on different MOF properties according to the work of Chu et al. [99].

Another significant element that influences the characteristics of produced MOFs is reaction temperature. Due to the high solubility of the reactants at high temperatures, increased crystallization is encouraged, leading to the creation of big, high-quality crystals [100]. The temperature of the reaction mixture has an influence on the speed of crystal nucleation and growth. The temperature of the reaction medium can also be changed to alter the morphology of synthesized MOFs. MOFs of Tm-succinate are produced at various temperatures using the same empirical formula but having monoclinic and triclinic shape. Two Ho-succinate MOFs were created by Bernini et al. [101, 102]. They observed that the MOF created using the hydrothermal approach at a higher temperature is thermally more stable than the MOF created at room temperature [103].

Table 2.1 Effect of pH on MOFs [1].

Sl. No	Properties of Co-MOFs	Complex-1	Complex-2	Complex-3
1	Chemical Formula	$[Co_2(L)(HBTC)_2(\mu_2\text{-}H_2O)(H_2O)_2].3H_2O$	$[Co_3(L)_2(BTC)_2].4H_2O$	$[Co_2(L)(BTC)(\mu_2\text{-}OH)(H_2O)_2].2H_2O$
2	Morphology	Monoclinic	Monoclinic	Orthorhombic
3	Space group	P2/c	C2/c	Pccn
4	pH	5	7	9
5	Color	Pink	Purple	Brown

2.7 Water Regeneration and Wastewater Treatment Using MOF Membranes

Membrane separation technology has shown a lot of potential for preserving the environment and saving energy in recent years. For instance, gas separation based on a membrane has a lot of potential in real-world applications [104]. However, a major problem with MOFs use in aqueous separation has always been its hydrothermal stability. It has a strong future in WWT since the dimensions of the pores found in the MOF membranes can be changed. The majority of the membranes of MOF are crystalline microporous membranes created by MOF crystals self-assembling on porous organic-inorganic substrates [105]. They possess excellent separation capacities because of their constantly varying pore diameters and chemical characteristics. The MOF microporous membrane is very customizable in terms of design and functionality. It can be rationally controlled via a variety of techniques or by interacting with organic ligands and inorganic metal ions [106]. The use of MOF membranes in WWT will be highlighted in this chapter, and it has gained significance as a result of the recent growth of MOFs [107].

2.8 Membrane Filtration

To filter substances down to the molecular level, a technique known as membrane filtration is necessary. In particular, the membrane acts as a barrier, preventing macromolecules and smaller particles from crossing into the cytoplasm but enabling small molecules, solvents, inorganic ions, and so on, to do so. A few factors that aid membrane filtration have been established via our research; the membranes of the MOF can further improve the filtration effectiveness of any filtering method. The applications of MOF in reverse osmosis (RO), MF, UF, and NF filtration processes are discussed here.

2.9 Microfiltration (MF)

Microfiltration (MF) uses membrane sieving to filter and separate by using the driving force behind the differential pressure across the membrane. Large colloids, bacteria, certain viruses, and suspended solids are among the particles between 0.1 and 1 mm that are retained by MF, whereas macromolecular organic substances and inorganic salts can pass through [108].

For this reason, membrane microfiltration usually makes use of MOF composite membranes, which consist of MOFs implanted into the microfiltration membrane to improve its performance. Micropollutants in water were filtered out using this membrane. The results showed that its permeability of water nearly doubled, and its capacity of adsorption improved by more than 40% [109]. ZIF-8's hydrophobicity is thought to have played a role, as it reduced the resistance water encountered upon contact with the membrane surface, increasing the MOF composite permeability membrane [110]. Hydrogen bonding between the MOF particles and the water also contributed to their hydrophilicity and, by extension, their permeability [111]. Numerous research depicts that microfiltration membranes of the MOF composite have enhanced permeability of the membrane, making them useful in a variety of settings, including wastewater treatment and membrane renewal of electronics and medical, food, and pharmaceutical industries. Thus, exploration in areas apart from water treatment has a favorable future [112].

2.10 Ultrafiltration (UF)

Ultrafiltration (UF) membranes, in contrast to microfiltration, contain smaller holes (varying from 0.001 to 0.1 mm) and involve greater compression as the permeation-inducing force. Additionally, because of the special characteristics of ultrafiltration, such as its greater separation efficiency, low working temperatures, and superior stability, it is efficient at eradicating compounds with high molecular weight, germs, and contaminated water containing organic particles. Notably the most popular technique for separating water for purification, ultrafiltration is also used extensively as a pretreatment for RO [113]. Due to the ultrafiltration membrane's superior efficiency in eliminating suspended bacteria, macromolecules, nanoparticles, and other contaminants, it has been extensively researched and used in the industry. As a result, ultrafiltration membrane modification is also receiving attention [114]. In the studies discussed next, existing materials can be used as filler for the polymeric ultrafiltration membrane in addition to the MOF being mixed with other substances to modify their pertinent properties. They discovered that the flux, rejection rate, antifouling characteristics, and hydrophilicity were all improved by 3.0 wt% of UiO-66 [115]. Results demonstrated that the hZIF8/PSF UF membrane maintained good rejection and antifouling capabilities while having water permeability that was around 2.8 times better than the membrane PSF [116]. The outcomes demonstrated that the CA/MOF@GO membrane had higher performance of antifouling and had a maximum water flux of 183.51 L/m^{-2} h 1 [117].

2.11 Nanofiltration (NF)

Nanofiltration (NF) is a relatively new membrane separation technology that is more commonly identified as low-pressure RO. This separation method lies between RO and ultrafiltration, and it allows certain solvents and inorganic salts to pass through the membrane. The NF technique was widely implemented in the water treatment industry because of its usefulness in ecological contexts. Additionally, the usage of membranes has been extended beyond just water purification to encompass the filtration and concentration of organic solutions with the advent of organic solvent nanofiltration (OSN), a recently discovered NF separation technology [118].

Golpour et al. developed a novel MOF membrane for nanofiltration that avoids the presence of the kinetic hydrate inhibitor (KHI). First, Yuan et al. developed a novel, pure ZIF-300 MOF membrane for use in the elimination of ions of heavy metals from sewage. The manufactured ZIF-300 membrane has an elimination rate of 99.21% for CuSO4, permeability of water of 39.2 L/m^2 h bar, and a high stability energy. It is anticipated that various pure ZIF-300 membranes would be utilized for heavy metal ion nanofiltration in wastewater [119]. By utilizing the Langmuir-Schaefer technique, Navarro et al. introduced MIL-101(Cr) nanoparticles to the thin-film nanocomposite (TFN) membrane. Methanol permeability was increased, while elimination of sunset yellow and rose bengal was maintained at above 90% [120, 121].

2.12 Reverse Osmosis (RO) and Forward Osmosis (FO)

RO is the most common membrane-based purification technique. With the increase in pressure above the osmotic pressure, water permeates in opposition to the osmotic gradient. FO, a newer membrane separation method, depends on spontaneous diffusion down the osmotic gradient [122].

Strong multivalent ion repulsion (99.0% for Al^{3+}, 86.3% for Ca^{2+}, and 98.0% for Mg^{2+}) and high permeability were both characteristics of these membranes (0.28 L/m^2 h 1 bar 1 lm) [123]. Hee et al. combined sulfuric acid-treated HKUST-1[Cu_3(BCT2)].MOF and a polysulfone (PSF) membrane to create a MOF composite membrane by RO. The experimental results demonstrated a 33% improvement in the flux of water and a modest rise in salt elimination to 96% due to high porosity of the structure, hydrophilicity, and antifouling characteristics of the MOF membrane [124].

Forward osmosis is an emerging filtration technology. RO is driven by hydraulic pressure and FO by osmotic pressure. FO is desirable since it operates at minimum pressure, uses less energy, and less fouls. MOF nanoscale membrane modification can lower mass transfer barriers and inner concentration polarization [125].

RO and FO are generally mixed with other filtration technologies or procedures to remove impurities. Song et al. and Wei et al. used this technology to treat hair color wastewater and shale gas reflux and produced water [126]. Coagulation as a pretreatment reduces fouling of membrane in ultrafiltration and RO. Ultrafiltration-reverse osmosis can better purify wastewater for reuse. It enables wastewater treatment and water regeneration [127].

2.13 Membrane Distillation (MD)

Membrane distillation is a thermal approach where only the vapor infiltrates the hydrophobic membrane. The earlier studies showed that MD is beneficial for water renewal and wastewater treatment, since it may be powered by waste heat and only vapor penetrated [128]. MD membranes should have low resistance to mass transfer, strong feed-side thermal efficiency, thermostability, and hydrophobicity. MOF functionalization added hydrophobicity to alumina's intrinsic permeability and stability. Fan et al. produced a superhydrophobic poly(vinylidenefluoride) nano membrane for direct contact MD for desalination [129]. According to the findings, the contact angle of water was increased to 138.06 2.18 degrees, the flux of water vapor was increased, and the rate of elimination of NaCl was 99.99 percent. The exclusion of heavy metals in the food industry are the applications for MD besides desalination [130]. By adopting a membrane distillation material that has been photothermally boosted, MD's energy efficiency can be increased even more. The inclusion of MOFs can also further improve the membrane's hydrophobicity, chemical stability, and thermal characteristics [131].

2.14 Membrane Pervaporation (PV)

A membrane separation process called pervaporation (PV) can separate a liquid mixture down to the molecular level. The feed solution containing the components are originally transferred through the membrane and condensed in the osmotic fluid during pervaporation [132]. The difference in vapor pressure between the solution and the permeate vapor drives the pervaporation, making it a more energy-efficient and less power-hungry process than MD. Typically, the permeate flux and separation factor are targeted while deciding on a membrane for pervaporation. One of pervaporation's most important applications is the regaining of organic solvents from wastewater [133]. There are two types of pervaporation membranes: organophilic and hydrophilic. The former was used for the recovery of solvents and the latter for elimination of water [134, 135].

Li et al. used the hydrophobic MAF-6 with PDMS to make a mixed matrix membrane (MMM) for the recovery of ethanol. MAF-6 improved the selectivity, permeability, and stability of the membrane due to its hydrophilic nature and high porous structure. Experiments showed that 15 wt% MAF-6 loading increased the hydrophobicity of the MMM membrane to its maximum, which led to the highest flow (1200 g/m^2h) and separation factor (14.9) [136].

Figure 2.6 shows the use of MOFs for wastewater treatment:

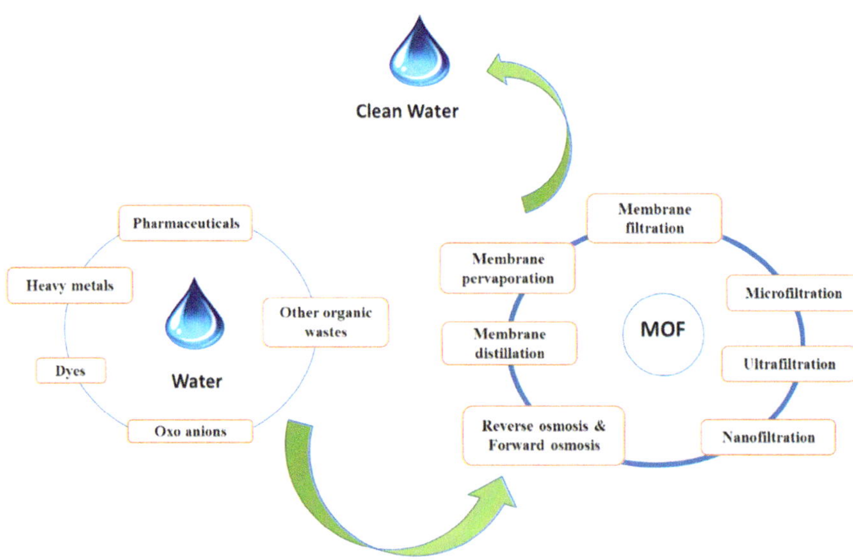

Figure 2.6 A schematic diagram for wastewater treatment by using MOFs.

2.15 Conclusion

As a result of their high adaptability and good compatibility, MOF membranes have seen increased usage in a wide variety of applications in recent years. Furthermore, the permeability and high selectivity of water-stable MOFs has led to an increase in liquid phase separation based on membranes, particularly with regard to wastewater treatment and water regeneration. There are several different categories of MOF membranes and their accompanying methods of synthesis are included in this chapter. MOFs' structural flaws serve a variety of useful roles as well. The many components of MOF membranes, such as MOF particles, membranes, and structural flaws, must be characterized as a means to advance the field of MOF membranes.

With MOFs, a new group of adsorbent materials has emerged. Due to the high capacity of adsorption, rapid adsorption kinetics, and excellent selectivity to harmful compounds, some MOFs are extremely useful for WWT. MOFs may effectively remove heavy metals, antibacterials, dyes, and other contaminants from wastewater when utilized as adsorbents. The primary process that causes adsorption is electrostatic contact. The elimination of contaminants from wastewater is accomplished by the net charge of the MOF surface distributed in the aqueous phase, and interaction with the cation or anion on the adsorbate leads to the formation of chemical bonds. Additionally, some MOFs that have been functionalized, such as those that have been sulfur- and amino-functionalized, are able to cooperate strongly with specific heavy metals, increasing their ability to bind those metals.

Water-stable MOFs are employed in membrane separation and adsorption for wastewater treatment. In practical applications, MOFs' high cost, weak stability, probable toxicity, and poor regeneration create hurdles for future study. All research discoveries must be applied to large-scale industry and be environmentally and socially friendly. Designing affordable and environmentally friendly MOFs is a challenge and opportunity for WWT. Post-modifications or novel MOF designs may reduce MOF toxicity, and fine sculpting may limit harmful chemical release during use. A number of MOFs appear to match the necessities for practical wastewater management, showing improvement for decontamination applications. For wastewater management and water restoration, MOF membranes are an important opportunity and challenge right now. There is still an extended way to go before MOF membranes could be used for large-level water treatment, but upcoming efforts in research are likely to considerably enhance the predictions of MOFs toward real-world application.

References

1 Soni, S., Bajpai, P.K., and Arora, C. (2020 September 20). A review on metal-organic framework: synthesis, properties and application. *Characterization and Application of Nanomaterials* 3 (2): 87–106.

2 Liu, X., Shan, Y., Zhang, S. et al. (2022 March 16). Application of metal-organic framework in wastewater treatment. *Green Energy & Environment (3).*

3 Shakoor, M.B., Nawaz, R., Hussain, F. et al. (2017 December 1). Human health implications, risk assessment and remediation of As-contaminated water: a critical review. *Science of the Total Environment* 601: 756–769.

4 Yang, Y., Ok, Y.S., Kim, K.H. et al. (2017 October 15). Occurrences and removal of pharmaceuticals and personal care products (PPCPs) in drinking water and water/sewage treatment plants: a review. *Science of the Total Environment* 596: 303–320.

5 Sharma, V.K. and Feng, M. (2019 June 15). Water depollution using metal-organic frameworks-catalyzed advanced oxidation processes: a review. *Journal of Hazardous Materials* 372: 3–16.

6 Jun, B.M., Al-Hamadani, Y.A., Son, A. et al. (2020 September 15). Applications of metal-organic framework based membranes in water purification: a review. *Separation and Purification Technology* 247: 116947.

7 Yu, S., Pang, H., Huang, S. et al. (2021 December 15). Recent advances in metal-organic framework membranes for water treatment: a review. *Science of the Total Environment* 800: 149662.

8 Li, W., Liu, Y.Y., Bai, Y. et al. (2020 August 5). Anchoring ZIF-67 particles on amidoximerized polyacrylonitrile fibers for radionuclide sequestration in wastewater and seawater. *Journal of Hazardous Materials* 395: 122692.

9 Liang, Z., Qu, C., Xia, D. et al. (2018 July 26). Atomically dispersed metal sites in MOF-based materials for electrocatalytic and photocatalytic energy conversion. *Angewandte Chemie International Edition* 57 (31): 9604–9633.

10 Wang, H.F., Chen, L., Pang, H. et al. (2020). MOF-derived electrocatalysts for oxygen reduction, oxygen evolution and hydrogen evolution reactions. *Chemical Society Reviews* 49 (5): 1414–1448.

11 Tsumori, N., Chen, L., Wang, Q. et al. (2018 April 12). Quasi-MOF: exposing inorganic nodes to guest metal nanoparticles for drastically enhanced catalytic activity. *Chem* 4 (4): 845–856.

12 Zhang, M., Zhou, W., Pham, T. et al. (2017 September 11). Fine tuning of MOF-505 analogues to reduce low-pressure methane uptake and enhance methane working capacity. *Angewandte Chemie* 129 (38): 11584–11588.

13 Zhu, R., Ding, J., Jin, L., and Pang, H. (2019 June15). Interpenetrated structures appeared in supramolecular cages, MOFs, COFs. *Coordination Chemistry Reviews* 389: 119–140.

14 Cao, Y., Wang, L., Shen, C. et al. (2019 March 15). An electrochemical sensor on the hierarchically porous Cu-BTC MOF platform for glyphosate determination. *Sensors and Actuators B: Chemical* 283: 487–494.

15 Chen, J., Xu, Q., Shu, Y., and Hu, X. (2018 July 1). Synthesis of a novel Au nanoparticles decorated Ni-MOF/Ni/NiO nanocomposite and electrocatalytic performance for the detection of glucose in human serum. *Talanta* 184: 136–142.

16 Kökçam-Demir, Ü., Goldman, A., Esrafili, L. et al. (2020). Coordinatively unsaturated metal sites (open metal sites) in metal-organic frameworks: design and applications. *Chemical Society Reviews* 49 (9): 2751–2798.

17 Kumar, P., Bansal, V., Kim, K.H., and Kwon, E.E. (2018 June 25). Metal-organic frameworks (MOFs) as futuristic options for wastewater treatment. *Journal of Industrial and Engineering Chemistry* 62: 130–145.

18 Candia-Onfray, C., Rojas, S., Zanoni, M.V., and Salazar, R. (2021 April 1). An updated review of metal-organic framework materials in photo (electro) catalytic

applications: from CO2 reduction to wastewater treatments. *Current Opinion in Electrochemistry* 26: 100669.

19 Dong, W., Wang, D., Wang, H. et al. (2019 February 1). Facile synthesis of In2S3/UiO-66 composite with enhanced adsorption performance and photocatalytic activity for the removal of tetracycline under visible light irradiation. *Journal of Colloid and Interface Science* 535: 444–457.

20 Xia, Q., Huang, B., Yuan, X. et al. (2018 November 15). Modified stannous sulfide nanoparticles with metal-organic framework: toward efficient and enhanced photocatalytic reduction of chromium (VI) under visible light. *Journal of Colloid and Interface Science* 530: 481–492.

21 Wang, C., Wang, H., Luo, R. et al. (2017 December 15). Metal-organic framework one-dimensional fibers as efficient catalysts for activating peroxymonosulfate. *Chemical Engineering Journal* 330: 262–271.

22 Wang, C., Kaneti, Y.V., Bando, Y. et al. (2018). Metal-organic framework-derived one-dimensional porous or hollow carbon-based nanofibers for energy storage and conversion. *Materials Horizons* 5 (3): 394–407.

23 Zhang, Y.B., Furukawa, H., Ko, N. et al. (2015 February 25). Introduction of functionality, selection of topology, and enhancement of gas adsorption in multivariate metal-organic framework-177. *Journal of the American Chemical Society* 137 (7): 2641–2650.

24 Li, M., Li, D., O'Keeffe, M., and Yaghi, O.M. (2014 January 22). Topological analysis of metal-organic frameworks with polytopic linkers and/or multiple building units and the minimal transitivity principle. *Chemical Reviews* 114 (2): 1343–1370.

25 Koros, W.J. and Zhang, C. (2017 March). Materials for next-generation molecularly selective synthetic membranes. *Nature Materials* 16 (3): 289–297.

26 Liu, J. and Yang, W. (2012 August 10). Water sustainability for China and beyond. *Science* 337 (6095): 649–650.

27 Shannon, M.A., Bohn, P.W., Elimelech, M. et al. (2010). Science and technology for water purification in the coming decades. *Nanoscience and Technology: A Collection of Reviews from Nature Journals* (452) 337–346.

28 Park, H.B., Kamcev, J., Robeson, L.M. et al. (2017 June 16). Maximizing the right stuff: the trade-off between membrane permeability and selectivity. *Science* 356 (6343): eaab0530.

29 Li, X., Liu, Y., Wang, J. et al. (2017). Metal-organic frameworks based membranes for liquid separation. *Chemical Society Reviews* 46 (23): 7124–7144.

30 Goh, P.S. and Ismail, A.F. (2018 May 15). A review on inorganic membranes for desalination and wastewater treatment. *Desalination* 434: 60–80.

31 Eddaoudi, M., Moler, D.B., Li, H. et al. (2001 April 18). Modular chemistry: secondary building units as a basis for the design of highly porous and robust metal– organic carboxylate frameworks. *Accounts of Chemical Research* 34 (4): 319–330.

32 Rowsell, J.L. and Yaghi, O.M. (2004 August 6). Metal-organic frameworks: a new class of porous materials. *Microporous and Mesoporous Materials* 73 (1–2): 3–14.

33 Yamada, T. and Kitagawa, H. (2009 May 13). Protection and deprotection approach for the introduction of functional groups into metal– organic frameworks. *Journal of the American Chemical Society* 131 (18): 6312–6313.

34 Horike, S., Shimomura, S., and Kitagawa, S. (2009). Soft porous crystals. *Nature Chemistry* 1 (9): 695–704. doi: 10.1038/nchem.444.

35 Férey, G. (2008). Hybrid porous solids: past, present, future. *Chemical Society Reviews* 37 (1): 191–214.

36 Lee, J., Farha, O.K., Roberts, J. et al. (2009). Metal-organic framework materials as catalysts. *Chemical Society Reviews* 38 (5): 1450–1459.

37 Wu, H., Zhou, W., and Yildirim, T. (2009 April 8). High-capacity methane storage in metal– organic frameworks M2 (dhtp): the important role of open metal sites. *Journal of the American Chemical Society* 131 (13): 4995–5000.

38 Zacher, D., Shekhah, O., Woll, W. et al. (2009). Thin films of metal-organic frameworks. *Chemical Society Reviews* 38: 1418–1429.

39 Yang, L.M., Vajeeston, P., Ravindran, P. et al. (2011). Revisiting isoreticular MOFs of alkaline earth metals: a comprehensive study on phase stability, electronic structure, chemical bonding, and optical properties of A–IRMOF-1 (A= Be, Mg, Ca, Sr, Ba). *Physical Chemistry Chemical Physics* 13 (21): 10191–10203.

40 Platero Prats, A.E., Iglesias, M., Snejko, N. et al. (2011 May 4). From coordinatively weak ability of constituents to very stable alkaline-earth sulfonate metal– organic frameworks. *Crystal Growth & Design* 11 (5): 1750–1758.

41 Platero Prats, A.E., de la Peña-o'shea, V.A., Iglesias, M. et al. (2010 February 8). Heterogeneous catalysis with alkaline-earth metal-based MOFs: a green calcium catalyst. *ChemCatChem* 2 (2): 147–149.

42 Serre, C. and Férey, G. (2002). Hydrothermal synthesis, thermal behaviour and structure determination from powder data of a porous three-dimensional europium trimesate: Eu 3 (H 2 O)(OH) 6 [C 6 H 3 (CO 2) 3]· 3H 2 O or MIL-63. *Journal of Materials Chemistry* 12 (10): 3053–3057.

43 Sabouni, R. Carbon dioxide adsorption by metal-organic frameworks (synthesis, testing and modeling).

44 Dinca, M. and Long, J.R. (2005 July 6). Strong H2 binding and selective gas adsorption within the microporous coordination solid Mg3 (O2C-C10H6-CO2) 3. *Journal of the American Chemical Society* 127 (26): 9376–9377.

45 Hamon, L., Llewellyn, P.L., Devic, T. et al. (2009). Co-adsorption and separation of CO2-CH4mixtures in the highly flexible MIL-53(Cr) MOF. *Journal of the American Chemical Society* 131: 17490–17499.

46 Serre, C.M., Mellot-Draznieks, C., Surblé, S. et al. (2007 March 30). Role of solvent-host interactions that lead to very large swelling of hybrid frameworks. *Science* 315 (5820): 1828–1831.

47 Choi, H.S. and Suh, M.P. (2009 September 1). Highly selective CO2 capture in flexible 3D coordination polymer networks. *Angewandte Chemie International Edition* 48 (37): 6865–6869.

48 Schneemann, A., Bon, V., Schwedler, I. et al. (2014). Flexible metal-organic frameworks. *Chemical Society Reviews* 43 (16): 6062–6096.

49 Liang, Z., Marshall, M., and Chaffee, A.L. (2009 May 21). CO2 adsorption-based separation by metal-organic framework (Cu-BTC) versus zeolite (13X). *Energy & Fuels* 23 (5): 2785–2789.

50 Millward, A.R. and Yaghi, O.M. (2005 December 28). Metal– organic frameworks with exceptionally high capacity for storage of carbon dioxide at room temperature. *Journal of the American Chemical Society* 127 (51): 17998–17999.

51 Demessence, A., D'Alessandro, D.M., Foo, M.L., and Long, J.R. (2009 July 1). Strong CO2 binding in a water-stable, triazolate-bridged metal– organic framework functionalized with ethylenediamine. *Journal of the American Chemical Society* 131 (25): 8784–8786.

52 Serre, C., Bourrelly, S., Vimont, A. et al. (2007 September 3). An explanation for the very large breathing effect of a metal-organic framework during CO2 adsorption. *Advanced Materials* 19 (17): 2246–2251.

53 Hu, J., Liu, J., Liu, Y., and Yang, X. (2016 May 19). Improving carbon dioxide storage capacity of metal-organic frameworks by lithium alkoxide functionalization: a molecular simulation study. *The Journal of Physical Chemistry C* 120 (19): 10311–10319.

54 Gordon, J., Kazemian, H., and Rohani, S. (2012 November 1). Rapid and efficient crystallization of MIL-53 (Fe) by ultrasound and microwave irradiation. *Microporous and Mesoporous Materials* 162: 36–43.

55 Xu, R., Pang, W., Yu, J. et al. (2007). Synthetic Chemistry of Microporous Compounds (I) Fundamentals and Synthetic Routes. *Chemistry of Zeolites and Related Porous Materials* (3)172–182.

56 Demazeau, G. (2011 April). Solvothermal and hydrothermal processes: the main physico-chemical factors involved and new trends. *Research on Chemical Intermediates* 37 (2): 107–123.

57 Schlesinger, M., Schulze, S., Hietschold, M., and Mehring, M. (2010 July 1). Evaluation of synthetic methods for microporous metal-organic frameworks exemplified by the competitive formation of [Cu2 (btc) 3 (H2O) 3] and [Cu2 (btc) (OH)(H2O)]. *Microporous and Mesoporous Materials* 132 (1–2): 121–127.

58 Yu, J. (2007 Jnauary 1). Synthesis of zeolites. *Introduction to Zeolite Science and Practice* 168: 39.

59 Morris, R.E. (2008 January 4). Ionic liquids and microwaves—making zeolites for emerging applications. *Angewandte Chemie International Edition* 47 (3): 442–444.

60 Parnham, E.R. and Morris, R.E. (2007 October 16). Ionothermal synthesis of zeolites, metal-organic frameworks, and inorganic–organic hybrids. *Accounts of Chemical Research* 40 (10): 1005–1013.

61 Liu, J., Zou, X., Liu, C. et al. (2016). Ionothermal synthesis and proton-conductive properties of NH 2-MIL-53 MOF nanomaterials. *CrystEngComm* 18 (4): 525–528.

62 Huang, L., Wang, H., Chen, J. et al. (2003 March 4). Synthesis, morphology control, and properties of porous metal-organic coordination polymers. *Microporous and Mesoporous Materials* 58 (2): 105–114.

63 Tranchemontagne, D.J., Hunt, J.R., and Yaghi, O.M. (2008 September 1). Room temperature synthesis of metal-organic frameworks: MOF-5, MOF-74, MOF-177, MOF-199, and IRMOF-0. *Tetrahedron* 64 (36): 8553–8557.

64 Cravillon, J., Münzer, S., Lohmeier, S.J. et al. (2009 April 28). Rapid room-temperature synthesis and characterization of nanocrystals of a prototypical zeolitic imidazolate framework. *Chemistry of Materials* 21 (8): 1410–1412.

65 Qian, J., Sun, F., and Qin, L. (2012 September 1). Hydrothermal synthesis of zeolitic imidazolate framework-67 (ZIF-67) nanocrystals. *Materials Letters* 82: 220–223.

66 Sánchez-Sánchez, M., Getachew, N., Díaz, K. et al. (2015). Synthesis of metal-organic frameworks in water at room temperature: salts as linker sources. *Green Chemistry* 17 (3): 1500–1509.

67 Nouar, F., Eckert, J., Eubank, J.F. et al. (2009 March 4). Zeolite-like metal– organic frameworks (zmofs) as hydrogen storage platform: lithium and magnesium ion-exchange and H2-(rho-ZMOF) interaction studies. *Journal of the American Chemical Society* 131 (8): 2864–2870.

68 Pichon, A. and James, S.L. (2008). An array-based study of reactivity under solvent-free mechanochemical conditions—insights and trends. *CrystEngComm* 10 (12): 1839–1847.

69 Pichon, A., Lazuen-Garay, A., and James, S.L. (2006). Solvent-free synthesis of a microporous metal-organic framework. *CrystEngComm* 8 (3): 211–214.

70 Friščić, T. and Fábián, L. (2009). Mechanochemical conversion of a metal oxide into coordination polymers and porous frameworks using liquid-assisted grinding (LAG). *CrystEngComm* 11 (5): 743–745.

71 Shinde, D.B., Aiyappa, H.B., Bhadra, M. et al. (2016). A mechanochemically synthesized covalent organic framework as a proton-conducting solid electrolyte. *Journal of Materials Chemistry A* 4 (7): 2682–2690.

72 Jhung, S.H., Chang, J.S., Hwang, J.S., and Park, S.E. (2003 October 3). Selective formation of SAPO-5 and SAPO-34 molecular sieves with microwave irradiation and hydrothermal heating. *Microporous and Mesoporous Materials* 64 (1–3): 33–39.

73 Kang, K.K., Park, C.H., and Ahn, W.S. (1999 May). Microwave preparation of a titanium-substituted mesoporous molecular sieve. *Catalysis Letters* 59 (1): 45–49.

74 Hwang, Y.K., Chang, J.S., Park, S.E. et al. (2005 January 14). Microwave fabrication of MFI zeolite crystals with a fibrous morphology and their applications. *Angewandte Chemie International Edition* 44 (4): 556–560.

75 Lu, C.M., Liu, J., Xiao, K., and Harris, A.T. (2010 January 15). Microwave enhanced synthesis of MOF-5 and its CO2 capture ability at moderate temperatures across multiple capture and release cycles. *Chemical Engineering Journal* 156 (2): 465–470.

76 Choi, J.-S., Son, W.-J., Kim, J. et al. (2008). Metal-organic framework MOF-5 prepared by microwave heating: factors to be considered. *Microporous Mesoporous Mater* 116 (1–3): 727–731.

77 Mueller, U., Schubert, M., Teich, F. et al. (2006). Metal-organic frameworks—prospective industrial applications. *Journal of Materials Chemistry* 16 (7): 626–636.

78 Mueller, U., Puetter, H., Hesse, M. et al., inventors; BASF SE, assignee (2011 June 28). Method for electrochemical production of a crystalline porous metal-organic skeleton material. United States patent US 7,968,739.

79 Suslick, K.S., Choe, S.B., Cichowlas, A.A., and Grinstaff, M.W. (1991 October). Sonochemical synthesis of amorphous iron. *Nature* 353 (6343): 414–416.

80 Gedanken, A. (2004). Sonochemical synthesis of amorphous iron. *Ultrason Sonochem* 11: 47–55.

81 Qiu, L.G., Li, Z.Q., Wu, Y. et al. (2008) Facile synthesis of nanocrystals of a microporous metal-organic framework by an ultrasonic method and selective sensing of organoamines. *Chemical Communications* 2008 (31): 3642–3644.

82 Li, Z.Q., Qiu, L.G., Xu, T. et al. (2009 January 15). Ultrasonic synthesis of the microporous metal-organic framework Cu3 (BTC) 2 at ambient temperature and pressure: an efficient and environmentally friendly method. *Materials Letters* 63 (1): 78.

83 Obeidli, A.A., Salah, H.F., Murisi, M.A., and Sabouni, R. (2022). Recent advancements in MOFs synthesis and their green applications. *International Journal of Hydrogen Energy* 47: 2561–2593.

84 Xuan, W., Zhu, C., Liu, Y., and Cui, Y. (2012). Mesoporous metal-organic framework materials. *Chemical Society Reviews* 41 (5): 1677–1695.

85 Zhao, D., Wan, Y., and Zhou, W. (2013). Chapter 2-synthesis approach of mesoporous molecular sieves. *Ordered Mesoporous Materials; Wiley-VCH Verlag & Co. KGaA: Weinheim, Germany* (2)5–54.

86 Pham, M.H., Vuong, G.T., Vu, A.T., and Do, T.O. (2011 December 20). Novel route to size-controlled Fe–MIL-88B–NH2 metal-organic framework nanocrystals. *Langmuir* 27 (24): 15261–15267.

87 Linares, N., Silvestre-Albero, A.M., Serrano, E. et al. (2014). Mesoporous materials for clean energy technologies. *Chemical Society Reviews* 43 (22): 7681–7717.

88 Sun, L.B., Li, J.R., Park, J., and Zhou, H.C. (2012 January 11). Cooperative template-directed assembly of mesoporous metal-organic frameworks. *Journal of the American Chemical Society* 134 (1): 126–129.

89 Zacher, D., Yusenko, K., Bétard, A. et al. (2011 February 1). Liquid-phase epitaxy of multicomponent layer-based porous coordination polymer thin films of [M (L)(P) 0.5] type: importance of deposition sequence on the oriented growth. *Chemistry – A European Journal* 17 (5): 1448–1455.

90 Shekhah, O. (2010 February 23). Layer-by-layer method for the synthesis and growth of surface mounted metal-organic frameworks (SURMOFs). *Materials* 3 (2): 1302–1315.

91 Horcajada, P., Serre, C., Grosso, D. et al. (2009 May 18). Colloidal route for preparing optical thin films of nanoporous metal-organic frameworks. *Advanced Materials* 21 (19): 1931–1935.

92 Aiyappa, H.B., Saha, S., Garai, B. et al. (2014 July 2). A distinctive PdCl2-mediated transformation of Fe-based metallogels into metal-organic frameworks. *Crystal Growth & Design* 14 (7): 3434–3437.

93 Yakovenko, A.A., Wei, Z., Wriedt, M. et al. (2014 November 5). Study of guest molecules in metal-organic frameworks by powder X-ray diffraction: analysis of difference envelope density. *Crystal Growth & Design* 14 (11): 5397–5407.

94 Banerjee, D. and Parise, J.B. (2011 October 5). Recent advances in s-block metal carboxylate networks. *Crystal Growth & Design* 11 (10): 4704–4720.

95 Huang, W.H., Yang, G.P., Chen, J. et al. (2013 January 2). Solvent influence on sizes of channels in three new Co (II) complexes, exhibiting an active replaceable coordinated site. *Crystal Growth & Design* 13 (1): 66–73.

96 He, Y.C., Guo, J., Zhang, H.M. et al. (2014). Tuning the void volume in a series of isomorphic porous metal-organic frameworks by varying the solvent size and length of organic ligands. *CrystEngComm* 16 (24): 5450–5457.

97 Akhbari, K. and Morsali, A. (2011). Effect of the guest solvent molecules on preparation of different morphologies of ZnO nanomaterials from the [Zn2(1,4-bdc)2(dabco)] metal-organic framework. *Journal of Coor-dination Chemistry* 64 (20): 352–3530.

98 Seetharaj, R., Vandana, P.V., Arya, P., and Mathew, S. (2019 March 1). Dependence of solvents, pH, molar ratio and temperature in tuning metal-organic framework architecture. *Arabian Journal of Chemistry* 12 (3): 295–315.

99 Yuan, F., Xie, J., Hu, H.M. et al. (2013). Effect of pH/metal ion on the structure of metal-organic frameworks based on novel bifunctionalized ligand 4′-carboxy-4, 2′: 6′, 4′′-terpyridine. *CrystEngComm* 15 (7): 1460–1467.

100 Chu, Q., Liu, G.X., Okamura, T.A. et al. (2008 February 6). Structure modulation of metal-organic frameworks via reaction pH: self-assembly of a new carboxylate containing ligand N-(3-carboxyphenyl) iminodiacetic acid with cadmium (II) and cobalt (II) salts. *Polyhedron* 27 (2): 812–820.

101 Wang, X.F., Liu, G.X., and Zhou, H. (2013 September 1). Syntheses, structures and physical properties of two zinc (II) coordination polymers with 1, 3, 5-tris (imidazol-1-ylmethyl)-2, 4, 6-trimethylbenzene and 1, 3, 5-benzenetricarboxylate. *Inorganica Chimica Acta* 406: 223–229.

102 de Oliveira, C.A., da Silva, F.F., Malvestiti, I. et al. (2013 January 1). Effect of temperature on formation of two new lanthanide metal-organic frameworks: synthesis, characterization and theoretical studies of Tm (III)-succinate. *Journal of Solid State Chemistry* 197: 7–13.

103 Zhang, K.L., Hou, C.T., Song, J.J. et al. (2012). Temperature and auxiliary ligand-controlled supramolecular assembly in a series of Zn (ii)-organic frameworks: syntheses, structures and properties. *CrystEngComm* 14 (2): 590–600.

104 Wang, C., Liu, X., Demir, N.K. et al. (2016). Applications of water stable metal-organic frameworks. *Chemical Society Reviews* 45 (18): 5107–5134.

105 Vaitsis, C., Kanellou, E., Pandis, P.K. et al. (2022 October 1). Sonochemical synthesis of zinc adipate metal-organic framework (MOF) for the electrochemical reduction of CO2: MOF and circular economy potential. *Sustainable Chemistry and Pharmacy* 29: 100786.

106 Dou, H., Xu, M., Wang, B. et al. (2021). Microporous framework membranes for precise molecule/ion separations. *Chemical Society Reviews* 50 (2): 986–1029.

107 Li, J., Wang, H., Yuan, X. et al. (2020 February 1). Metal-organic framework membranes for wastewater treatment and water regeneration. *Coordination Chemistry Reviews* 404: 213116.

108 Teng, J., Chen, Y., Ma, G. et al. (2020 April 1). Membrane fouling by alginate in polyaluminum chloride (PACl) coagulation/microfiltration process: molecular insights. *Separation and Purification Technology* 236: 116294.

109 Dharupaneedi, S.P., Nataraj, S.K., Nadagouda, M. et al. (2019 February 8). Membrane-based separation of potential emerging pollutants. *Separation and Purification Technology* 210: 850–866.

110 Suen, S.Y. (2015 January 1). Mixed matrix membranes for adsorption application. *Journal of Chemical Engineering & Process Technology* 6 (1): 1.

111 Nayak, V., Jyothi, M.S., Balakrishna, R.G. et al. (2015 June). Preparation and characterization of chitosan thin films on mixed-matrix membranes for complete removal of chromium. *ChemistryOpen* 4 (3): 278–287.

112 Ragab, D., Gomaa, H.G., Sabouni, R. et al. (2016 September 15). Micropollutants removal from water using microfiltration membrane modified with ZIF-8 metal-organic frameworks (MOFs). *Chemical Engineering Journal* 300: 273–279.

113 Campbell, J., Burgal, J.D., Szekely, G. et al. (2016 April 1). Hybrid polymer/MOF membranes for Organic Solvent Nanofiltration (OSN): chemical modification and the quest for perfection. *Journal of Membrane Science* 503: 166–176.

114 Yang, S., Zou, Q., Wang, T., and Zhang, L. (2019). Effects of GO and MOF@GO on the permeation and antifouling properties of cellulose acetate ultrafiltration membrane. *Journal of Membrane Science* 569: 48–59.

115 Ma, J., Guo, X., Ying, Y. et al. (2017 April 1). Composite ultrafiltration membrane tailored by MOF@ GO with highly improved water purification performance. *Chemical Engineering Journal* 313: 890–898.

116 Sun, H., Tang, B., and Wu, P. (2017 June 28). Development of hybrid ultrafiltration membranes with improved water separation properties using modified superhydrophilic metal-organic framework nanoparticles. *ACS Applied Materials & Interfaces* 9 (25): 21473–21484.

117 Yin, N., Wang, K., Wang, L., and Li, Z. (2016 December 15). Amino-functionalized MOFs combining ceramic membrane ultrafiltration for Pb (II) removal. *Chemical Engineering Journal* 306: 619–628.

118 Sarango, L., Paseta, L., Navarro, M. et al. (2018 March 25). Controlled deposition of MOFs by dip-coating in thin film nanocomposite membranes for organic solvent nanofiltration. *Journal of Industrial and Engineering Chemistry* 59: 8–16.

119 Yuan, J., Hung, W.S., Zhu, H. et al. (2019 February 15). Fabrication of ZIF-300 membrane and its application for efficient removal of heavy metal ions from wastewater. *Journal of Membrane Science* 572: 20–27.

120 Xu, Y., Gao, X., Wang, X. et al. (2016 October 26). Highly and stably water permeable thin film nanocomposite membranes doped with MIL-101 (Cr) nanoparticles for reverse osmosis application. *Materials* 9 (11): 870.

121 He, Y., Tang, Y.P., Ma, D., and Chung, T.S. (2017 November 1). UiO-66 incorporated thin-film nanocomposite membranes for efficient selenium and arsenic removal. *Journal of Membrane Science* 541: 262–270.

122 Kadhom, M., Yin, J., and Deng, B. (2016 December 6). A thin film nanocomposite membrane with MCM-41 silica nanoparticles for brackish water purification. *Membranes* 6 (4): 50.

123 Park, H.M., Jee, K.Y., and Lee, Y.T. (2017 November 1). Preparation and characterization of a thin-film composite reverse osmosis membrane using a polysulfone membrane including metal-organic frameworks. *Journal of Membrane Science* 541: 510–518.

124 Duan, J., Pan, Y., Pacheco, F. et al. (2015 February 15). High-performance polyamide thin-film-nanocomposite reverse osmosis membranes containing hydrophobic zeolitic imidazolate framework-8. *Journal of Membrane Science* 476: 303–310.

125 Wang, F., Zheng, T., Xiong, R. et al. (2019 October 1). Strong improvement of reverse osmosis polyamide membrane performance by addition of ZIF-8 nanoparticles: effect of particle size and dispersion in selective layer. *Chemosphere* 233: 524–531.

126 Yin, J., Kim, E.S., Yang, J., and Deng, B. (2012 December 15). Fabrication of a novel thin-film nanocomposite (TFN) membrane containing MCM-41 silica nanoparticles (NPs) for water purification. *Journal of Membrane Science* 423: 238–246.

127 Kadhom, M., Hu, W., and Deng, B. (2017 June 14). Thin film nanocomposite membrane filled with metal-organic frameworks UiO-66 and MIL-125 nanoparticles for water desalination. *Membranes* 7 (2): 31.

128 Cheng, D., Zhao, L., Li, N. et al. (2019 October 1). Aluminum fumarate MOF/PVDF hollow fiber membrane for enhancement of water flux and thermal efficiency in direct contact membrane distillation. *Journal of Membrane Science* 588: 117204.

129 Tan, Y.Z., Han, L., Chew, N.G. et al. (2018 December 1). Membrane distillation hybridized with a thermoelectric heat pump for energy-efficient water treatment and space cooling. *Applied Energy* 231: 1079–1088.

130 Lalia, B.S., Guillen, E., Arafat, H.A., and Hashaikeh, R. (2014 January 2). Nanocrystalline cellulose reinforced PVDF-HFP membranes for membrane distillation application. *Desalination* 332 (1): 134–141.

131 Kebria, M.R., Rahimpour, A., Bakeri, G., and Abedini, R. (2019 January 15). Experimental and theoretical investigation of thin ZIF-8/chitosan coated layer on air gap membrane distillation performance of PVDF membrane. *Desalination* 450: 21–32.

132 Li, Q., Liu, Q., Zhao, J. et al. (2017 December 15). High efficient water/ethanol separation by a mixed matrix membrane incorporating MOF filler with high water adsorption capacity. *Journal of Membrane Science* 544: 68–78.

133 Shen, J., Liu, G., Huang, K. et al. (2016 September 1). UiO-66-polyether block amide mixed matrix membranes for CO2 separation. *Journal of Membrane Science* 513: 155–165.

134 Li, S., Chen, Z., Yang, Y. et al. (2019 May 15). Improving the pervaporation performance of PDMS membranes for n-butanol by incorporating silane-modified ZIF-8 particles. *Separation and Purification Technology* 215: 163–172.

135 Wu, X., Wei, W., Jiang, J. et al. (2018 November 19). High-flux high-selectivity metal-organic framework MIL-160 membrane for xylene isomer separation by pervaporation. *Angewandte Chemie International Edition* 57 (47): 15354–15358.

136 Wang, X., Chen, J., Fang, M. et al. (2016 May 1). ZIF-7/PDMS mixed matrix membranes for pervaporation recovery of butanol from aqueous solution. *Separation and Purification Technology* 163: 39–47.

3

Application of MOFs in the Removal of Pharmaceutical Waste from Aquatic Environments

Gagandeep Kaur, Parul Sood, Lata Rani, and Nitin Verma*

Chitkara University School of Pharmacy, Chitkara University, Himachal Pradesh, India
** Corresponding author*

3.1 Introduction

An increase in pollution from both industrial and non-industrial sources has recently endangered water supplies. There is a daily outflow of around two million tonnes of sewage into pure water [1]. According to the United Nations' annual report, the average amount of wastewater generated is 1.5 billion cubic meters per day. This amount is about equivalent to six times the volume of all the rivers in the globe today [2]. About 90% of urban water supplies are contaminated, with sewage from homes accounting for 70% and industrial effluent for 33%. Neither source receives any treatment before being dumped into lakes and rivers [3].

Researchers from all around the world have brought attention to the presence of medicines in water bodies over the past few decades, citing the inadequacy of traditional wastewater treatment units in removing these substances. Because of this, several investigations have centered on finding these pharmacological ingredients [4–6]. As the world's population continues to grow at an alarming rate, so does the need for pharmaceutically active substances. Fortunately, this rise in demand, in part driven by changes in people's diets and exercise habits, has spurred the development of increasingly potent [7–9]. Although no confirmed side effects have been linked to these pharmaceutically active compounds, it is hypothesized that prolonged exposure causes endocrine disruption [10]. Moreover, there are currently no recognized limits for these compounds in aquatic environments. Scientists all over the world are worried about the difficulty of removing endocrine-disrupting compounds from water systems [11–13].

Consumption of numerous novel pollutants has expanded alongside modern lifestyle and economic progress. Medicinal items, plasticizers, additives, cosmetics, flame retardants, detergents, and so on, all fall under the category of endocrine disruptors, which are organic substances. Because of their widespread use, high production volumes, and general importance to modern living, pharmaceuticals

Metal Organic Frameworks for Wastewater Contaminant Removal, First Edition. Edited by Arun Lal Srivastav, Lata Rani, Jyotsna Kaushal, and Tien Duc Pham.

and personal care products (PPCPs) have been a focal point of research into environmental contaminants (ECs). To be specific, PPCPs include analgesics, antibiotics, diuretics, nonsteroidal anti-inflammatory drugs (NSAIDs), different antiseptics, stimulating medicines, beta-blocking drugs, beauty products including sunscreen agents and perfumes, dietary supplements, and their metabolites. A study accomplished on priority PPCPs identified some drugs, namely carbamazepine (CBZ) and iopamidol, NSAIDS including diclofenac sodium and ibuprofen, clofibric acid, triclosan (TCS), bisphenol A (BPA), and phthalates.

Antibacterial TCS is one of the choices on the list, and it was estimated that roughly 350 tonnes of TCS was being used in European countries. Like other phenolic compounds, BPA is used as a monomer in the plastics industry. Endocrine-disrupting effects have been observed even at extremely low concentrations of this substance. Research has demonstrated that the two most common chemicals in human wastewater, caffeine (CAFF) and cannabidiol (CBZ), are detrimental to the environment. These PPCPs have the potential to bioaccumulate on plants and animals in the environment, which could lead to antibiotic resistance and even cancer. However, these compounds' complex physicochemical properties have made treatment thus far elusive (e.g., nonbiodegradability). CIP, or ciprofloxacin hydrochloride, is a powerful antibiotic in the fluoroquinones family [14]. Because of this, its introduction into the environment is mostly attributable to industrial- and consumer-scale manufacturing and use. A little amount of the active constituent escapes during industrial production and purification and ends up in the waste stream. Ciprofloxacin removal from wastewater has been the subject of numerous research that has been published in the literature on the topic of alternative treatment technologies [15, 16]. Drug residues are a relatively new type of emerging pollutant that has been found in a wide variety of aquatic and sedimentary habitats around the globe [17, 18]. Because they tend to build up in sea organisms, they provide a serious ecological threat [19, 20]. For this reason, cleansing the environment of pharmaceutical waste is a crucial step in ensuring the well-being of future generations and protecting the planet. Due to their varied physicochemical features and low concentrations, medicines in the aquatic system have proven difficult to remove effectively [21, 22].

Several approaches like adsorption, ozonation, photo-degradation, and coagulation and sedimentation [16] have been applied for wastewater treatment [23–30]. Adsorption is an important technology for contamination remediation owing to its minimal cost of operation, high elimination efficacy, and low production of secondary hazardous pollutants [31]. Recovering a pollutant has the potential to be accomplished under certain conditions. Activated carbon (AC) [32], silicate mesoporous materials [33], porous natural polymers [34, 35], and MOFs are only some of the materials of porous nature that have seen rapid growth due to the current focus on adsorption [36]. In recent years, researchers have started examining MOFs commonly known as porous coordination polymers (PCPs) [37]. Self-linkage of organic chelates with metal ions or metal ion clusters creates a patterned porous structure on the external surface of most MOFs. Larger specific surface area (the Langmuir specific surface area was greater than 6550 m^2/g), ultra-high porosity [38], a flexible

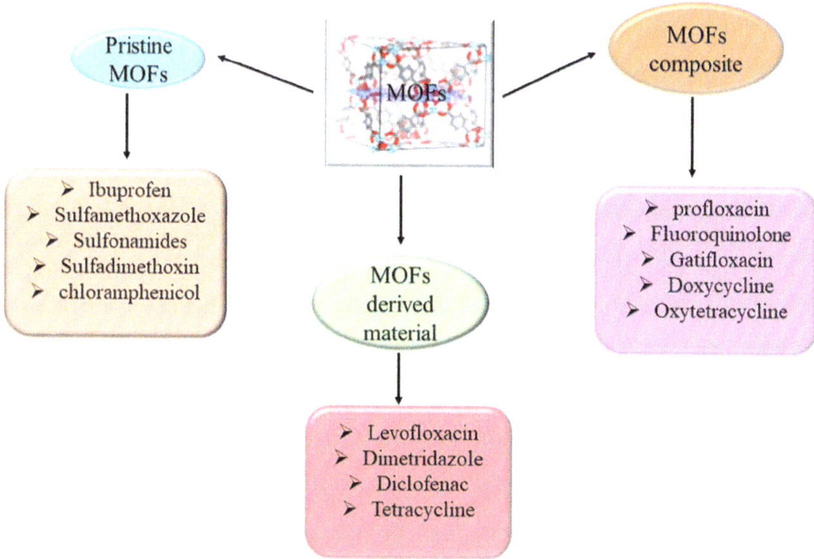

Figure 3.1 Application of MOFs for the removal of pharmaceutical waste from water.

structure, and simple preparation are just some of the wonderful properties of MOFs. Gas storage or separation [39], catalysis with or without radiations [40], sensing [41], and drug delivery are just a few of the applications that stand to benefit greatly from MOFs' unique properties [42]. MOFs and their derivatives have demonstrated great application promise for wastewater treatment because of their high flexibility and tunable physicochemical features, especially in the adsorptive removal of metal ions and developing organic pollutants [37, 43–50]. Despite the recent explosive growth in the use of MOFs for the removal of pharmaceuticals, there has not been a comprehensive evaluation of the connection between structure and adsorption efficacy. Figure 3.1 shows the application of MOFs for the removal of pharmaceutical waste from water.

3.2 The Potential of MOFs and Their Analogs to Resist Water Stability

For widespread commercialization, it is crucial that MOFs and their variants like metal oxides, porous, or metal-carbon can survive water. MOFs and their ability to withstand water have been the subject of several reviews [36, 51]. Key characteristics postulated to define the aqueous stability of MOFs include the extent and strength of coordination bonding amongst metal and ligands as well as the shielding of functional groups that exist at coordination sites [51]. Stronger coordination bonds are found between hard Lewis bases and acids (or soft Lewis acids and bases), according to the HSAB theory rather than between soft Lewis bases and hard acids [52]. The formation of water-stable MOFs is thereby achieved through the combination of carboxylate-based ligands with hard acidic metal ions such as Zr^{4+}, Cr^{3+}, Al^{3+}, and Fe^{3+}. Several representative carboxylate-based MOFs show exceptional water stability; these include the MIL series from the Material Institute Lavoisier (MIL-53, 100, and 101) [53, 54] and the UiO series from the University of Oslo (UiO-66, 67, and 68) [55].

Zr-based MOFs like PCN-222 and UiO-66 exhibit excellent water stability even in acidic or alkaline environments, while iron-based MOFs i.e., MIL-100 and chromium-based MIL-101, may be successfully kept in freshwater for numerous months [54, 56, 57]. MOFs that are resistant to water can also be created by combination of metal ions formed from soft acids like zinc, copper, nickel, manganese, or silver ion with that of ligands of soft bases (such as triazole, tetrazole, imidazole and pyrazoles) [58]. Metal azolate frameworks (MAFs) include the zeolitic imidazolate framework (ZIF) [59]. Stable even in a highly basic environment, modern MOFs are based on triazoles and pyrazoles [60]. Improved water stability is achieved by adding hydrophobic functionalities to MOFs, such as alkyl and different fluorinated groups, which also increases the metal-ligand interactions. Crystal lattices can be protected against water damage by providing a water repellant pore surface or by modifying metal ion clusters. A recent study by Yang et al., [61] for instance, showed that methyl addition significantly improved MOF durability [61]. Alterations in the powder X-ray diffraction (PXRD) patterns of MOF-5 were observed after a day of exposure to air. However, both methyl-modified CH_3MOF-5 and DiCH_3MOF-5 displayed structural stability for up to four days. CALF-25 possesses non-polar alkyl substituents inside the pores and displays more structural stability at high humidity [62]. Hydrophobic pores made by ethyl ester groups encircled the metal's core, preventing water molecules from penetrating [62]. This feature was exploited by Yang et al. who fabricated fluorinated MOFs by incorporating fluorine into the structure, rendering them very hydrophobic [63]. According to Li et al. [64] it is possible to increase hydrostability by incorporating multiple metal ions into a single MOF structure. Comparatively, nickel-doped MOF-5 retained more moisture than the unadulterated version. Enhanced surface area and pore dimensions of nickel-doped MOF-5 aided in the cleaning up of pollution as well.

3.3 Methods for the Development and Design of Aqueous-Stable Composites of Metal-Organic Frameworks

High adsorption capacity and favorable kinetics are the two key factors that make the MOFs excellent adsorbents. The following are a few examples of commonly used metal-organic frameworks: MOFs clubbed with graphene oxide (GO/MOFs) [65], carbon nanotubes (CNT/MOFs) [66], and polydimethylsiloxane (PDMS/MOFs) [67, 68]. As an illustration, consider Cu-BTC@GO, which retains its pore structure and hydrophobicity for up to 10 hours [69]. When exposed to water, the surface area of the purest form of copper-BTC diminished from 1188 m^2/g to 20.7 m^2/g, but the surface area of contaminated forms remained relatively unchanged. The findings showed that GO's hydrophobic properties can shield metallic sites' coordination bonds from ionic attacks. Coating the surface of MOFs is an intriguing method to increase their water stability. In order to create composites that are impervious to moisture and water a simple vapor deposition method has been proposed. Notably, after just one hour in ambient air, the structure of MOF-5 changed and collapsed completely. A substantial improvement in the composite resistance to moisture was noticed after a PDMS coating was applied. All of the PDMS-coated samples, surprisingly, maintained their crystallite size and shape and porosity structure for at least three months [68].

3.4 Synthesis and Design of Water-Stable MOF-Derived Materials

Many other types of porous materials, such as porous carbons and metal oxides, have been developed through controlled pyrolysis from MOFs. The higher strength, as well as chemical, physical, and thermal stability of these materials, make them ideal for a variety of applications [51]. To make highly stable porous carbon [70], came up with the novel idea of using hygroscopic MOF-5 as a sacrificial precursor [70]. Specific surface area can reach up to 2872 m2/g in the final product. Controlling the heat treatment, Yang et al. found significantly improved MOF-5's resistance to moisture [71]. The product was encased in a layer of amorphous carbon, which protected the framework from being degraded by water. Heat-treated MOF-5 retained its crystalline structure and porous characteristics after 14 days in the air. Using glucose and sucrose as supplementary carbon sources, MOF-5-derived porous carbon was synthesized [72]. After absorbing water, the resultant porous carbon still had a high degree of porosity. In contrast, MOF-5 showed a dramatic drop in its specific surface area. In addition to the aforementioned methods, MOFs are also excellent starting materials for creating water-resistant metal oxides. Many metal oxides, including CuO, Fe_2O_3, ZnO, ZrO_2, Mn_3O_4, $CoMn_2O_4$, etc., with desirable properties including higher adsorption capacity, chemical stability, and ecological friendliness, have been generated by direct calcification of MOFs in air and put to use in environmental applications. Nickel mixed iron-oxide nanocubes were made by utilizing a Prussian blue analog as a mold [73]. The resulting substance proved highly effective at removing harmful levels of As(V) and Cr(VI) from water. Although the adsorption capacity of NiFe-oxide nanocubes did drop slightly when eluted with 0.5 M NaOH solution, this was largely attributable to the remarkable aqueous stability of nanocubes. Hierarchically structured porous carbon nanohybrids have shown considerable improvements in aqueous stability and adsorption efficiency, similar to metal oxides and porous carbons derived from MOF-derived. For instance, porous iron-containing MOFs are shown experimentally to be useful as a novel category of precursors for producing ferric oxide-carbon composites (MOFs) [74]. The resulting composite showed remarkable adsorption properties for the removal of several contaminants. Moreover, it was incredibly recyclable and had the added benefit of being magnetically recoverable.

3.5 MOFs and Their Hybrids as Versatile Adsorbents for Capturing Pharmaceutical Drugs

Recognizing the underlying interaction mechanisms is crucial for developing materials with high adsorption capacity. Studies have shown that organic pollutants are mostly adsorbed onto MOFs by acid-base, hydrophobic, π-π-hydrogen bonds, electrostatic, and coordinate bond interactions [43]. Moreover, the size and form of MOFs would have a major impact on them. The most recent developments in the use of MOFs and their different hybrids for drug elimination are discussed here.

3.6 MILs and Their Derived Compounds

Materials of Institute Lavoisier (MILs) possess various good specific features like strong porosity, higher stability, and extremely large surface area, ranking them as ideal adsorbents for wastewater treatment. Historically, core metals were linked to organophosphorus acid or succinic acid for MIL synthesis. Carboxylate and trivalent cations like Fe^{3+}, Al^{3+}, Ga^{3+}, In^{3+}, Vn^{3+}, and Cr^{3+} are the main building blocks of most modern MILs. Due to their strong and powerful structure, MILs and their derivatives have found widespread usage in the adsorptive clearance of medicine residues in recent studies. That is true for both unmodified MILs (like MIL-101, 100, 53, 88, 68, etc.) and MIL composites (like carbon nanotubes-MOF, GO-MOF, and magnetic MIL, etc.)

3.7 Pristine MILs

MILs are among the best MOFs for pharmaceuticals remediation because of their exceptional hydrolytic stability. PDMS coating on MOF surfaces might increase their resistance to moisture and water [68]. Exploring the adsorption capability of MOFs toward dugs like furosemide (FS) and sulfasalazine remarkably revealed their practical application (SAZ). MIL-100(Cr) showed the highest water stability when compared to MOF-177(Zn), MOF-5(Zn), HKUST-1(Cu), MOF-505(Cu), UMCM-150(Cu), ZIF-8(Zn), and MIL-100(Zn). As a result of the synergistic binding relationship, MIL-100(Cr) was able to efficiently bind and collect FS and SAZ at very low concentrations. MIL-100 (iron) and MIL-101 (chromium) were also found to be effective in removing the drug clofibrate and naproxen from the body [11]. The adsorption effectiveness of MIL-101 and 100 were significantly better than that of AC in terms of time and extent of adsorption that occurred due to the strong electrostatic bond existing between naproxen and MOFs.

In comparison to MIL-100 (Fe), MIL-101(Cr) performed better because of its larger surface area, more porous structure, and more readily accessible active sites. The metallic center is an essential component for adsorption according to the findings of various scientists who looked into the adsorption of different drugs like acetylsalicylic acid and roxarsone, zeolite Y, goethite, and MILs of series numbered 100(Fe),

53(Cr), 101, 100(Cr), and 100(Al) [75]. Since organic arsenic compounds have a lower adsorption energy than water molecules, MIL-100(Fe) was found to have the highest adsorption capacity and fastest kinetic. It has been suggested that a mesoporous structure, incorporation of specific functionalities, and the use of a flexible framework can all improve adsorption efficacy. Recent research has shown that mesoporous MIL53(Al) has 4.4 times faster adsorption rates of triclosan (TCS) at a concentration of 1 mg/l compared to microporous MIL53(Al) [76]. Functional groups such as hydroxyl [77–79], primary amines [80], and sulphonic acid have also been proposed for insertion [81]. Therefore, the coordinatively unsaturated sites on the MIL-101 skeleton were used to integrate aminomethanesulfonic acid (AMSA) and ethylenediamine (ED) (CUSs). NAP and CA were effectively removed using the resulting materials with -SO_3H (AMSA-MIL-101) and -NH_2 (ED-MIL-101) [82]. The acid-base relationship between the acidic and basic functional groups may explain the reason for more removal effectiveness of ED-MIL-101 than pure MIL-101 and AMSA-MIL-101. The breathing effect motivated researchers to look at how the flexibility and stiffness of MOFs affected the adsorption behavior of carbamazepine (CBZ), dimetridazole (DMZ), and sulfamethoxazole (SMZ) [83, 84]. When comparing the adsorption capacities of rigid MIL-101(Cr) (13.5 mg/g) and flexible MIL-53(Cr) (101.1 mg/g) [85]. The malleability of MIL-53(Cr) is a possible explanation. However, MIL-101 (Cr adsorption) performance was hindered by its rigidity, immutable structure, and hydrophilicity. Combining PXRD data with a computational model, we find that the adsorption process causes the pores of flexible MIL-53(Cr) to increase from narrower to wide, decreasing binding efficiency and increasing hydrophobicity.

3.8 MILs Composites

Improvements in water stability and adsorption capability make the development and research of MOFs composites particularly exciting. Sodium alginate (SA) [86], chitosan (CS), GO [87–89], multiwall carbon nanotubes (MWCNTs) [61, 90], metal nanoparticles [91], and Fe_3O_4 have all been mixed with MILs for this purpose [92–94]. Particular composites, MIL-101(Cr)/SA and MIL-101(Cr)/CS, were made by including MIL-101(Cr) into an SA and CS matrix, respectively [86]. The MIL-101(Cr)/CS hybrid outperformed SA, CS, and MIL-101(Cr)/SA in terms of adsorption capacity. Recently, the double-solvent method has been used with the overpowering reduction approach to confine copper and cobalt bimetallic nanoparticles within the pores of MIL-101 (Cr) [91]. Increases in tetracycline (TC) adsorption capacity of roughly 140% have been seen for Cu-Co/MIL-101 when compared to that of pure MOFs. The increase was traced back to the chemistry between the Cu and Co mixed metal nanoparticles and the TC molecular structures. Other interactions playing crucial roles include the pore dimension selective adsorption process, electrostatic repelling forces, acid-base ionic exchange reactions, π-π stack arrangements, and hydrogen bonding. Researchers are interested in magnetic MOFs because of the fast response of magnetic materials combined with the beneficial effects of MOFs. For the purpose of ciprofloxacin elimination, Fe_3O_4@MIL-100(Fe) and Fe_3O_4@MOF-235(Fe) were developed [95]. The obtained magnetic composites outperformed both the reported AC and CNTs in terms of their superparamagnetic property and their ability to collect CIP (322.6 mg/g). The success of MILs-based composites in capturing medicines was further validated by these findings.

3.9 MILs-Derived Materials

Related derivatives, such as MIL-88 and 53 (Fe), have been produced using a wide variety of MILs as starting materials. For the adsorption of ciprofloxacin, for instance, it was disclosed that carbonization of a mixed-valence MIL-53(Fe) produces magnetic mesoporous carbon (Fe_3O_4/C) [96]. Well-dispersed nano molecules of ferric oxide (30–40 nm) were found in the carbon matrix of Fe_3O_4-based carbonic structures, demonstrating a mesoporous structure with a significant specific surface area (908.2 m^2/g). The composite may be easily recycled through magnetic separation because of its high magnetic power (22.81 emu/g). Incredible stability of ciprofloxacin as a removal medium is evidenced by the fact that its adsorption affinity did not alter after five cycles. A study showed that magnetic carbon with iron can be produced through direct pyrolysis of MIL-53(Fe) doped with zinc [97]. Due to its greater specific surface area, smaller pore diameter, and greater total pore volume, iron-based magnetic carbons owe a higher adsorption value of 453 mg/g than pure magnetic carbon. Recent investigations on adsorption have implicated ionic interaction and the pores-filling effect in achieving efficient capture. It turns out that even trace amounts of TC can be successfully eliminated, and its amount can be lowered to the safest level.

3.10 ZIFs and Their Derived Compounds

ZIFs have been proposed as a means of removing harmful substances from water because of their stability and hydrophobicity.

3.11 Pristine ZIFs

It was investigated whether ZIF-8 nanoparticles, synthesized in methanol at room temperature, could adsorb TC and remove it from an aqueous system [98]. The remarkable adsorption capacity (930 mg/g) of the produced material can be attributed to the Pi-Pi stacking and electrostatic interactions among ZIF-8 and TC. Investigation was done on the adsorption behavior of TC on a number of ZIFs with different chemical and textural components, including ZIF-67-NO_2, ZIF-67-Cl, ZIF-67-SO_4, ZIF-8-Octahedron, ZIF-8-Leaf, and ZIF-8-Cuboid [99]. Adsorption capacity was highest for ZIF-67-OAC at 446.9 g/l. Contrary to expected based on results of surface area calculated by Langmuir and volume of pores, the sequence of adsorption capabilities for various ZIFs is not constant. By using polyelectrolytes as a structural guiding agent, hierarchically porous ZIF-8 (HpZIF-8) with mesopores and macropores were created [100] to investigate the advantages of the enlarged pore and streamline the production process. The advantages of the stratified porous structure, which can provide a large number of attainable adsorption sites and promote mass transfer, were demonstrated by the observation that HpZIF-8–10(D) (D = 1.0, 1.5, 2.0) had higher adsorption capacities toward TC and chloramphenicol (CP) than mZIF-8–10. This study's findings provide credence to the idea that pore widths and functional group compositions both play important roles in the efficient elimination of pharmaceuticals.

3.12 ZIFs Composites

The regular structure of the material is preserved in composites formed from ZIFs, and the low cost and high recycling rate of the ZIFs are further additional benefits. A number of composites based on ZIFs have been produced and put to use as the favorite adsorbents for purging pharmaceuticals. This category includes materials like metal-organic framework resins (MOF-resins), ZIF-8-carbon-sulfur (CS), ZIF-8-konjac glucomannan (KGM), and ZIF-8-GO [101–103]. Through the incorporation of common MOFs e.g. Fe-BTC, ZIF-8, 67, and HKUST-1 into a CS matrix, a wide variety of MOFs-CS composites have been created [103]. In tests against natural polymer TC adsorbents, the ZIF-8-CS composite was shown to have the highest adsorption performance of 495.04 mg/g. Ionic and hydrogen bonding along with π-π stacking were shown to be crucial through both the experimental characterizations and density functional theory (DFT) simulations. This significant finding demonstrates the promise of simulation techniques based on molecular dynamics as one of the influential methodologies for investigating interaction mechanisms at the atomic scale. Self-assembled ZIF-8@SiO_2@Fe_3O_4, produced with sodium laurate and cetyltrimethylammonium bromide (CTAB) as template agents has a high recycling efficiency [104]. By altering the concentration of template agents, pore size could be adjusted. The adsorption capacity for ceftazidime (CAZ) under optimal conditions was 74.26 mg/g, but for unprocessed ZIF-8 was only 39.1 mg/g.

In order to go further, scientists developed platinum-free magnetic micromotors ZIF-8-C-Fe_2O_3/c-Al_2O_3/MnO_2 (ZIF-8-M). Because of its high surface area and porosity, ZIF-8-M has demonstrated outstanding adsorption for congo red of 394 mg/g and 242 mg/g for doxycycline (DOC). These micromotors based on the ZIF-8 were able to propel themselves through real wastewater, including lake and tap water, and successfully absorb harmful substances.

3.13 Materials Derived from ZIFs

Two of the most extensively researched newly generated ZIFs are ZIF-8 and 67. In particular, using rare hybrid materials in a contained setting allows for the production of hollow carbonized derivatives. Using multi-metallic MOFs containing two or more metal ions, ZIFs have been used to produce compounds with multi-metals or metal oxides. In order to design porous carbons, scientists used ZIF-8 as a self-template [105]. At 1000 degrees Celsius, the surface area of carbons generated from MOFs was remarkably higher than that of pure ZIF-8 (1073 m2/g). It is worth noting that the optimum adsorption values obtained for isobutyl methyl ether (IBU) and diclofenac sodium (DCF) were 320 and 400 mg/g, respectively. Although hydrogen bonding is suspected to play a major role, hydrophobic and π-π interactions are other possible contributors to this remarkable performance. After comparing ZIF-8-generated porous carbons to SMZ-derived porous carbons, it was revealed that both types were equally effective at removing CIP [102, 106]. Adsorption performance was predicted to be enhanced for ionizable liquid-filled ZIF-8-derived carbon (IMDCs) compared to that of carbons derived from ZIF (MDCs). Hydrogen bonding with the desired pollutant can only form if there is a sufficient amount of nitrogen in the MDC, which is the root cause of its superior performance. The adsorption capacity of MOF-derived materials was enhanced in another study by creating hollow structures [107, 108] in order to remove TC by adsorption.

To create N-doped porous carbons (NPC), NHPC-1 and NHPC-2 molds were made from ZIF-8 that had been coated with resorcinol and formaldehyde (ZIF-8@RF) or ZIF-8 that had been treated with tannic acid alone (ZIF-8@TA). Adsorption performance of TC (518.1 mg/g) was 2.9 times higher for NHPC-2 when compared to NPC (180.2 mg/g) because of its excessive nitrogen content, a large number of defects, and quite well pore size distribution. Several types of interactions, including hydrogen bonding, hydrophobicity, and π-π stacking arrangements, were proposed as being important. Even after eight cycles, the adsorption capability of NHPC-2 for TC was unchanged, demonstrating its high stability and robust recycling potential. Because of its simplicity of use, magnetic technology has shown great promise for real-world applications. To this end, used an MNPC adsorbent family comprised of strong magnetic nitrogen-doped nanoporous carbon to get rid of CIP in water [109]. Precursors including ZnO, $Co(OH)_2$, and dimethylimidazole were pyrolyzed in a single step to produce MNPCs [110].

3.14 UiOs Composite Materials

Many different UiO-66 analogs, including those containing the functional groups like sulphonyl, amino, amide, and carboxylic have been utilized to purge medicines [111, 112]. The structural properties of the resulting MOFs are significantly impacted by not only the alteration of functional groups, but also the synthesis condition and activation technique, which in turn affects the surface area, hydrophobicity, and adsorption capacity of the MOFs. A considerable increase in sulfachloropyradazine (SCP) adsorption capacity (417 mg/g) was seen between chloroform-activated UiO-66 and untreated UiO-66, as reported [113]. MOFs with controlled pore diameters were created by either modulating the linkers or inserting a defect in an effort to improve the adsorption performance. After two minutes at a dose of 100 mg/L, 95% of CBZ was collected over UiO-67, but only 35% was taken over UiO-66 [114]. In particular, the high concentration of modulators allowed for the fabrication of UiO-66 with continuous missing linker flaws [115]. Improved ROX adsorption capacity and adsorption rate were made possible by the creation of Zr-OH, which in turn benefited from the insertion of cracks, which increased porosity. The adsorption capacity went up to 730 mg/g, and the adsorption equilibrium period went down to 30 min from its original 240 min. In order to effectively remove medicines, the data revealed that both pore size and a wide surface area were essential.

Another interesting and promising strategy is metal-doped modification, which involves the deliberate introduction of another metal that might be Ce, Co, or Mn, etc., in order to construct a bimetallic MOF, which has a unique structure and enhanced activity [116, 117]. Adsorptive TC removal by Co-doped UiO-66 was studied, for instance, by Cao et al. Co-doped UiO-66, the end product, showed 7.6 times the capacity of pure UiO-66 (224.1 mg/g) [116]. Likewise, it was found that Mn-doped UiO-66 significantly improved upon bare UiO-66 in terms of TC adsorption [117]. Increases in Langmuir surface area and pore size with Mn insertion were found to be conducive to TC removal via π-π and hydrogen bond couplings.

Wrinkled UiOs composites are intriguing since they can be used to better increase surface area and expose defective areas. The interaction between the adsorbent and the desired pollutant is aided by the wrinkle structure. Based on this idea, it was found that wrinkle-designed UiO-66@Fe_3O_4 absorbed almost twice as much salicylic acid per unit gram as virgin UiO-66 [118]. Because of the increased speed with which the coordination contact between salicylic acid and wrinkled MOFs formed, this enhancement was mostly due to the creation of latticed defects. According to the findings, the MOF wrinkle structure plays a critical role in improving adsorption efficiency.

3.15 UiOs-Derived Materials

Recently, it was demonstrated that the compound UiO-66-NH_2 can be used as a template to create carbon composites with pores of different sizes [119]. The use of CTAB and sodium laurate-containing templates allowed for the control of pore size. UC-0.1 (pore size 5.38 nm) had the highest absorption capacity at 84.23 mg/g. Thermodynamic, isothermal, and kinetic analyses of adsorption led scientists to infer that electrostatic and hydrophobic interactions were crucial for the adsorption process.

3.16 Pharmaceutical Drug Resistance

Antibiotic residues have accumulated in the environment due to varying patterns of antibiotic usage, most typically for treating bacterial infections and raising animals. Long half-lives, inability to degrade in the body, and high lipophilicity are hallmarks of medicinal compounds [120]. The pharmaceutical residues themselves are the cause of this increased environmental persistence. The sluggish rate of auto-degradation of all of these pharmaceutical remnants, along with the sustained discharge of considerable quantities, causes contamination to grow from nano to micro levels [121]. Antimicrobial use may also have the unintended consequence of selecting for resistance. When antimicrobial concentrations drop below the minimum inhibitory concentration, antibiotic-resistant bacteria tend to proliferate rather than susceptible ones. Therefore, the resistant bacteria can endure antibiotic doses that would ordinarily be deadly to microorganisms of identical species [122]. Resistance genes to antimicrobials can be transferred among bacteria that are merely related to one another, in spite to the germs that are typically present in humans [123].

Protein adsorption at the solid-liquid interface is essential to a wide variety of natural processes and technological applications [124]. The study of protein–solid adsorption interactions is crucial to many fields, including nanoscience, healthcare, analysis, biotechnology, molecular genetics, and biophysics. There have been many attempts to find an adsorbent that can effectively remove organic pollutants from water, and many have failed [125]. Because of its advantageous qualities including increased bacterial internalization, flexible structural control, excellent selectivity, and controlled drug release, new materials like MOFs are promising prospects for antibacterial therapy to combat antibiotic resistance. Nano silica has been used in a variety of environmental circumstances, not just for the treatment of medicinal compounds [126]. MOFs are a potential solution for treating wastewater [127] because of their versatility.

3.17 Conclusion

Contamination due to pharmaceutical wastes in water has increased rapidly over the last two decades due to a fast increase in industrialization, urbanization, and population. These contaminants cause serious health issues; for instance, carcinogenicity, cardiovascular disease, and neurological disorders. Therefore, the remediation of heavy metals from water is necessary. There are numerous techniques like precipitation, coagulation, advanced oxidation process, and adsorption, but these techniques suffer some limitations, like higher cost, production of secondary pollution, and complex operation. Adsorption is the optimized technique due to easy operation, cost effectiveness, and production of lesser secondary pollution. The selection of the adsorbent is the key factor in adsorption. MOFs is a promising adsorbent due to higher surface area, high porosity, and high efficiency. In this chapter, authors have discussed the removal efficiency of different MOFs for the removal of pharmaceutical waste.

References

1 Dulio, V. and Slobodnik, J. (2009 August). NORMAN—network of reference laboratories, research centers and related organizations for monitoring of emerging substances. *Environmental Science and Pollution Research* 16 (1): 132–135.

2 UNESCO WWAP (2003). Water for people, water for life: 3rd world water forum in Kyoto, Japan. http://www.unesco.org/new/en/natural%20sciences/environment/water/wwap/wwdr/wwdr1-2003.

3 Geissen, V., Mol, H., Klumpp, E. et al. (2015). Emerging pollutants in the environment: a challenge for water resource management. *The International Soil and Water Conservation* 3: 57–65.
4 Mompelat, S., Le Bot, B., and Thomas, O. (2009). Occurrence and fate of pharmaceutical products and by-products, from resource to drinking water. *Environment International* 35: 803.
5 Daughton, C.G. (2003). Pollution from the combined activities, actions, and behaviors of the public: pharmaceuticals and personal care products. *NoeCal SETAC News* 14: 5.
6 Kümmerer, K. (2001). Drugs in the environment: emission of drugs, diagnostic aids and disinfectants into wastewater by hospitals in relation to other sources: a review. *Chemosphere* 45: 957.
7 Gadipelly, C., Rathod, V.K., and Marathe, K.V. (2014). Pharmaceutical industry waste water: review of the technologies for water treatment and reuse. *Industrial & Engineering Chemistry Research* 53 (29): 11571.
8 Huerta, B. and Barceló, D. (2012). Pharmaceuticals in biota in the aquatic environment: analytical methods and environmental implications. *Analytical and Bioanalytical Chemistry* 404: 2611.
9 Jones, J.N., Voulvoulis, O.A.H., and Lester, N. (2010). Human pharmaceuticals in the aquatic environment a review. *Environmental Technology* 22: 1383.
10 Rivera-Utrilla, J., Sánchez-Polo, M., Ferro-García, M.Á. et al. (2013). Pharmaceuticals as emerging contaminants and their removal from water. a review. *Chemosphere* 93: 1268.
11 Hasan, Z., Jeon, J., and Jhung, S.H. (2012). Adsorptive removal of naproxen and clofibric acid from water using metal-organic frameworks. *Journal of Hazardous Materials* 209–210: 151.
12 Barceló, D. (2010). Water quality and assessment under scarcity: prospects and challenges in Mediterranean watersheds. *Journal of Hydrology* 383: 1.
13 Rivera-Jiménez, S.M. and Hernández-Maldonado, A.J. (2008). Nickel(II) grafted MCM-41: a novel sorbent for the removal of Naproxen from water. *Microporous Mesoporous Materials* 116: 246.
14 Rakshit, S., Sarkar, D., Elzinga, E.J. et al. (2013). Mechanisms of ciprofloxacin removal by nano-sized magnetite. *Journal of Hazardous Materials* 246–247: 221.
15 Mestre, A.S., Pires, J., Nogueira, J.M.F., and Carvalho, A.P. (2007). Activated carbons for the adsorption of ibuprofen. *Carbon N. Y.* 45: 1979.
16 Boyd, G.R., Reemtsma, H., Grimm, D.A., and Mitra, S. (2003). Pharmaceuticals and personal care products (PPCPs) in surface and treated waters of Louisiana, USA and Ontario, Canada. *The Science of the Total Environment* 311: 135.
17 Liu, W., Sutton, N.B., Rijnaarts, H.H.M., and Langenhoff, A.A.M. (2016). Pharmaceutical removal from water with iron- or manganese-based technologies: a review. *Critical Reviews in Environmental Science and Technology* 46: 1584–1621.
18 Kairigo, P., Ngumba, E., Sundberg, L.-R. et al. (2020). Occurrence of antibiotics and risk of antibiotic resistance evolution in selected Kenyan wastewaters, surface waters and sediments. *Science of the Total Environment* 720: 137580.
19 Marchant, J. (2018). When antibiotics turn toxic. *Nature* 555: 431–433.

20 Lee, C.C., Hsieh, C.Y., Chen, C.S., and Tien, C.J. (2020). Emergent contaminants in sediments and fishes from the Tamsui river (Taiwan): their spatial-temporal distribution and risk to aquatic ecosystems and human health. *Environmental Pollution* 258: 113733.

21 Pavithra, K.G., Senthil Kumar, P., Sundar Rajan, P. et al. (2017). Sources and impacts of pharmaceutical components in wastewater and its treatment process: a review. *The Korean Journal of Chemical Engineering* 34: 2787–2805.

22 Cristovao, M.B., Janssens, R., Yadav, A. et al. (2020). Predicted concentrations of anticancer drugs in the aquatic environment: what should we monitor and where should we treat? *The Journal of Hazardous Materials* 392: 122330.

23 Ahmed, M.B., Zhou, J.L., Ngo, H.H., and Guo, W. (2015). Adsorptive removal of antibiotics from water and wastewater: progress and challenges. *Science of the Total Environment* 532: 112–126.

24 de Andrade, J.R., Oliveira, M.F., da Silva, M.G.C., and Vieira, M.G.A. (2018). Adsorption of pharmaceuticals from water and wastewater using nonconventional low-cost materials: a review. *Industrial & Engineering Chemistry Research* 57 (9): 3103–3127.

25 Yu, F., Li, Y., Han, S., and Ma, J. (2016). Adsorptive removal of antibiotics from aqueous solution using carbon materials. *Chemosphere* 153: 365–385.

26 Santamaría, L., Devred, F., Gaigneaux, E.M. et al. (2020). Effect of the surface properties of Me2+/Al layered double hydroxides synthesized from aluminum saline slag wastes on the adsorption removal of drugs. *Microporous Mesoporous Mater* 309: 110560.

27 Kumar, A., Sharma, S.K., Sharma, G. et al. (2019). Wide spectral degradation of norfloxacin by ag@BiPO4/BiOBr/BiFeO3 nano-assembly: elucidating the photocatalytic mechanism under different light sources. *The Journal of Hazardous Materials* 364: 429–440.

28 Sharma, G., Gupta, V.K., Agarwal, S. et al. (2018). Fabrication and characterization of trimetallic nano-photocatalyst for remediation of ampicillin antibiotic. *The Journal of Molecular Liquids* 260: 342–350.

29 Oberoi, A.S., Jia, Y., Zhang, H. et al. (2019). Insights into the fate and removal of antibiotics in engineered biological treatment systems: a critical review. *Environmental Science & Technology* 53: 7234–7264.

30 Kanakaraju, D., Glass, B.D., and Oelgem¨oller, M. (2018). Advanced oxidation processmediated removal of pharmaceuticals from water: a review. *The Journal of Environmental Management* 219: 189–207.

31 Xu, Y., Liu, T., Zhang, Y. et al. (2017). Advances in technologies for pharmaceuticals and personal care products removal. *Journal of Materials Chemistry* 5: 12001–12014.

32 Mansour, F., Al-Hindi, M., Yahfoufi, R. et al. (2018). The use of activated carbon for the removal of pharmaceuticals from aqueous solutions: a review. *Reviews in Environmental Science and Biotechnology* 17: 109–145.

33 Alothman, Z.A. (2012). A review: fundamental aspects of silicate mesoporous materials. *Materials* 5: 2874–2902.

34 Huang, L., Mao, N., Yan, Q. et al. (2019). Magnetic covalent organic frameworks for the removal of diclofenac sodium from water. *ACS Applied Nano Materials* 3: 319–326.

35 Huang, L., Shen, R., Liu, R., and Shuai, Q. (2020). Thiol-functionalized magnetic covalent organic frameworks by a cutting strategy for efficient removal of Hg2+ from water.

36 Dhaka, S., Kumar, R., Deep, A. et al. (2019). Metal-organic frameworks (MOFs) for the removal of emerging contaminants from aquatic environments. *Coordination Chemistry Reviews* 380: 330–352.

37 Gao, Q., Xu, J., and Bu, X.H. (2019a). Recent advances about metal-organic frameworks in the removal of pollutants from wastewater. *Coordination Chemistry Reviews* 378: 17–31.

38 Wang, T.C., Bury, W., G´omez-Gualdr´on, D.A. et al. (2015). Ultrahigh surface area zirconium MOFs and insights into the applicability of the BET theory. *Journal of the American Chemical Societyc.* 137: 3585–3591.

39 Lin, R.-B., Xiang, S., Xing, H. et al. (2019). Exploration of porous metal-organic frameworks for gas separation and purification. *Coordination Chemistry Reviews* 378: 87–103.

40 Dhakshinamoorthy, A., Li, Z., and Garcia, H. (2018). Catalysis and photocatalysis by metal-organic frameworks. *Chemical Society Reviews* 47: 8134–8172.

41 Wang, H., Lustig, W.P., and Li, J. (2018). Sensing and capture of toxic and hazardous gases and vapors by metal-organic frameworks. *Chemical Society Reviews* 47: 4729–4756.

42 Wu, M.X. and Yang, Y.W. (2017). Metal-organic framework (MOF)-based drug/cargo delivery and cancer therapy. *Advanced Materials* 29 (23): 1606134.

43 Hasan, Z. and Jhung, S.H. (2015). Removal of hazardous organics from water using metal-organic frameworks (MOFs): plausible mechanisms for selective adsorptions. *The Journal of Hazardous Materials* 283: 329–339.

44 Khan, N.A., Hasan, Z., and Jhung, S.H. (2013). Adsorptive removal of hazardous materials using metal-organic frameworks (MOFs): a review. *Journal of Hazardous Materials* 244–245: 444.

45 Huang, L., He, M., Chen, B., and Hu, B. (2016). A mercapto functionalized magnetic Zr-MOF by solvent-assisted ligand exchange for Hg2+ removal from water. *Journal of Materials Chemistry A* 4: 5159–5166.

46 Huang, L., He, M., Chen, B.B., and Hu, B. (2018). Magnetic Zr-MOFs nanocomposites for rapid removal of heavy metal ions and dyes from water. *Chemosphere* 199: 435–444.

47 Alqadami, A.A., Naushad, M., Alothman, Z.A., and Ghfar, A.A. (2017). Novel metal–organic framework (mof) based composite material for the sequestration of U(VI) and Th(IV) metal ions from aqueous environment. *ACS Appliled Mater Interfaces* 9: 36026–36037.

48 Kobielska, P.A., Howarth, A.J., Farha, O.K., and Nayak, S. (2018). Metal–organic frameworks for heavy metal removal from water. *Coordination Chemistry Reviews* 358: 92–107.

49 Li, J., Wang, X., Zhao, G. et al. (2018a). Metal-organic framework-based materials: superior adsorbents for the capture of toxic and radioactive metal ions. *Chemical Society Reviews* 47: 2322–2356.

50 Wei, X., Wang, Y., Chen, J. et al. (2020). Adsorption of pharmaceuticals and personal care products by deep eutectic solvents-regulated magnetic metal-organic framework adsorbents: performance and mechanism. *The Chemical Engineering Journal* 392: 124808.

51 Wang, C., Kim, J., Tang, J. et al. (2020). New strategies for novel MOF-derived carbon materials based on nanoarchitectures. *Inside Chemistry* 6: 19–40.

52 Devic, T. and Serre, C. (2014). High valence 3p and transition metal based MOFs. *Chemical Society Reviews* 43: 6097–6115.

53 F′erey, G., Latroche, M., Serre, C. et al. (2003). Hydrogen adsorption in the nanoporous metal-benzenedicarboxylate M (OH) (O2C–C6H4–CO2)(M= Al3+, Cr3+), MIL-53. *Chemical Communications* (24): 2976–2977.

54 F′erey, G., Mellot-Draznieks, C., Serre, C. et al. (2005). A chromium terephthalate-based solid with unusually large pore volumes and surface area. *Science* 309: 2040.

55 Kandiah, M., Nilsen, M.H., Usseglio, S. et al. (2010). Synthesis and stability of tagged UiO-66 Zr-MOFs. *Chemistry of Materials* 22: 6632–6640.

56 Cavka, J.H., Jakobsen, S., Olsbye, U. et al. (2008). A new zirconium inorganic building brick forming metal organic frameworks with exceptional stability. *Journal of the American Chemical Societyc.* 130: 13850–13851.

57 Feng, D., Gu, Z.Y., Li, J.R. et al. (2012). Zirconium-metalloporphyrin PCN-222: mesoporous metal-organic frameworks with ultrahigh stability as biomimetic catalysts. *Angewandte Chemie International Edition* 51: 10307–10310.

58 Banerjee, R., Phan, A., Wang, B. et al. (2008). High-throughput synthesis of zeolitic imidazolate frameworks and application to CO2 capture. *Science* 319: 939–943.

59 Zhang, J.P., Zhang, Y.B., Lin, J.B., and Chen, X.M. (2012). Metal azolate frameworks: from crystal engineering to functional materials. *Chemical Reviews* 112: 1001–1033.

60 Colombo, V., Galli, S., Choi, H.J. et al. (2011). High thermal and chemical stability in pyrazolate-bridged metal–organic frameworks with exposed metal sites. *Chemical Science* 2: 1311–1319.

61 Yang, J., Grzech, A., Mulder, F.M., and Dingemans, T.J. (2011). Methyl modified MOF-5: a water stable hydrogen storage material. *Chemical Communications* 47 (18): 5244–5246.

62 Taylor, J.M., Vaidhyanathan, R., Iremonger, S.S., and Shimizu, G.K.H. (2012). Enhancing water stability of metal–organic frameworks via phosphonate monoester linkers. *Journal of the American Chemical Societyc.* 134: 14338–14340.

63 Yang, C., Wang, X., and Omary, M.A. (2007). Fluorous metal-organic frameworks for high density gas adsorption. *Journal of the American Chemical Societyc.* 129: 15454–15455.

64 Li, H., Shi, W., Zhao, K. et al. (2012). Enhanced hydrostability in Nidoped MOF-5. *Inorganic Chemistry* 51: 9200–9207.

65 Li, L., Liu, X.L., Geng, H.Y. et al. (2013). A MOF/graphite oxide hybrid (MOF: HKUST-1) material for the adsorption of methylene blue from aqueous solution. *Journal of Materials Chemistry* 1: 10292–10299.

66 Xiong, W., Zeng, G., Yang, Z. et al. (2018a). Adsorption of tetracycline antibiotics from aqueous solutions on nanocomposite multi-walled carbon nanotube functionalized MIL-53 (Fe) as new adsorbent. *Science of the Total Environment* 627: 235–244.

67 Li, X., Yuan, H., Quan, X. et al. (2018b). Effective adsorption of sulfamethoxazole, bisphenol A and methyl orange on nanoporous carbon derived from metal-organic frameworks. *Journal Environmental Science* 63: 250–259.

68 Zhang, W., Hu, Y., Ge, J. et al. (2014). A facile and general coating approach to moisture/water-resistant metal–organic frameworks with intact porosity. *Journal of the American Chemical Societyc.* 136: 16978–16981.

69 Zu, D.D., Lu, L., Liu, X.Q. et al. (2014). Improving hydrothermal stability and catalytic activity of metal–organic frameworks by graphite oxide incorporation. *The Journal of Physical Chemistry C* 118: 19910–19917.

70 Liu, B., Shioyama, H., Akita, T., and Xu, Q. (2008). Metal-organic framework as a template for porous carbon synthesis. *Journal of the American Chemical Societyc.* 130: 5390–5391.

71 Yang, S.J. and Park, C.R. (2012). Preparation of highly moisture-resistant black-colored metal organic frameworks. *Advanced Materials* 24: 4010–4013.

72 Sun, X., Wu, T., Yan, Z. et al. (2019). Novel MOF-5 derived porous carbons as excellent adsorption materials for n-hexane. *Journal of Solid State Chemistry* 271: 354–360.

73 Xie, Y., Chen, C., Lu, X. et al. (2019). Porous NiFe-oxide nanocubes derived from prussian blue analogue as efficient adsorbents for the removal of toxic metal ions and organic dyes. *The Journal of Hazardous Materials* 379: 120786.

74 Banerjee, A., Gokhale, R., Bhatnagar, S. et al. (2012). MOF derived porous carbon–Fe3O4 nanocomposite as a high performance, recyclable environmental superadsorbent. *Journal of Materials Chemistry* 22: 19694.

75 Jun, J.W., Tong, M., Jung, B.K. et al. (2015). Effect of central metal ions of analogous metal–organic frameworks on adsorption of organoarsenic compounds from water: plausible mechanism of adsorption and water purification. *Chemistry—A European Journal* 21: 347–354.

76 Dou, R., Zhang, J., Chen, Y., and Feng, S. (2017). High efficiency removal of triclosan by structure-directing agent modified mesoporous MIL-53(Al). *Environmental Science and Pollution Research* 24: 8778–8789.

77 Sarker, M., Song, J.Y., and Jhung, S.H. (2017). Adsorption of organic arsenic acids from water over functionalized metal-organic frameworks. *The Journal of Hazardous Materials* 335: 162–169.

78 Song, J.Y. and Jhung, S.H. (2017). Adsorption of pharmaceuticals and personal care products over metal-organic frameworks functionalized with hydroxyl groups: quantitative analyses of H-bonding in adsorption. *The Chemical Engineering Journal* 322: 366–374.

79 Seo, P.W., Bhadra, B.N., Ahmed, I. et al. (2016). Adsorptive removal of pharmaceuticals and personal care products from water with functionalized metalorganic frameworks: remarkable adsorbents with hydrogen-bonding abilities. *Scientific Reports* 6: 34462.

80 Lv, Y., Zhang, R., Zeng, S. et al. (2018). Removal of p-arsanilic acid by an amino-functionalized indium-based metal–organic framework: adsorption behavior and synergetic mechanism. *The Chemical Engineering Journal* 339: 359–368.

81 Chai, F., Zhao, X., Gao, H. et al. (2019). Effective removal of antibacterial drugs from aqueous solutions using porous metal–organic frameworks. *Journal of Inorganic and Organometallic Polymers and Materials* 1305–1313.

82 Hasan, Z., Choi, E.J., and Jhung, S.H. (2013). Adsorption of naproxen and clofibric acid over a metal–organic framework MIL-101 functionalized with acidic and basic groups. *The Chemical Engineering Journal* 219: 537–544.

83 Peng, Y., Zhang, Y., Huang, H., and Zhong, C. (2018). Flexibility induced high-performance MOF-based adsorbent for nitroimidazole antibiotics capture. *The Chemical Engineering Journal* 333: 678–685.

84 Gao, Y., Kang, R., Xia, J. et al. (2019b). Understanding the adsorption of sulfonamide antibiotics on MIL-53s: metal dependence of breathing effect and adsorptive performance in aqueous solution. *Journal of Colloid and Interface Science* 535: 159–168.

85 Gao, Y., Liu, K., Kang, R. et al. (2018). A comparative study of rigid and flexible MOFs for the adsorption of pharmaceuticals: kinetics, isotherms and mechanisms. *The Journal of Hazardous Materials* 359: 248–257.

86 Zhuo, N., Lan, Y., Yang, W. et al. (2017). Adsorption of three selected pharmaceuticals and personal care products (PPCPs) onto MIL-101 (Cr)/natural polymer composite beads. *Separation and Purification Technology* 177: 272–280.

87 Jia, M., Wang, F., Bian, Y. et al. (2013 May 1). Effects of pH and metal ions on oxytetracycline sorption to maize-straw-derived biochar. *Bioresource Technology* 136: 87–93.

88 Yu, L.L., Cao, W., Wu, S.C. et al. (2018). Removal of tetracycline from aqueous solution by MOF/graphite oxide pellets: preparation, characteristic, adsorption performance and mechanism. *Ecotoxicology and Environmental Safety F.* 164: 289–296.

89 Yu, L.L., Luo, Z.F., Zhang, Y.Y. et al. (2019). Contrastive removal of oxytetracycline and chlortetracycline from aqueous solution on Al-MOF/GO granules. *Environmental Science and Pollution Research* 26: 3685–3696.

90 Xiong, W., Zeng, Z., Li, X. et al. (2018b). Multi-walled carbon nanotube/amino-functionalized MIL-53(Fe) composites: remarkable adsorptive removal of antibiotics from aqueous solutions. *Chemosphere* 210: 1061–1069.

91 Jin, J., Yang, Z., Xiong, W. et al. (2019). Cu and Co nanoparticles co-doped MIL-101 as a novel adsorbent for efficient removal of tetracycline from aqueous solutions. *Science of the Total Environment* 650: 408–418.

92 Bayazit, S.S., Danalioglu, S.T., Salam, M.A., and Kuyumcu, O.K. (2017). Preparation of magnetic MIL-101 (Cr) for efficient removal of ciprofloxacin. *Environmental Science and Pollution Research* 24: 25452–25461.

93 Naeimi, S. and Faghihian, H. (2017). Application of novel metal organic framework, MIL-53 (Fe) and its magnetic hybrid: for removal of pharmaceutical pollutant, doxycycline from aqueous solutions. *Environ. Toxicol. Pharmacol.* 53: 121–132.

94 Zhang, Y.Y., Liu, Q., Yang, C. et al. (2019). Magnetic aluminum-based metal organic framework as a novel magnetic adsorbent for the effective removal of minocycline from aqueous solutions. *Environmental Pollution* 255: 113226.

95 Moradi, S.E., Haji Shabani, A.M., Dadfarnia, S., and Emami, S. (2016). Effective removal of ciprofloxacin from aqueous solutions using magnetic metal–organic framework sorbents: mechanisms, isotherms and kinetics. *Journal of the Iranian Chemical Society* 13: 1617–1627.

96 Jiang, C., Zhang, X., Xu, X., and Wang, L. (2016). Magnetic mesoporous carbon material with strong ciprofloxacin adsorption removal property fabricated through the calcination of mixed valence Fe based metal-organic framework. *The Journal of Porous Materials* 23: 1297–1304.

97 Xiong, W., Zeng, Z., Zeng, G. et al. (2019). Metal-organic frameworks derived magnetic carbon-alpha Fe/Fe3C composites as a highly effective adsorbent for tetracycline removal from aqueous solution. *The Chemical Engineering Journal* 374: 91–99.

98 Wu, C.S., Xiong, Z.H., Li, C., and Zhang, J.M. (2015). Zeolitic imidazolate metal organic framework ZIF-8 with ultra-high adsorption capacity bound tetracycline in aqueous solution. *RSC Advances* 5 (100): 82127–82137.

99 Dehghan, A., Zarei, A., Jaafari, J. et al. (2019 February 1). Tetracycline removal from aqueous solutions using zeolitic imidazolate frameworks with different morphologies: a mathematical modeling. *Chemosphere* 217: 250–260.

100 Chen, X., Jiang, X., Yin, C. et al. (2018b). Facile fabrication of hierarchical porous ZIF-8 for enhanced adsorption of antibiotics. *The Journal of Hazardous Materials* 194–204.

101 Yuan, Y., Yang, D., Mei, G. et al. (2018 May 5). Preparation of konjac glucomannan-based zeolitic imidazolate framework-8 composite aerogels with high adsorptive capacity of ciprofloxacin from water. *Colloids and Surfaces A: Physicochemical and Engineering Aspects* 544: 187–195.

102 Song, Z., Ma, Y.L., Li, C.E. et al. (2019 March 18). Molecular sieving film prepared by vacuum filtration for the efficient removal of tetracycline antibiotics from pharmaceutical wastewater. *Advances in Materials Science and Engineering* 2019.

103 Zhao, R., Ma, T., Zhao, S. et al. (2020 February 15). Uniform and stable immobilization of metal-organic frameworks into chitosan matrix for enhanced tetracycline removal from water. *Chemical Engineering Journal* 382: 122893.

104 Duan, H., Hu, X., and Sun, Z. (2020 February 15). Magnetic zeolite imidazole framework material-8 as an effective and recyclable adsorbent for removal of ceftazidime from aqueous solution. *Journal of Hazardous Materials* 384: 121406.

105 Bhadra, B.N., Ahmed, I., Kim, S., and Jhung, S.H. (2017 April 15). Adsorptive removal of ibuprofen and diclofenac from water using metal-organic framework-derived porous carbon. *Chemical Engineering Journal* 314: 50–58.

106 Ahmed, I., Bhadra, B.N., Lee, H.J., and Jhung, S.H. (2018 March 1). Metal-organic framework-derived carbons: preparation from ZIF-8 and application in the adsorptive removal of sulfamethoxazole from water. *Catalysis Today* 301: 90–97.

107 Liang, C., Tang, Y., Zhang, X. et al. (2020 March 1). ZIF-mediated N-doped hollow porous carbon as a high performance adsorbent for tetracycline removal from water with wide pH range. *Environmental Research* 182: 109059.

108 Liang, C., Zhang, X., Feng, P. et al. (2018 July 15). ZIF-67 derived hollow cobalt sulfide as superior adsorbent for effective adsorption removal of ciprofloxacin antibiotics. *Chemical Engineering Journal* 344: 95–104.

109 Tang, Y., Chen, Q., Li, W. et al. (2020 April 15). Engineering magnetic N-doped porous carbon with super-high ciprofloxacin adsorption capacity and wide pH adaptability. *Journal of Hazardous Materials* 388: 122059.

110 Huang, L., He, M., Chen, B.B. et al. (2017 January 25). Highly efficient magnetic nitrogen-doped porous carbon prepared by one-step carbonization strategy for Hg2+ removal from water. *ACS Applied Materials & Interfaces* 9 (3): 2550–2559.

111 Sarker, M., Song, J.Y., and Jhung, S.H. (2018). Carboxylic-acid-functionalized UiO-66-NH2: a promising adsorbent for both aqueous- and non-aqueous-phase adsorptions. *The Chemical Engineering Journal* 331: 124–131.

112 Lin, S., Zhao, Y., and Yun, Y.-S. (2018). Highly effective removal of nonsteroidal anti-inflammatory pharmaceuticals from water by Zr(IV)-based Metal–Organic Framework: adsorption performance and mechanisms. *ACS Applied Materials & Interfaces* 10: 28076–28085.

113 Azhar, M.R., Abid, H.R., Periasamy, V. et al. (2017 August 15). Adsorptive removal of antibiotic sulfonamide by UiO-66 and ZIF-67 for wastewater treatment. *Journal of Colloid and Interface Science* 500: 88–95.

114 Akpinar, I. and Yazaydin, A.O. (2017 December 27). Rapid and efficient removal of carbamazepine from water by UiO-67. *Industrial & Engineering Chemistry Research* 56 (51): 15122–15130.

115 Li, B., Zhu, X., Hu, K. et al. (2016). Defect creation in metal-organic frameworks for rapid and controllable decontamination of roxarsone from aqueous solution. *The Journal of Hazardous Materials* 302: 57–64.

116 Cao, J., Yang, Z.H., Xiong, W.P. et al. (2018 December 1). One-step synthesis of co-doped UiO-66 nanoparticle with enhanced removal efficiency of tetracycline: simultaneous adsorption and photocatalysis. *Chemical Engineering Journal* 353: 126–137.

117 Yang, Z., Pei, H., Hou, Q. et al. (2018 January 15). Algal biofilm-assisted microbial fuel cell to enhance domestic wastewater treatment: nutrient, organics removal and bioenergy production. *Chemical Engineering Journal* 332: 277–285.

118 Yin, Y., Shi, M., Ren, Y. et al. (2020 May 1). Wrinkle structure on multifunctional MOFs to facilitate PPCPs adsorption in wastewater. *Chemical Engineering Journal* 387: 124196.

119 Hu, X., Sun, T., Jia, L. et al. (2020 May 15). Preparation of metal-organic framework based carbon materials and its application to adsorptive removal of cefepime from aqueous solution. *Journal of Hazardous Materials* 390: 122190.

120 Patel, P., Hanini, A., Shah, A. et al. (2019). Surface modification of nanoparticles for targeted drug delivery. In: *Surface Modification of Nanoparticles for Targeted Drug Delivery* 19–31. Cham: Springer.

121 Dinh, T.D., Phan, M.N., Nguyen, D.T. et al. (2022). Removal of beta-lactam antibiotic in water environment by adsorption technique using cationic surfactant functionalized nanosilica rice husk. *Environmental Research* 210: 112943.

122 Kovalakova, P., Cizmas, L., McDonald, T.J. et al. (2020). Occurrence and toxicity of antibiotics in the aquatic environment: a review. *Chemosphere* 251: 126351.

123 Vu, T.N., Le, P.H.P., Pham, D.N.P. et al. (2022). Highly adsorptive protein inorganic nanohybrid of Moringa seeds protein and rice husk nanosilica for effective adsorption of pharmaceutical contaminants. *Chemosphere* 307: 135856.

124 Mitra, S.P. (2020). Protein adsorption on biomaterial surfaces: subsequent conformational and biological consequences–A review. *Journal of Surface Science and Technology* 36 (1–2): 7–38.

125 Tchinsa, A., Hossain, M.F., Wang, T., and Zhou, Y. (2021). Removal of organic pollutants from aqueous solution using metal organic frameworks (MOFs)-based adsorbents: a review. *Chemosphere* 284: 131393.

126 Salami, B.A., Oyehan, T.A., Gambo, Y. et al. (2022). Technological trends in nanosilica synthesis and utilization in advanced treatment of water and wastewater. *Environmental Science and Pollution Research* 1–41.

127 Ezugwu, C.I., Sonawane, J.M., and Rosal, R. (2021). Redox-active metal-organic frameworks for the removal of contaminants of emerging concern. *Separation and Purification Technology* 120246.

4

Efficiency of MOFs in Water Treatment Against the Emerging Water Contaminants Such as Endocrine Disruptors, Pharmaceuticals, Microplastics, Pesticides, and Other Contaminants

*Jogindera Devi and Ajay Kumar**

Department of Chemistry, School of Basic and Applied Sciences, Maharaja Agrasen University, Baddi (H.P.)-174103, India
** Corresponding author*

4.1 Introduction

Advancements in the industrial field have led to a generation of various harmful effluents. These include dyes, personal care products, drugs, heavy metals, surfactants, and pesticides, etc., which are being released to the environment without any pretreatment. Toxins that have a significant adverse effect on humans, animals, and marine life are included on a priority pollutant list that has been developed by the the United States Environmental Protection Agency (USEPA), World Health Organization (WHO), and other international governmental entities [1]. Only a few of the toxins recently discovered in wastewater treatment facility effluents include sunscreens, analgesics, antibiotics, plasticizers, antiseptics, and stimulants of the central nervous system (CNS) [2]. This suggests that these micropollutants are really not properly removed using traditional wastewater treatment methods, which is quite alarming. The overall scenario has led to a decay of water quality, which poses various health issues to both flora and fauna. Thus, treatment of wastewater has become a global challenge. Recently, many scientists have developed new methods and materials to deal with the problem [3, 4]. However, finding eco-friendly and cost effective strategies for waste treatment is, again, a challenge. A novel class of porous material called MOFs is composed of organic linkers and nodes that contain metal. MOF-based materials are of high importance for wastewater remediation because of their distinctive qualities such as large surface area, large adsorption capacity, high stability, etc [5]. The porous structure of MOFs is regarded as an important feature because we can alter the pore size to some extent, according to our requirement for wastewater treatment plants (WWTPs) [6]. The most significant techniques which are commonly used for wastewater treatment are sedimentation, soxhlet extraction, centrifugation, oxidation, and membrane separation, photocatalysis, microwave catalysis, photocatalytic ozonation,

Metal Organic Frameworks for Wastewater Contaminant Removal, First Edition. Edited by Arun Lal Srivastav, Lata Rani, Jyotsna Kaushal, and Tien Duc Pham.

coagulation, adsorption, ion exchange, and electrolysis, etc [7, 8]. Contrarily, adsorption and advanced oxidationprocesses offer low cost and simple implementation strategies. Both photocatalytic and adsorption techniques demonstrated great potential for the reduction or elimination of various contaminants.

Adsorption is a method with good efficiency for the effective elimination of contaminants from the aqueous phase. On the other hand, photocatalysis has the ability to degrade trace organic pollutants in mild conditions with minimal byproduct formation [9]. Thus, the present chapter summarizes the exploration of MOFs for the photocatalaytic and adsorption removal of different impurities.

4.2 Chemical Contaminants: Those Mysterious Ingredients in Ground and Surface Water

4.2.1 Endocrine Disruptors (EDs)

The careless handling of industrial waste or effluents has deteriorated the quality of both surface as well as ground water resources. Ions of various heavy metals like iron, manganese, arsenic, and cadmium, etc., are found in groundwater at higher levels than the permitted concentration. The other class of contaminants such as endocrine disruptors, potential carcinogens, etc., are also being continuously detected in groundwater. EDs are disorders that adversely affect the normal functioning of metabolism in animal cells. These chemicals are responsible for the disorder, creating a negative effect on metabolism [10]. There are various chemicals that may act as EDs: estradiol, mifepristone, estriol, ethinylestradiol, and bisphenol A, etc [11]. These can penetrate the cell and interfere with the endocrine system and, ultimately, alter hormonal release. Some of the adverse effects are obesity, thyroid problems, cardiovascular disease, Alzheimer's, cancer, and reproductive system problems [12]. The detection of EDs in various water resources are of serious concern. Saha et al. reported the presence of di (2-ethylhexyl) phthalate in the concentration 1005.490, 3203.33 & 2706.135 $\mu g\ L^{-1}$ in agricultural soils, groundwaters, and river waters, respectively. They also report the presence of other EDs such as bisphenol A (13.99–228.03 $\mu g\ L^{-1}$), di-n-butyl phthalate (117.492–182.29 $\mu g\ L^{-1}$), and 4-tert octylphenol (180.680–829.93 $\mu g\ L^{-1}$) [13]. Another study by Jian et al. also revealed the presence of atrazine, acetochlor, alachlor, heptachlor, and chlorpyrifos in various water samples [14]. Similarly, Idowu et al. studied a sample of water and sediment from a river. They found BPA concentrations in river waters varied from 0.41 to 5.19 μg/L, and in river sediments from 0.64 to 10.6 μg/L [15]. Thus, there is a need for efficient strategies to eliminate such substrate from water resources to minimize the causing detrimental effects.

4.2.2 Microplastics (MPs)

Contamination from microplastics (MPs) such as plasticizers, stabilizers, and flame retardants mainly arise from industry. Wang et al. and Li et al. found that small MPs can enter in to flora and fauna, which resulted in abnormal growth [16, 17]. When MPs enter animals or the human body, it can result in intestinal injury, loss of weight, liver necrosis, and many other health complications. In a study by Zhou et al., the frequency of microplastics found in fish gills or gastrointestinal tracts was shown as C.idella (93.33% and 86.67%), H.molitrix (91.67% and 100%), C. Molitorella (87.50% and 87.5%), O.niloticus (92.86% and 100%), and P. fulvidraco (90.91% and 81.82%) [18]. Another work by Zhang et al. reports the source of MPs in the soil. The concentration of MPs in compost were found in chicken manure (CM)14,720 ± 2,468 kg^{-1}, sludge (SC) 8600 ± 1,428 kg^{-1}, and domestic waste (DW)11,640 ± 3565 kg^{-1} [19]. The detection of MPs in soil and aquatic habitats is alarming because both sites are the base of an ecosystem. If preventive measures were not taken to check the MPs spill to these sites, there will definitely be serious consequences because soil and water are the backbones of the entire food chain on Earth.

4.2.3 Contaminants from the Agriculture Sector

Pesticides, insecticides, and herbicides are frequently used chemicals that are used to control insects and pests for the protection of crops. Thus, these are being frequently detected in water sources as well as in soil. Roshanak et al. studied a water sample from the Marun river and found the presence of two pesticides (1, 3-dichloropropene and alachlo) at the concentration 58 .0 & 2.44 g/L, respectively [20]. Tolgyesi et al. studied the presence of phenoxy herbicides in a different water sample. 40% of the examined samples were found to contain phenoxy herbicides in various concentrations ranging from 1.91 to 40.5 $\mu g\ L^{-1}$ [21]. Olisah et al. reported the greatest concentration of organophosphate pesticides, found in leaves of the Phragmites australis plant (16.41–31.39 $\mu g\ kg^{-1}$ dw), followed by sediments (3.30–8.07 $\mu g\ kg^{-1}$ dw), and roots (13.92–30.88 $\mu g\ kg^{-1}$ dw) [22]. The presence of these pollutants at these sites indicate that they are quite resistant to biotic or abiotic breakdown. Therefore, the complete eradication of agricultural contaminants is essential to safeguard environmental safety as well as public health.

4.2.4 Pharmaceutical Effluents

However, wastewater produced by the pharmaceutical industries contains significant quantities of a number of drugs, such as analgesics, antibiotics, anti-diabetics, beta-blockers, contraceptives, lipid regulators, and antidepressants, etc [23]. Koreje et al. reported the presence of the antiretroviral drug nevirapine at a high concentration in the range of 1–2 $g\,L^{-1}$ in underground water near septic tanks that were utilized as sources for drinking water [24]. Similarly, other researchers have reported the presence of various pharmaceutical effluents in surface- as well as groundwater such as the antibacterial agent josamycin (100 $\mu g\,L^{-1}$), clindamycin (1.1 $g\,L^{-1}$), erythromycin in spring (1.60 $\mu g\,L^{-1}$), summer (0.772 $\mu g\,L^{-1}$), and winter (0.546 $\mu g\,L^{-1}$), hydrochlorothiazide (0.07 $\mu g\,L^{-1}$), and losartan potassium

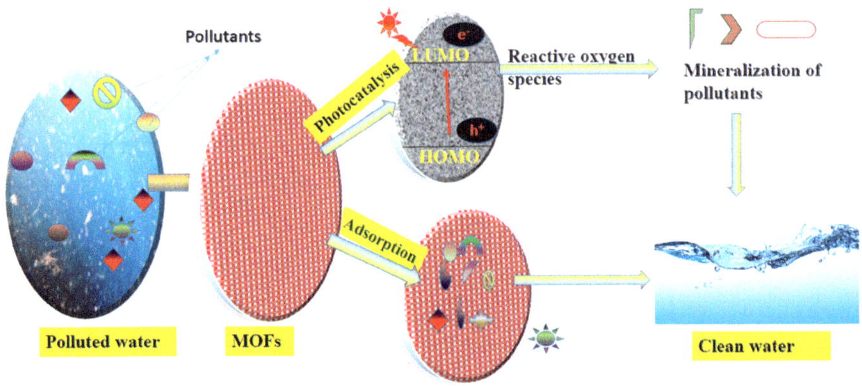

Figure 4.1 Adsorption and photocatalysis of pollutants by MOFs.

($0.09\,\mu g\,L^{-1}$), respectively [25, 26]. The presence of these illicit substrates in surface- and groundwater are alarming, and they have shown a major impact on the living system [27]. Thus, the elimination of such substrates has become the priority for a sustainable environment. Various processes and materials have been continuously explored for wastewater remediation. Advanced oxidation processes such as photocatalysis, ozonolysis, and photo-Fenton processes and materials such as semiconductor-based photocatalysts, MOFs, activated carbon, etc., have been studied for wastewater remediation [28–30]. Figure 4.1 is a pictorial representation of MOFs and their application for wastewater treatment via photocatalysis and adsorption.

4.3 MOFs

Metal complexes or metallic ions connected with chelates by consciousness are called metal-organic frameworks (MOFs), which may be crystalline or amorphous [31]. The abundance of porous matrices, high surface area, and various functionalities make MOFs a good potential candidate for the adsorption of pollutants as well as for the anchoring sites of nanoparticles [32]. Due to the adjustable nature of their pore structure and the simplicity with which they can be modified without changing the morphology of the structure, MOFs have received a lot of interest. They are used to remove numerous hazardous compounds through adsorptive action from both aqueous and non-aqueous environments [33, 34]. It has been demonstrated that MOFs can adsorb more As, Cr, and Se oxyanions than metal hydroxides. Additionally, material modification and functionalization strategies can address the issue of light harvesting properties and may enhance surface qualities which might be beneficial for MOF-based photocatalytic operations [35]. However, there are certain areas that require research for the large-scale exploration of MOFs, such as robustness in aqueous acid/base solutions, reusability, recyclability, and the working mechanism in complex wastewater effluents [36].

4.3.1 MOF Stability in the Aqueous Phase

The scaffold stability in aqueous mediums is one of the most significant factors when using MOFs for the remediation of aqueous contaminants [37]. It was determined that MOF stability was determined by how much their typical chemical properties changed as a result of their exposure to an aquatic environment. Due to MOF's susceptibility to moisture, the separation membranes in water are unstable due to the interaction of water molecules with the framework. Water molecules may quickly assault the relatively weak metal-organic coordinate bonds, which may lead to MOF structural damage or crystal phase change [38]. Therefore, it is crucial to accurately determine how water molecules affect the fundamental characteristics of MOFs (for example, coordination of metal ligands and structural integrity), especially for the creation of water-stable MOFs [39]. Thus, strengthening the metal-organic coordination links can improve MOF water stability.

4.3.2 Improving the Water Stability of MOFs: General Enhancement Strategies

The metal-ligand bond of MOFs, which is stronger than the interaction between the central metal ion and hydrogen bonds, is largely responsible for water solubility. This prevents engulfing and hydrolysis of the M–L bond from water [40]. The use of metal ions with high valency (such as Ce, Ti, Fe, Cr, and Zr) rather than metal ions with low valency (such as Cu, Zn, Ni, and Pb) can accomplish this. The aforementioned requirements are essential to improve MOFs' thermodynamic stability, but they are not the only aspect that determines the MOFs stability in water. Kinetic stability and a high activation energy barrier are the other key factors to take into account. As MOFs do not have strong thermodynamic stability by nature (water might undergo hydrolysis of the coordination bond if it reaches the metal center), consistency can be attained via dynamic variables, such as steric effects and hydrophobicity, that create an activation energy barrier [41].

Pure MOFs often have a poor ability to remove contaminants; thus, modification of pure MOFs is the utmost requirement. The incorporation of various functional groups of MOFs and loading with other materials like hydrotalcite and graphene oxide can increase MOF capacity for pollutant removal [42]. In order to selectively remove particular contaminants, MOFs with specific pore diameters may be the area of research. But as of yet, there is no specific synthesis process that can precisely adjust the diameter and density of MOFs. However, various researchers have approaches and strategies to attain mesoporous and microporous structures of MOFs [43].

In general, high valence metal ions produce a more stiff structure that is resistant to hydrolysis due to their large size. Thus, metal ions with high valency like Ni^{2+}, Zr^{4+}, Cr^{3+}, and Fe^{3+} are widely utilized in conjunction with carboxyl groups linking ions [44].The metal-ligand coordination bond strongly influences the stability of MOFs, which is made by using metal ions as nodes and multidentate organic ligands containing O or N donors as linkers. As a result, these bond coordination numbers can be used as a predictor of the relative stability of MOFs [45]. The Cr^{3+} cation, which is simple to connect with carboxylate, can effectively inhibit the dissociation of the metal-ligand bond, which may be partially responsible for the stability of MIL-101 [46]. For instance, the Fe-based MOF types MIL-100, MIL-101, and UiO-66 give excellent chemical stability in both aqueous conditions and a variety of solvent media [47]. Water-stable MOF structures can also emerge when soft divalent metal ions and softer azolate ligands are combined. Hydrophobic characteristics can also be used to impart durability in wastewaters [48].

The ligands present in MOFs determine their consistency. Many MOFs remain unstable because of the metal and ligand's relatively weak connections [49]. The substantial correlation between the structural characteristics of MOFs results from the various topologies and the effects of the ligands present [50]. Whereas the 2-D layer ligands are connected to the metal sites by carboxylate groups in each structure, the pillaring ligands are coordinated by nitrogen bonds. Because of these variations in ligand flexibility, it may be possible to create highly porous and mechanically or chemically stable MOFs by using a ligand rigidification method [51].

For photocatalysis and water splitting processes, ligand functionalization and the introduction of co-catalysts can increase light harvesting properties. The separation of photo-generated charge transfer has recently been improved by using precious elements including Pt, Pd, and Au as co-catalysts [52]. Zhou et al. used 4-(1,2,4-triazol-4-yl)-phenylphosphonic acid (1,2,4-triazole, pKa=13.9) as the intermediate linker interacting with nickel (II) to generate the trinuclear metal phosphonate, [Ni3(Hptz)6(H2O)6]9H2O [53]. When the precursor was used to make a 2D MOF, it was discovered that the structure is remarkably stable for a week in acid solution, heat, and several volatile chemicals. The MOF possessed selectivity for the sorption of CO_2 from the mixture of N_2 and CO_2 [54]. Therefore, the above listed factors such as hydrophobization, ligand functionalization, and charge on metal cluster, etc., are the crucial points that should be considered as important during the modification/fabrication of MOFs.

4.4 Possibilities for Wastewater Treatment Applications Using MOFs

MOFs' material-based strategies are the recent approaches that are being looked into for the removal of water contamination. The inorganic moieties such as clusters, atom chains, and organic linkers such as phosphonate-, azolate-, and carboxylate-based MOFs are recognized as a unique variety of porous conjugated polymers that self-assemble into multifunctional irregular crystal lattices. These materials have been suggested for a variety of uses, such as medication administration, fuel cells, light harvesting, photocatalysis, adsorption, separation, sensing, and catalysis [55]. MOFs can be molded into monoliths, pellets, membranes, columns, or other shapes that are appropriate for decontamination devices [56]. In order to design materials that improve adsorbents and/or catalysts for the treatment of contaminated water, a lot of study has been invested in to modify MOFs by increasing their pore diameter (creating defects by adding functional groups, etc.) or by synthesizing MOF composites (graphite oxide associated with metals, perovskites, etc.) [57].

4.4.1 MOF-Supported Adsorption & Photocatalysis

A lot of research has been done on MOF-based adsorption and photocatalysis treatments of wastewater due to its low cost, simplicity of use, and cheap energy requirements. The following factors can increase the adsorptive removal effectiveness of MOFs for pollutants from wastewater: i) The size-dependent interaction where the pore size of MOFs and the molecular dimension of wastewater pollutant removal are identical; ii) the hydrogen bonding (H bonding) interaction, which can be enhanced by adding heteroatom-containing functional groups to MOFs to bind the removal of pollutants; iii) the electrostatic interaction, which can be controlled by modifying the surface energy of MOFs [58, 59]. For photocatalytic degradation of MOFs, however, there are a number of challenges to solve. These include decreasing the recombination rate of excited carriers to effectively slow down degradation by decreasing the band gap for easily accessible visible light-induced degradation and more effective solar harnessing. MOFs can be modified with organic linkers, porphyrin, metal-oxoclusters, and other substances to increase their stability, surface area, and light responsiveness, among other things [60, 61]. The photocatalysis mechanism can be considered as the excitation of an electron from the HOMO (highest occupied molecular orbital) to the LUMO (lowest unoccupied molecular orbital) when light is reflected on the MOFs, leaving a hole in the HOMO [62]. The organic molecules can be oxidized by this hole by interacting with OH to produce an $OH^{\bullet}$ radical [63]. In order to perform photocatalysis, the photoactive MOFs must produce pairs of electron holes in their valence bands and conduction, respectively. The released electrons (e) mix with oxygen in an aqueous medium to produce oxygen radicals, which then convert into hydroxyl radicals ($OH^{\bullet}$). The target pollutant could be sufficiently attacked by the $OH^{\bullet}$ and h^+ active species in both instances, which could then break every bond in the analyte and ultimately transform into a non-toxic aliphatic species or completely mineralize [64]. Therefore, evaluating MOFs for photocatalytic applications requires consideration of their capacity to capture direct sunlight. MOFs with Ti, Cr, Zr, and Fe metal ions are stable in water and have the capacity to capture direct solar energy [65].

Utilizing MOFs to combine adsorption and photocatalysis processes is regarded as more advantageous over sole adsorption and photocatalysis processes. The porous structure of MOFs allow quick adsorption of the target pollutants. The process leads to a rise in the concentration of pollutant molecules on the surface, which encourages the ensuing photocatalytic destruction. In order to improve the accessibility of MOFs to pollutants, the photocatalytic active surface of MOFs can encourage the production of reactive oxygen species (ROS), which effectively breaks down huge contaminant molecules into smaller particles that can more easily enter the pores of MOFs [66].

The adsorption mechanism is, however, completely dependent on hydrogen bonds, electrostatic interactions, or a combination of these between the adsorbate and adsorbents. Among other factors, it has been observed that the surface energy of the framework strongly influences its adsorption abilities [67]. To control MOFs adsorption capabilities, a variety of ways have been followed to control their surface charge.

Generally, adsorption is the technique where a pollutant molecule is transferred from one place to another place and may remain in its original form. But photocatalysis offers a way to mineralize or completely destroy the original pollutant molecule. Thus, photocatalysis is the preferred technique over the conventional adsorption technique [68]. There are some key factors that are essentially required for efficient working of a photocatalyst, such as a high tendency to harvest solar light, an appropriate band gap, a low rate of electron/hole pair charge recombination, a high surface area for the photochemical reaction to occur, and a good support material if working in the nano range [69]. The scaffold of MOFs solely offers mostly all of these advantages. In the context of the mentioned advantages, various researchers have worked on the fabrication of MOF-based photocatalysts.

Moreover, the overall efficacy of both the adsorption and photocatalysis processes is based on surface contact between the pollutant molecule and photocatalyst/adsorbent. It is essential for the pollutant molecule that it should face a maximum encounter so that the ongoing remediation process can undergo smoothly. If we consider MOF-based scaffolds for adsorption and photocatalysis, following are some factors that play a crucial role in governing the effective interaction between the pollutant molecule and the MOF's scaffold. Figure 4.2 represents the various interactions between the MOF's surface and the pollutant molecule.

4.4.2 π-π Interactions

π-π interactions are a fundamental mechanism in the adsorption process, which arises as a result of the interaction between π electrons that may present in the targeted pollutant molecule as well as in MOFs [70]. Qin et al. studied the adsorption removal of bisphenol A (BPA) based on Fe MIL-101-Cr and MIL-100 MOFs, and they compared the obtained result for the same pollutant with activated carbon. Maximum adsorption removal was obtained for MIL-101-Cr. Actually, the pollutant moiety which contain π electrons had more affinity toward MIL-101, which resulted in a π-π interaction [71]. However, these are weak interactions, even though they contributed to the encounter between the MOF's scaffold and the pollutant molecule. For the MOF-based adsorption of other pollutants with benzene rings, such as 17-ethinylestradiol, sulfamethoxazole, tetracycline, and tyrosine etc., π-π interactions are one main factor among others [72, 73].

4.4.3 Electrostatic Interactions

Electrostatic interactions are another main factor that are responsible for the surface–surface interaction between pollutant molecules and the MOF's scaffold [74]. Generally, the surface charge on the adsorbent surface and net charge on the pollutant molecule govern the type of interaction. These interactions may be attractive or repulsive. Attractive interactions have +ve effect on the encounter while repulsive interaction causes – ve effects. Thus, the fate of the interaction is purely dependent on the isoelectric point of both adsorbent and adsorbate. In return, the effective surface charge is dependent on the working pH on which the ongoing adsorption process is monitored [75]. The pH factor is highly responsible for the alteration of

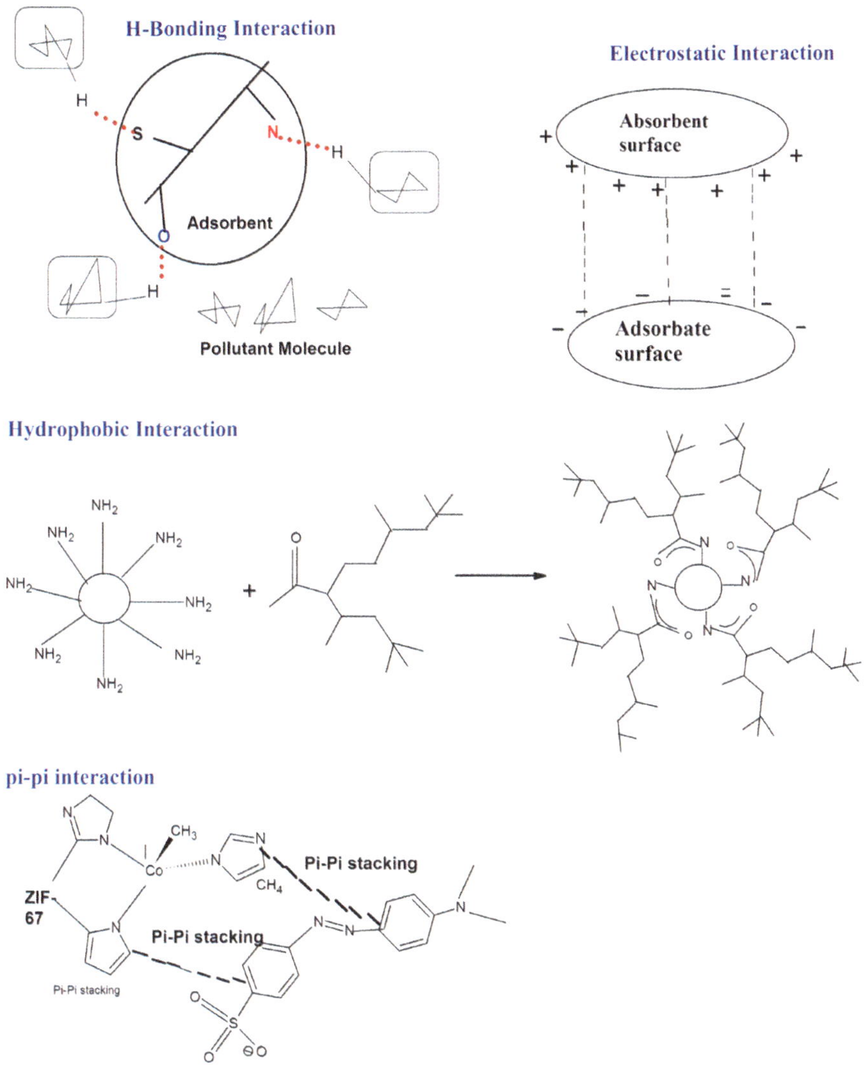

Figure 4.2 Various interactions responsible for pollutant adsorption onto the MOFs surface.

the surface charge. As the pH value goes below the point of zero charge (PZC) or isoelectric point (pH < PZC), the surface becomes positively charged. On the contrary, the surface becomes negatively charge as the pH value goes above the point of zero charge (PZC) or isoelectric point (pH > PZC) [76]. Thus, electrostatic interactions have high influence on the adsorption process and can be kept in mind while we fabricate MOFs for specific application.

4.4.4 Hydrophobic Interactions

Hydrophobic compounds often contain long carbon chains that are nonpolar and have little water solubility [77]. When organic materials are adsorbing from aqueous mediums, hydrophobic interactions are frequently seen. As versatile adsorbents, MOFs of hydrophobic nature were used for spilt oil adsorption [78]. Yang et al. [79] initially described the works in a variety of MOFs (FMOFs-1) composed of silver (I) 3,5-bis(trifluoromethyl)-1,2,4-triazolate for cleaning up oil spills. FMOFs-1 has a high hydrophobic interaction due to its inner surface that has been perfluorinated. Because of its hydrophobic interactions with nonpolar adsorbates, FMOFs-1 has a high affinity and adsorption efficiency for C_6–C_8 hydrocarbons and no discernible water adsorption. Soybean oil droplets were removed from water over Cu-BTC at a rapid rate (about six times that of a conventional AC) according to Lin et al [80]. The hydrophobic interactions between the benzene rings in Cu-BTC and soybean oil led to significant adsorption.

4.4.5 H-Bonding

To eliminate polar organic pollutants from water, the hydrogen bond is highly beneficial or effective because hydrogen bonding helps to interact the pollutant molecule with the adsorbent [81]. MOFs containing amino groups are reported as good adsorbents for the adsorption of cationic dyes due to the strong hydrogen bonding [82]. Thus, the functionalization of MOFs with such functional groups, which have high potential for hydrogen bonding, may contribute more for the adsorption removal of pollutants.

The outcome demonstrated by Z. Hasan et al. for the adsorption of pyridine was preferred over NH_2-UiO-66 in comparison to pyridine and UiO-66 in terms of both capacity and adsorption efficiency. With a binding energy of -74.2 kJ mol^{-1}, the amino group on pyridine's N is exposed to form a H-bond with the amino group on NH2-UiO-66. Instead, the optimized geometry predicted that the N on pyridine would move away once again to form a H-bond after initially placing the N on pyridine to face the N on the NH2-UiO-66. As a result, it was hypothesized that H-bonding had been essential for increasing pyridine binding above basic NH2-UiO-66 [83]. The quantity of pollutant adsorbed per square meter of surface grew as the number of hydrogen bonding sites increased. The solution's pH is also required to promote hydrogen bonding and subsequently adsorption. The various applications of MOFs as photocatalysts and adsorbents are shown in Table 4.1 and Table 4.2.

4.5 Use of MOFs for Water Remediation: Issues & Perspectives

Through selective adsorption and/or photocatalytic destruction, MOFs show significant promise for the removal of several contaminants from wastewater. When employing MOFs for water treatment or other applications, the material's stability under working conditions is an important factor in deciding the fate of practical applicability. MOF-based composites used in water remediation contain toxic metals (such as cadmium, lead, mercury, arsenic, chromium, and cobalt) and/or noxious chemical moieties (such as pyridine and 4,4-bipyridine). Thus, these areas need

Table 4.1 MOF-based photocatalyst applications.

Sr. no.	MOF-based photocatalyst	Utilization	Photocatalytic efficiency	References
1)	g-C_3N_4/ NH_2-MIL-125(Ti)	Photocatalytic CO_2 reduction to CO and CH_4	383.79 μmol g^{-1} & 13.8 μmol g^{-1}	[84]
2)	Ag_3PO_4/ TpPa-1-COF	Photocatalytic degradation of pesticide pymetrozine (PYM) and Rhodamine (Rh.B)	99.3% & 96.7%	[85]
3)	NH_2-MIL-125(Ti)	Photocatalytic degradation of dichlorophen and trichlorophenol	93.28% and 92.19%	[86]
4)	Fe_3O_4/CuO@C	Photocatalyst degradation of ciprofloxacin antibiotic	98.5%	[87]
5)	Hetero-ZnO/ Fe_2O_3	Photocatalytic degradation of methylene blue (MB) dye	100%	[88]
6)	2D Cd/Co-based MOFs	Photoreduction of Cr(VI) and degradation of methyl orange dye	100% & 85%	[89]
7)	Cu-BTC MOFs	Degradation of ofloxacin (OFX) antibiotic	72%	[90]
8)	Ag-MOF	Degrdation of 2-methyl-4-chlorophenoxyacetic acid (MCPA) and 2,4-dichlorophenoxyacetic acid (2,4-D) herbicides	98.0% & 96.0%	[91]
9)	NiAlCe LDH/ RGO	Photocatalysis of ciprofloxacin	94%	[92]
10)	g-CN@TiO_2/ MoS_2	Degradation methylene blue	97.5%	[93]
11)	NH_2-UiO-66	Photodegradation of ketorolac tromethamine and tetracycline	68.3% & 71.8%	[94]
12)	GO@UiO-66	Photodegradation of carbamazepine	(> 90%)	[95]
13)	TCS@MOF	Monocrotophos (MCP)	~98.79%	[96]
14)	Zn-MOF @AC	Photocatalytic degradation of brilliant green & methyl orange dye	86.4% &77.5%	[97]
15)	Ag/ZIF-67/TiO_2/ Cu	Photocatalytic degradation of 4-aminothiophenol pesticide	100%	[98]

Table 4.2 MOF-based adsorbents and their applications.

S.No.	MOF-based adsorbent	Utilization as adsorbent	Adsorption efficiency	References
1	Al@Fe-MOF	Selenite (Se(1V))	75.33 mg g^{-1}	[99]
2	Fe_3O_4/Cu-BTC@ CNT (FCuC)	Methylene blue	152 mg g^{-1}	[100]
3	MOFs-UIO-67	Organophosphorus insecticides, dichlorvos and metrifonate	571.43 mg g^{-1} & 378.78 mg g^{-1}	[101]
4	Zr-MOFs (TCPP@ MOF-808s)	ECs compounds BPA, 17β-E2, 17α-E2 and17α- E1	94.34, 104.17, 109.89 & 121.95 mg g^{-1} respectively	[102]
5	3-D CaFu MOFs	Imidacloprid pesticides and cadmium ions	467.23 & 781.2 mg g^{-1}	[103]
6	Cu-BTC@CA	Dimethoate pesticide	543.2 mg g^{-1}	[104]
7	MOF-on-COF (MCA/UiO-67)	2,4-dichlorphenoxyacetic acid (2,4-D) and glyphosate (GLP) herbicides	615.12 & 675.48 mg g^{-1}	[105]
8	MOF@ cottonfibre (HCF)	Uranium(VI)	241.28 mg g^{-1}	[106]
9	IL@ZIF-8-derived carbon (IMDC)	Diuron and 2,4-D	284 mg g^{-1} & 448 mg g^{-1}	[107]
10	[Sn(II)-BDC MOF]	Eosin Y (EY)	208.3 mg/g	[108]
11	MOF-808	Diclofenac	833 mg g^{-1},	[109]
12	Al-MOF	EDs bisphenol A, 17α-ethynylestradiol and perfluorooctanoic acid	138.4, 200.4 & 169.2 mg g^{-1}	[110]
13	ZIF-67	Microplastics Polystyrene	11.6 mg g^{-1}	[111]
14	UiO-66-OH@ MF-3	Microplastics	95.5%	[112]
15	MOF-5	Hexavalent chromium (VI)	78.6 mg g^{-1}	[113]

more research to explore more biocompatible moieties. Determining the potential leaching of ligand or metal from MOFs and employing complicated water compositions or continuous flow conditions to measure the performance of MOFs under more realistic settings is therefore essential. The majority of the presented research only takes short-term recyclability (10 cycles) into account; the long-term reuse and integrity of MOF materials is mainly unexplored. Recovering the adsorbed ECs or other pollutants for further purification and reuse may also be financially advantageous. So that novel materials can be created and the adsorption capabilities of existing MOFs can be increased, more study is required to understand the interactions between MOFs and pollutants. Several MOFs, including HKUST-1, MIL-101-Cr, MIL-100-Fe, UiO-66, ZIF-8, MOF-5, MIL-53,NU-1000, and others, have so far been manufactured at the tone scale and are offered for sale on the market. Despite this, some MOFs are still in their early stages and need more in-depth analysis of important factors such as reusability, safety, efficacy, lifetime, cost, and industrial conditions, etc. In addition, MOFs are viewed as multipurpose platforms in the context of water remediation because they serve as pollutant detectors, microextraction and separation tools for ECs, detoxifying agents for overdose treatments, and, therefore, more.

4.6 Future

MOFs have demonstrated multidirectional application in the area of wastewater treatment as well as for other fields such as photocatalytic water splitting, hydrogen production, and CO_2 reduction, etc. Hence, MOF-based materials possess high potential for future research. The advantageous properties offered by MOFs may be explored for various fields such as biomedical and energy storage, etc.

4.7 Conclusions

The present chapter summarizes the vast application and utilization of MOFs in various fields. The potential of MOF-based composites is not only applicable for wastewater treatment but can also be explored for the fabrication of other advanced materials such as energy storage devices and drug delivery applications, etc. The incorporation of biocompatible metals or ligands in MOFs may be the area of research that can be explored to minimize secondary pollutants. The present study highlights the applicability of MOFs as photocatalysts and adsorbents. Due to high practical ease of these two processes, more advancement in research will be needed for the removal of trace contaminants. Thus, the insight of the overall literature suggests that MOFs possess a great potential that can be further explored for various environmental applications.

References

1 Parida, V.K., Saidulu, D., Majumder, A. et al. (2021 October 1). Emerging contaminants in wastewater: a critical review on occurrence, existing legislations, risk assessment, and sustainable treatment alternatives. *Journal of Environmental Chemical Engineering* 9 (5): 105966.

2 Rathi, B.S., Kumar, P.S., and Show, P.L. (2021 May 5). A review on effective removal of emerging contaminants from aquatic systems: current trends and scope for further research. *Journal of Hazardous Materials* 409: 124413.

3 Chandel, M., Thakur, M., Sharma, A. et al. (2022 October 1). Chlorophyll sensitized (BiO) 2CO3/CdWO4/rGO nano-hybrid assembly for solar assisted photo-degradation of chlorzoxazone. *Chemosphere* 305: 135472.
4 Kumar, A., Pathania, D., Gupta, N. et al. (2020 December 1). Photo-degradation of noxious pollutants from water system using Cornulaca monacantha stem supported ZnFe2O4 magnetic bio-nanocomposite. *Sustainable Chemistry and Pharmacy* 18: 100290.
5 Far, H.S., Hasanzadeh, M., Nashtaei, M.S. et al. (2020 May 13). PPI-dendrimer-functionalized magnetic metal-organic framework (Fe3O4@ MOF@ PPI) with high adsorption capacity for sustainable wastewater treatment. *ACS Applied Materials & Interfaces* 12 (22): 25294–25303.
6 Li, J., Wang, H., Yuan, X. et al. (2020 February 1). Metal-organic framework membranes for wastewater treatment and water regeneration. *Coordination Chemistry Reviews* 404: 213116.
7 Kumar, A., Sharma, G., Thakur, M., and Pathania, D. (2019 August). Sol–gel synthesis of polyacrylamide-stannic arsenate nanocomposite ion exchanger: binary separations and enhanced photo-catalytic activity. *SN Applied Sciences* 1 (8): 1-1.
8 Kumar, A., Sharma, K., Thakur, M. et al. (2022 August 1). Fabrication of high visible light active LaFeO3/Cl-g-C3N4/RGO heterojunction for solar assisted photo-degradation of aceclofenac. *Journal of Environmental Chemical Engineering* 10 (4): 108098.
9 Kumar, A., Chandel, M., Rana, A. et al. Recent Innovation and advances in utilization of graphene oxide based photocatalysis.
10 Ghosh, A., Tripathy, A., and Ghosh, D. (2022 March 6). Impact of endocrine disrupting chemicals (EDCs) on reproductive health of human. In *Proceedings of the Zoological Society*, 1–15. India: Springer.
11 Xu, J., Li, K., Zhang, S. et al. (2020 March 27). Removal of endocrine-disrupting chemicals from environment using a robust platform based on metal-organic framework nanoparticles. *ACS Applied Nano Materials* 3 (4): 3646–3651.
12 Kasonga, T.K., Coetzee, M.A., Kamika, I. et al. (2021 January 1). Endocrine-disruptive chemicals as contaminants of emerging concern in wastewater and surface water: a review. *Journal of Environmental Management* 277: 111485.
13 Saha, S., Narayanan, N., Singh, N., and Gupta, S. (2022 January). Occurrence of endocrine disrupting chemicals (EDCs) in river water, ground water and agricultural soils of India. *International Journal of Environmental Science and Technology* 22: 1–6.
14 Jian, Y., Yunting, X., Xianghong, T. et al. (2022 July). Endocrine disrupting compounds (EDCs) in source water, finished water, and tap water from drinking water treatment plants and its human risk assessment in Chengdu Plain, China. *Human and Ecological Risk Assessment: An International Journal* 23: 1–6.
15 Idowu, G.A., David, T.L., and Idowu, A.M. (2022 March 1). Polycarbonate plastic monomer (bisphenol A) as emerging contaminant in Nigeria: levels in selected rivers, sediments, well waters and dumpsites. *Marine Pollution Bulletin* (176) 113444.
16 Fu, L., Li, J., Wang, G. et al. (2021 July 1). Adsorption behavior of organic pollutants on microplastics. *Ecotoxicology and Environmental Safety* 217: 112207.

17 Li, S., Wang, P., Zhang, C. et al. (2020 April). Influence of polystyrene microplastics on the growth, photosynthetic efficiency and aggregation of freshwater microalgae Chlamydomonas reinhardtii. *Science of the Total Environment* 20 (714): 136767.

18 Aiguo, Z., Di, S., Chong, W. et al. (2022 April 1). Characteristics and differences of microplastics ingestion for farmed fish with different water depths, feeding habits and diets. *Journal of Environmental Chemical Engineering* 10 (2): 107189.

19 Zhang, J., Wang, X., Xue, W. et al. (2022 July 1). Microplastics pollution in soil increases dramatically with long-term application of organic composts in a wheat–maize rotation. *Journal of Cleaner Production* 356: 131889.

20 Kalantary, R.R., Barzegar, G., and Jorfi, S. (2022 January 1). Monitoring of pesticides in surface water, pesticides removal efficiency in drinking water treatment plant and potential health risk to consumers using Monte Carlo simulation in Behbahan city, Iran. *Chemosphere* 286: 131667.

21 Tölgyesi, Á., Korozs, G., Tóth, E. et al. (2022 January 1). Automation in quantifying phenoxy herbicides and bentazon in surface water and groundwater using novel solid phase extraction and liquid chromatography tandem mass spectrometry. *Chemosphere* 286: 131927.

22 Olisah, C., Rubidge, G., Human, L.R., and Adams, J.B. (2021 December 1). A translocation analysis of organophosphate pesticides between surface water, sediments and tissues of common reed Phragmites australis. *Chemosphere* 284: 131380.

23 Adeleye, A.S., Xue, J., Zhao, Y. et al. (2022 February 15). Abundance, fate, and effects of pharmaceuticals and personal care products in aquatic environments. *Journal of Hazardous Materials* 424: 127284.

24 K'oreje, K.O., Vergeynst, L., Ombaka, D. et al. (2016 April 1). Occurrence patterns of pharmaceutical residues in wastewater, surface water and groundwater of Nairobi and Kisumu city, Kenya. *Chemosphere* 149: 238–244.

25 Meffe, R. and de Bustamante, I. (2014 May 15). Emerging organic contaminants in surface water and groundwater: a first overview of the situation in Italy. *Science of the Total Environment* 481: 280–295.

26 Obando, M.A., Estela, J.M., and Cerdà, V. (2008 September 10). Multi-syringe chromatography (MSC) system for the on-line solid-phase extraction and determination of hydrochlorothiazide and losartan potassium in superficial water, groundwater and wastewater outlet samples. *Journal of Pharmaceutical and Biomedical Analysis* 48 (1): 212–217.

27 Du, C., Zhang, Z., Yu, G. et al. (2021 June 1). A review of metal-organic framework (MOFs)-based materials for antibiotics removal via adsorption and photocatalysis. *Chemosphere* 272: 129501.

28 Krishnan, R.Y., Manikandan, S., Subbaiya, R. et al. (2021 August 1). Removal of emerging micropollutants originating from pharmaceuticals and personal care products (PPCPs) in water and wastewater by advanced oxidation processes: a review. *Environmental Technology & Innovation* 23: 101757.

29 Lozano, I., Pérez-Guzmán, C.J., Mora, A. et al. (2022 March 4). Pharmaceuticals and personal care products in water streams: occurrence, detection, and removal by electrochemical advanced oxidation processes. *Science of the Total Environment* (827) 154348.

30 Rivera-Utrilla, J., Sánchez-Polo, M., Ferro-García, M.Á. et al. (2013 October 1). Pharmaceuticals as emerging contaminants and their removal from water. a review. *Chemosphere* 93 (7): 1268–1287.

31 Wibowo, A., Marsudi, M.A., Pramono, E. et al. (2021 August 30). Recent improvement strategies on metal-organic frameworks as adsorbent, catalyst, and membrane for wastewater treatment. *Molecules* 26 (17): 5261.

32 Yu, F., Bai, X., Liang, M., and Ma, J. (2021 February 1). Recent progress on metal-organic framework-derived porous carbon and its composite for pollutant adsorption from liquid phase. *Chemical Engineering Journal* 405: 126960.

33 Liu, D., Gu, W., Zhou, L. et al. (2022 January 1). Recent advances in MOF-derived carbon-based nanomaterials for environmental applications in adsorption and catalytic degradation. *Chemical Engineering Journal* 427: 131503.

34 Chen, J., Abazari, R., Adegoke, K.A. et al. (2022 October 15). Metal-organic frameworks and derived materials as photocatalysts for water splitting and carbon dioxide reduction. *Coordination Chemistry Reviews* 469: 214664.

35 Abdi, J., Vossoughi, M., Mahmoodi, N.M., and Alemzadeh, I. (2017 October 15). Synthesis of metal-organic framework hybrid nanocomposites based on GO and CNT with high adsorption capacity for dye removal. *Chemical Engineering Journal* 326: 1145–1158.

36 Abdollahi, N., Moussavi, G., and Giannakis, S. (2022 February 15). A review of heavy metals' removal from aqueous matrices by metal-organic frameworks (MOFs): state-of-the art and recent advances. *Journal of Environmental Chemical Engineering* (10) 107394.

37 Dhaka, S., Kumar, R., Deep, A. et al. (2019 February 1). Metal-organic frameworks (MOFs) for the removal of emerging contaminants from aquatic environments. *Coordination Chemistry Reviews* 380: 330–352.

38 Zhang, S., Wang, J., Zhang, Y. et al. (2021 December 15). Applications of water-stable metal-organic frameworks in the removal of water pollutants: a review. *Environmental Pollution* 291: 118076.

39 Liu, B., Vikrant, K., Kim, K.H. et al. (2020). Critical role of water stability in metal-organic frameworks and advanced modification strategies for the extension of their applicability. *Environmental Science: Nano* 7 (5): 1319–1347.

40 He, T., Kong, X.J., and Li, J.R. (2021 July 14). Chemically stable metal-organic frameworks: rational construction and application expansion. *Accounts of Chemical Research* 54 (15): 3083–3094.

41 Lee, Y.J., Chang, Y.J., Lee, D.J., and Hsu, J.P. (2018 December 1). Water stable metal-organic framework as adsorbent from aqueous solution: a mini-review. *Journal of the Taiwan Institute of Chemical Engineers* 93: 176–183.

42 Bagherzadeh, S.B., Kazemeini, M., and Mahmoodi, N.M. (2021 November 15). Preparation of novel and highly active magnetic ternary structures (metal-organic framework/cobalt ferrite/graphene oxide) for effective visible-light-driven photocatalytic and photo-Fenton-like degradation of organic contaminants. *Journal of Colloid and Interface Science* 602: 73–94.

43 Feng, M., Zhang, P., Zhou, H.C., and Sharma, V.K. (2018 October 1). Water-stable metal-organic frameworks for aqueous removal of heavy metals and radionuclides: a review. *Chemosphere* 209: 783–800.

44 Subramaniyam, V., Thangadurai, T.D., and Lee, Y.I. Zirconium based metal-organic framework for the adsorption of Cu (II) ions in real water samples. Cleane.

45 Rego, R.M., Kuriya, G., Kurkuri, M.D., and Kigga, M. (2021 February 5). MOF-based engineered materials in water remediation: recent trends. *Journal of Hazardous Materials* 403: 123605.

46 Liu, Q., Wu, X., Xie, Z., and Zhang, K. (2022 July). Construction of PPSU-MoS2/PA-MIL-101 (Cr) membrane with highly enhanced permeance and stability for organic solvent nanofiltration. *Membranes* 12 (7): 639.

47 huang, H., Zhang, W., Wang, L. et al. (2021 September 24). Vapor deposition-prepared MIL-100 (Cr)-and MIL-101 (Cr)-supported iron catalysts for effectively removing organic pollutants from water. *ACS omega* 6 (39): 25311–25322.

48 Hamisu, A.M., Ariffin, A., and Wibowo, A.C. (2020 October). Cation exchange in metal-organic frameworks (MOFs): the hard-soft acid-base (HSAB) principle appraisal. *Inorganica Chimica Acta* 1 (511): 119801.

49 Shi, W., Li, W., Nguyen, W. et al. (2022 August 1). Advances of metal-organic frameworks in analytical applications. *Materials Today Advances* 15: 100273.

50 Cui, P.P., Zhang, X.D., Kang, Y.S. et al. (2021 January 25). Cobalt-based metal-organic frameworks for adsorption of CO2 and C2 hydrocarbons: effect of auxiliary ligands with different functional groups. *Inorganic Chemistry* 60 (4): 2563–2572.

51 Li, X.Y., Duan, H.Y., and He, C. Engineering a series of isoreticular pillared-layer ultramicroporous MOFS for gas and vapor uptakes. Available at SSRN 4090823.

52 Hwang, I., Mazare, A., Will, J. et al. (2022). Inhibition of H2 and O2 recombination: the key to a most efficient single-atom co-catalyst for photocatalytic H2 evolution from plain water. *Advanced Functional Materials* (32) 2207849.

53 Zhai, F., Zheng, Q., Chen, Z. et al. (2013). Crystal transformation synthesis of a highly stable phosphonate MOF for selective adsorption of CO 2. *CrystEngComm* 15 (11): 2040–2043.

54 Li, N., Xu, J., Feng, R. et al. (2016). Governing metal-organic frameworks towards high stability. *Chemical Communications* 52 (55): 8501–8513.

55 Desai, A.V., Sharma, S., Let, S., and Ghosh, S.K. (2019 September 15). N-donor linker based metal-organic frameworks (MOFs): advancement and prospects as functional materials. *Coordination Chemistry Reviews* 395: 146–192.

56 Jang, S., Song, S., Lim, J.H. et al. (2020 February 6). Application of various metal-organic frameworks (MOFs) as catalysts for air and water pollution environmental remediation. *Catalysts* 10 (2): 195.

57 Oladoye, P.O., Adegboyega, S.A., and Giwa, A.R. (2021 December 1). Remediation potentials of composite metal-organic framewrks (MOFs) for dyes as water contaminants: a comprehensive review of recent literatures. *Environmental Nanotechnology, Monitoring & Management* 16: 100568.

58 Mondol, M.M. and Jhung, S.H. (2021 October 1). Adsorptive removal of pesticides from water with metal-organic framework-based materials. *Chemical Engineering Journal* 421: 129688.

59 Ibrahim, A.O., Adegoke, K.A., Adegoke, R.O. et al. (2021 July 1). Adsorptive removal of different pollutants using metal-organic framework adsorbents. *Journal of Molecular Liquids* 333: 115593.

60 Wang, Q., Gao, Q., Al-Enizi, A.M. et al. (2020). Recent advances in MOF-based photocatalysis: environmental remediation under visible light. *Inorganic Chemistry Frontiers* 7 (2): 300–339.

61 Khodayari, A. and Sohrabnezhad, S. (2021 May 1). Fabrication of MIL-53 (Al)/Ag/AgCl plasmonic nanocomposite: an improved metal-organic framework based photocatalyst for degradation of some organic pollutants. *Journal of Solid State Chemistry* 297: 122087.

62 Chen, C.X., Xiong, Y.Y., Zhong, X. et al. (2022 January 17). Enhancing photocatalytic hydrogen production via the construction of robust multivariate Ti-MOF/COF composites. *Angewandte Chemie* 134 (3): e202114071.

63 Kim, S. and Choi, W. (2002 May 1). Kinetics and mechanisms of photocatalytic degradation of (CH3) n NH4-n+(0≤ n≤ 4) in TiO2 suspension: the role of OH radicals. *Environmental Science & Technology* 36 (9): 2019–2025.

64 Xu, X., Wang, J., Chen, T. et al. (2021 November 5). Deep insight into ROS mediated direct and hydroxylated dichlorination process for efficient photocatalytic sodium pentachlorophenate mineralization. *Applied Catalysis B: Environmental* 296: 120352.

65 Wang, S. and Serre, C. (2019 June 16). Toward green production of water-stable metal-organic frameworks based on high-valence metals with low toxicities. *ACS Sustainable Chemistry & Engineering* 7 (14): 11911–11927.

66 Pi, Y., Li, X., Xia, Q. et al. (2018 April 1). Adsorptive and photocatalytic removal of persistent organic pollutants (POPs) in water by metal-organic frameworks (MOFs). *Chemical Engineering Journal* 337: 351–371.

67 Li, T.T., Liu, Y.M., Wang, T. et al. (2018 December 1). Regulation of the surface area and surface charge property of MOFs by multivariate strategy: synthesis, characterization, selective dye adsorption and separation. *Microporous and Mesoporous Materials* 272: 101–108.

68 Chenab, K.K., Sohrabi, B., Jafari, A., and Ramakrishna, S. (2020 June 1). Water treatment: functional nanomaterials and applications from adsorption to photodegradation. *Materials Today Chemistry* 16: 100262.

69 Ndolomingo, M.J., Bingwa, N., and Meijboom, R. (2020 May). Review of supported metal nanoparticles: synthesis methodologies, advantages and application as catalysts. *Journal of Materials Science* 55 (15): 6195–6241.

70 Wang, C.C., Sheu, G.B., Ke, S.Y. et al. (2015). A 3D porous supramolecular architecture via π–π assembly of 2D metal-organic frameworks (MOFs): structure-versus-luminescence reversibility and gas adsorption properties. *CrystEngComm* 17 (6): 1264–1272.

71 Qin, F.X., Jia, S.Y., Liu, Y. et al. (2015 April 3). Adsorptive removal of bisphenol A from aqueous solution using metal-organic frameworks. *Desalination and Water Treatment* 54 (1): 93–102.

72 Zhang, Z., Chen, Y., Wang, Z. et al. (2021 March 15). Effective and structure-controlled adsorption of tetracycline hydrochloride from aqueous solution by using Fe-based metal-organic frameworks. *Applied Surface Science* 542: 148662.

73 Yu, K., Ahmed, I., Won, D.I. et al. (2020 July 1). Highly efficient adsorptive removal of sulfamethoxazole from aqueous solutions by porphyrinic MOF-525 and MOF-545. *Chemosphere* 250: 126133.

74 Cao, Y., Chen, X., Li, X., and Wang, B. (2021 May 20). Tuning surface functionalization and pore structure of UiO-66 metal-organic framework nanoparticles for organic pollutant elimination. *ACS Applied Nano Materials* 4 (5): 5486–5495.

75 Zhu, B., Xia, P., Ho, W., and Yu, J. (2015 July 30). Isoelectric point and adsorption activity of porous g-C3N4. *Applied Surface Science* 344: 188–195.

76 Sharma, A., Pathania, D., and Kumar, A. (2020 March 27). Bio-polymer based tragacanth gum (TG) loaded Fe3O4 nanocomposite for the sequestration of tenacious congo red dye from waste water. *Journal of Material Science and Technology Research* 7: 92–100.

77 Jayaramulu, K., Geyer, F., Schneemann, A. et al. (2019 August). Hydrophobic metal-organic frameworks. *Advanced Materials* 31 (32): 1900820.

78 Hasan, Z. and Jhung, S.H. (2015 February11). Removal of hazardous organics from water using metal-organic frameworks (MOFs): plausible mechanisms for selective adsorptions. *Journal of Hazardous Materials* 283: 329–339.

79 Yang, C., Kaipa, U., Mather, Q.Z. et al. (2011 November 16). Fluorous metal-organic frameworks with superior adsorption and hydrophobic properties toward oil spill cleanup and hydrocarbon storage. *Journal of the American Chemical Society* 133 (45): 18094–18097.

80 Lin, K.Y., Yang, H., Petit, C., and Hsu, F.K. (2014 August 1). Removing oil droplets from water using a copper-based metal-organic frameworks. *Chemical Engineering Journal* 249: 293–301.

81 Uddin, M.J., Ampiaw, R.E., and Lee, W. (2021 December 1). Adsorptive removal of dyes from wastewater using a metal-organic framework: a review. *Chemosphere* 284: 131314.

82 Ahmed, I., Hasan, Z., Lee, G. et al. (2022 February 15). Contribution of hydrogen bonding to liquid-phase adsorptive removal of hazardous organics with metal-organic framework-based materials. *Chemical Engineering Journal* 430: 132596.

83 Hasan, Z., Tong, M., Jung, B.K. et al. (2014 September 11). Adsorption of pyridine over amino-functionalized metal-organic frameworks: attraction via hydrogen bonding versus base–base repulsion. *The Journal of Physical Chemistry C* 118 (36): 21049–21056.

84 Ikreedeegh, R.R. and Tahir, M. (2021 August 1). Indirect Z-scheme heterojunction of NH2-MIL-125 (Ti) MOF/g-C3N4 nanocomposite with RGO solid electron mediator for efficient photocatalytic CO2 reduction to CO and CH4. *Journal of Environmental Chemical Engineering* 9 (4): 105600.

85 Liu, J., Feng, C., Li, Y. et al. (2022 May 1). Photocatalytic detoxification of hazardous pymetrozine pesticide over two-dimensional covalent-organic frameworks coupling with Ag3PO4 nanospheres. *Separation and Purification Technology* 288: 120644.

86 Zhang, S., Du, M., Kuang, J. et al. (2019 October 15). Surface-defect-rich mesoporous NH2-MIL-125 (Ti)@ Bi2MoO6 core-shell heterojunction with improved charge separation and enhanced visible-light-driven photocatalytic performance. *Journal of Colloid and Interface Science* 554: 324–334.

87 Tran, V.A., Nguyen, T.L., and Doan, V.D. (2021 May 1). Fabrication of Fe3O4/ CuO@ C composite from MOF-based materials as an efficient and magnetically separable photocatalyst for degradation of ciprofloxacin antibiotic. *Chemosphere* 270: 129417.

88 Zaman, F.U., Xie, B., Zhang, J. et al. (2021 November 29). MOFs derived hetero-ZnO/Fe2O3 nanoflowers with enhanced photocatalytic performance towards efficient degradation of organic dyes. *Nanomaterials* 11 (12): 3239.

89 Yi, X.H., Wang, F.X., Du, X.D. et al. (2018 September 15). Highly efficient photocatalytic Cr (VI) reduction and organic pollutants degradation of two new bifunctional 2D Cd/Co-based MOFs. *Polyhedron* 152: 216–224.

90 Kaur, R., Kaur, A., Kaur, R. et al. (2021 May 1). Cu-BTC metal-organic framework (MOF) derived Cu-doped TiO2 nanoparticles and their use as visible light active photocatalyst for the decomposition of ofloxacin (OFX) antibiotic and antibacterial activity. *Advanced Powder Technology* 32 (5): 1350–1361.

91 Hayati, P., Mehrabadi, Z., Karimi, M. et al. (2021). Photocatalytic activity of new nanostructures of an Ag (i) metal-organic framework (Ag-MOF) for the efficient degradation of MCPA and 2, 4-D herbicides under sunlight irradiation. *New Journal of Chemistry* 45 (7): 3408–3417.

92 Gao, Z., Liang, J., Yao, J. et al. (2021 August 1). Synthesis of Ce-doped NiAl LDH/RGO composite as an efficient photocatalyst for photocatalytic degradation of ciprofloxacin. *Journal of Environmental Chemical Engineering* 9 (4): 105405.

93 Karpuraranjith, M., Chen, Y., Rajaboopathi, S. et al. (2022 January 1). Three-dimensional porous MoS2 nanobox embedded g-C3N4@ TiO2 architecture for highly efficient photocatalytic degradation of organic pollutant. *Journal of Colloid and Interface Science* 605: 613–623.

94 Kaur, M., Mehta, S.K., and Kansal, S.K. (2022 February 8). Construction of multifunctional NH2-UiO-66 metal-organic framework: sensing and photocatalytic degradation of ketorolac tromethamine and tetracycline in aqueous medium. *Environmental Science and Pollution Research* (30)1–21.

95 Heu, R., Ateia, M., Awfa, D. et al. (2020 May 12). Photocatalytic degradation of organic micropollutants in water by Zr-MOF/GO composites. *Journal of Composites Science* 4 (2): 54.

96 Vigneshwaran, S., Sirajudheen, P., Karthikeyan, P. et al. (2021 March 15). Immobilization of MIL-88 (Fe) anchored TiO2-chitosan (2D/2D) hybrid nanocomposite for the degradation of organophosphate pesticide: characterization, mechanism and degradation intermediates. *Journal of Hazardous Materials* 406: 124728.

97 Govindaraju, S., Arumugasamy, S.K., Chellasamy, G., and Yun, K. (2022 January 5). Zn-MOF decorated bio activated carbon for photocatalytic degradation, oxygen evolution and reduction catalysis. *Journal of Hazardous Materials* 421: 126720.

98 Shi, C., Qin, L., Wu, S. et al. (2021 October 15). Highly sensitive SERS detection and photocatalytic degradation of 4-aminothiophenol by a cost-effective cobalt metal-organic framework-based sandwich-like sheet. *Chemical Engineering Journal* 422: 129970.

99 Wang, R., Xu, H., Zhang, K. et al. (2019 February 15). High-quality Al@ Fe-MOF prepared using Fe-MOF as a micro-reactor to improve adsorption performance for selenite. *Journal of Hazardous Materials* 364: 272–280.

100 Jabbari, V., Veleta, J.M., Zarei-Chaleshtori, M. et al. (2016 November 15). Green synthesis of magnetic MOF@ GO and MOF@ CNT hybrid nanocomposites with high adsorption capacity towards organic pollutants. *Chemical Engineering Journal* 304: 774–783.

101 Jamali, A., Shemirani, F., and Morsali, A. (2019 December 25). A comparative study of adsorption and removal of organophosphorus insecticides from aqueous solution by Zr-based MOFs. *Journal of Industrial and Engineering Chemistry* 80: 83–92.

102 Han, L., Liu, X., Zhang, X. et al. (2022 February 15). Preparation of multivariate zirconia metal-organic frameworks for highly efficient adsorption of endocrine disrupting compounds. *Journal of Hazardous Materials* 424: 127559.

103 Singh, S., Kaushal, S., Kaur, J. et al. (2021 June 1). CaFu MOF as an efficient adsorbent for simultaneous removal of imidacloprid pesticide and cadmium ions from wastewater. *Chemosphere* 272: 129648.

104 Abdelhameed, R.M., Abdel-Gawad, H., and Emam, H.E. (2021 April 1). Macroporous Cu-MOF@ cellulose acetate membrane serviceable in selective removal of dimethoate pesticide from wastewater. *Journal of Environmental Chemical Engineering* 9 (2): 105121.

105 Luo, X., Wang, C., Huang, G. et al. (2022 October 1). Bio-inspired chitosan aerogel decorated with MOF-on-COF heterostructure hybrid as recyclable scavenger of herbicides in water. *Separation and Purification Technology* 298: 121616.

106 Yang, A., Wang, Z., and Zhu, Y. (2020 November 6). Facile preparation and adsorption performance of low-cost MOF@ cotton fibre composite for uranium removal. *Scientific Reports* 10 (1): 1–0.

107 Sarker, M., Ahmed, I., and Jhung, S.H. (2017 September 1). Adsorptive removal of herbicides from water over nitrogen-doped carbon obtained from ionic liquid@ ZIF-8. *Chemical Engineering Journal* 323: 203–211.

108 Ghosh, A. and Das, G. (2020 May 1). Green synthesis of Sn (II)-BDC MOF: preferential and efficient adsorption of anionic dyes. *Microporous and Mesoporous Materials* 297: 110039.

109 Prasetya, N. and Li, K. (2021 August 1). MOF-808 and its hollow fibre adsorbents for efficient diclofenac removal. *Chemical Engineering Journal* 417: 129216.

110 Jun, B.M., Hwang, H.S., Heo, J. et al. (2019 November 25). Removal of selected endocrine-disrupting compounds using Al-based metal-organic framework: performance and mechanism of competitive adsorption. *Journal of Industrial and Engineering Chemistry* 79: 345–352.

111 Wan, H., Wang, J., Sheng, X. et al. (2022 February 4). Removal of polystyrene microplastics from aqueous solution using the metal-organic framework material of ZIF-67. *Toxics* 10 (2): 70.

112 Chen, Y.J., Chen, Y., Miao, C. et al. (2020). Metal-organic framework-based foams for efficient microplastics removal. *Journal of Materials Chemistry A* 8 (29): 14644–14652.

113 Babapour, M., Dehghani, M.H., Alimohammadi, M. et al. (2022 January 1). Adsorption of Cr (VI) from aqueous solution using mesoporous metal-organic framework-5 functionalized with the amino acids: characterization, optimization, linear and nonlinear kinetic models. *Journal of Molecular Liquids* 345: 117835.

5

Metal-Organic Frameworks for Wastewater Contaminants Removal

*Khushbu Sharma, Priyanka Devi, and Prasann Kumar**

Department of Agronomy, School of Agriculture, Lovely Professional University, Phagwara, 144411 (Punjab), India
** Corresponding author*

5.1 Introduction

Water, as the source of life, is critical to human survival and development. However, water contamination has gotten worse in recent decades as a result of the release of various pollutants. According to the Environmental Protection Agency of the United States, there are some inorganic contaminants, particularly metal oxoanions, that are prioritized by the EPA as pollutants in water. Heavy metal oxoanions, for example, are contaminants in water. Industrial effluent containing CrO_4/Cr_2O_7 as well as $HAsO_4/H_2AsO_4/H_3AsO_3$ poses a risk. Furthermore, some radioactive substances endanger human health and the natural environment. Metal oxoanions (SeO_3) are contaminants (SeO_4^{-2}) that have been discovered in groundwater. All of the current scenarios have raised concerns about the water supplies we consume [1]. As a result, efficient strategies for removing them are being researched immediately to remove pollutants from water and then meet our demand for safe drinking water. Several technologies have been investigated in recent decades to mitigate the concentration of metal ions in drinking water. Adsorption is one method for removing these highly hazardous metal oxoanion compounds from water. Biological treatments, chemical oxidation/reduction, membrane filtration, and other methods are used. However, the majority of these solutions have drawbacks of their own [2]. For example, biological treatments tend to generate secondary contaminants and a lot of waste sludge. Adsorption is thought to be the most effective approach when compared to other methods. Because of its easy operation and ultra-low energy consumption, this technology has the potential for application [3]. To guarantee a highly efficient system with minimal secondary pollutants, adsorption heavily relies on the selection of an adequate sorbent. So far, unique to the removal of metal oxoanions, many adsorbent materials have been developed to remove contaminants

Metal Organic Frameworks for Wastewater Contaminant Removal, First Edition. Edited by Arun Lal Srivastav, Lata Rani, Jyotsna Kaushal, and Tien Duc Pham.

from wastewater, such as nanosized and layer-based activated carbons, sulfides, polymers, etc.

Adsorbent materials are often limited by the following drawbacks:

i) (i) Low adsorption capacity, chemical stability, and thermal stability (LMS); (ii) Comparatively low sorption capacity (AC) (iii); (iii) Sluggish sorption kinetics; (iv) Limited selectivity (Layered double hydroxides [LDHs] and NMO) and; (iv) Weak reusability (PAER).

These disadvantages are induced by inherent adsorbent characteristics such as their tiny pore size along with low volume, insufficient external area, and poor adsorption [4]. As a result, research on novel adsorbent materials that can be modified is currently underway to address the aforementioned issues and efficiently remove metal oxoanions. MOFs are new porous materials composed of metals and organic compounds that form metal ions and organic ligands via coordination interactions. MOFs are made up of alkaline earth metal ions such as Mg^{2+} and Ca^{2+}, transition metal ions [5], lanthanide metal ions (Pr^{3+} and La^{3+}), and other metal ions (Bi^{3+}). Coordination geometries ranging from 2 to 12 can be used. The organic ligands of MOFs are mostly quadrilateral, tetrahedral, and octahedral and include N- and carboxylate-containing ligands that can modify their coordination. MOF architecture is designed to suit the geometric requirements of metal ions. This is varied and designable. Numerous MOFs with substantially bigger surface areas have been discovered so far; (1000–10000 m^2g^1) than standard materials have been recorded, which is crucial because pore sizes and characteristics could be adjusted to meet needs [6].

The following characteristics distinguish MOFs as adsorbents: For starters, many pores can provide routes to permit target access and transfer. Second, the broad surface area can expose enough sites to capture targets. Third, it is adjustable. The structure can be used to embellish functional groups in order to give extra binding sites for the removal of goals. Because of their superior stability, regulated porosity, and huge surface area, MOFs, with their adjustable structure, have received a wide range of interest as next-generation adsorbents. Charged MOFs have emerged as an intriguing subclass. The charge-induced functionalities caused a great deal of concern. Normally, guest ions with opposing charges, which often occupy the pores or channels in MOFs, are required to balance the charged host framework. As a result, cationic MOFs provide ion exchanges that provide an opportunity to catch the anion pollutant.

5.2 Aqueous Phase MOF Stability

The structural stability of MOFs must be considered when using them to treat aqueous pollutants. The key factor considered is the presence of MOFs in aqueous media. According to studies showing that MIL-100, MIL-101, ZIF-67, and others are water stable under direct exposure because of the strong coordinated connections, water is not a problem for Zr-based MOFs. Due to their high coordination number, they are known as the most stable in water and secondary building units (SBUs) [7]. Unfortunately, due to poor cooperation, some MOFs in water are responsive to metal and organic ligands, while others are not appropriate for application in water treatment. For example, the MOF-5 is water-sensitive, and its construction has been entirely ruined by water. Due to the presence of oxygen in water, molecules immediately attack the tetrahedron of ZnO_4 SBUs and release organic ligands. Low et al. proposed that ligands' major structural degradation mechanisms are movement and hydrolysis. MOFs deform and collapse in water [8]. So far, four effective approaches for producing stable MOFs in water have been documented:

- Metal carboxylate frameworks are composed of high-valence metal ions (e.g., Zr^{4+}, Fe^{3+}) and metal frameworks with nitrogen donor ligands (e.g., ZIFs).
- The addition of hydrophobic groups to organic ligands, as well as the presence of highly hydrophobic groups (ethyl eater and fluoro methoxy groups), can make MOFs hydrophobic and prevent them from degrading.

5.3 MOF Degradation in Water

- The introduction of various metal ions into the MOF. Because of this, the exchange of various metal cations can boost the stability of MOFs in water through the strengthened metal-ligand connection.
- MOF composites, in combination with other hydrophobic media, can create a hydrophobic environment for MOFs, preventing water molecules from entering pores and preserving them.

5.4 Influence of MOF Structure

The pH of wastewater varies from area to area and habitat. As a result, it is critical to guarantee that MOFs used as adsorbents are steady and perform well in terms of resistance to proton or hydroxide ions. MOFs outperform a neutral water solution [4] because the ions proton and hydroxide are sensitive in acids and basic aqueous solutions. As previously stated, MOFs based on high-valent metals are often detrimental to MOFs because their ions and carboxylate ligands have high acid resistance. MOFs are considered to have great performance, which is demonstrated by structures made of low-valent metal ions and isolate-based ligands. Fundamental solution stability is determined by the pKa. The organic ligands' values and the strength of their coordination bonds (Liu et al., 2017). Furthermore, to increase their stability at varying pH levels, ligand modification, such as the use of porphyrin-containing compound rings, can improve their stability in both acidic and basic situations.

5.5 2D Nanostructured Coating

Platelets, sheets, plates, membranes, and films are examples of 2D nanostructures, which are distinguished by their great lateral dimensions and atomic thickness in the third dimension. The unique 2D architecture provides electron confinement within two dimensions in the ultrathin region, a well-developed surface-to-volume ratio, outstanding mechanical properties (strength and flexibility), as well as considerable surface exposure, which favors surface functionalization for the alteration of properties. As a result, materials containing 2D nanostructures are preferred. Typically, differentiated electrochemical performance is expected as compared to their bulk counterparts. The thickness of 2D ultrathin Co-MOFs was controlled to 2 nm via a surfactant-assisted solvothermal process. CO-MOFs are relatively inactive when compared to micronano and bulk-electrocatalytic OER activity. One of the most prominent 2D MOF composite architectures is the coating or loading of MOFs on 2D materials to support the ZIF-8 coating on as-prepared graphene oxide (GO) and reduced GO. At ambient temperatures, NSs were synthesized utilizing a liquid-liquid synthesis technique. The ZIF-8 uniform could also be wrapped around the Pt/NPs hybrid NSs, which all have 2D morphologies. MOF development in situ on 2D supports such as GO NSs can also produce MOF particle shapes. For instance, a bimetallic ZnCo ZIFs (Co/Zn =3/6). The NPS was loaded with ZnCo-ZIF-GO composite electrocatalyst, which was synthesized on GO NSs using in-situ MOF growth. The scanning electron microscope picture of the rhombic dodecahedral ZIF-67 was revealed by ZnCo-ZIF, GO, and O, averaging 60 nm in size and being uniformly distributed over the 2D sheet. A covalent construction of graphene and MOF via amide bonds was recently described [9]. Before reacting with the amine, the graphene material acid (GA) was first created. UiO-66-$_2$ is a functionalized, Zr-based MOF. The unbalanced supercapacitor made of GA-UiO-66-NH_2 composites and Ti3C2Tx MXene was highlighted for its exceptional power density (16 Wkg^{-1}) as well as energy density (73 Wkg^{-1}). The synergistic impact of the two 2D components provided the high capacitance and long cycle life of the composite. A supercapacitor made on MXene/MPFs electrodes demonstrated a 20.4 W h/cm2 areal energy density as well as 7000-cycle stability, owing to its symmetrical, ultrathin, and flexible nature.

5.6 3D Nanostructure of MOF

Building a 3D MOF composite superstructure with preserved porosity has been both difficult and intriguing with embedded ultrathin NiFe MOF NSs in the 3D organized macropores of NiFe hydroxide (NFH) created using an in-situ growing method. NFH precursors were pushed into the interstices of as-assembled polystyrene [PS] sphere templates for coprecipitation. After the PS templates were removed, the 3D-ordered macroporous OM-NFH was discovered. NiFe composites were created. This was accomplished using OM-NFH as both a metal source and a template in a solvothermal process. The majority of NiFe is a hydroxide with no organized macroporous structure. For comparison, NiFe was synthesized and hybridized. The hierarchical NiFe composite performed well and improved electrocatalytic performance in electrocatalytic applications. This was accomplished using OM-NFH as both a metal source and a template in a solvothermal process. The outcomes stressed the importance of NiFe composition and hierarchical structure in electrocatalysis. Another appealing method for creating 3D hierarchical structures

[10] found that MOF composites can be used to disperse nanoarrays on supports or substrates with high conductivity or surface area. The conductive supports are used in array structures to provide effective electron transport and composite structural toughness. The directed arrangement of nanostructured materials helps to avoid restacking and ensures that active sites receive adequate exposure. The porous construction could improve electrolyte, reactant, and product mass transfer.

Park and Joo (2007) suggested that ZIF-67 was grown directly on noble-metal-sensitized ZnO nanorods to obtain nanoarrays of ZnO, M, and ZIF-67 (M = Au, Pt, and Ag) [11]. The hydrothermal reaction is used to create an indium tin oxide (ITO) glass substrate. Following that, Au NPs were dispersed onto the ZnO nanorods, followed by in-situ ZIF-67 nucleations on the surface of ZnO, Au without the use of additional metal sources. The obtained arrays had a rough surface and were well-maintained in a one-dimensional nanostructure. The TEM scan also showed this composite's sandwich shape, with an 84 nm ZnO inner core and a ZIF-67 outer shell measuring 17 nm. The composition played a role in electron-hole separation and electron transfer over the ZIF-67 shell toward the ZnO, Au core in favor of long-term photoelectrochemical stability. Splitting water performance, a versatile synthetic technique for ultrathin bimetal-MOF NS (BMNS) arrays was also developed in another paper and carried out on self-supported 3D macroporous conductive substrates with in-situ transformation of LDHs). NiCo-BDC BMNSs are used in electrocatalytic OER. The array of BDC (1,4-dicarboxybenzene) on nickel foam (NF). Overpotentials of 230 mV were required to deliver a current density of 10 mA/cm^2. In comparison, the NiCo-BDC dissociated on NF and BMNS powder was reloaded with Nafion as a binder. It lacked activity and stability under identical test conditions. The improved performance of in-situ prepared arrays could be attributed to nanostructures with 2D NS arrays, which have increased ion diffusion, more accessible active sites, and greater proximity to the electrolyte. Insulating polymeric binders, for example, Nafion, were frequently used to glue materials to substrate electrodes, which could reduce conductivity and induce corrosion in the region of interaction between catalytic active sites and electrolytes. As a result, when employing in-situ growth or transformation strategies, the intimate contact between the MOFs and the conductive substrates promoted electron transport between the two components.

5.7 MOF-Based Materials' Adsorption Processes for Heavy Metal Oxyanion

The removal of heavy metal oxyanions using MOF-based materials is mostly due to MOF adsorptive behaviors. For SeO_3, SeO_4, and $HAsO_4/H_2AsO_4/H_3AsO_3$, thermodynamic studies demonstrate the spontaneous properties of these adsorption processes, and several adsorption isotherm models have been utilized to simulate adsorption behavior. It has been revealed that the Langmuir model often fits adsorption onto MOF-based materials well, illustrating the uniform monolayer adsorption process. The pseudo-second-order kinetic equation could be used to model kinetics in general, indicating that coordination effects (chemisorption) are dominant in capture [9]. In general, researchers proposed heavy metal oxyanion adsorption methods by MOFs:

i) Between MOFs and heavy metal oxoanions with coordinately unsaturated sites (CUSs), coordination bonds (M–O-Se and M–O-As; M = metal centers in MOFs) are formed.
ii) The primary method of adsorption of Cr (VI) oxoanions on pristine MOFs is due to their highly organized structures. MOF requires extra-anion storage in the cavities or pores to balance the charged host framework. It carries out ion exchange for heavy metal oxyanions [12].
iii) The acid-base notion plays a significant role in the attraction of heavy metal oxyanions. Heavy metal oxoanions, as Lewis bases, can interact with Lewis acidic sites in MOFs via orbital overlap, increasing adsorption capacity.
iv) Bonding between the inorganic chain in MOF and heavy metal oxoanions has been observed and proposed as a mechanism for efficient adsorptive properties, which can be improved by introducing more -OH, -NH_2, and other groups.
v) Interactions that occur as a result of ionic contacts between target anions and positively charged anions. Charged sites in MOFs are also thought to aid in heavy metal adsorption by metal anion.
vi) Adsorption-reduction combined adsorption via increased oxidation activity from metastable active regions of the amorphous MOF structure (from highly poisonous As(III) to As(V)) or by exposing photoactive MOFs to light.

The lighting (from Cr (IV) to Cr (III)) helped the polluted water [13]. Pollutant interaction processes with adsorbents have a significant impact on cleanup and have an important effect on linkers, which can prevent this. Adsorbent desorption and regeneration, as well, turn the desorption and regeneration behavior of the depleted adsorbent aid in the identification of potential contact mechanisms. With formal channels of communication, adsorption processes can cause medical responses. If that happens, the obtained effectiveness will be very low in the irreversible process and unrecoverable active sites, resulting in poor cycling performance. The regeneration adsorption sites may function as solutions with an electrolyte solution and new adsorbent with satisfactory recycling performance. This is a strategy shown to be effective in increasing the pore size of MOFs, resulting in oxyanions, which hasten the metal oxoanions' diffusion of longer MOFs. However, the longer, more expensive organic ligands limit their commercial availability. MOFs in which functions are located at the terminals of metallic atoms or organic linkers, suffer structural instability when their organic ligands are too long. In addition, new functional groups are introduced. MOFs can be formed by assembling organic ligands. These functional groups include NH2 and NH3. -SH, can serve as adsorption sites to interact with targets, hence increasing adsorption pristine MOFs' capacity and selectivity [14].

Similar outcomes can be obtained by introducing functional groups at the metal nodes because of the steric barrier. As a result, functional groups at metal nodes are more active than those at organic ligands. Typically, postsynthetic modification procedures are used to bring functional groups together. With designable-free anion MOFs, uncoordinated anions are required to balance the cationic host MOFs, as previously stated. These free anions in the environment, MOF pores, or channels, can be easily swapped to capture the metal oxoanion targets. Anions with the same particular shape as the target metal are thus produced. Oxoanion is meant to make substitution easier [9]. For example, Dai et al. (2017) reported a Ni-based MOF with exchangeable tetrahedral shape SO4 and no coordination anions that has a strong affinity for MnO_4 ions with comparable geometry CUSs. The CUSs or open metal sites are revealed by excavating terminally MOF-coordinated solvent molecules. CUSs function as Lewis acids, interacting with Lewis bases like -N(CH)2 in malachite green. Furthermore, ions can replace terminal solvent molecules as a result of improved coordination by being targeted to associate with metal ions of MOFs. It has been discovered that SeO_3 as well as SeO4. Instead, two ions can coordinate with Zr nodes to produce an inner sphere. As a result, the adsorption performance of H_2O molecules is outstanding. This is an example of how CUSs are active sites in an irreversible process. Defects will be produced in MOFs by adding modulators to the precursors, which will increase pore size and create additional adsorption sites for targets. For example, flaws can be seen in UIO-66 when it is heated.

Using monocarboxylic acid modulators, the Brunauer-Emmett-Teller (BET) surface area increased from 1174 to 1776 m2g1 [2]. Rather than preparing MOFs, this technique is more convenient and cost-effective because it uses organic linkers that are longer. Furthermore, the flaws increase the hydrophilicity of MOFs, which makes them advantageous for use in aqueous solutions. The emergence of composites based on MOFs can be synthesized to create hybrid materials from various useful materials; for example, low-cost supports (silica) [6] and magnetic materials (Fe_3O_4). MOF-based composites outperform conventional composites in general due to the synergistic action of their constituents. Furthermore, MOF-based composites, because of their larger size, are considerably easier to remove

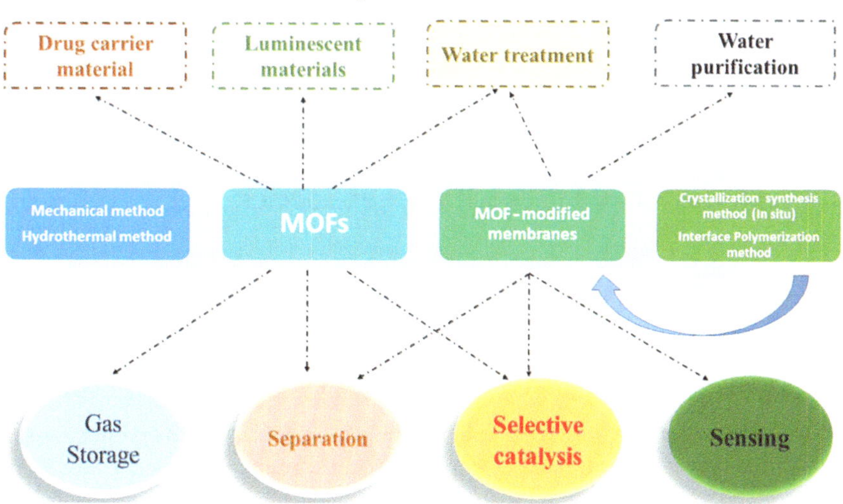

Figure 5.1 Adsorption mechanisms of heavy metal oxoanions by MOF-based materials.

from water than individual MOFs of their huge sizes, which solves the disadvantage of MOF recycling (Figure 5.1).

5.8 Remediation Through Perfect MOFs

The c-MOF requires extra anions in the voids to balance the charged host framework. It uses ion exchange in practical applications. In 1990, Robson and Hoskins published the main ion-exchange study on a coordination polymer (of which MOFs are a subclass), a 3D copper-based coordination polymer with huge holes [6]. There are a lot of BF4 anions in the pores, which are easily swapped for PF6 anions. Since then, ion exchanges have been studied from every angle. BUT-39, a Zr-based 3D MOF that recently demonstrated fast adsorption kinetics and strong adsorption capacity to Cr2O7 [15], was able to stabilize due to a balance between physicsorption and chemisorption. BUT-39 reached equilibrium after adsorption at 116.5 mg/g/minute Cr(VI). The coordination of the Zr6 cluster and Cr2O7 may result in chemisorption. Meanwhile, BUT-39 (500 ppm) demonstrated superior objectionable adsorption for chromium derivatives, carbonate, boron, phosphate, sulfate, and nitrate anions at high concentrations (0.5 mM). As previously stated, organic ligands adorned with specific groups and all CUSs can increase strong binding affinity for specific ions, resulting in high sorption capacity and a fast sorption rate. Manos's squad began by decorating the previous stiff terephthalate scaffold with a bulky functional group (-NH-CH2-py). MOR-2 is similar to MOR-1 but includes an extra-framework Cl anion that can be easily replaced with a Cr_2O_7 anion. Positively charged pyridinium-methylammonium moieties were added. They were expected to interact strongly with Cr(VI) oxoanions. Furthermore, the Cr_2O_7 anions can bind with Zr_6 clusters to form their terminal OH or H_2O ligands, which are replaced. As a result, the MOR-2 displayed admirably high Cr_2O_7-based sorption capacities and astonishingly fast sorption kinetics [9]. It significantly outperforms known MOF sorbents like ZJU-101 and ABT_2ClO_4. MOR-1 (30–130 mg/g1) with its high Cr (VI) sorption capacity (193.7 mg Cr (VI) per g) [3]. The process of adsorption of MOR-2 on Cr_2O_7, even after 1000 times, and the concentration of two anion species remains nearly unchanged, creating a surplus of competitive anion species such as carbonate, boron, phosphate, sulfate, and nitrate.

5.9 Interaction of MOFs with Other Species

The greater charge of Cr_2O_7 can be attributed to MOR-2's selectivity for two anion states and significant interactions with the material's structure [16].

Recent studies suggest that low-dimensional MOFs (1D MOFs) have higher electrical conductivity and more accessible active sites, which accelerate interactions with targets in the water. Cai's group reported iron-gallic acid hybrids (Fe-GA), a classic example of a 1D MOF containing infinite chains of trans-corner-sharing FeO_6 octahedra. In the presence of 1 g/L1 Fe-GA, the removal efficiency for Cr (VI) was in the range of 85–100%, while the initial concentration of Cr (VI) varied from 100–1500 mg/L. The interaction between Cr (VI) and Fe-GA reached equilibrium within ten minutes, with an adsorption capacity of 820 mg Cr (VI)/g. As a result, the exceptional performance is attributed to the combined effects of reduction, metal substitution, and coprecipitation.

Among them, CrO_4 adsorption onto 3D MOFs through anion-exchanged single-crystal-to-single-crystal (SC-SC) transformation is notable in 19 cases. The 3D report demonstrated a high adsorption capacity (42 mg Cr (VI) per g) for CrO_4 with high selectivity via SC-SC ion exchange. It is ascribed to the fact that the small channel size enhances the interaction between the CrO_4 ions and the matrix [17]. In this way, CrO_4 ions are fixed in the channels based on the electrostatic interaction and the crystal lattice. In addition, rich hydrogen bonds and CHO bonds between CrO_4 ions and the framework also play an important role in the capture of CrO_4 ions. Importantly, it showed excellent repeatability. This MOF was immersed in a KNO_3 aqueous solution after reaching adsorption equilibrium with CrO_4 ions. With a high concentration of NO_3 ions as triggers, a fast exchange dynamic was observed. After 9 hours, 95.5% of the CrO42 ions were released, which is the maximum efficiency. If the ion-exchange process were maintained, Cr_2O_7 2 ions would replace NO_3 ions and return to the framework. After five cycles, the exchange efficiency was slightly reduced to about 91.4%. The renewability of these MOFs is of great importance for practical applications.

5.10 With the Use of MOF Composites

To enhance the separation performance of MOF adsorbents from the water after adsorption, embedding ferrites into MOFs to make them magnetic is an effective approach to achieving a convenient separation of this material from water. The magnetic properties of this material leads to rapid and easy separation from solution with a recovery yield of close to 100%. They usually exhibit enhanced properties compared with their counterparts owing to the synergistic effects among their 20 different components. The HPU-13 and Fe_3O_4 composites were created with the criteria in mind, and they have maximum capacities of 190.7 and 210.4 mg Cr (VI)/g for Cr2O7 2 ions, respectively. Because of the synergistic reaction of Cr (VI) reduction and adsorption, these adsorption capacities are much higher than those of most previously reported cationic MOFs. The adsorption of Cr2O7 2 ions by HPU-13, Fe_3O_4 composites is a two-step process that includes the following steps: (i) The large cationic channels provide active sites for Cr (VI) and (ii) HPU-13, Fe_3O_4 composites partially reduce the adsorbed Cr(VI) due to the reducibility of Cu(I) as the metal centers in HPU-13.

a) Hierarchical UiO-66-NH_2, silica buildup, and column packing for $(Cr_2O_7)^{2-}$ removal.
b) Column efficiency of UiO-66-NH_2 and silica in the presence of competing ions for Cr_2O_7 sorption.

To realize the real-life water treatment application, continuous-flow ion-exchange columns with specific requirements regarding materials as the active 21 sorbents are demanded, which need to show fast and selective sorption performance toward targeted ions and have an appropriate particle size to ensure a continuous flow of treated water at a reasonable rate. This demonstration has excellent mechanical properties to withstand the high pressure applied by the water passing through the column. A novel methodology that can integrate the high surface area with excellent chemical and physical durability is needed [18].

Researchers have used an ion-exchange column with a UiO-66-NH_2 silica composite as the active solid phase to remove Cr (VI) ions. The column made from this composite has a high uptake of Cr2O7 2 ions, with a capacity of 132.5 mg Cr (VI)/g. This ion-exchange column can remove Cr (VI) ions even in the presence of competing anions such as Cl, Br, NO_3, and SO4 [19]. Surprisingly, cement wastewater containing 3.8 ppm Cr_2O_7 was used to test this composite's decontamination capabilities via column ion exchange. The results showed that after running the column three times, the total Cr(VI) concentration of the eluted solution decreased to 0.15 ppm, which is well within the permitted discharge limits for industrial wastewater [20]. Although various ion exchanges in MOFs between Cr_2O_7 and other anions were investigated, their Cr adsorption capacity is still not satisfactory. New materials with high adsorption capacity should be developed to capture Cr(VI). Through the exploration of the synergic effect by anchoring additional functional components, MOFs such as ZIFs, UiO-66,6, and MILs have been utilized to remove As from water.

However, their adsorption efficiency is insufficient. Researchers attempted to create bimetallic MOFs by incorporating additional metal ions into the pure framework. This is expected to have a large impact on crystal structure, particle size, and specific surface area, resulting in modified adsorption performance. Based on the previously reported MIL-88B (Fe) [9], we created the bimetallic Fe/Mg-MIL-88B by incorporating Mg ions into MIL-88B (Fe). The Fe/Mg-MIL-88B had a remarkable adsorption capacity of 303 mg/g^{-1} for H_3AsO_3, which was significantly greater than that of the monometallic MIL-88B (Fe) and other reported MOF materials. The lattice characteristics vary, and the SSA of Fe/Mg-MIL-88B increases as some Mg^{2+}

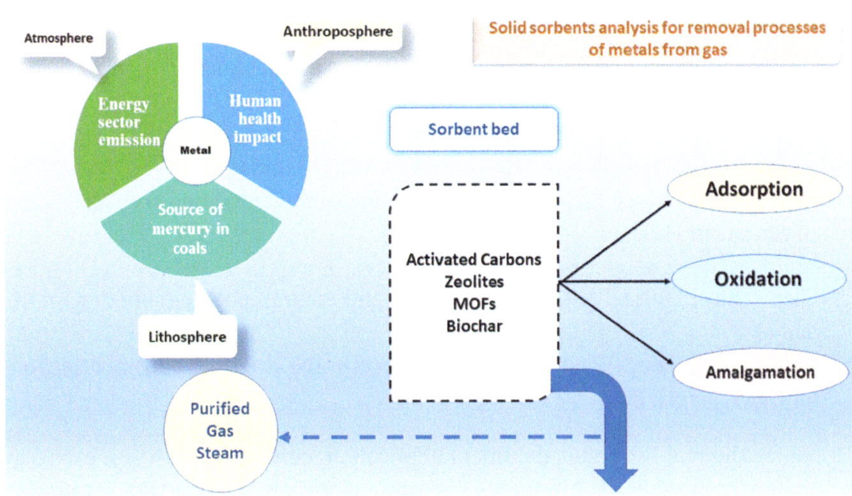

Figure 5.2 Solid sorbents analysis for removal processes of metal from gas.

ions replace Fe^{3+} ions. The high SSA suggests that the addition of Mg can increase the exposure of active areas to oxoanion capture.

Figure 5.2 shows solid sorbent analysis for the removal processes of metal from gas.

5.11 Removal of Metal Ions through Adsorption

Due to the strong interaction between H_2AsO_4 and metal ions (Fe^{3+} and Mg^{2+}) via the creation of Mg-O-As and Fe-O-As bonds, Fe/Mg-MIL-88B (0.5) could swiftly approach adsorption equilibrium within 30 min. Furthermore, Fe/Mg-MIL-88B (0.5) adsorbents have outstanding regenerative ability and stability, retaining 84% of the H_2AsO_4 removal efficiency even after five adsorption-desorption rounds. Surprisingly, Fe/Mg-MIL-88B (0.5) adsorbents can achieve a 99.6% removal rate in five minutes when the adsorbent dosage is increased to 0.05 g/L in the 0.5 ppm H_2AsO_4 solution. So after the adsorption, the residual [5] as concentration is below 10 ppb, which reaches the drinking water standard. Furthermore, Liao et al. effectively produced nanostructured Fe-Co bimetallic MOF-74 to expose additional active sites created by two distinct metal ions in the lattice. The Fe/Mn-MOF-74, with an acceptable Fe/Mn ratio of 1.96, also demonstrated a high experimental adsorption capacity for H_3AsO_3 of 161.6 mg/g1 (Wang et al., 2014). They used XPS to examine the adsorption process of the Fe/Mn-MOF-74 for H_3AsO_3. The results showed that the oxidation of As(III) to As(V) occurred on the surface of Fe/Mn-MOF-74 rather than as a result of adsorption.

Overall, the removal mechanism of Fe/Mn-MOF-74 from H_3AsO_3 is as follows: The coordination bonds transport the H_3AsO_3 molecules to the solid-liquid interface, where they are adsorbed onto the surface of Fe/Mn-MOF-74 (Fe-O-As). At the solid-liquid interfaces, redox reactions with low-coordinated manganese (Mn^{3+} and Mn^{4+}) oxidize As(III) species to As(V) species. Because of their aqueous solution stability and wide surface area, modified MOFs such as UiO-66, a well-known Zr-based MOF, are utilized. As previously reported, pristine UiO-66 may take up to 85 mg/g of H_2AsO_4 [20]. Although this adsorption capacity is greater than that of many other adsorbents, it is significantly lower than expected given its ultrahigh porosity and large surface area. The result shows that the pore width of UiO-66 is approximately 6, which is less than that of hydrated arsenate. As a result, hydrated arsenate is unable to enter the inner channels of UiO-66, decreasing its adsorption capability.

The synthesis of MOF-76 (Y) and MOF-76 (Y)-Ac(b) ASV removal from simulated gold mining wastewater (c) alkaline As(V) adsorption mechanism is proposed on 37 MOF-76 (Y)-Ac. Because of the extra-active M-OH sites, coordinated modulation is another interesting technique for improving the adsorption performance of MOFs [8]. This is a simple solvothermal approach to create yttrium-based MOFs, MOF-76(Y), and an acetate-modulated one, MOF-76(Y)-Ac [9]. When compared to MOF-76(Y), coordinated modulation with sodium acetate as the capping reagent gives MOF-76(Y)-Ac more porosity, surface area, and CUSs. All of this points to MOF-76(Y)-Ac's adsorption capacity. As a result, the maximum adsorption capacity of HAsO4-2 by MOF-76(Y)-Ac within 30 minutes is 201.5 (mg/g), which is greater than that of MOF-76(Y) (187.8 mg/g) in alkaline conditions.

Furthermore, its high reusability and durability make it suitable for use in alkaline gold smelting wastewater (pH = 10). The major adsorption mechanism, As, is the production of Y–O-As via ligand exchange between arsenate and the organic ligand. Furthermore, the ligand change with Y–OH sites and the chemical precipitation mechanism cannot be overlooked.

5.12 MOF Composites are Used for Removal

Because of their magnetic-responsive qualities for easy harnessing and separation from treated water solutions, magnetic composites are chosen for practical applications. Because of the strong and irreversible interaction between As and iron oxide, ferrous magnetic nanoparticles (MNPs) are suitable adsorbents for As removal. MNP adsorbents, on the other hand, typically have limited adsorption capacity, poor selectivity, or sluggish adsorption kinetics [5]. These drawbacks can be mitigated by coating porous materials, such as MOFs, on the surface of the MNPs. A number of magnetic MOF composites based on Fe_3O_4 have recently been investigated for As-oxoanion capture. Folens et al. developed a hybrid nanomaterial, Fe_304, MIL-101(Cr), with adsorption capacities of 121.5 and 80 mg/g/s H_2AsO_4, respectively, and phosphate ions have little effect on removal efficiency or selectivity. The structure of Fe_304, MIL-101(Cr) was very well preserved following desorption, with no substantial leaching of Cr or Fe components observed [21]. Hasan et al. (2012) created magnetic-responsive Fe_3O_4 and ZIF-8 composites for H_3AsO_3 capture from aqueous solutions. Due to strong surface complex interaction, Fe_3O_4, ZIF-8 with the core-shell structure (Fe_3O_4 as core and ZIF-8 as shell) displayed a large surface area of 1133 m^2/ g, which was monolayer adsorption with a maximum capacity of 100 mg/gl for H_3AsO_3 [22]. Haque et al. (2011) also investigated a magnetic Fe3O4, UiO-66 composite with a unique core-shell structure for efficient H_2AsO_4 removal from wastewater. Because of the magnetic-responsive characteristic, the composites have an adsorption capacity of 73.2 mg/g and can be easily isolated from purified water solutions.

The other type of magnetic MOF composite is $CoFe_2O_4$, MIL-100 (Fe), which has a core-shell and mesoporous structure and has a good adsorption capacity for H_3AsO_3 and H_2AsO_4 [9]. The greatest adsorption capacities for H_2AsO_4 were 114.8 mg/g and 143.6 mg/g for H_3AsO_3. Furthermore, in two minutes, CoFe2O4, MIL-100(Fe) removed 100 39 ppb H_3AsO_3 and H_2AsO_4. Furthermore, owing to the size-exclusion influence of the MOF shell and electrostatic repulsion, the adsorption efficiency of $CoFe_2O_4$, MIL-100(Fe) did not decrease in the presence of inorganic ions. To generate Fe-O-As complexes, the authors recommended that H_3AsO_3 and H_2AsO_4 substitute terminal hydroxyl groups on $CoFe_2O_4$, MIL-100 (Fe). H_2AsO_4's very high adsorption capacity could be attributed to hydrogen bonding, which resulted in multi-layer adsorption of neutral As species (H_3AsO_3). Adsorption of ionic As species (H_2AsO_4) occurred in a monolayer reaction mechanism.

Building MOFs on 1D nanowire is an efficient way to overcome separation challenges and produce excellent As removal [9]. Zheng et al. (2017) successfully synthesized 1D-MnO_2, ZIF-8 composites with a core-shell structure. H_2AsO_4 adsorption capacity on MnO_2, ZIF-8 was 140.3 mg/g, which was significantly higher than that of pure ZIF-8 nanoparticles (90.8 mg/g). It was attributed to the ability of the core's -MnO_2 nanowires to oxidize as well as the electrostatic interaction between oxidized As(V) and ZIF-8. Importantly, the distinct 1D nanostructure effectively inhibited ZIF-8 particle aggregation in aqueous solutions, allowing for easy separation from aqueous media and integration of ZIF-8 nanoparticles into chitosan-grafted poly(N-vinyl caprolactam) (chitosan-g-PNVCL) nanofibers. The greatest experimental adsorption capacity of chitosan-g-PNVCL/ZIF-8 nanofibers for H_2AsO_4 adsorption was 258.5 mg/g1, with equilibrium reached in 30 minutes. The huge specific area of chitosan-g-PNVCL/ZIF-8 nanofibers and the electrostatic interaction of copolymer functional groups such as carboxylic and amide with metal ions were ascribed to the high adsorption capacity.

5.13 COFs are a New Class of Materials that Have Similar MOF Structures

As a consequence, they can also serve as a carrier for iron oxide, resulting in composites that trap pollutants. The use of bimetallic MOFs to date, MOFs such as ZIFs, UiO-66, and MILs have been utilized to remove As from water. However, the adsorption efficiency is insufficient. Researchers attempted to create bimetallic MOFs by incorporating additional metal ions into the pure framework. This is expected to have a large impact on crystal structure, particle size, and specific surface area, resulting in modified adsorption performance. Based on the previously reported MIL-88B (Fe), Zhang et al. (2017) created the bimetallic Fe/Mg-MIL-88B by incorporating Mg ions into MIL-88B (Fe) [19]. The Fe/Mg-MIL-88B (0.5, the Fe3+: Mg2+ feeding ratio) had a remarkable adsorption capacity of 300 mg/g for H_2AsO_4, which was significantly greater than that of the monometallic MIL-88B (Fe) and other reported MOF materials [23]. The lattice characteristics vary, and the SSA of Fe/Mg-MIL-88B increases as some Mg^{2+} ions replace Fe^{3+} ions. The substantial SSA implies that the addition of Mg can enhance sufficient exposure of active areas for oxoanion capture.

Due to the strong interaction between H_2AsO_4 and metal ions (Fe^{3+} and Mg^{2+}) via the creation of Mg-O-As and Fe-O-As bonds, Fe/Mg-MIL-88B (0.5) could swiftly approach adsorption equilibrium within 30 min. Furthermore, Fe/Mg-MIL-88B (0.5) adsorbents have outstanding regenerative ability and stability, retaining 84% of the H_2AsO_4 removal efficiency even after five adsorption-desorption rounds. Surprisingly, Fe/Mg-MIL-88B (0.5) adsorbents may reach a 99.6% elimination rate in five minutes when the absorbent dosage is increased to 0.05 g/l in the 0.5 ppm H_2AsO_4 solution. As a result of the adsorption, the residual As concentration is less than 10 ppb, which meets the drinking water requirement.

Furthermore, effectively produced nanostructured Fe-Co bimetallic MOFs [4] expose additional active sites created by two distinct metal ions in the lattice. Within 12 hours, the highest adsorption capabilities of Fe/Co-MOF-74 (molar ratio of Fe/Co = 2:1) toward H3AsO3 and H_2AsO_4 were 265.52 and 290.29 mg/g1, respectively. Furthermore, the presence of sulfate and carbonate ions had no discernible effect on adsorption. Electrostatic contact and coordination connections between metal centers and O (from H_3AsO_3 and H_2AsO_4) were important in the As adsorption process. Excitingly, Fe/Co-MOF-74 adsorbents were effectively regenerated in NaOH solution and demonstrated great durability, with only a 6.6% and 8.7% decline in H_3AsO_3 and H_2AsO_4 removal after five cycles, respectively.

The Fe/Mn-MOF-74, with an acceptable Fe/Mn ratio of 1.96, also demonstrated a high experimental adsorption capacity for H_3AsO_3 of 161.6 mg/g 1. Wang et al. used XPS to examine the adsorption process of the Fe/Mn-MOF-74 for H_3AsO_3. The results demonstrated that the oxidation of As(III) to As(V) occurred on the surface of Fe/Mn-MOF-74 rather than as a result of adsorption. The increased oxidation efficiency of Fe/Mn-MOF-74 was attributed to the amorphous structure's large proportion of Mn nodes and metastable active sites. The consumption of high-valence manganese in Fe/Mn-MOF-74 was attributed to the electron transfer from the oxidation of As(III) at the solid-liquid interfaces [24]. Overall, the removal mechanism of Fe/Mn-MOF-74 from H_3AsO_3 is as follows:

i) The coordination bonds transport the H_3AsO_3 molecules to the solid-liquid interface, where they are adsorbed onto the surface of Fe/Mn-MOF-74 (Fe-O-As).
ii) At the solid-liquid interfaces, redox reactions with low-coordinated manganese (Mn^{3+} and $Mn^{4+)}$ oxidize As(III) species to As(V) species.CUSs are the primary active sites for H_2AsO_4 adsorption. In normal UiO-66, however, Zr atoms are covalently saturated with ligands.

To improve UiO-66's H_2AsO_4 adsorption ability, pore diameters must be increased or the number of CUSs must be increased. Through ligand-selective thermolysis, Song et al. (2016) found that diameters must be increased or the number of CUSs must be increased. As previously said, the introduction of faults can cure both of the above concerns at the same time. Through ligand-selective thermolysis, [1] created hierarchically porous UiO-66 (HP-UiO-66) with tunable oxygen vacancies and mesopores [25]. Controlling the breakdown temperature selectively removes NH2-BDC ligands, resulting in numerous CUSs. HP-UiO-66 produced an exceptional adsorption capacity of 248.8 mg/g1 for H2AsO4 with a fast adsorption rate by establishing Zr-O-As bonds due to the synergetic impact of hierarchical pores and CUSs from oxygen vacancies. In spite of the extra-active M-OH sites, coordinated modulation is another interesting technique for improving the adsorption performance of MOFs. Researchers have used a simple solvothermal approach to create yttrium-based MOFs, MOF-(Y), and an acetate-modulated one, MOF-(Y)-Ac [26]. When compared to MOF-(Y), coordinated modulation with sodium acetate as the capping reagent gives MOF-(Y)-Ac more porosity, surface area, and CUSs.

5.14 Application of MOF Composites

For MOF water buoyancy, the existence of water vapour in any industrial stream must be considered when selecting adsorbents for purification and adsorption systems. Despite their exceptional properties and wide range of applications, some unfavorable circumstances (such as metal ions, organic ligands, metal-ligand coordination geometry, pore surface hydrophobicity, and so on) have led to the instability of MOFs. As a result, the stability of MOFs has been a major concern for teams of researchers all over the globe. Chemical stability, in this context, refers to the ability of MOFs to maintain their structure under adverse conditions. This is usually established following some specific treatments of MOFs using a powdered X-ray diffraction (PXRD) pattern to characterize the samples' crystallinity and BET surface area based on N2-sorption isotherms and porosity probing studies. The chemical stability of MOFs is primarily controlled by two factors: the external component (the operating environment) and the internal factor (i.e., MOF structures). Many significant MOFs have been documented as unstable because their frameworks become compromised when exposed to moisture in air or water, both of which are always present components in most industrial processes. Because of their structural shortcomings, most MOFs are unsuitable for a broad range of applications. The hydrophobicity, organic linkers, metal ion clusters, geometry, surface area, and other working conditions of MOFs can all be ascribed to the stability or instability of these frameworks. The susceptibility of the bonds that coordinate the framework has been widely reported to add significantly to its stability. Researchers have created the Fe3O4(Cr) hybrid nanomaterial, which had adsorption capacities of 121.5 and 80 mg/gl for H2AsO4, respectively.

5.15 Gas Separation and Adsorption

The effective porous materials produced as conventional porous solid materials are needed in industrial processes for adsorptive removal, isolation, or separation (Mueller et al., 2006). The porous metal-organic frameworks with their customised structures and tunable surface properties are now suitable materials for gas adsorption and separation. They have also demonstrated that they are thermally stable, and most of the time, their structures are large enough to suit the removal of some guest species, which leads to permanent porosity. MOFs can thus be said to be suitable for research and actual applications in gas separation and purification as adsorbents for industry (Yang et al., 2012). A porous MOF—Co2(ad)2(CO2CH3) 2 DMF0.5H2O (bio-MOF-11)—was created by An et al. in 2010 using the solvothermal technique of synthesis and adenine in DMF. This MOF had a high thermal capacity for CO_2 adsorption, a high CO_2 capacity, and a high preference for adsorption of CO_2 over N_2. With an 8.9 wt% dynamic capacity and open magnesium sites that considerably compete with adsorbents in CO_2 adsorption, Mg-MOF-74, a MOF already in use, can release adsorbed CO_2 at a significantly lower temperature of 80 °C. Mg-MOF-74 has a remarkable capacity for adsorption and release of the adsorbed substance. This study demonstrates how superb CO_2 adsorbent or storage materials can be made from MOFs with open metal sites.

5.16 MOF Composites

Because MOFs have widely dispersed active sites, MOF-based composites with other embedded functional catalysts have great potential in tandem reactions [27]. This could be improved by strategically engineering synergic catalysis of the MOF matrix and embedded materials. During catalysis, designable functional organic ligands almost always serve as additional Lewis acid or base reactive sites. Among these catalysts, amino-modified MOFs have received a lot of attention because the -NH2 groups not only stabilised the NPs but also acted as active sites in some reactions. When it came to formic acid dehydrogenation at room temperature, the catalyst with amino-functionalized MIL-125 embedding Pd NPs performed better than Pd NPs loaded in unadulterated MIL-125. This is due to the weakly basic -NH2, which could accept a proton under reaction conditions to produce -+HNH2 and facilitate O-H bond dissociation. Meanwhile, the Pd format endures -hydride elimination to produce CO2 while leaving the Pd hydride species to produce H2. Because of its slightly alkaline sites, the amino group is commonly used in basic reactions such as Knoevenagel condensation. Uniform core-shell Pd@IRMOF-3 nanostructures, for example, have been effectively prepared with single Pd nanoparticles surrounded by amino-functionalized IRMOF-3. In a cascade reaction with 4-nitrobenzaldehyde as the reactant, the as-obtained composite catalyst demonstrated very high catalytic efficacy. The nitro group was then hydrogenated on embedded Pd NPs to create the final product after the Knoevenagel condensation happened at the -NH2 group. Aside from the -NH2 group, some acidic groups can also serve as second active sites in tandem reactions by collaborating with embedded functional materials. By using a simple thermal decomposition, our lab was able to create highly distributed Pd NPs supported on a sulfonic acid functionalized MOF catalyst, Pd/MIL-101-SO3H. This catalyst did exceptionally well in the one-pot synthesis of ethyl tetrahydro-2-furoate (ETF) from furoic acid due to the cooperation of Pd NPs and -SO3H acidic sites in Pd/MIL-101-SO3H. (FA).

Some metal-sites fixed on organic linkers may collaborate with embedded species in addition to organic linkers alone. Lin's group used in situ photoreduction to incorporate Pt NPs into stable, porous, and phosphorescent metal-organic frameworks composed of [Ir(ppy)2(bpy)]+ linkers and Zr-based secondary construction units. Because of the synergistic photoexcitation of MOFs and electron acceptors of embedded Pt NPs, the as-prepared hybrid catalysts demonstrated high activity for photocatalytic hydrogen generation. And also affixed ReI(CO)3(BPYDC)Cl (BPYDC = 2,2′-bipyridine-5,5′-dicarboxylate) into zirconium MOFs, UiO-67 (Ren-MOF) near the surface of Ag nanocubes [28]. Under visible light, intensified near-surface electric fields at the Ag nanocube surface increased CO2-to-CO conversion by about 7-fold.

Metal node unsaturated coordinated sites could also interact with embedded materials to enable tandem reactions. In general, MOF open metal sites function as Lewis acid sites. MIL-101 is a popular and extensively used MOF as a support in cooperative work with embedded materials due to its abundance of open chromium centres [29]. To accomplish efficient one-step methyl isobutyl ketone (MIBK) synthesis, a tandem catalyst with Pd NPs supported on MIL-101(Cr) was developed. Acetone was first transformed to mesityl oxide (MO) for catalysis via high density unsaturated Cr Lewis acid sites. At Pd active sites, the resulting MO molecule was further hydrogenated to create the final product MIBK. Multifunctional catalysts have also been used in tandem processes involving nitroarene reduction and carbonyl compound reductive amination. Nitroarene is first reduced to an aromatic amine on noble metal NPs, then further reacted with aldehyde or ketone at Cr Lewis acid sites to create the final secondary arylamine products.

COFs are a new class of materials that have similar structures to MOFs [30]. As a result, they can also serve as a carrier for iron oxide, resulting in composites that trap pollutants. The embedding of Fe203 was created.

5.17 Agrochemical Adsorption and Removal

Long-term residual deposition of organic agrochemicals (e.g., herbicides, insecticides, and fungicides) produces a variety of major environmental and health hazards. Until recently, research MOFs have been used to remove hazardous agrochemicals from water despite being quite uncommon and the target objects being limited to 2,4-D, which is a pesticide that is commonly used in agriculture. It is, however, a probable carcinogen and mutagen. The first illustration of MOFs' adsorptive elimination of 2,4-D from water was investigated. Jhung reported on this in 2013; MIL-53 is mentioned in this submission and demonstrated a significantly greater adsorption capacity and bigger adsorption than typical porous compounds like activated carbon as well as USY zeolite. The fundamental adsorption mechanism was discovered. MIL-53's zeta potential and adsorption behavior at different pH and temperature levels are discussed. The findings revealed host π-π stacking and electrostatic interactions. This system relies heavily on guests. Jhang returned a few years later, and coworkers reported the first instance of the adsorptive. The findings revealed that UiO-66 displayed a higher MCPP adsorption rate and capacity than activated carbon, with a bigger pore size, particularly at low MCPP concentrations. According to the zeta potential at various pH levels, a potential mechanism was discussed, revealing that both weak electrostatic and π-π stacking interactions are crucial functions in this system.

MOFs have efficient cationic herbicide adsorption. NKU-101's anionic framework is made up of tetrapodal mesoporous cage structural blocks and a plethora of uncoordinated oxygen atoms, allowing for enough room and active sites for guest adsorption. The adsorption isotherms revealed that it adsorbed capacities of 100 and 200 mg/g for methyl viologen and diquat, respectively. The residual concentrations of these two herbicides were calculated to be around 18–20 ppb (MV) and 9–10 ppb (DQ) using the concentration-peak area of LC-MS/MS, which is substantially lower than the Codex Alimentarius Commission's standard (200 ppb).

5.18 Pharmaceutical and Personal Care Adsorption Removal Products (PPCPs)

PPCPs have been identified as a new contaminant discovered in a variety of water sources, including not only surface water but also groundwater and drinking water. Because of its tenacity, the accumulation of PPCPs in nature has become a serious threat to public health and ecosystems. UiO-66, as well as its functionalized derivative, was used by Jhung (specifically, UiO-66 with SO3H/NH2) [13]. Jung et al. (2015) demonstrated the first use of MOFs in the elimination of diclofenac sodium (DCF). Contrasts between the adsorption of DCF by various adsorbents revealed that UiO-66 was the most effective, and UiO-66s performed significantly better than activated carbon (AC) in terms of both kinetics and capacities. UiO-66's pseudo-second-order rate constant was 0.014, which was the highest. The adsorption capacity was 189 mg/g, which was 3.5 and 2.5 times greater than that of AC. According to scientists, the increased adsorption rate and capacity of UiO-66 should be attributable to specific factors.

Weak interactions, such as electrostatic and π-π stacking interactions, exist between the adsorbate and UiO-66. Furthermore, base-base repulsion and acid-base attraction are to blame for the shift in the functionalized UiO-66's adsorption capacity. Li and colleagues (2015) recently developed a topological design technique. BUT-12 is an isostructural Zr-MOF obtained by coworkers as well as BUT-13. The H3CTTA and H3TTNA luminous ligand acids, which were redesigned and introduced, are used in the self-assembly method. Because of Zr-O bonding, the Zr6 cluster, which has a high degree of BUT-12 and BUT-13, may have strong water stabilities.

It also increases the overall hydrophobicity of the framework. The BUT-12 and BUT-13 were found to be stable in tests. After being submerged in water and HCl solution, the crystalline framework of NaOH solution (pH = 10) and 2 M, 6 M, and concentrated HCl remains even after 24 hours of boiling water. Because of their exceptional stability, these MOFs were used to adsorb a variety of antibiotics (NFs, NMs, SAM, and CAP) in aqueous solutions. The outcomes showed that BUT-12 had high adsorption rates toward BUT-13, which rapidly absorbed NZF and NFT, whereas NZF and NFT were slowly adsorbed by BUT-13. The adsorption behaviors of CAP and SAM differed significantly. This is most likely due to their differing pore diameters. According to the findings, the authors clarified that the necessary control experiments for antibiotics rely on pore size and hydrophobic pore surface adsorption in water. As a result, the higher adsorption capabilities of the two MOFs were compared to typical porous materials. The presence of methyl groups in the ligands was attributed to improving the hydrophobicity of the resultant MOFs and inhibiting the competitive adsorption of water and antibacterial molecules.

5.19 MOFs for Photocatalytic Elimination of Organic Pollutants

Photocatalytic degradation of organic contaminants is a more complete method of water purification than adsorptive removal. For instance, in-situ synthesis of highly reactive transitory species (for example, H_2O2, OH, and O_2) is capable of converting harmful organic pollutants into less hazardous or non-poisonous compounds. According to recent findings, MOFs are new and promising photocatalysts for the degradation of organic contaminants in wastewater. Photocatalytic dye removal (2.2.2.1) MOF central metals are thought to have a large influence on photocatalytic capabilities due to their distinct electronic structures. In order to testify to this, researchers created a series [M3(L)2(H2O)6] H2On (M = Mn) of isomorphous MOFs, Mn 0.7Co 0.3 (2), Mn 0.5Co 0.5 (3), Mn 0.3Co 0.7 (4), and Co (5) through the use of 1-aminobenzene.

5.20 Conclusion

The recent five-year development of MOF-based materials indicates that they are strong adsorbents for heavy metal oxoanions derived from water. This chapter thoroughly examines adsorption performance and describes adsorption methods including anion exchange, electrostatic interactions, hydrogen bonding, van der Waals interactions (which are examples of attractive forces, anion interactions, and so on), redox processes, and coordination between them. We can vary the pore diameters of metal oxoanions, modify the pores of MOFs, and add suitable functional groups into MOFs to speed up the spread of target ions and improve the interaction between MOFs and target ions. A huge number of MOFs, particularly those with a very large surface area, are unstable in water, which limits their use in water treatment. The pH will also have an effect on the charge of functional groups, which has varying affinities to target ions. Furthermore, some MOFs' structures will be damaged by strong acids or basic conditions. As a result, the active pH range of particular MOFs is often relatively limited.

Some approaches to overcoming the aforementioned challenges could be investigated by developing environmentally friendly methods of synthesizing MOFs, such as microwave, electrochemical, and mechanical approaches; adding functional groups to improve MOF stability; changing the architectures of ultrahigh porosity MOFs to fully utilize the adsorption sites abundantly distributed throughout the tiny structure; and fabrication. MOF-based hybrids are being developed to improve their recyclability and reusability. In a nutshell, MOF-based compounds have outstanding adsorption performance and stability, as well as notable reusability, and are prepared on a low-cost, large-scale basis. Eco-friendly strategies are projected to be used in practical applications for removing heavy metal oxyanions. the pore diameters of MOFs.

It will cause the situation to deteriorate. Some approaches to overcoming the aforementioned challenges could be investigated by developing environmentally friendly methods of synthesizing MOFs, such as microwave, electrochemical, and mechanical approaches; adding functional groups to improve MOF stability; changing the architectures of ultrahigh porosity MOFs to fully utilize the adsorption sites abundantly distributed throughout the tiny structure; and fabrication. MOF-based hybrids are being developed to improve their recyclability and reusability. In a nutshell, MOF-based compounds have outstanding adsorption performance and stability, as well as notable reusability, and are prepared on a low-cost, large-scale basis. Eco-friendly strategies are projected to be used in practical applications for removing heavy metal oxyanions. Multiple active materials with different functions could improve their performance in ECS applications. When compared to a simple mixture of nanostructures, well-designed hierarchical nanostructures could enable synergistic interactions for improved performance. The constituent MOFs and functionalities chosen with care, combined with reasonable dimensional designs, have been proven to provide adequate ECS performance improvements. MOF composites have demonstrated consistent capacity and potential in energy conversion technologies such as photovoltaics, fuel electrocatalysis, solar cells, fuel cells, LEDs, and energy storage technologies such as SCs, LIBs, LSBs, and MABs. MOFs and integrating functional materials may be used based on their relative functionalization in composite locations and distinctive roles when functional materials (carbon-based materials, polymers, etc.) are used as shelters or other structures. They can improve substrates' chemical and electrochemical stability, electric conductivity, and mechanical strength. MOFs can also be used as porous supports or coatings to allow for functional materials (metal NPs, QDs, POMs, etc.) in order to sift, block, or prevent electrolytes or reactants from aggregating and leaching functional materials. It should be noted that both MOFs and functional materials in ECS systems could use active sites. Novel product development MOF and integrating material functions provide feasible approaches for expanded applications and improved ECS performance.

Acknowledgment

It is with great appreciation that the authors would like to acknowledge the Department of Agronomy at Lovely Professional University for their constant support and encouragement throughout the research process.

Author Contributions

In addition to contributing to the outline, the author is responsible for leading the draft and editing of the manuscript. It was K.S., P.D., and P.K who conducted an extensive literature search and contributed to writing sections and constructing figures and tables. In addition to providing professional advice, P.K. helped revise the final version of the document and participated in its revision. In the end, all authors read and approved the final version of the manuscript.

Conflicts of Interest

None

References

1 Song, L.L., Chen, C., Luo, F. et al. (2016). Isoreticular MOFs functionalized in the pore wall by different organic groups for high-performance removal of uranyl ions. *Journal of Radioanalytical and Nuclear Chemistry* 310 (1): 317–327.
2 Niu, Z., Liu, L., Zhang, L., and Chen, X. (2014). Porous graphene materials for water remediation. *Small* 10 (17): 3434–3441.
3 Li, C.P., Chen, J., Liu, C.S., and Du, M. (2015). Dynamic structural transformations of coordination supramolecular systems upon exogenous stimulation. *Chemical Communications* 51 (14): 2768–2781.
4 Dias, E.M. and Petit, C. (2015). Towards the use of metal-organic frameworks for water reuse: a review of the recent advances in the field of organic pollutants removal and degradation and the next steps in the field. *Journal of Materials Chemistry A* 3 (45): 22484–22506.
5 Moreno-Castilla, C. (2004). Adsorption of organic molecules from aqueous solutions on carbon materials. *Carbon* 42 (1): 83–94.
6 Ho, Y.S. (2006). Review of second-order models for adsorption systems. *Journal of Hazardous Materials* 136 (3): 681–689.
7 Qin, Q., Wang, Q., Fu, D., and Ma, J. (2011). An efficient approach for Pb (II) and Cd (II) removal using manganese dioxide formed in situ. *Chemical Engineering Journal* 172 (1): 68–74.
8 Ahmed, I. and Jhung, S.H. (2017). Applications of metal-organic frameworks in adsorption/separation processes via hydrogen bonding interactions. *Chemical Engineering Journal* 310: 197–215.
9 Zheng, T., Yang, Z., Gui, D. et al. (2017). Overcoming the crystallization and designability issues in the ultrastable zirconium phosphonate framework system. *Nature Communications* 8 (1): 1–11.
10 Dai, R., Peng, F., Ji, P. et al. (2017). Electron crystallography reveals atomic structures of metal-organic nanoplates with M12 (μ3-O) 8 (μ3-OH) 8 (μ2-OH) 6 (M= Zr, Hf) secondary building units. *Inorganic Chemistry* 56 (14): 8128.
11 Ding, Y., Chen, Y.-P., Zhang, X. et al. (2017). Controlled intercalation and chemical exfoliation of layered metal-organic frameworks using a chemically labile intercalating agent. *Journal of the American Chemical Society* 139: 9136.
12 Yee, K.K., Reimer, N., Liu, J. et al. (2013). Effective mercury sorption by thiol-laced metal-organic frameworks: in strong acid and the vapor phase. *Journal of the American Chemical Society* 135 (21): 7795–7798.
13 Jung, B.K., Jun, J.W., Hasan, Z., and Jhung, S.H. (2015). Adsorptive removal of p-arsanilic acid from water using mesoporous zeolitic imidazolate framework-8. *Chemical Engineering Journal* 267: 9–15.

14 Drever, J.I. (1988). *The Geochemistry of Natural Waters*, vol. 437. Englewood Cliffs: Prentice hall.

15 Yang, R.T., Hernández-Maldonado, A.J., and Yang, F.H. (2003). Desulfurization of transportation fuels with zeolites under ambient conditions. *Science* 301 (5629): 79–81.

16 Li, S., Chen, Y., Pei, X. et al. (2016). Water purification: adsorption over metal-organic frameworks. *Chinese Journal of Chemistry* 34 (2): 175–185.

17 Aguila, B., Banerjee, D., Nie, Z. et al. (2016). Selective removal of cesium and strontium using porous frameworks from high level nuclear waste. *Chemical Communications* 52 (35): 5940–5942.

18 Cristaldi, A., Fiore, M., Zuccarello, P., Oliveri Conti, G., Grasso, A., Nicolosi, I., ... & Ferrante, M. (2020). Efficiency of wastewater treatment plants (Wwtps) for microplastic removal: A systematic review. *International Journal of Environmental Research and Public Health* 17 (21): 8014.

19 Vorosmarty, C.J., Green, P., Salisbury, J., and Lammers, R.B. (2000). Global water resources: vulnerability from climate change and population growth. *Science* 289 (5477): 284–288.

20 Wang, S. and Peng, Y. (2010). Natural zeolites as effective adsorbents in water and wastewater treatment. *Chemical Engineering Journal* 156 (1): 11–24.

21 Hasan, Z., Jeon, J., and Jhung, S.H. (2012). Adsorptive removal of naproxen and clofibric acid from water using metal-organic frameworks. *Journal of Hazardous Materials* 209: 151–157.

22 Haque, E., Jun, J.W., and Jhung, S.H. (2011). Adsorptive removal of methyl orange and methylene blue from aqueous solution with a metal-organic framework material, iron terephthalate (MOF-235). *Journal of Hazardous Materials* 185 (1): 507–511.

23 Baudron, S.A., Avarvari, N., Batail, P. et al. (2003). Singular crystalline β '-layered topologies directed by ribbons of self-complementary amide⊙⊙⊙ amide ring motifs in [EDT-TTF-(CONH2) 2] 2X (X= HSO4-, ClO4-, ReO4-, AsF6-): coupled activation of ribbon curvature, electron interactions, and magnetic susceptibility. *Journal of the American Chemical Society* 125 (38): 11583–11590.

24 Ahluwalia, S.S. and Goyal, D. (2007). Microbial and plant derived biomass for removal of heavy metals from wastewater. *Bioresource Technology* 98 (12): 2243–2257.

25 Rapti, S., Pournara, A., Sarma, D. et al. (2016). Rapid, green and inexpensive synthesis of high quality UiO-66 amino-functionalized materials with exceptional capability for removal of hexavalent chromium from industrial waste. *Inorganic Chemistry Frontiers* 3 (5): 635–644.

26 Wang, C.C., Li, J.R., Lv, X.L. et al. (2014). Photocatalytic organic pollutants degradation in metal-organic frameworks. *Energy & Environmental Science* 7 (9): 2831–2867.

27 Qin, Yuhuan, Hao,Mingming, Wang, Jiaqi, Yuan, Rusheng, Li, Zhaohui. (2022). Rational Design of a Core–Shell Structured Plasmonic Au@MIL-100(Fe) Nanocomposite for Efficient Photocatalysis. *ACS Applied Materials & Interfaces* 14 (51): 56930–56937.

28 Zhao, Meiting, Deng, Ke, He, Liangcan, Liu, Yong, Li, Guodong, Zhao, Huijun, and Tang, Zhiyong (2014). Core–shell palladium nanoparticle@ metal–organic frameworks as multifunctional catalysts for cascade reactions. *Journal of the American Chemical Society* 136 (5): 1738–1741. doi: 10.1021/ja411468.

29 Xie, L. H., Liu, X. M., He, T., & Li, J. R. (2018). Metal-organic frameworks for the capture of trace aromatic volatile organic compounds. *Chem* 4 (8): 1911–1927.

30 Kolpin, D.W., Furlong, E.T., Meyer, M.T. et al. (2002). Pharmaceuticals, hormones, and other organic wastewater contaminants in US streams, 1999– 2000: a national reconnaissance. *Environmental Science & Technology* 36 (6): 1202–1211.

6

"Green Applications of Metal-Organic Frameworks for Wastewater Treatment"

Ankita Saini[1,], Sunil Kumar Saini[2], and Parul Lakra[1]*

[1] *Department of Chemistry, Faculty of Physical Sciences, P.D.M. University, Bahadurgarh, Haryana - 124507, India*
[2] *Department of Zoology, Faculty of Life Sciences, P.D.M. University, Bahadurgarh, Haryana - 124507, India*
[*] *Corresponding author*

6.1 Introduction

Metal-organic frameworks (MOFs) are a versatile class of advanced organic and inorganic blended porous crystalline materials that contain regular arrangement of cations surrounded by organic linker molecules [1–3]. The cations form nodes which bind together all arms of the linker molecules to construct repeating cage-like structures called coordinating polymers or MOFs. Owing to the presence of hollow porous structures, MOFs display a large internal surface area [4–9]. MOFs were first prepared by Yaghi et al. in 1995 [10]. It was investigated how MOFs can be used in various applications in wastewater treatment [11–17]. They are also used for gas purification [18], chemical and biosensors [19, 20], phase separation [21], bio-imaging, electrocatalysis and photocatalysis [22–24], light capture and energy conversion [25, 26], and in drug delivery [26–28]. In this chapter, firstly, we will discuss types of MOFs and methods and procedures involved in the synthesis of MOFs. Further, we have briefed about the use of green chemistry in the preparation of MOFs using sustainable approaches followed by their application in the treatment of wastewater by the removal of different contaminants.

MOFs showcase various applications because of their great properties like large surface area, high porosity, abundant active adsorption sites, stability, and flexible design as displayed in Figure 6.1. The physicochemical and electronic properties of MOFs can be engineered easily by employing varying lengths of linker moiety. They are nontoxic and have liquid-like density. They can be recycled and demand low energy for preparation [9, 29].

The term MOF is the most frequently used for the coordinated class of frameworks followed by a numeral. The abbreviation MOF not only depicts the presence of a porous structure but also a well-defined geometry resulting in a rigid framework. However, a number of frameworks are clubbed into classes with common letter

Metal Organic Frameworks for Wastewater Contaminant Removal, First Edition. Edited by Arun Lal Srivastav, Lata Rani, Jyotsna Kaushal, and Tien Duc Pham.

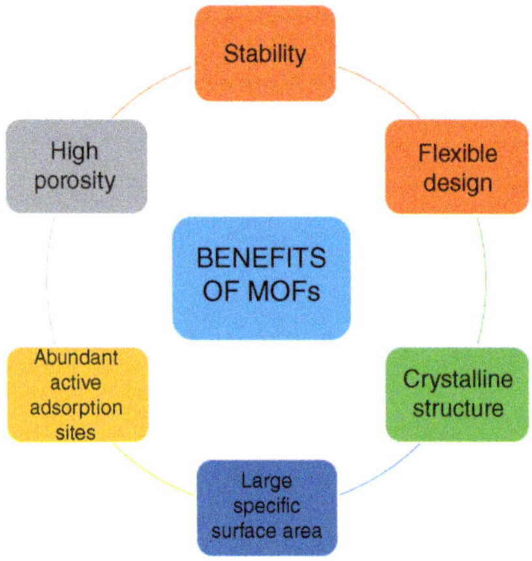

Figure 6.1 Benefits of using MOFs which make them efficient for wastewater treatment.

designations according to the place of origin of their discovery, which is not based on similarity in their structure. UiO (University of Oslo- UiO-66, UiO-67, UiO-68); MIL (Materials of Institut Lavoisier- MIL-101 (Cr), MIL-100 (Al), MIL-43 (Fe)); NU (Northwestern University- NU-1000); HKUST (Hong Kong University of Science and Technology- HKUST-1, Co-HKUST); ZIFs (Zeolitic Imidazolate Frameworks- ZIF-8, ZIF-67); and JUC (Jilin University China- JUC-165, JUC-167, JUC-199); MOF-N (metal-organic frameworks- MOF-5, MOF-177, MOF-199) are some of such classes. Apart from these, the most prominent class of MOFs is abbreviated as ZIF with a number since they have a zeolite-like topology and framework.

MOFs have been prepared by two different procedures, which are:

- Solvothermal synthesis
- Non-solvothermal synthesis

Solvothermal synthesis involves the use of a solvent in a chemical reaction at the boiling point of the solvent. It is also called hydrothermal because mostly water is used as the solvent. The choice of solvent is a crucial step in the synthesis of MOFs since it has a significant impact on how a coordinated environment forms. The solvent assists in the formation of coordinate bonds with metal ions in the synthesis of MOFs [30]. As a result, the final lattice structure is significantly influenced by the solvent [31].

Non-solvothermal synthesis refers to the preparation of MOFs without the use of solvents or a lot less of the solvent is used. Under atmospheric pressure, the procedure is carried out in an open flask below the solvent's boiling point. Without the use of complicated equipment, non-solvothermal synthesis can be performed both at room temperature and by heating [32–35]. Usually, in the synthesis of MOFs, conventional electric heating is done through convection from the oven, and the heat source is used for maintaining the temperature, ranging from the required temperature to 250°C, which is approximately the boiling point of the solvent that is in use [36].

Different processes that are used for heating are:

1) **Microwave synthesis**
   Microwaves (MW) are electromagnetic radiations that have frequencies ranging from 300- to 300,000 MHz. These radiations have two effects on the substrates: action on polar molecules and free ions that lead to heating. There are several factors that affect the heating efficiency, such as the dipole moment of polar molecules and their free rotation [33]. Since the solvent plays a significant role in the synthesis of MOFs, the use of a particular solvent is required for MW radiation.
   Microwave radiation was first used for the synthesis of MOFs in 2005 [37]. The microwave process helps in the decrement of time taken to synthesize the MIL-100 framework from 96 hours to 4 hours. It was synthesized in an aqueous solution of chromium metal, benzene-1,3,5-tricarboxylic acid (H_3BTC), and hydrofluoric acid (HF). The reactants were stirred well by using MW radiations and then they were placed in an autoclave and slowly heated to 220°C [37, 38]. The MIL-100 was synthesized within one hour. It was amazing to see that, after a 24-fold reduction in synthesis time, the yield of the product remains the same as traditional synthesis.
2) **Electrochemical synthesis**
   This method of MOF synthesis includes metals ions that are introduced through the electrochemical process. Particularly, the metal ions are allowed to pass through the dissolution of the anode into a reaction mixture that contains linker molecules and electrolytes. This process should be continuous and the creation of ions during the reaction should be avoided, which will result in a good yield of MOFs [36]. The electrochemical method was first used for the synthesis of HKUST-1 MOF [39].
3) **Mechanochemical synthesis**
   The mechanochemical mechanism involves the introduction of mechanical energy such as ball milling in solids. This technique has been popular for high yield and fast responses with or without a small amount of solvent. The presence of liquid components during a mechanochemical reaction provides additional benefits such as easier crystallization, greater mobility of the reactants, a higher degree of crystallinity of the reaction product, and a good yield of the target product [40–43]. The mechanochemical method was first used for the synthesis of MOFs in 2006 [41]. This process involved mixing copper acetate with isonicotinic acid for a short period of time in a ball mill, resulting in a product that was well-crystallized with the formula of $Cu(INA)_{2x}H_2O.yAcOH$, whereas HKUST-1 was synthesized in 2010 by the mechanochemical method [44]. In both studies, copper acetate and H_3BTC have been used in solvent-free situations. Yuan et al. focus on the fact that the $Cu(INA)_2$ is built almost naturally while the formation of HKUST-1 requires more time for mixing. Moreover, a small amount of acetic acid added during the reaction rapidly increases the synthesis rate of $Cu(INA)_2$ instead of HKUST-1 [44, 45]. Due to the potential of carrying out chemical reactions at room temperature and a large reduction in solvent, mechanochemical synthesis is a particularly appealing technique.
4) **Sonochemical synthesis**
   Cavitation is primarily responsible for how ultrasound affects liquid and colloids [46–48]. The sonochemical method was first used in 2008 for the synthesis of ($Zn_3(BTC)_2$) MOF [49]. Both of the substrates, zinc acetate and H_3BTC, were mixed in 20% ethanol and subjected to sonication, yielding 75.3% of the product

after five minutes of sonication. Furthermore, Schlesinger et al. compared six different methods of synthesis of HKUST-1, which were: synthesis on atmospheric pressure with heating under reflux, solvothermal synthesis, electrochemical synthesis, microwave, mechanical synthesis, and sonication [50]. All procedures resulted into a pure phase of the product, and it was observed that the maximum yield using minimal synthesis was achieved through microwave radiation.

6.2 Role of Green Chemistry in Preparation of MOFs

Sustainable methods for the synthesis of MOFs are of supreme importance for energy conservation. Apart from the amazing properties of MOFs, their physical applications are limited because of conventional preparation methods that are harmful for nature. But, now there are many green approaches that are being considered for the sustainable synthesis of MOFs.

Greening the synthetic approach for MOFs has grabbed the interest of many researchers as a critical step toward practical commercial applications. The green synthesis of MOFs is done by keeping 12 principles of green chemistry. The prime focus for green synthesis of MOFs is based on: [51, 52]

i) **Sustainable metal ions-**
There is a need for the use of safer and less toxic sustainable metals that can be used for the formation of MOFs, which are: zinc, iron, copper, calcium, magnesium, potassium, sodium, zirconium, manganese, etc. [53].

ii) **Safer solvents / auxiliary solvents-**
Solvent selection is a fundamental part of the synthesis of MOFs. Solvents play a pivotal role in the formation of coordination. It has been found that recoverable solvents that are derived from reusable waste help in the reduction of toxicity [54, 55]. Green, alternative techniques and strategies are employed for the replacement of toxic and non-environmentally friendly solvents by recommended green solvents such as water, isopropyl alcohol, acetone, ionic liquids, supercritical liquids, ethyl acetate/ethyl lacrate, and bio-derived solvents [56–60].

1) **Water**
The safest, most accessible, and most affordable solvent is water. It is simple to recycle and purify produced materials [61], and it would be perfect to develop a scalable water-based synthesis alternative for MOFs. The easiest, least expensive, and safest posttreatment approach is regarded as water-based synthesis. The majority of recent work on the synthesis of MOFs, including ZIFs, concentrates on methods that use water at room temperature and atmospheric pressure [62] and metal salts and chiral, fluorinated, and N-linked ligands such as oxides and sulfates.

2) **Supercritical liquids**
Another excellent choice of medium for an environmentally friendly reaction is supercritical liquids. Water and supercritical carbon dioxide ($scCO_2$) are two possible solvents that are strong, environmentally friendly solvents for the production of MOFs [63]. Depending on the synthesis conditions applied, they can

be made using green techniques because of their remarkable properties such as polarity, viscosity, and surface tension. These eco-friendly solvents permit recycling and do not require distillation [64, 65].

3) **Ionic liquids**
Ionic liquids (ILs), which consist of organic/inorganic cations and anions are new and emerging choices for the replacement of harmful solvents where 4° ammonium/phosphonium or imidazolium will make a positively charged cationic component. The related anionic species may be a halogen, hexafluorophosphate, or triflate [66, 67]. Compounds that are environmentally friendly solvents and have melting points below water include ILs. It is necessary to take care of some ILs since they are also toxic, such as 1-butyl-3- methylimidazolium, perfluorinated anions, and several imidazoles [68, 69].

4) **Bio-derived solvents**
Biomass is a renewable feedstock that can be catalyzed into usable energy and biochemicals. Biomass is a viable raw material for the synthesis of compounds since it demonstrates a number of intriguing traits such as oxygenation tendency, stereochirality, and easy availability [70]. A number of biomass resources including logs, corn cobs, bagasses, wood, and press cake can be used to produce cyrene, also known as 6,8-dioxabicyclo [3.2.1] octanone, a dipolar aprotic green solvent. Cellulose is the most productive biomass source for the synthesis of levoglucosenone LGO i.e., LGO and (-)cyrene. Two processes are required to produce 2H-LGO: first, solid biomass is converted into LGO, and then LGO is reduced to 2H-LGO using a hydrogenation technique [71]. Water is the only by-product of this green conversion of solid biomass.

iii) **Organic linkers from biomass / biomolecule-derived organic linkers**
The choice of organic linkers is also a crucial step for ensuring the desired structure of MOFs. Exploring environmentally friendly linker substitutes is necessary [72]. Due to the favorable coordination reactivity with metal nodes, carboxylic acids are the most prevalent and traditional linkers accessible for MOF production. Almost all the commercially available linkers are expensive because of their petrochemical origins [73]. For sustainable production, there is a requirement for a linker moiety that produces a nontoxic intermediate and byproduct. Green linkers based on biomolecules as multifunctional ligands for MOF production, amino acids, cyclodextrins, nucleobases, saccharides, and peptides can be used. These bioactive compounds feature reactive chemical groups that mix well with a range of metals to create BioMOFs, which are long-lasting frameworks. Their inherent properties including the presence of low symmetry and great flexibility, self-assembly, and biocompatibility have boosted interest in the synthesis of BioMOFs [74]. However, they also have a number of drawbacks. The most often-used biomolecules in the creation of BioMOFs are nucleobases. Recently, the nontoxic chemicals adenine, butanedioic acid, and isobutyric acid were used to create two green biomimctic Co-MOFs [75].

Currently, the most affordable, long-lasting, and secure metal salts that provide ecologically friendly anions for MOF synthesis include metal hydroxides, acetates, and oxides. According to Julien et al. [51], the benign anions that are effective for MOF synthesis include hydroxides, carbonates, sulfates, acetates, oxides, and acetylacetonate since they produce water as a byproduct. Metal salts based on sulfate anion, which are water soluble, can be used as metal nodes for the synthesis of MOFs in a green protocol. In fact, researchers exploring for environmentally friendly methods of MOF synthesis are finding interest in the examination of various salts displaying characteristics of water solubility and stability. However, the porosity, stiffness, and crystalline structure of sulfate anions inside the framework topology can be impacted by significant interfacial interactions [76–78]. It provides higher catalytic efficiency because the coordination of linkers with metal ions results in Bronsted acidity [79]. For example, metal acetylacetonate salts have been regarded as green reagents in several industrial processes as a desirable option for alternative metals using the hydrothermal synthesis of MOFs. For instance, employing acetylacetonate salts at ambient temperatures, various MOFs from the UiO family (MOF-801 and UiO66-NH_2) and MIL-88A have been synthesized with high yields of 60–80% [80]. However, when protonated linkers like carboxylic acids or imidazoles were initially utilized for MOF production, a pool of hazardous byproducts were formed. These more environmentally friendly methods for MOF synthesis at room temperature are quicker, more inventive, and more broadly applicable than hydrothermal synthesis.

The most plentiful and renewable sources of carbon are biopolymers like cellulose, lignin, and chitin, which are generated from living organisms and are used as biomass linkers. Numerous functional groups found in carbon-based materials enable the creation of a wide range of chemical entities [81, 82]. The usefulness of molecules derived from biomass has the potential to increase the use of sustainable chemistry in numerous fields, including the food, environmental, and medical industries [83]. It has been extensively investigated how biomass can be converted into bioactive substances like 5-hydroxyl methyl furfural (HMF) and saccharides [84]. The conversion of various biomasses into useful organic acids including formic, lactic, and acetic acid has also received tremendous study and research [85–87]. One instance is the direct synthesis of luminous nanosized porous zinc phosphates from waste polyethylene terephthalate (PET) bottle waste as a source of linker by Huang et al., who created a sustainable method for MOF linkers [88]. Additionally, Cr-MOF has been created utilizing formic acid and waste PET [89]. Additionally, PET waste has been used as a linker in the hydrothermal synthesis of MIL-family MOFs such as MIL-47, MIL-53, and MIL-101 [90, 91].

6.3 Green Application of MOFs in the Removal of Contaminants from Wastewater

Water is one of the most crucial parts of everyone's lives [92]. Heedless use of technology and the available resources contribute to many environmental difficulties. Water pollution is a major issue in contemporary times as the addition of organic dyes, pesticides, microplastics, etc. Inorganic contaminants including heavy metals, non-metals, and various other pollutants contaminate the water and deteriorate its quality, which further makes it unfit for everyone's use. These contaminants come from naturally occurring and man-made sources mainly from industrial, municipal, and hospital wastewater treatment plants, drainage systems, and landfills [93].

MOFs are widely used for the elimination of toxic pollutants from wastewater because they possess a remarkable crystalline structure, large surface area, high porosity, abundant adsorption sites, and a flexible design. Due to these properties, MOFs are observed to be a good and possible fit for use in wastewater treatment. Water treatment can be done by biological treatments, advanced oxidation processes, and phase changing [94, 95]. According to the chemical composition, the contaminated substances can be divided into organic and inorganic categories, which can be removed by adsorption, degradation, or the reduction method [96]. Adsorption and photodegradation are the most sustainable techniques used for water purification because these techniques are economical, simple, and have easily operated designs, which can be used for treating wastewater.

For the removal of inorganic contaminants particularly from agricultural practices, home sewage and industries including mining, battery, nuclear, and textile dyes contaminated with heavy metals is made possible by the use of MOFs as adsorbents. Heavy metals like Cd, Zn, Pb, Fe, Cu, Hg, Ni, Mn, and Co are typically present but only in trace amounts; nevertheless, when they are present in large quantities in wastewater effluent they are considered harmful [97, 98].

Moreover, MOFs can also act as photocatalysts for the photodegradation of organic pollutants present in wastewater. Theoretically, photodegradation is a better method for wastewater treatment than adsorption since it results in the total eradication of the pollutant rather than just a simple phase transition, and no additional treatment is needed. Because MOFs contain organic linkers, they offer a rather broad absorption spectrum that enables the creation of a charge-separated state, which supports photocatalysis [99, 100]. Alvaro et al. described the ZnO cluster-containing MOF-5's behavior as a semiconductor. The material was discovered to have a 3.4 eV band gap, an absorption band centered at 450 nm and stability toward light exposure. In the photodegradation of phenol, the activity of MOF-5 was evaluated and contrasted with that of TiO_2 nanoparticles and another semiconductor, such as ZnO [101].

So, nevertheless it is of no doubt to state that MOFs can be applied for the removal of both organic and inorganic contaminants from industrial and domestic wastewater, including the removal of heavy metal ions, radioactive elements, poly-aromatic hydrocarbons (PAH), microplastics, pesticides, pharmaceuticals, and dyes. A few of the composites/frameworks used for the removal of inorganic and organic pollutants from wastewater that are discussed in this chapter are summarized and listed in Table 6.1.

6.3.1 MOFs for the Removal of Inorganic Contaminants

Now, in this particular section of the chapter, we will discuss the use of MOFs as adsorbents for heavy metals in detail, owing to the high gravimetric and volumetric densities of MOFs arising from their porous structure. MOFs that have been synthesized in green protocols are studied for the adsorption of various heavy metal ions, particularly As, Cd, Cr, Pb, and Hg, from wastewater. These are described below:

Table 6.1 Summarized list of various composites/frameworks discussed in this chapter that are used for the removal of inorganic and organic contaminants from wastewater.

S. No.	Composite Porous Material	Method of Synthesis	Type of Pollutant	Removal of Species	Removal Capacity (mg/g)	Ref. No.
1)	UiO-66	Mechanochemical	Inorganic	Arsenic (As) V	303	[102]
2)	Fe-BTC MOF	Solvothermal	Inorganic	Arsenic (As) V	57.7	[103]
3)	ZIF-8	Mechanochemical	Inorganic	Arsenic (As) III, V	2.02, 1.42	[104]
4)	ZIF 8	Hydrothermal	Inorganic	Arsenic (As) III, V	49.49, 60.03	[105]
5)	MIL-53(Fe)	Solvothermal	Inorganic	Arsenic (As) V	21	[106]
6)	MIL-53(Al)	Solvothermal	Inorganic	Arsenic (As) V	106	[107]
7)	MOF-808	Microwave	Inorganic	Arsenic (As) V	25	[108]
8)	UiO-66	Solvothermal	Inorganic	Arsenic (As)	303	[102]
9)	UiO-66-$(SH)_2$	Solvothermal	Inorganic	Arsenic (As) III, V	40, 10	[109]
10)	$Cu_3(BTC)_2$-SO_3H	Mechanochemical	Inorganic	Cadmium (Cd) II	88.7	[110]
11)	AMOF-1	Solvothermal	Inorganic	Cadmium (Cd) II	41	[111]
12)	PCN-100	Solvothermal	Inorganic	Cadmium (Cd) II	163	[112]
13)	UiO-66-NHC(S)NHMe	Ultrasonic	Inorganic	Cadmium (Cd) II	49	[113]
14)	3D Co(II) MOF	Ultrasound	Inorganic	Cadmium (Cd) II	2.3	[114]
15)	HS-mSi@MOF-5	Hydrothermal	Inorganic	Cadmium (Cd) II	98	[115]
16)	TMU-5	Mechanochemical	Inorganic	Cadmium (Cd) II	43	[116]
17)	Cu-terephthalate MOF	Solvothermal	Inorganic	Cadmium (Cd) II	100	[117]
18)	HKUST-1-MW	Microwave	Inorganic	Cadmium (Cd) II	32	[118]
19)	MnO_2-MOF	Solvothermal	Inorganic	Cadmium (Cd) II	176	[119]

20)	UiO-66-NHC(S)NHMe	Ultrasonic	Inorganic	Chromium (Cr) III	117	[113]
21)	TMU-5	Mechanochemical	Inorganic	Chromium (Cr) III	123	[116]
22)	Fe_3O_4@MIL-100	Hydrothermal	Inorganic	Chromium (Cr) II	18	[120]
23)	chitosan-MOF (UiO-66)	Microwave	Inorganic	Chromium (Cr) IV	94	[121]
24)	Cu-BTC	Hydrothermal	Inorganic	Chromium (Cr) VI	48	[122]
25)	Silver triazolo-MOF	Mechanochemical	Inorganic	Chromium (Cr) VI	37	[123]
26)	ZJU-101	Solvothermal	Inorganic	Chromium (Cr) VI	245	[124]
27)	TMU-30	Electrothermal	Inorganic	Chromium (Cr) VI	145	[125]
28)	MOR-1-HA	Solvothermal Reflux	Inorganic	Chromium (Cr) VI	280	[126]
29)	UiO-66-NHC(S)NHMe	Ultrasonic	Inorganic	Lead (Pb) II	232	[113]
30)	3D Co(II) MOF	Ultrasound	Inorganic	Lead (Pb) II	3.1	[114]
31)	HSmSi@MOF-5	Hydrothermal	Inorganic	Lead (Pb) II	312.5	[115]
32)	TMU-5	Mechanochemical	Inorganic	Lead (Pb) II	251	[116]
				Cobalt (Co) II	63	
				Copper (Cu) II	57	
33)	MIL-53(Al)	Mechanochemical	Inorganic	Lead (Pb) II	492.4	[127]
34)	AMOF-1	Solvothermal	Inorganic	Lead (Pb) II	71	[111]
35)	Cu-terephthalate MOF	Solvothermal	Inorganic	Lead (Pb) II	80	[117]
36)	HKUST-1 MW	Microwave	Inorganic	Lead (Pb) II	98	[118]
37)	Dy(BTC)(H_2O)(DMF)	Sonochemical	Inorganic	Lead (Pb) II	5	[128]
38)	MnO_2-MOF	Solvothermal	Inorganic	Lead (Pb) II	917	[119]

(Continued)

Table 6.1 (Continued)

S. No.	Composite Porous Material	Method of Synthesis	Type of Pollutant	Removal of Species	Removal Capacity (mg/g)	Ref. No.
39)	MOF-5/ $Zn_4O(BDC)_3$	Solvothermal	Inorganic	Lead (Pb) II	659	[129]
40)	Thiol-HKUST-1	Solvothermal	Inorganic	Mercury (Hg) II	714	[130]
41)	UiO-66-NHC(S)NHMe.	Ultrasonic	Inorganic	Mercury (Hg) II	769	[113]
42)	3D Co(II) MOF	Ultrasound	Inorganic	Mercury (Hg) II	5.17	[114]
43)	FJI-H12	Solvothermal	Inorganic	Mercury (Hg) II	440	[131]
44)	LMOF-263	Solvothermal	Inorganic	Mercury (Hg) II	380	[132]
45)	SH@SiO_2/$Cu(BTC)_2$	Solvothermal Encapsulation	Inorganic	Mercury (Hg) II	210	[133]
46)	MIL-101-Thymine	Solvothermal	Inorganic	Mercury (Hg) II	52	[134]
47)	BioMOF CaII.CuII 6[(S,S)-methox]3(OH)2(H_2O)·16H_2O	Solvohydrothermal	Inorganic	Mercury (Hg) II	900	[135]
48)	MOF-74-Zn	Solvohydrothermal	Inorganic	Mercury (Hg) II	63	[136]
49)	MIL-53(Al)	Solvothermal	Inorganic	Silver (Ag) I	183	[137]
50)	3D Co(II) MOF	Ultrasound	Inorganic	Iron (Fe) III	2.5	[114]
				Aluminum (Al) III	4.22	
51)	Cu-terephthalate MOF	Solvothermal	Inorganic	Iron (Fe) III	175	[117]
				Manganese (Mn) II	150	
				Zinc (Zn) II	115	
52)	MOF-5	Electrochemical	Inorganic	Copper (Cu) II	290	[138]

53)	Cd-MOF-74	Solvothermal	Inorganic	Copper (Cu) II	189.5	139]
54)	ZIF-8@PAM	Hydrothermal	Inorganic	Copper (Cu) II	224	[140]
55)	Sn(II)-TMA	Solvothermal	Inorganic	Fluoride	30.86	[141]
56)	UiO-66, NU-1000 MOF-525	Ultrasonic	Organic Drug	Tetracycline (TC)	145, 356, 807	[142]
57)	MIL 53(Fe–Cu)	Solvothermal	Organic Drug	Ciprofloxacin (CF)	190	[143]
58)	V_2O_5@Ch/Cu-TMA	Solvothermal	Organic Drug and Inorganic	Levofloxacin (LEVO) and Chromium (Cr) VI	91.99%–97.20%, 92.43–96.95%%	[144]
59)	MIL-53 (Fe)_UTS-2	Sonochemical	Organic Dye	Methyl Orange (MO) Methylene Blue (MB) bisphenol A (BPA)	494.2 221.9 160.1	[145]
60)	$Zn_2[(S,S)$-serimox$](H_2O)_2$	Hydrothermal	Organic Dye	Brilliant Green (BG)	100%	[146]
61)	Zn L-Asp bioMOF	Solvothermal	Organic Dye	Direct Red 81 (DR-81)	27.14, 95.3%	[147]
62)	Zn-MOF nano-cubes/ CZM	One pot Solvothermal	Organic Dye	Malachite Green (MG)	953.14	[148]
63)	Cu MOFs/Fe_3O_4	Solvothermal	Organic Dye and Inorganic	Malachite Green and (MG) Lead (Pb) II	113.67, 219	[149]
64)	CAU-1	Hydrothermal	Organic Product	Nitrobenzene	970	[150]

(Continued)

Table 6.1 (Continued)

S. No.	Composite Porous Material	Method of Synthesis	Type of Pollutant	Removal of Species	Removal Capacity (mg/g)	Ref. No.
65)	MIL-68 (Al)	Hydrothermal	Organic Product	Nitrobenzene	1130	[150]
66)	UiO-66-OH	Solvothermal	Organic Micro/ Nano Polymers	Microplastics, polyvinylidiene fluoride (PVDF), polymethyl methacrylate (PMMA), and polystyrene (PS)	PVDF = 90.1 ± 2.1%;. PMMA = 88.2 ± 1.7% PS = 85.7 ± 4.8%	[151]
67)	ZIF-8@Aerogel	Solvothermal	Organic Micro/ nano Polymers	Microplastics, poly(1,1-difluoroethylene) (PDFE) and polystyrene (PS)	PDFE = 91.4%, PS = 85.8%,	[152]

1) **Removal of Arsenic**

Wastewater from domestic households has been found to have traces of arsenic in it. There have been many studies conducted for the selective removal of arsenic through different MOFs from wastewater.

Wang et al. synthesized a water-stable, Zr-based MOF (UiO-66), which was prepared for arsenic adsorption for first time. UiO-66 MOF acted as an adsorbent and performed admirably over a pH range of 1 to 10. In this MOF, benzene dicarboxylate ligand and hydroxyl group are the arsenic species binding sites. It displayed a significant As-V removal capacity of 303 mg/g at a pH of 2 [102]. Zhu et al. examined the production and adsorption efficiency by Fe-BTC MOF where the MOF contained iron nodes and 1,3,5-benzenetricarboxylic (BTC) acid linkers by solvothermal autogenous pressure synthesis. With an altered pH, it was discovered that the removal effectiveness of As-V by Fe-BTC polymer was better than 96% between pH 2 and 10 with the maximum capacity of 57.7 mg/g [103]. According to Jian et al., ZIF-8, which was seen at neutral pH, has moderate capacities for both As(V) and As(III). The sources of the metal ions for As(III) and As(V), respectively, were $NaAsO_2$ and $Na_3AsO_412.H_2O$, which were used to calculate the adsorption capacity of the MOF at room temperature to be 2.02 and 1.42 mg/g [104]. Jian et al. demonstrated ZIF-8 as efficient As (III) and As (V) adsorbents with maximal adsorption capacities of 49.49 and 60.03 mg/g, respectively. The ZIF was synthesized at room temperature in aquesous conditions by the hydrothermal method [105].

In 2015, Vu et al. provided a detailed description of the adsorption of As(V) in MIL-53(Fe) demonstrating an adsorption capability of 21 mg/g. This MIL-53 MOF was synthesized using HF-free solvothermal methods [106]. Furthermore, Li et al. demonstrated the utilization of a related MOF called MIL-53(Al), which demonstrated a maximum As(V) adsorption capacity of 106 mg/g at a pH of 8 [107]. Li et al. provided MOF-808 for the adsorption of As(V). In this study, the MOF was produced by microwave irradiation. When compared to crystals produced using a proper solvothermal synthesis, the crystals obtained via microwave synthesis were nanosized, in the range of 150 to 200 nm. However, when the pH of the solution was lowered to 4 using HCl, it was discovered that the MOF had an adsorption capability of 25 mg/g. Rapid adsorption made it possible to remove 95% of As(V) from the solution in 30 minutes at an initial concentration of 5 ppm only [108].

Wang et al. produced UiO-66 at a pH of 2 that had a 303 mg/g capacity. Over a wide range of pHs, the MOF showed encouraging adsorption abilities [102]. Using the thiolated derivative of UiO-66, i.e., UiO-66-$(SH)_2$, Audu et al. reported the simultaneous capture of As(III) and As(V) species. In this instance, it was discovered that the As(V) species interacted with the node of the MOF while the As(III) species were entrapped by a thiolated linker with an adsorption capacity of 40 mg/g and 10 mg/g, respectively, after six hours of attainment of equilibrium [109].

2) **Removal of Cadmium**

For the adsorption of cadmium, Wang et al. constructed a $Cu_3(BTC)_2$-SO_3H framework. $Cu_3(BTC)_2$ was modified postsynthesis and then oxidized with sulfonic acid to synthesize it. The pH value of 6 was discovered to be ideal for Cd(II) adsorption.

Chelation between the sulfonic groups of the MOF and Cd(II) caused the adsorption process to take place with a maximum adsorption capacity of 88.7 mg/g. The functionalization of the framework with sulfonic acid significantly boosted selectivity for Cd(II) over other metal ions, which could be a result of SO_3H numerous bonding sites and coordination modes [110].

AMOF-1 was invented by Chakraborty et al. This substance was made from flexible tetracarboxylate linkers and Zn(II) metal ions. When equilibrium had been reached after 24 hours, the maximum Cd-ion absorption was discovered to be 41 mg/g [111].

Fang et al. reported on the existence of PCN-100, another Cd(II) removing MOF. Through the linkers used to insert metal ions, it was discovered that the MOF displayed a chelating coordination mode [112]. Saleem et al. worked on UiO-66-NHC(S)NHMe, a MOF with a maximum capacity of 49 mg/g that can adsorb Cd(III) among other ions [113]. Another MOF for Cd(II) removal was created by Abbasi et al. It was a 3D cobalt and TATAB (TATAB = 4,4',4"-s-triazine-1,3,5-triyltri-paminobenzoate) based MOF (3D Co(II) MOF) and nanosized particles using hydrothermal synthesis or ultrasound irradiation. The 3D Co(II) MOF was compared to its nanostructured counterpart, which showed improved adsorption capabilities [114]. Another MOF, called HS-mSi@MOF-5, was produced by Zhang et al. using a hydrothermal approach. It was discovered that HS-mSi@MOF-5 attained equilibrium for Cd(II) adsorption in about 30 minutes, like MOF-5, with an estimated adsorption capacity of 98 mg/g larger than MOF-5's with 43.6 mg/g [115]. Tahmasebi et al. reported azine-based TMU-5 MOF. The maximum adsorption capacity for this MOF for Cr (III) was found to be 43 mg/g [116]. Further, Rahimi et al. also examined the magnetic Cu-terephthalate MOF manufactured using solvothermal techniques for its adsorption properties. The maximal Cd (II) removal capability of this MOF was reported to be 100 mg/g [117]. The HKUST-1-MW MOF's synthesis process was revealed by Zou et al. through microwave irradiation. Chemical adsorption was used to remove Cd(II), and the framework's capacity was discovered to be 32 mg/g with equilibrium attained within 80 minutes [118].

3) Removal of Chromium

UiO-66-NHC(S)NHMe, an UiO-66-NH_2 framework with S-functionalization, was synthesized by Saleem et al. A set of modified MOFs were framed when UiO-66 was reacted with a number of diphosgene, thiophosgene, or isothiocyanates. When compared to the unmodified MOF, UiO-66-NHC(S)NHMe showed up to a 25-fold increase in metal ion adsorption capability with a value of 49 mg/g [113]. Tahmasebi et al. synthesized three azine and imine functionalized MOFs using the mechanochemical procedure for Cr removal [116]. The application of Fe_3O_4@MIL-100 Fe magnetic microspheres was proven by Yang et al. An in-situ hydrothermal reaction was employed to build the MOF shell from Fe_3O_4 crystal seeds. At a pH of 2, the material's maximal adsorption capacity was found to be 18 mg/g [120].

A bio-derived chitosan-MOF (UiO-66) composite synthesized by Wang et al. was able to adsorb 94 mg/g of Cr(VI) in the form of $Cr_2O_7^{2-}$. The composite was synthesized by grinding and green energy source i.e. microwave irradiation [121]. Another Cr-adsorbing MOF, Cu-BTC, a copper-benzene-tricarboxylate-based MOF, was reported by Maleki et al. It was discovered that Cr(VI) (in the form of $Cr_2O_7^{2-}$) has an adsorption capability of 48 mg/g. It was demonstrated that the highest adsorption took place at a neutral pH [122]. Li et al. synthesized MOFs based on a silver triazolo core by the hydrothermal method. Cr-VI was moderately adsorbed and reached equilibrium after four hours with a maximum capacity of 37 mg/g, and the MOF was discovered to be recyclable [123]. A zirconium-based MOF called ZJU-101 that has the ability to adsorb substances with a capacity of 245 mg/g was designed by Zheng et al. [124] TMU-30, an isonicotinate N-oxide based MOF, was made by Aboutorabi et al. to effectively remove Cr(VI) as CrO_4^{2-} from water. This MOF was particularly efficient at removing Cr(VI) with a capacity of 145 mg/g in a short amount of time [125]. The greener, quick, and secure synthesis technique for MOR-1-HA, an UiO-66-based amino-functionalized MOF, with an alginic acid coating was developed by Rapti et al. The synthesis of MOR-1-HA MOF was carried via the reflux method. Through an ion exchange mechanism, Cr(VI) was adsorbed by substituting $Cr_2O_7^{2-}$ from the solution for Cl^- anions of the MOF. The highest adsorption capacity, which was less than the unmodified MOR-1, was estimated to be 280 mg/g. This led to the conclusion that the MOF's ion exchange capacities were changed by the alginic acid coating [126, 153].

4) **Removal of Lead**

A Zr-based MOF, UiO-66-NHC(S)NHMe, was employed by Saleem et al. as a lead adsorbent (II). For lead, the highest reported adsorption capacity was 232 mg/g [113]. Co(II) MOF was introduced by Abbasi et al. to remove Pb ions from wastewater. The results showed that adsorption values were strongly correlated with a solution pH of 6 [114]. Zhang et al. investigated HSmSi@MOF-5, a silica-coated, thiolated MOF-5 derivative for Pb(II) elimination. This substance had a high capacity for adsorption at 312 mg/g, and it reached equilibrium rather quickly—within half an hour [115]. Tahmasebi et al. reported MOF TMU-5 and demonstrated it to be a successful Pb(II) adsorbent after being synthesized using a mechanochemical approach. The maximum adsorption capacity was 251 mg/g, and equilibrium was established in under 15 minutes [116]. According to Ricco et al., MIL-53 and iron-oxide nanoparticles were combined to create a series of magnetic framework composites called MIL-53. It was found that the amount of Pb(II) that MOF could adsorb increased significantly when there were 50% or more NH_2 moieties present in the structure [127]. Chakraborty et al. presented another novel framework for the elimination of Pb(II) ions from water. An equilibrium was reached in 24 hours for this anionic, Zn(II), and tetracarboxylate-based MOF (AMOF-1) with a maximum Pb ion absorption of 71 mg/g [111]. The Pb ion uptake in Cu-terephthalate MOF synthesized by Rahimi et al. ensured a maximum adsorption capacity of 80 mg/g at an ideal pH of 7 [117].

Zou et al. demonstrated an efficient method for the synthesis of HKUST-1 with functionalization with a polyoxometalate. This particular functionalized MOF had favorable selectivity and uptake for Pb(II) ions with a maximum adsorption of 98 mg/g over a short time of ten minutes to reach equilibrium [118]. Using a lanthanide-based, rod-like MOF called Dy(BTC)(H_2O)(DMF), Jamali et al. found that it could reach equilibrium in under ten minutes and had moderate adsorption values of 5 mg/g of Pb(II). The MOF was also discovered to be reusable, retaining 94–98% of its adsorption capability after five consecutive cycles with an adsorption capacity of 492.4 mg/g [128]. Further, Qin et al. provided MnO_2-MOF, which was highly efficient at adsorbing Pb(II) ions in aqueous solutions. The MOF's metal ion adsorption was characterized by quick equilibration within one hour, and lead(II) ion uptake capacity was measured at 917 mg/g [119]. Rivera et al. also showed that MOF-5 resulted in excellent Pb(II) uptake from wastewater with a maximum capacity of 659 mg/g at 45°C [129].

5) **Removal of Mercury**

Thiol-HKUST-1, a modified HKUST-1 MOF, displayed a profound capacity to bind Hg(II) ions synthesized by Ke et al. The thiol functionalization was carried by treating unfunctionalized HKUST-1 MOF with dithioglycol during the solvothermal process of its synthesis. However, the thiol-modified version of the HKUST-1 MOF displayed exceptionally good adsorption values of 714 mg/g with almost 100% removal capacity with equilibrium attainment within the first 120 minutes. This is interesting because the unmodified HKUST-1 MOF showed no affinity toward Hg(II) ions [130]. By performing covalent postsynthetic alterations on the basis UiO-66-NH_2, Saleem et al. synthesized UiO-66-NHC(S)NHMe, and when Hg(II) adsorption capabilities of this MOF were examined, it was discovered that the modified framework significantly increased metal ion uptake up to 25-fold, with adsorption reaching 99% after 240 minutes with a removal capacity of 769 mg/g [113]. Moreover, another 3D Co(II) MOF was designed by Abbasi et al. The MOF adsorbed 70% of Hg(II) ions within an equilibrium reached in 100 minutes at an optimal pH of 6 [114].

Liang et al. synthesized S-modified MOF, FJI-H12, which displayed an efficient ability to adsorb mercury (II). The highest adsorption capacity was estimated to be 440 mg/g with a visible lowering of the adsorption rate after 60 minutes. At an ideal pH of 7, submerging the MOF in KSCN solution for 24 hours resulted in 86% of the MOF regeneration [131]. Furthermore, LMOF-263, a luminous MOF, was discovered by Rudd et al. Hg(II) quenched the material's photoluminescence in water by 84%, making it a high-selective heavy metal detection MOF. It attained equilibrium in 30 minutes and was estimated to have a maximum adsorption removal capacity of 380 mg/g [132]. In a subsequent work, Sohrabi et al. suggested using a SH@SiO_2/$Cu(BTC)_2$ for Hg adsorption. The HKUST-1 and thiol-modified Si-nanoparticles were used to synthesize the composite. Here, it was determined that the highest adsorption capacity was 210 mg/g, with the equilibrium being attained in 60 minutes at an ideal pH of 6 [133].

MIL-101-Thymine, a postsynthesis modification of MIL101 containing thiol functional groups, was synthesized by Luo et al. The maximal adsorption capacity of MIL-101-Thymine was discovered to be 52 mg/g, with equilibrium attained in 200 minutes at an ideal pH of 6. Both more basic and acidic pH levels were associated with a decrease in adsorption capability. Mercury was discovered to be more preferentially selected by the framework than other heavy metal ions [134]. In 2016, Mon et al. presented a different MOF, referred to as bioMOF, as a stable Hg-adsorbent with an exceptional adsorption capability. Using $HgCl_2$ and CH_3HgCl in water and water-methanol environments, the maximal $HgCl_2$ adsorption capacity was determined to be 900 mg/g [135]. Lastly, Xiong et al. studied Hg(II) adsorption from water using MOF-74-Zn. Solvohydrothermal synthesis was used to generate this framework. The maximum adsorption capacity at pH 6 was

determined to be 63 mg/g, with attainment of equilibrium after 90 minutes [136].

6) **Removal of other metals (Ag, Fe, Al, Zn, Ni, Mn, Co, Cu, and radioactive elements)**

In this section, we have included, in brief, about the application of MOFs synthesized by green methods for the removal of remaining metals such as Ag, Fe, Al, Zn, Ni, Mn, Co, Cu, and radioactive elements.

MIL-53(Al), a MOF having an affinity for Ag(I) ion adsorption, was designed by Cheng et al. [137] The framework was made by adding thiol groups to MIL-53 after it had already been synthesized. Surprisingly, the thiol groups stabilized the silver nanoparticles and adsorbed silver ions within the framework. Within three hours of contact, the maximal adsorption capacity of 183 mg/g was observed.

The 3D Co(II) MOF was synthesized by Abbasi et al. to have the ability to remove Fe(III) and Al(III) from aqueous solutions. The substance was visible to the naked eye for adsorption detection because it changed color when impregnated with metal ions. The MOF absorbed 100% Fe(III) and 90% Al(III), reaching equilibrium in 80 and 100 minutes and having an ideal pH of 6 with different metal ion concentrations of 10–40 ppm. [114]. Rahimi et al. used Cu-terephthalate MOF as a reliable Fe(III), Mn(II), and Zn(II) adsorbent from water. At an ideal pH of 7, the maximal adsorption capacities for Mn(II), Zn(II), and Fe(III) were found to be around 175, 150, and 115 mg/g, respectively, with equilibrium being reached for all three in about 120 minutes [117].

The Zn-based MOF-5 was structured by Bakhtiari et al. as a potent Cu(II) adsorbent. The framework was shown to have a remarkable 290 mg/g adsorption capacity with equilibrium being attained after 30 minutes [138]. With a remarkable adsorption of 189.5 mg/g of Cu(II) ions, Cd-MOF-74 was demonstrated to not only adsorb Cu(II) ions from the water sample, but it was also able to detect them via photoluminescence quenching. With the presence of additional metal ions in solutions, such as Co(II) or Ni(II), the framework was extremely selective for Cu(II) ions when H_2O, MeOH, and DMF were used as the solvents [139].

The selective adsorption capabilities of numerous Zr-based MOFs i.e., ZIF-8@PAM for Cu(II) ions over Ni(II) ions, were revealed by Zheng et al. [140] Cu(II) adsorption was possible using ZIF-8 MOF. The MOF had an exceptional, superior Cu(II) adsorption capacity compared to other metal ions. The adsorption process was quick, as it attained an equilibrium within 20 minutes with a maximum adsorption capacity of 224 mg/g [140].

Further, Ghosh et al. synthesized Sn(II)-based MOFs by the solvothermal synthesis method using benzene-1,3,5-tricarboxylic acid as an organic linker for the adsorption of F^- from wastewater. With an initial concentration of 12 mg/L, the equilibrium F^- concentration in a batch adsorption experiment was obtained at 1.6 mg/L, indicating an 84% removal effectiveness over a pH range of 3 to 10. The maximum F^- adsorption capacity of 30.86 mg/g was determined at an ambient temperature. This work shed light on the environmentally friendly production of Sn(II)-TMA MOF and its potential as an effective adsorbent for F^- removal from aqueous solutions [141].

6.3.2 MOFs for the Removal of Organic Contaminants

Organic contaminants involve a variety of molecules such as dyes, drugs, pesticides, antibiotics, and microplastics [154]. Herein, we will firstly discuss the removal of pharmaceutical contaminants from various wastewater sources. MOFs can be helpful in the removal of these organic contaminants from wastewater. Following are some of the studies done in order to remove pharmaceutical contaminants from wastewater.

For the removal of tetracycline (TC) from aqueous solutions, a group of Zr-MOFs such as UiO-66, NU-1000, and MOF-525 were synthesized by the ultrasonic method by Xia et al. They displayed excellent pore topologies and properties, making them effective MOFs for TC elimination. UiO-66, NU-1000, and MOF-525 each had adsorption capacities of 145 mg/g, 356 mg/g, and 807 mg/g, which were all much greater than those of typical adsorbents [142]. Jana et al. employed MOF MIL-53(Fe–Cu) by single-step solvothermal synthesis for the treatment of pharmaceutical waste that contains different compounds by using a combined methodology of adsorption and photocatalytic degradation. The MOF prepared had a maximum adsorption capacity of 190 mg/g for ciprofloxacin and a maximum photodegradation of 74.48% and 57.88% in the presence of UV and visible radiations, respectively. This experiment showed that MIL-53 (Fe–Cu) is helpful and can be used in the treatment of wastewater containing pharmaceuticals in good yield [143]. By incorporating nanoscale MOFs (Cu-TMA) into vanadium pentoxide that is embedded in a chitosan matrix, a nano-biosorbent V_2O_5@Ch/Cu-TMA was synthesized by the encapsulation method for the effective removal of levofloxacin (LEVO) and Cr (VI) from water by Mahmoud et al. While LEVO was shown to be spontaneous and to advance by exothermic processes, adsorption of Cr(VI) was endothermically driven and spontaneous. With a percentage range between 92.43% and 96.95% for the removal of inorganic contaminant Cr(VI) from all natural water samples and 91.99% to 97.20% for the removal of the organic contaminant LEVO from tap and wastewater, the MOF V_2O_5@Ch/Cu-TMA proved to be an exceptional nano-biosorbent [144].

Secondly, in this section of the chapter, we will briefly highlight the application of MOFs for the removal of dyes from wastewater [155]. An iron-based MOF, MIL-53 (Fe), was synthesized by Lee et al. It was made using a sonochemical process and the effectiveness of its adsorption for organic dye viz methyl orange (MO), methylene blue (MB), and bisphenol A (BPA) was evaluated. The sonochemical synthesis of MIL- 53(Fe) particles was quicker than solvothermal synthesis. An amazing maximal adsorption capacity for MO, MB, and BPA elimination from aqueous solutions was demonstrated by MIL-53(Fe)_UTS-2 with an absorption capacity of 494.2, 221.9 and 160.1 mg/g, respectively [145]. A novel eco-friendly and water-stable Zn-based MOF, $Zn_2[(S,S)\text{-serimox}](H_2O)_2$ where (*S,S*)-serimox is bis[(*S*)-serine] oxalyl diamide, was synthesized using a derivative of the amino acid L-serine. It was heavily embellished with hydroxylated L-serine residues, which are extremely flexible. The synthesized MOF photodegraded the organic dye, brilliant green (BG), within 120 minutes. This MOF provided highly effective wastewater treatment by trapping more than 90% of the dye, even at very low quantities like 10 ppm, which are comparable to those typically seen in industrial wastewaters [146].

Furthermore, a Zn-based L-aspartic acid bioMOF (Zn L-Asp bioMOF) was synthesized by Salama et al. It effectively adsorbed Direct Red 81 (DR-81), which is an anionic organic dye. The synthesized bioMOF had a large surface area, chemical inertness, selectivity, and thermal and mechanical stability according to physicochemical characterization. The adsorption studies demonstrated that Zn L-Asp bioMOF is a potent adsorbent for DR-81 from aqueous solutions, showing 27.14 mg/g with 95.3% removal, which is recyclable for up to eight cycles as well as for simple sorbent regeneration [147].

It is well known that the production of colors, polymers, insecticides, explosives, pharmaceuticals, and lubricating oil all make extensive use of nitrobenzene, which is one of the prominent contaminants in wastewater from industries. In this regard, two aluminum-based MOFs, named CAU-1 and MIL-68 (Al), were synthesized by Xie et al. using the hydrothermal method and employed them for the first time to capture nitrobenzene from wastewater. Their respective adsorption capabilities were 970 mg/g and 1130 mg/g. Methanol might also be used to completely regenerate CAU-1 and MIL-68 (Al) without causing secondary pollutants. The findings show that MOFs are effective adsorbents for capturing nitrobenzene from wastewater [150].

Malachite green is another dye frequently used in industries. A Zn-based MOF was made by Khalil et al. It was made via a one-pot synthesis at ambient temperature, which is considered among one of the 12 principles of green chemistry. When it was used to remove the examined wastewater samples containing the dye malachite green, Zn-MOF nano-cubes demonstrated a remarkable adsorption capability of 953.14 mg/g. An efficient approach for the management of industrial wastewater and water filtration was demonstrated by the microporous cubic Zn MOF as an adsorbent [148]. Shi et al. framed Cu MOFs/Fe_3O_4 based MOF as an adsorbent for the removal of Pb(II) and malachite green from wastewater. By developing Cu-MOFs in-situ with doping Fe_3O_4 nanoparticles, Cu-MOFs/Fe_3O_4 composite was synthesized quickly by the solvothermal method. The results of the adsorption studies showed that Pb(II) and MG may be removed simultaneously using Cu-MOFs/Fe_3O_4 as an adsorbent. The adsorption capabilities were quite high and were determined as 113.67 mg/g for MG and 219.00 mg/g for Pb^{2+}. The Fe_3O_4/Cu-MOFs, which were investigated as viable adsorbents for wastewater treatment, were discovered to be recyclable for the removal of both inorganic ion Pb(II) and organic dye, MG [149].

After discussing numerous green applications of MOFs in the removal of organic contaminants such as pharmaceuticals and organic dyes, let us now shed light on how MOFs are capable of flushing out microplastics present in wastewater. Microplastics are recently discovered pollutants that are widespread, nowadays. They move from one area to another via various mediums and are little in size. One of the most significant pathways for the movement of microplastics is through water. Microplastic particles are bad for the ecosystem as a whole, as well as for humans, animals, and plants. Microplastic traces have been found in wastewater, which is a public health problem. In order to guarantee the removal of microplastics, various MOFs have been employed. Because of their great durability, proper framework, and adequate porosity, they may effectively trap pollutants. The use of MOFs in the removal of microplastics from wastewater is discussed briefly in the following review of studies.

Chen et al. loaded Zr-MOFs on melamine foam, which was synthesized using 1,4-dicarboxybenzene ligand with various functional groups. Zr-based UiO-66-X where X=H, NH_2, OH, Br, and NO_2 MOFs were synthesized using the solvothermal method. These frameworks were then utilized to test the removal of microplastics in a solution containing microplastics of polymethyl methacrylate (PMMA), polyvinylidiene fluoride (PVDF), and polystyrene (PS). The particle size and zeta potential of different microplastics influenced removal performance. This demonstrates that removal could be accomplished using the synthesized MOF, which has a high recyclability. Among various composites, UiO-66-OH@MF-3 displayed a high elimination efficiency of 90.1 ± 2.1% for PVDF, 88.2 ± 1.7% for PMMA, and 85.7 ± 4.8% for PS [151]. You et al. generated a long-lasting zinc MOF-based ZIF-8@Aerogel material by developing ZIF-8 in situ on wood aerogel fibers using the solvothermal method, which successfully removed microplastics from the sample water and seawater. Removal rates on micro/nano polymers such as poly(1,1-difluoroethylene), PDFE, and polystyrene were 91.4% and 85.8%, respectively. This study demonstrated a novel and efficient method for removing small-sized microplastic particles from the environment using MOFs synthesized from biomass linkers [152].

There is still a great deal of interest in studying MOF materials for organic and inorganic contaminant adsorption and photodegradation. As a result, the research area of photocatalysis is gradually becoming the focus of research, particularly in the field of wastewater treatment using MOFs. Although several MOFs have been demonstrated to be effective for the removal of contaminants, there is still a dearth of MOF materials that are quite stable as well as complying with the guidelines of green chemistry for a sustainable environment. In conclusion, MOFs are still being used in the field of green applications, but more work needs to be done to enhance its inherent properties and streamline the reaction pathway for a more straightforward and effective synthesis.

6.4 Conclusion and Future Prospects

In this chapter, we have highlighted the application of various MOFs synthesized by amalgam of green chemistry and sustainable technology for wastewater treatment. MOFs are widely used for the elimination of toxic contaminants from wastewater because they possess outstanding properties such as high porosity, large specific surface area, crystalline structure, abundant adsorption sites, flexible frameworks, and stability. It is best to utilize the lowest quantity of solvent possible while synthesizing, activating, or purifying the MOF. Other organic solvents may be used in green syntheses that are stable under synthesis conditions and sufficiently recyclable as considered under the 12 principles of green chemistry. Based on the potential hazards the counter anion may cause, the metal source should be chosen. Moreover, in order to yield MOFs on a large scale for commercial use, it is important to develop straightforward and efficient production techniques. A lot of work has been put into developing novel methodologies for quick and easy synthesis of MOFs during the past few decades. These include microwave synthesis and electrochemical, sonochemical, and mechanochemical methods, which are discussed in this chapter. These methods not only decrease the amount of energy required during the production process, they simultaneously accelerate the formation and easy purification of synthesized MOFs. This is because green approaches are typically evaluated, and the synthesis process must first and foremost be designed to give the desired product. Recently, the focus of MOF research has shifted to the identification of synthesis conditions and methodologies that employ non-hazardous reactants at relatively milder reaction conditions while generating less toxic waste than traditional methods for sustainability.

6.5 Conflict of Interest

The authors declare no conflict of interest.

References

1 Kim, K., Culp, J., Ohodnicki, P. et al. (2021). Synthesis of high-quality Mg-MOF-74 thin film via vapor assisted crystallization. *Applied Materials & Interfaces* 13 (29): 35223–35231.

2 Wang, R., Xu, H., Zhang, K. et al. (2019). High-quality Al@Fe-MOF prepared using Fe-MOF as a micro-reactor to improve adsorption performance for selenite. *Journal of Hazardous Materials* 364: 272–280.

3 Chen, L., Zhang, X., Cheng, X. et al. (2020). The function of metal-organic frameworks in the application of MOF-based composites. *Nanoscale Advances* 2: 2628–2647.

4 Gelfand, B. and Shimizu, G. (2016). Parameterizing and grading hydrolytic stability in metal organic frameworks. *Dalton Transactions* 45: 3668–3678.

5 Wang, G., Zhang, Z., Wang, J. et al. (2015). Study of the influence of pore width on the disposal of benzene employing tunable OMCs. *Industrial & Engineering Chemistry Research* 54: 1074–1080.

6 Hu, G., Zhang, Z., Zhang, X., and Li, T. (2021). Size and shape effects of MnFe2O4 nanoparticles as catalysts for reductive degradation of dye pollutants. *Frontiers of Environmental Science & Engineering* 15 (108).

7 Hu, T., Jia, Q., He, S. et al. (2017). Novel functionalized metal-organic framework MIL-101 adsorbent for capturing ox tetracycline. *Journal of Alloys and Compounds* 727: 114–122.

8 Zhao, Z., Ma, X., Li, Z., and Lin, Y. (2011). Synthesis, characterization and gas transport properties of MOF-5 membranes. *Journal of Membrane Science* 382 (1–2): 82–90.

9 Bian, Y., Xiong, N., and Zhu, G. (2018). Technology for the remediation of water pollution: a review on the fabrication of metal-organic frameworks. *Processes* 206 (122).

10 Yaghi, O., Li, G., and Li, H. (1995). Selective binding and removal of guests in a microporous metal-organic framework. *Nature* 378: 703–706.

11 Hasan, Z., Jeon, J., and Jhung, S. (2012). Adsorptive removal of naproxen and clofibric acid from water using metal-organic frameworks. *The Journal of Hazardous Materials* 209–210: 151–157.

12 Seo, P., Khan, N., Hasan, Z., and Jhung, S. (2016). Adsorptive removal of artificial sweeteners from water using metal-organic frameworks functionalized with urea or melamine. *ACS Applied Materials & Interfaces* 8: 29799–29807.

13 Wu, Z., Yuan, X., Zhong, H. et al. (2016). Enhanced adsorptive removal of p-nitrophenol from water by aluminum metal-organic framework/reduced graphene oxide composite. *Scientific Reports* 6: 25638.

14 Akpinar, I. and Yazaydin, A. (2017). Rapid and efficient removal of carbamazepine from water by UiO-67. *Industrial & Engineering Chemistry Research* 56: 15122–15130.

15 Martinez, J., Leyva, R., and Padilla, E. (2018). Sorption of diclofenac from aqueous solution on an organobentonite and adsorption of cadmium on organobentonite saturated with diclofenac. *Clays and Clay Minerals* 66: 515–528.

16 Daliran, S., Ghazagh-Miri, M., Oveisi, A. et al. (2020). A pyridyltriazol functionalized zirconium metal-organic framework for selective and highly efficient adsorption of palladium. *ACS Applied Materials & Interfaces* 12: 25221–25232.

17 Hu, C., Hu, X., Li, R., and Xing, Y. (2020). MOF derived ZnO/C nanocomposite with enhanced adsorption capacity and photocatalytic performance under sunlight. *The Journal of Hazardous Materials* 385: 121599.

18 Trickett, C., Helal, A., Al-Maythalony, B. et al. (2017). The chemistry of metal-organic frameworks for CO_2 capture, regeneration and conversion. *Nature Reviews Materials* 2: 17045.

19 Kreno, L., Leong, K., Farha, O. et al. (2012). Metal-organic framework materials as chemical sensors. *Chemical Reviews* 112: 1105–1125.

20 Palakollu, V., Chen, D., Tang, J.N. et al. (2022). Recent advancements in metal-organic frameworks composites based electrochemical (bio)sensors. *Microchim Acta* 189 (161).

21 Yaghi, O., Kalmutzki, M., and Diercks, C. (2019). Liquid- and gas-phase separation in MOFs. *Introduction to Reticular Chemistry* (1) 365–393.

22 Dhakshinamoorthy, A., Asiri, A., and Garcia, H. (2019). 2D metal-organic frameworks as multifunctional materials in heterogeneous catalysis and electro/photocatalysis. *Advanced Materials* 31: 1900617.

23 Yang, D. and Gates, B. (2019). Catalysis by metal-organic frameworks: perspective and suggestions for future research. *ACS Catal* 9: 1779–1798.

24 Wang, Z., Huang, J., Mao, J. et al. (2020). Metal-organic frameworks and their derivatives with graphene composites: preparation and applications in electrocatalysis and photocatalysis. *Journal of Materials Chemistry* A8: 2934–2961.

25 Heo, D., Do, H., Ahn, S., and Kim, S. (2020). Metal-organic framework materials for perovskite solar cells. *Polymers* 12 (9): 2061.

26 Rojas, S., Arenas-Vivo, A., and Horcajada, P. (2019). Metal-organic frameworks: a novel platform for combined advanced therapies. *Coordination Chemistry Reviews* 388: 202–226.

27 Abánades, I. and Forgan, R. (2019). Application of zirconium MOFs in drug delivery and biomedicine. *Coordination Chemistry Reviews* 380: 230–259.

28 Lawson, H., Walton, S., and Chan, C. (2021). Metal-organic frameworks for drug delivery: a design perspective. *ACS Applied Materials & Interfaces* 13 (6): 7004–7020.

29 Park, K., Ni, Z., Cote, A. et al. (2006). Exceptional chemical and thermal stability of zeolitic imidazolate frameworks. *Proceedings of the National Academy of Sciences* 103: 10186–10191.

30 Seetharaj, R., Vandana, P., Arya, P., and Mathew, S. (2019). Dependence of solvents, pH, molar ratio and temperature in tuning metal-organic framework architecture. *The Arabian Journal of Chemistry* 12: 295–315.

31 Belyakov, S., Krukle-Berzina, K., Mishnev, A., and Shubin, K. (2020). Stability and phase transitions of nontoxic c-cyclodextrin-K+ metal-organic framework in various solvents. *Crystals* 10 (1): 37.

32 Li, H., Eddaoudi, M., O'Keeffe, M., and Yaghi, O. (1999). Design and synthesis of an exceptionally stable and highly porous metal-organic framework. *Nature* 402: 276–279.

33 Butova, V., Soldatov, M., Guda, A. et al. (2016). Metal- organic frameworks: structure, properties, methods of synthesis and characterization. *Russian Chemical Reviews* 85: 280–307.

34 Furukawa, H., Cordova, K., O'Keeffe, M., and Yaghi, O. (2013). The chemistry and applications of metal-organic frameworks. *Science* 341: 1230444–1230444.

35 Lillerud, K., Olsbye, U., and Tilset, M. (2010). Designing heterogeneous catalysts by incorporating enzyme-like functionalities into MOFs. *Topics in Catalysis* 53: 859–868.

36 Stock, N. and Biswas, S. (2012). Synthesis of metal-organic frameworks (MOFs): routes to various MOF topologies, morphologies, and composites. *Chemical Reviews* 112: 933–969.

37 Jhung, S., Lee, J., and Chang, J. (2005). Microwave synthesis of a nanoporous hybrid material, chromium trimesate, bulletin of the korean chemical society. 26: 880–881.

38 Férey, G., Serre, C., Draznieks, C. et al. (2004). Solid with giant pores prepared by a combination of targeted chemistry, simulation, and powder diffraction. *Angewandte Chemie International Edition* 43 (46): 6296–6301.

39 Mueller, U., Schubert, M., Teich, F. et al. (2006). Metal- organic frameworks—prospective industrial applications. *Journal of Materials Chemistry* 16 (7): 626–636.

40 James, S., Adams, C., Bolm, C. et al. (2012). Mechanochemistry: opportunities for new and cleaner synthesis. *Chemical Society Reviews Chemical Society Reviews* 41 (1): 413–447.

41 Pichon, A., Garay, A., and James, S. (2006). Solvent-free synthesis of a microporous metal-organic framework. *Crystal. Engineering. Communication.* 8 (3): 211.

42 Do, J. and Frišcic, T. (2017). Mechanochemistry: a force of synthesis. *ACS Central Science* 3: 13–19.

43 Tan, D. and García, F. (2019). Main group mechanochemistry: from curiosity to established protocols. *Chemical Society Reviews* 48: 2274–2292.

44 Yuan, W., Garay, A., Pichon, A. et al. (2010). Study of the mechanochemical formation and resulting properties of an archetypal MOF: cu3(BTC)2 (BTC = 1,3,5-benzenetricarboxylate). *Crystal. Engineering. Communication* 12 (12): 4063.

45 Lv, D., Chen, Y., Li, Y. et al. (2017). Efficient mechanochemical synthesis of MOF-5 for linear alkanes adsorption. *Journal of Chemical & Engineering Data* 62: 2030–2036.

46 Bigdeli, F., Ghasempour, H., Tehrani, A. et al. (2017). Ultrasound-assisted synthesis of nano-structured Zinc(II)-based metal- organic frameworks as precursors for the synthesis of ZnO nano-structures. *Ultrasonics Sonochemistry* 37: 29–36.

47 Abuzalat, O., Wong, D., Elsayed, M. et al. (2018). Sonochemical fabrication of Cu(II) and Zn(II) metal-organic framework films on metal substrates. *Ultrasonics Sonochemistry* 45: 180–188.

48 Mojtabazade, F., Mirtamizdoust, B., Morsali, A., and Talemi, P. (2018). Ultrasonic-assisted synthesis and the structural characterization of novel the zig-zag Cd(II) metal-organic polymer and their nanostructures. *Ultrasonics Sonochemistry* 42: 134–140.

49 Qiu, L., Li, Z., Wu, Y. et al. (2008). Facile synthesis of nanocrystals of a microporous metal-organic framework by an ultrasonic method and selective sensing of organoamines. *Chemical Communications* 31: 3642.

50 Schlesinger, M., Schulze, S., Hietschold, M., and Mehring, M. (2010). Evaluation of synthetic methods for microporous metal-organic frameworks exemplified by the competitive formation of [Cu2(btc)3(H_2O)3] and [Cu2(btc)(OH)(H_2O)]. *Microporous and Mesoporous Materials* 132 (1–2): 121–127.

51 Julien, P., Mottillo, C., and Frišcic′, T. (2017). Metal-organic frameworks meet scalable and sustainable synthesis. *Green Chemistry* 19 (12): 2729–2747.

52 Kumar, S., Jain, S., Nehra, M. et al. (2020). Green synthesis of metal-organic frameworks: a state-of-the-art review of potential environmental and medical applications. *Coordination Chemistry Reviews* 420: 213407.

53 Reinsch, H., Vermoortele, F., Stassen, I. et al. (2015). Green synthesis of zirconium-MOFs. *CrystEnggComm* 17: 4070–4074.

54 Tao, Y., Huang, G., Li, H., and Hill, M. (2019). Magnetic metal-organic framework composites: solvent-free synthesis and regeneration driven by localized magnetic induction heat. *ACS Sustainable Chemistry & Engineering* 7: 13627–13632.

55 Lomba, L., Zuriaga, E., and Giner, B. (2019). Solvents derived from biomass and their potential as green solvents. *Current Opinion in Green and Sustainable Chemistry* 18: 51–56.

56 Tobiszewski, M., Namies′nik, J., and Pena-Pereira, F. (2019). Environmental risk-based ranking of solvents using the combination of a multimedia model and multicriteria decision analysis. *Green Chemistry* 19: 1034–1042.

57 Camargo, A., Bueno, M., Alfonso, F. et al. (2019). Hansen. Solubility parameters for selection of green extraction solvents. *Trends in Analytical Chemistry* 118: 227–237.

58 Prat, D., Wells, A., Hayler, J. et al. (2015). CHEM21 selection guide of classical- and less classical-solvents. *Green Chemistry* 18: 288–296.

59 Alde, C.C., Hayler, J., Henderson, R. et al. (2016). Updating and further expanding GSK's solvent sustainability guide. *Green Green Chemistry* 18: 3879–3890.

60 Byrne, F., Jin, S., Paggiola, G. et al. (2016). Sherwood J.Tools and techniques for solvent selection: green solvent selection guides. *Sustainable Chemical Processes* 4 (7).

61 Chen, J., Shen, K., and Li, Y. (2017). Greening the processes of metal-organic framework synthesis and their use in sustainable catalysis. *ChemSusChem* 10: 3165–3187.

62 Cadot, S., Veyre, L., Luneau, D. et al. (2014). A waterbased and high space-time yield synthetic route to MOF Ni2(dhtp) and its linker 2,5-dihydroxyterephthalic acid. *Journal of Materials Chemistry A* 2: 17757–17763.

63 Guan, H., LeBlanc, R., Xie, S., and Yue, Y. (2018). Recent progress in the syntheses of mesoporous metal-organic framework materials. *Coordination Chemistry Reviews* 369: 76–90.

64 Shehada, S., Clark, J., Paggiola, G., and Sherwood, J. (2016). Tunable solvents: shades of green. *Chemical Engineering and Processing: Process Intensification* 99: 88–96.

65 Capuzzo, A., Maffei, M., and Occhipinti, A. (2013). Supercritical fluid extraction of plant flavors and fragrances. *Molecules* 18: 7194–7238.

66 Dašic′, M., Stankovic′, I., and Gkagkas, K. (2019). Molecular dynamics investigation of the influence of the shape of the cation on the structure and lubrication properties of ionic liquids. *Physical Chemistry Chemical Physics* 21: 4375–4386.

67 Hanamertani, A., Pilus, R., Idris, A. et al. (2018). Ionic liquids as a potential additive for reducing surfactant adsorption onto crushed Berea sandstone. *Journal of Petroleum Science and Engineering* 480–490 (162).

68 Figoli, A., Marino, T., Simone, S. et al. (2014). Towards non-toxic solvents for membrane preparation: a review. *Green Chemistry* 16: 4034–4059.

69 Isosaari, P., Srivastava, V., and Sillanpää, M. (2019). Ionic liquid-based water treatment technologies for organic pollutants: current status and future prospects of ionic liquid mediated technologies. *Science of the Total Environment.* 690: 604–619.

70 Comba, M., Tsai, Y., Sarotti, A. et al. (2018). Levoglucosenone and its new applications: valorization of cellulose residues. *European Journal of Organic Chemistry* 2018: 590–604.

71 Huang, X., Kudo, S., and Hayashi, J. (2019). Two-step conversion of cellulose to levoglucosenone using updraft fixed bed pyrolyzer and catalytic reformer. *Fuel Processing Technology* 191: 29–35.

72 Bousfield, T., Pearce, K., Nyamini, S. et al. (2019). Synthesis of amides from acid chlorides and amines in the bio-based solvent CyreneTM. *Green Chemistry* 21: 3675–3681.

73 DeSantis, D., Mason, J., James, B. et al. (2017). Techno-economic analysis of metal-organic frameworks for hydrogen and natural gas storage. *Energy Fuels* 31: 2024–2032.

74 An, H., Li, M., Gao, J. et al. (2019). Incorporation of biomolecules in metal-organic frameworks for advanced application. *Coordination Chemistry Reviews* 384: 90–106.

75 Li, N., Liu, J., Liu, J. et al. (2019). Adenine components in biomimetic metal-organic frameworks for efficient CO2 photoconversion. *Angewandte Chemie* 131: 5280–5285.

76 Reinsch, H., Bueken, B., Vermoortele, F. et al. (2015). Green synthesis of zirconium-MOFs. *CrystEngComm* 17:4070–.

77 Lenzen, D., Bendix, P., Reinsch, H. et al. (2018). Scalable green synthesis and fullscale test of the metal-organic framework CAU-10-H for use in adsorption driven chillers. *Advanced Materials* 30.

78 Reinsch, H. (2016). Green synthesis of metal-organic frameworks. *European Journal of Inorganic Chemistry* 2016: 4290–4299.

79 Jiang, J., Gándara, F., Zhang, Y. et al. (2014). Superacidity in sulfated metal-organic framework-808. *Journal of the American Chemical Society* 136: 12844–12847.

80 Camur, C., Carvajal, J., Imaz, I., and Maspoch, D. (2018). Metal acetylacetonates as a source of metals for aqueous synthesis of metal-organic frameworks. *ACS Sustainable Chemistry & Engineering* 6: 14554–14560.

81 Wyman, C., Cai, C., and Kumar, R. (2019). *Bioethanol from Lignocellulosic Biomass, In: Energy from Org. Mater.* New York: Springer. 997–1022.

82 Yan, N. and Chen, X. (2015). Sustainability: don't waste seafood waste. *Nature* 524 (2015): 155–157.

83 Fermoso, F., Serrano, A., Fariñas, B. et al. (2018). Valuable compound extraction, anaerobic digestion, and composting: a leading biorefinery approach for agricultural wastes. *Journal of Agricultural and Food Chemistry* 66: 8451–8468.

84 Wang, Q., Fu, M., Li, X. et al. (2019). Aluminum alkoxycatalyzed biomass conversion of glucose to 5-hydroxymethylfurfural: mechanistic study of the cooperative bifunctional catalysis. *Journal of Computational Chemistry* 40: 1599–1608.

85 Pileidis, F. and Titirici, M. (2016). Levulinic acid biorefineries: new challenges for efficient utilization of biomass. *ChemSusChem* 9: 562–582.

86 Morthensen, S., Zeuner, B., Meyer, A. et al. (2018). Membrane separation of enzyme-converted biomass compounds: recovery of xylose and production of gluconic acid as a value-added product. *Separation and Purification Technology* 194: 73–80.

87 Wang, Y., Cao, W., Luo, J. et al. (2019). One step open fermentation for lactic acid production from inedible starchy biomass by thermophilic Bacillus coagulans IPE2. *Bioresource Technology* 272: 398–406.

88 Huang, Y., Lai, Y., Lin, C., and Wang, S. (2011). Direct use of waste PET as unfailing source of organic reagents in the synthesis of intrinsic white/yellow luminescent nanoporous zincophosphates. *Green Chemistry* 13: 2000–2003.

89 Ren, J., Dyosiba, X., Musyoka, N. et al. (2016). Green synthesis of chromium-based metal-organic framework (CrMOF) from waste polyethylene terephthalate (PET) bottles for hydrogen storage applications, Int. *International Journal of Hydrogen Energy* 41: 8141–18146.

90 Deleu, W., Stassen, I., Jonckheere, D. et al. (2016). Waste PET (bottles) as a resource or substrate for MOF synthesis. *Journal of Materials Chemistry A* 4: 9519–9525.

91 Lo, S., Raja, D., Chen, C. et al. (2016). Waste polyethylene terephthalate (PET) materials as sustainable precursors for the synthesis of nanoporous MOFs, MIL-47, MIL-53(Cr, Al, Ga) and MIL-101(Cr). *Dalton Transactions* 45: 9565–9573.

92 Westall, F. and Brack, A. (2018). The importance of water for life. *Space Science Reviews* 214 (50).

93 Madhav, S., Ahamad, A., Singh, A. et al. (2019). Water pollutants: sources and impact on the environment and human health.

94 Dias, E. and Petit, C. (2015). Towards the use of metal-organic frameworks for water reuse: a review of the recent advances in the field of organic pollutants removal and degradation and the next steps in the field. *Journal of Materials Chemistry A* 3: 22484–22506.

95 Cheremisinoff, N. (2007). *Handbook of Water and Wastewater Treatment Technologies*. Waltham: Butterworth-Heinemann.

96 Fijalkowski, K., Rorat, A., Grobelak, A., and Kacprzak, M. (2017). The presence of contaminations in sewage sludge – the current situation. *Journal of Environmental Management* 203 (3): 1126–1136.

97 Zhou, Q., Yang, N., Li, Y. et al. (2020). Total concentrations and sources of heavy metal pollution in global river and lake water bodies from 1972 to 2017. *Global Ecology and Conservation* 22.

98 Sankaran, R., Show, P., Ooi, C. et al. (2020). Feasibility assessment of removal of heavy metals and soluble microbial products from aqueous solutions using eggshell wastes. *Clean Technologies and Environmental Policy* 22.

99 Hariganesh, S., Vadivel, S., Maruthamani, D. et al. (2019). Facile large scale synthesis of CuCr 2 O 4 /CuO nanocomposite using MOF route for photocatalytic degradation of methylene blue and tetracycline under visible light. *Applied Organometallic Chemistry* 34 (2).

100 Wang, C., Wang, X., and Liu, W. (2020). The synthesis strategies and photocatalytic performances of TiO2/MOFs composites: a state-of-the-art review. *The Chemical Engineering Journal* 391: 123601.

101 Alvaro, M., Carbonell, E., Ferrer, B. et al. (2007). Semiconductor behavior of a metal-organic framework (MOF). *Chemistry* 13: 5106–5112.

102 Wang, C., Liu, X., Chen, J., and Li, K. (2015). Superior removal of arsenic from water with zirconium metal-organic framework UiO-66. *Scientific Reports* 5: 16613.

103 Zhu, B., Yu, X., Jia, Y. et al. (2012). Iron and 1,3,5-benzenetricarboxylic metal-organic coordination polymers prepared by solvothermal method and their application in efficient As(V) removal from aqueous solutions. *The Journal of Physical Chemistry C* 116 (15): 8601–8607.

104 Jian, M., Liu, B., Zhang, G. et al. (2014). Comparing adsorption of arsenic and antimony from single-solute and bi-solute aqueous systems onto ZIF-8. *Colloids and Surfaces A: Physicochemical and Engineering Aspects* 465: 67–76.

105 Jian, M., Liu, R., Yao, J., and Zhang, X. (2015). Adsorptive removal of arsenic from aqueous solution by zeolitic imidazolate framework-8 (ZIF-8) nanoparticles. *Colloids and Surfaces A: Physicochemical and Engineering Aspects* Colloid Surf. A Physicochem. Eng. Asp 465: 67–76.

106 Vu, T., Le, G., Dao, C. et al. (2015). Arsenic removal from aqueous solutions by adsorption using novel MIL-53(Fe) as a highly efficient adsorbent. *RSC Advances* 5 (7): 5261–5268.

107 Li, J., Wu, Y., Li, Z. et al. (2014). Characteristics of arsenate removal from water by metal-organic frameworks (MOFs). *Water Science and Technology* 70 (8): 1391–1397.

108 Li, Z., Yang, J., Sui, K., and Yin, N. (2015). Facile synthesis of metal-organic framework MOF-808 for arsenic removal. *Materials Letters* 160: 412–414.

109 Audu, C., Nguyen, H., Chang, C. et al. (2016). The dual capture of AsV and AsIII by UiO-66 and analogues. *Chemical Science* 7 (10): 6492–6498.

110 Wang, Y., Ye, G., Chen, H. et al. (2015). Functionalized metal-organic framework as a new platform for efficient and selective removal of cadmium(ii) from aqueous solution. *Journal of Materials Chemistry A* 3: 15292–15298.

111 Chakraborty, A., Bhattacharyya, S., Hazra, A. et al. (2016). Post-synthetic metalation in an anionic MOF for efficient catalytic activity and removal of heavy metal ions from aqueous solution. *Chemical Communications* 52: 2831–2834.

112 Fang, Q., Yuan, D., Sculley, J. et al. (2010). Functional mesoporous metal–organic frameworks for the capture of heavy metal ions and size-selective catalysis. *Inorganic Chemistry* 49: 11637–11642.

113 Saleem, H., Rafique, U., and Davies, R. (2016). Investigations on post-synthetically modified UiO-66-NH_2 for the adsorptive removal of heavy metal ions from aqueous solution. *Microporous and Mesoporous Materials* 221: 238–244.

114 Abbasi, A., Moradpour, T., and Hecke, K. (2015). A new 3D cobalt (II) metal-organic framework nanostructure for heavy metal adsorption. *Inorganica Chimica Acta* 430: 261–267.

115 Zhang, J., Xiong, Z., Li, C., and Wu, C. (2016). Exploring a thiol-functionalized MOF for elimination of lead and cadmium from aqueous solution. *The Journal of Molecular Liquids* 221: 43–50.

116 Tahmasebi, E., Masoomi, M., Yamini, Y., and Morsali, A. (2014). Application of mechanosynthesized azine-decorated zinc(II) metal-organic frameworks for highly efficient removal and extraction of some heavy-metal ions from aqueous samples: a comparative study. *Inorganic Chemistry* 54: 425–433.

117 Rahimi, E. and Mohaghegh, N. (2015). Removal of toxic metal ions from sungun acid rock drainage using mordenite zeolite, graphene nanosheets, and a novel metal organic framework. *Mine Water and the Environment* 35: 18–28.

118 Zou, F., Yu, R., Li, R., and Li, W. (2013). Microwave-assisted synthesis of HKUST-1 and functionalized HKUST-1-@$H_3PW_{12}O_{40}$: selective adsorption of heavy metal ions in water analyzed with synchrotron radiation. *Chemphyschem.* 14: 2825–2832.

119 Qin, Q., Wang, Q., Fu, D., and Ma, J. (2011). An efficient approach for Pb(II) and Cd(II) removal using manganese dioxide formed in-situ. *The Chemical Engineering Journal* 172: 68–74.

120 Yang, Q., Zhao, Q., Ren, S. et al. (2016). Fabrication of core-shell Fe_3O_4@MIL-100(Fe) magnetic microspheres for the removal of Cr(VI) in aqueous solution. *Journal of Solid State Chemistry* 244: 25–30.

121 Wang, K., Tao, X., Xu, J., and Novel Chitosan MOF, Y.N. (2016). Composite Adsorbent for the Removal of Heavy Metal Ions. *Chemistry Letters* 45 (2016): 1365–1368.

122 Maleki, A., Hayati, B., Naghizadeh, M., and Joo, S. (2015). Adsorption of hexavalent chromium by metal organic frameworks from aqueous solution. *Journal of Industrial and Engineering Chemistry* 28: 211–216.

123 Li, L., Feng, X., Han, R. et al. (2017). Cr(VI) removal via anion exchange on a silver-triazolate MOF. *The Journal of Hazardous Materials* 321: 622–628.

124 Zhang, Q., Yu, J., Cai, J. et al. (2015). A porous Zr-cluster-based cationic metal organic framework for highly efficient $Cr_2O_7^{2-}$ removal from water. *Chemical. Communication* 51: 14732–14734.

125 Aboutorabi, L., Morsali, A., Tahmasebi, E., and Buyukgungor, O. (2016). Metal organic framework based on isonicotinate N-oxide for fast and highly efficient aqueous phase Cr(VI) adsorption. *Inorganic Chemistry* 55: 5507–5513.

126 Rapti, S., Pournara, A., Sarma, D. et al. (2016). Rapid, green and inexpensive synthesis of high quality UiO-66 amino-functionalized materials with exceptional capability for removal of hexavalent chromium from industrial waste. *Inorganic Chemistry Frontiers* 3: 635–644.

127 Ricco, R., Konstas, K., Styles, M. et al. (2015). Lead(ii) uptake by aluminium based magnetic framework composites (MFCs) in water. *Journal of Materials Chemistry A* 3: 19822–19831.

128 Jamali, A., Tehrani, A., Shemirani, F., and Morsali, A. (2016). Lanthanide metal organic frameworks as selective microporous materials for adsorption of heavy metal ions. *Dalton Transactions* 45 (2016): 9193–9200.

129 Rivera, J., Rincon, S., Youssef, C., and Zepeda, A. (2016). Highly efficient adsorption of aqueous Pb(II) with mesoporous metal-organic framework-5: an equilibrium and kinetic study. *Journal of Nanomaterials* 9.

130 Ke, F., Qiu, L., Yuan, Y. et al. (2011). Thiol-functionalization of metal-organic framework by a facile coordination-based postsynthetic strategy and enhanced removal of Hg^{2+} from water. *The Journal of Hazardous Materials* 196: 36–43.

131 Liang, L., Chen, Q., Jiang, F. et al. (2016). In situ large-scale construction of sulfur-functionalized metal organic framework and its efficient removal of Hg(II) from water. *Journal of Materials Chemistry A* 4: 15370–15374.

132 Rudd, N., Wang, H., Fuentes-Fernandez, E. et al. (2016). Highly efficient luminescent metal organic framework for the simultaneous detection and removal of heavy metals from water. *ACS Applied Materials & Interfaces* 8: 30294–30303.

133 Sohrabi, M. and Acta, M. (2013). Preconcentration of mercury(II) using a thiol-functionalized metal-organic framework nanocomposite as a sorbent. *Springer Link* 181: 435–444.

134 Luo, X., Shen, T., Ding, L. et al. (2016). Novel thymine-functionalized MIL-101 prepared by post-synthesis and enhanced removal of Hg^{2+} from water. *The Journal of Hazardous Materials* 306: 313–322.

135 Mon, M.,.F., Ferrando-Soria, L., Marti-Gastaldo, C. et al. (2016). Selective and efficient removal of mercury from aqueous media with the highly flexible arms of a BioMOF. *Chemie – International Edition* 55: 11167–11172.

136 Xiong, Y., Li, J., Gong, L. et al. (2017). Using MOF-74 for Hg^{2+} removal from ultra-low concentration aqueous solution. *Journal of Solid State Chemistry* 246: 16–22.

137 Cheng, X., Liu, M., Zhang, A. et al. (2015). Size-controlled silver nanoparticles stabilized on thiol-functionalized MIL-53(Al) frameworks. *Nanoscale* 7: 9738–9745.

138 Bakhtiari, N. and Azizian, S. (2015). Adsorption of copper ion from aqueous solution by nanoporous MOF-5: a kinetic and equilibrium study. *The Journal of Molecular Liquids* 206: 114–118.

139 Zheng, T., Zhao, J., Fang, Z. et al. (2017). A luminescent metal organic framework with high sensitivity for detecting and removing copper ions from simulated biological fluids. *Dalton Transactions* 46: 2456–2461.

140 Zhao, Y., Pan, Y., Liu, W., and Zhang, L. (2015). Removal of heavy metal ions from aqueous solutions by adsorption onto ZIF-8 nanocrystals. *Chemistry Letters* 44: 758–760.

141 Ghosh, A. and Das, A. (2020). Green synthesis of a novel water stable- Sn(II) – TMA metal- organic framework (MOF): efficient adsorbent for fluoride in the aqueous medium in the wide pH range. *The Royal Society of Chemistry New Journal of Chemistry (44) 1354–1361.*

142 Xia, J., Gao, Y., and Yu, G. (2021). Tetracycline removal from aqueous solution using zirconium-based metal-organic frameworks (Zr-MOFs) with different pore size and topology: adsorption isotherm, kinetic and mechanism studies. *Journal of Colloid and Interface Science* 590: 495–505.

143 Jana, A., Chatterjee, A., and Basu, J. (2021). A binary MOF of iron and copper for treating ciprofloxacin-contaminated wastewater by an integrated technique of adsorption and photocatalytic degradation. *New Journal of Chemistry* 45: 17196–17210.

144 Mohamed, E., Amira, M., Azab, M., and Abdelfattah, A. (2021). Effective removal of levofloxacin drug and Cr(VI) from water by a composed nanobiosorbent of vanadium pentoxide@ chitosan@ MOFs. *International Journal of Biological Macromolecules* 188: 879–891.

145 Lee, J., Ahn, J., and Kwak, S. (2022). Facile sonochemical synthesis of flexible Fe based metal organic frame works and their efficient removal of organic contaminants from aqueous solutions. *ACS Omega* 7: 23213–23222.

146 Escamilla, P., Viciano-Chumillas, M., Bruno, R. et al. (2021). Photodegradation of brilliant green dye by a zinc bioMOF and crystallographic visualization of resulting CO_2. *Molecules Basel, Switzerland* 26: 4098.

147 Salama, E., Ghanim, M., Hassan, H. et al. (2022). Novel aspartic-based bio-MOF adsorbent for effective anionic dye decontamination from polluted water. *RSC Advances* 12: 18363–18372.

148 El-Sewify, I., Radwan, A., Shahat, A. et al. (2022). Superior adsorption and removal of aquaculture and bio-staining dye from industrial wastewater using microporous nanocubic Zn-MOFs. *Microporous and Mesoporous Materials* 329: 111506.

149 Shi, Z., Xu, C., Guan, H. et al. (2018). Magnetic metal organic frameworks (MOFs) composite for removal of lead and malachite green in wastewater. *Colloids and Surfaces A: Physicochemical and Engineering Aspects* 539: 382–390.

150 Xie, L., Liu, D., Huang, H. et al. (2014). Efficient capture of nitrobenzene from wastewater using metal organic frameworks. *Chemical Engineering Journal* 246: 142–149.

151 Chen, Y., Chen, Y., Miao, C. et al. (2020 8). Metal organic framework based foams for efficient micro plastic removal. *Journal of Material Chemistry A* 8: 14644–14652.

152 You, D., Zhao, Y., Yang, W. et al. (2022). Metal-organic framework-based wood aerogel for effective removal of micro/nano plastics. *Chemical Research in Chinese Universities* 38: 186–191.

153 Wang, C., Du, X., Li, J. et al. Photocatalytic Cr(VI) reduction in metal-organic frameworks: a mini-review. *Applied Catalysis B: Environmental*. 2016; 193: 198–216.

154 Li, X. Gao X, Ai L, Jiang (2015). Magnetic cobalt nanoparticles embedded in hierarchically porous nitrogen-doped carbon frameworks for highly efficient and well-recyclable catalysis. *The Journal of Environmental Chemical Engineering* 274: 238–246.

155 Aquino, J., Rocha-Filho, R., Ruotolo, L. et al. (2014). Electrochemical degradation of a real textile wastewater using β-PbO2 and DSAR anodes. *Chemical Engineering Journal* 251: 138–145.

7

Case Studies (Success Stories) on the Application of Metal-Organic Frameworks (MOFs) in Wastewater Treatment and Their Implementations; Review

Arpit Kumar[1,], Mahesh Rachamalla[2], and Akshat Adarsh[3]*

[1] *Science Section, Bihar Bal Bhawan Kilkari, Department of Education Government of Bihar, India*
[2] *Department of Biology, University of Saskatchewan, Canada*
[3] *Science Section, Bihar Bal Bhawan Kilkari, Department of Education Government of Bihar, India*
[*] *Corresponding author*

7.1 Introduction

The rising concentration of new toxins in water bodies is a key worry in this decade. Water is life's most powerful solvent [1]. Furthermore, this is not the case for people. Due to its high solubility, it is used for all daily cleaning duties and the disposal of various soluble wastes. Wastewater is the accumulation of all of the above. It is often caused by the use of water. It is not that big of a deal, to a certain extent. Nature can, without a doubt, manage it. However, surface water availability is largely unchanged, and its quality is degrading because of the continued release of chemicals into the environment, notably fertilizers resulting from the rise in agricultural productivity [2–4]. It now seems to have numerous significant direct and indirect effects. Emerging contaminants are substances that have emerged during the last several decades, and they are often detected in low amounts in water streams, are poorly regulated, and pose harm to the environment, human health, and economic activity [5, 6]. All of these water contaminants have several input sources, including industrial waste dumping, hospital effluents, home wastewater, sewage treatment plants (STPs), and water treatment plants (WTPs) [7]. Historically, raw sewage poured into a body of water may naturally self-purify and clean the stream. However, due to population growth and widespread urbanization, sewage discharge has increased at a far quicker pace than natural filtering. The water body gets eutrophicated as a result of the excess nutrients generated, resulting in a progressive decline in water quality. Ammonia (NH_3)/ammonium (NH+4), nitrate (NO-3)/nitrite (NO-2), hazardous metals, crude oil, phosphorous (P)/phosphate (PO4+3), and sulfate/sulfide minerals are common water pollutants. Depending on their amounts, trace elements and minerals in the water may be useful or harmful. Minerals derived from groundwater are advantageous at low quantities but harmful at high doses [8]. Therefore, cheap water treatment solutions are in

Metal Organic Frameworks for Wastewater Contaminant Removal, First Edition. Edited by Arun Lal Srivastav, Lata Rani, Jyotsna Kaushal, and Tien Duc Pham.

great demand. Several wastewater treatment methods, such as the enhanced oxidation process [9], adsorption [10, 11], oxidation-reduction [11, 12], membrane filtration [13, 14], chemical treatment, mechanical [15, 16], and incineration, have been widely used [9]. According to the need for space-intensive, expensive-to-maintain buildings and facilities with advanced technology, these treatment procedures are not better.

Sewage Treatment Policies and State Implementation Strategies

> *Wastewater quality is improved through its treatment, and numerous physical, chemical, and biological techniques are used to remove contaminants from wastewater* Figure 7.1.

As the first federal legislation of its kind, the Water (Prevention and Control of Pollution) Act of 1974 sought to reduce water pollution and increase water conservation efforts throughout the country. Wastewater discharge is a major source of pollution, and this law aims to fix that. To further limit water pollution, the Water (Prevention and Control of Pollution) Cess Act, 1977 became effective in 1992. Taxes paid by business owners and municipal leaders bolster the budgets of the Water Act's central and state boards. As of the end of 2016, India has a water treatment capacity of 27.3% and a sewage treatment capacity of 18.6% (with an additional 5.2% capacity being built), as reported by the Central Pollution Control Board (March 2021). Considering the severity of the issue, India's waste and sewage treatment capacity is far from adequate, and without immediate action, there might be significant issues.

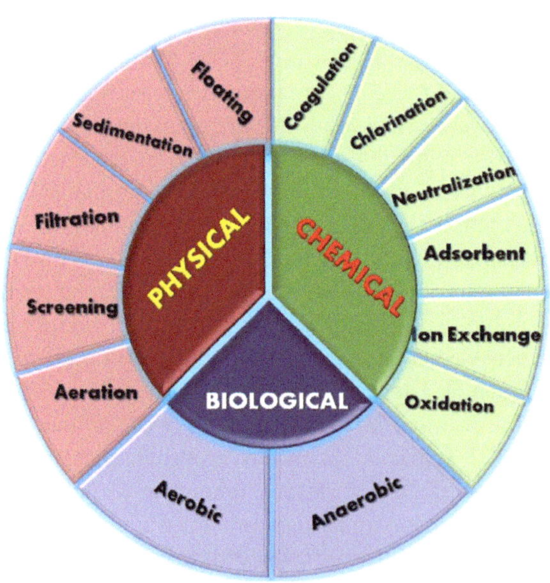

Figure 7.1 Schematic depiction of the different wastewater treatment processes, from primary treatment to secondary treatment (with permission) [17].

It is greater than the global average of about 20%. An individual treats an aggregate of more than four billion gallons of sewage daily or small, clustered wastewater systems that serve more than one in five houses in the United States. In order to deliver effective, efficient wastewater treatment, EPA wants to be a resource for decision-makers in rural, exurban, and suburban areas across the nation. Implementing agencies are aware of the challenges associated with managing wastewater in developed areas that already have old, ineffective, or damaged septic systems, as well as in more recent construction that calls for high-performance treatment technology along an effective supply chain to protect nearby groundwater lakes, rivers, streams, wetlands, and coastal waters. To enhance the water quality of the river Ganga to an acceptable degree and prevent pollution from entering the river, India has successfully established large-scale sewage treatment policies and methods. The Ganga Action Plan has implemented several sewage treatments, interception, and diversion systems (GAP) [18].

The GAP and Yamuna Action Plan (YAP) have resulted in more than 70 sewage treatment facilities since 1985. These facilities use a variety of technologies with differing degrees of mechanism, energy inputs, land requirements, skilled labor requirements, and so on. Early on, the choice of technology was made based on prior knowledge and how well it was thought to function. Additionally, several technologies have been tested on a pilot size at various phases of these action plans, and some of them have been scaled up for higher capacity facilities.

Within the nation, in this industry, there has been significant experience and knowledge with unique approaches for wastewater treatment application with rising trends over the past 20 years. The sewage treatment methods have been categorized based on a thorough review of performance and cost data from several sewage treatment facilities in the Ganga River basin and abroad that use all the technical alternatives. Four categories have been established for therapeutic technology. The classification may not be universal, but it is unquestionably accurate given India's experiences with wastewater treatment facilities over the previous two to three decades, particularly in the Ganga basin. Implementing agencies are aware of the challenges associated with managing wastewater in developed areas that already have old, ineffective, or damaged septic systems, as well as in more recent construction that calls for high-performance treatment technology along an effective supply chain to protect nearby groundwater lakes, rivers, streams, wetlands, and coastal waters [18].

Alternative methods are required owing to the limitations, the need to enhance the processes above, case studies, and the desire for a more economical and efficient treatment strategy. MOF-driven treatment techniques are widely renowned for their capacity to remove and mineralize all dangerous contaminants at a reduced cost. This chapter discusses the application of metal-organic frameworks (MOFs) as adaptive, highly effective case studies that may be implemented and their method for removing hazardous organic/inorganic pollutants from wastewater without using materials. The material supplied by the MOFs on their procedures has been summarized, and different types of wastewater treatment applications and case studies have been reviewed.

7.2 Metal-Organic Framework (MOF)

Metal-organic frameworks (MOFs), polymers with three-dimensional pore structures, have been rapidly evolving during the last decade [19, 20]. MOFs' primary molecular structure consists of metal nodes connected by organic linkers, which may be used to fabricate one-, two-, and three-dimensional structures with extraordinarily high porosity. In a three-dimensional MOF structure, metal ions and organic linkers produce a polymeric structure known as porous coordinate polymers [21]. MOFs have the highest surface areas for porous materials, up to 6200 m2/g [19]. Up to 2 nm wide pores in MOFs may allow small molecules to pass through. However, large molecules are rarely incorporated because of the pores (e.g., proteins and enzymes). There have been efforts to reduce the size of the crystal to the nanometer scale and to increase the pore size to the mesopore regime (2–50 nm) [22]. The kind of functional groups and their guest-host chemistry are the most significant properties of MOFs. Nitrogen and oxygen are examples of electron-donor molecules that function as organic ligands to connect the metal ions in the framework [23–25]. The kind of functional groups and their guest-host chemistry are the most significant properties of MOFs. Nitrogen and oxygen are examples of electron-donor molecules that function as organic ligands to connect the metal ions in the framework Figure 7.2. Because of the covalent interactions formed between the ligands and the clusters/metal ions, a three-dimensional crystalline hierarchy is formed throughout the synthesis procedure [26–28]. An effective option for generating a MOF structure is shown in Figure 7.2. Several SBU geometries are often used for describing MOF structure. The size and form of the ligand have a considerable effect on the structure's characteristics and its use. By changing the metal ions and organic linkers in MOFs, it is possible to develop a variety [26].

Recently, metal oxide and MOFs for the photocatalytic degradation of dye, antibiotic elimination, and pharmaceuticals have been stabilized by researchers [29, 30]. Khan et al. provided a metal oxide framework for the removal of heavy metal ions from wastewater in a separate evaluation. As soon as it was found that MOFs possessed a highly porous structure and outstanding catalytic activity, several applications for their usage in wastewater treatment were apparent. MOFs are ideal adsorbents for extracting poisonous dyes and other organic pollutants from wastewater due to their porosity, excellent properties, and high specific surface area [31].

7.2.1 Properties and Applications of MOFs

According to studies, MOFs' programmable molecular structure allows easy modification to produce bespoke MOFs with various predetermined uses (Table 7.1.) Despite their intriguing properties, MOFs have drawbacks compared to other materials like zeolites, such as poor water stability and limited temperature stability. MOFs have been regarded as promising materials for such a wide variety of

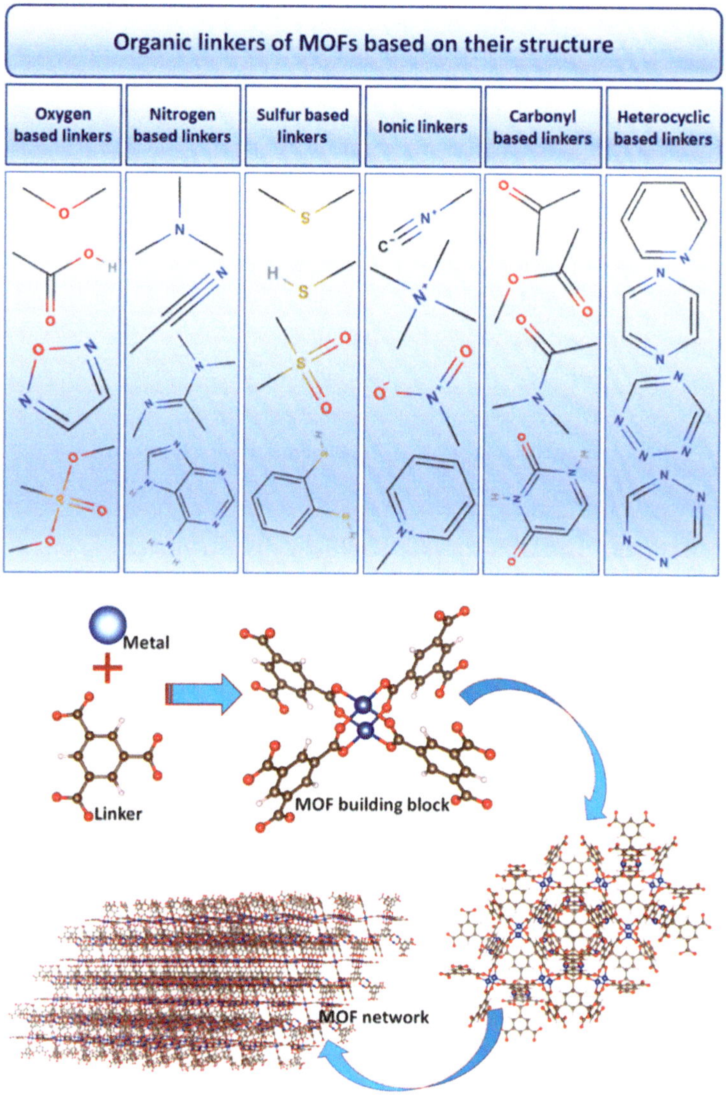

Figure 7.2 Different organic functional groups are categorized according to their chemical properties, as well as the basic approach to creating MOF structures (with permission; adopted from [26].)

applications that the number of applications has increased. This has contributed to their development and success in incorporation into various wastewater treatment and dye removal practices, as well as subsequent laboratory-scale research and numerous studies into wastewater treatment, heavy metal removal, hazardous dye removal, sensing, catalytic processes, gas separation, adsorption, storage, magnetic separation, and magnesite removal.

Table 7.1 A compendium of MOF synthesis methods, characteristics, and applications (with permission) [26].

Metal-Organic Frameworks		
Properties	**Applications**	**Synthesis Approaches**
High surface area	Gas/energy storage	Hydro/solvothermal
High porosity	Separation	Electrochemical
Tunable topography	Waste disposal	Mechanochemical
Crystallinity	Catalysis	Diffusion
Multiple affinity	Sensing	Microwave
Easy preparation	Drug delivery	Solvent evaporation
Stability	Electrode materials	Sonochemical

7.3 Applications of MOFs in Wastewater Treatment: Case Studies

Numerous industrial-scale implementations and academic studies of these crystalline porous MOFs have been conducted. The effectiveness of photocatalysis in purifying tainted water has given it increased popularity lately. The [32] study was analyzed as a nonprecious metal Ppotocatalyst for visible light-driven wastewater treatment. There is still a need for clean water, which can be achieved by removing the accumulated harmful organic compounds produced by industrial or medical waste, it was expected that the photocatalytic decontamination of wastewater using visible-light sensitive materials would be of significant concern. Researchers have developed a novel photocatalyst based on the crystallization of size-optimized colloidal MOF crystallites of hematite (Fe_2O_3) and titanium dioxide (TiO_2) (anatase). In their study, Fe_2O_3/TiO_2@MIL-101 displayed more visible-light-driven antibiotic degradation activity than bare Fe_2O_3 (hematite), TiO_2 (anatase), and commercially available TiO_2. It is a reusable photocatalyst with the characteristics of reusability and stability (P25). In addition, we observed that the Fe_2O_3 and TiO_2 support on colloidal MIL-101 crystallites enhanced photocatalytic activity; no other additives were necessary for oxidative pollutant removal. This study focuses on the component that tackles the critical issue of clinical wastewater treatment and indicates whether the development of noble-metal-free materials for photocatalytic water purification systems is advantageous in this case. Figure 7.3

Companies such as Panasonic [33] utilize photocatalysts and ultraviolet (UV) radiation to eliminate toxic substances in water, demonstrating that other sectors were able to overcome the challenge and deploy photocatalytic technologies based on MOFs. In this technological application, TiO_2 photocatalysts are bound to commercial adsorbents and a catalyst termed zeolite, allowing for efficient separation and recovery of photocatalysts from the water for reuse. In this research, TiO_2 was used to mineralize several chemical molecules into innocuous byproducts. To succeed in this endeavor,

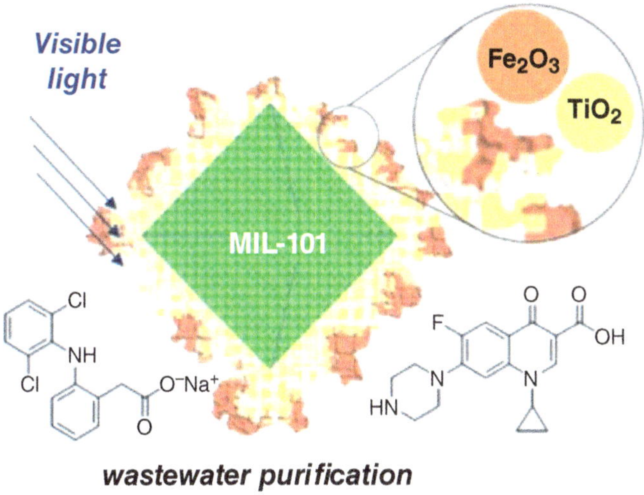

Figure 7.3 A figure depicting the core-shell configuration of the Fe_2O_3/TiO_2@MIL-101 photocatalyst. The MOF-supported Fe_2O_3/TiO_2 junction system is an effective technology for treating wastewater when exposed to visible light [32].

a catalyst requires the utilization of ultraviolet (UV) light, either naturally occurring or artificially produced. The range of organic materials it degraded included oestrogens, pesticides, dyes, crude oil, microorganisms like viruses and chlorine-resistant diseases, and inorganic chemicals like nitrous oxide. Figure 7.4 Industrial wastewater that has been heavily contaminated with organic chemicals or metals may be cleaned up using a photocatalytic water treatment system.

Figure 7.4 Panasonic's demo unit with their photocatalytic water filtration technology, adapted with their permission [33].

MOF-based membrane technologies for wastewater treatment offer tremendous potential and should be widely employed to reject contaminants via exclusion size [34]. With the properties of porousness, high surface area, smaller particle size, aspect ratio control, tuneability, compatibility with a polymeric network, and availability of functional groups, MOFs are finding widespread use as membrane functionalizing agents. Figure 7.5 shows how [35, 36] applied the same techniques to swiftly include new species or functionalities without altering their fundamental makeup. Because of its ease of use and minimal chemical requirements, membrane separation has the potential to replace conventional wastewater treatment systems [37]. Numerous membrane technologies, including FO, RO, NF, and UF, make use of MOF-enhanced membranes. Recent studies have analyzed implementation strategies at the ground level [34, 38–40]. Depending on chemical properties, water chemistry, and membrane features, MOF-based membranes may remove organic and inorganic pollutants [41]. Thin-film nafion (TFN) [42], porous mixed matrix [43], solvothermal method [44], electrodeposition method [45], vacuum filtration [46], and thin-film composite (TFC) [38], evaluate and enhance MOF-based membranes for liquid separation, especially desalination [47]. Figure 7.5: In-situ preparatory, mixing, and interfacial polymerization techniques for MOF-based membranes in liquid separations [41].

Forward osmosis (FO), reverse osmosis (RO), ultrafiltration (UF), and nanofiltration (NF) membranes are the four major categories of filtration membranes based on their outer surfaces and pore diameters. Nanoscale and angstrom-scale holes [48, 49] are necessary for molecules and ions to undergo NF and RO, respectively. Filtration is a time-consuming and power-hungry process because of the high total mixed phase (TMP) involved in their production. Standard UF/MF membranes, in contrast to extreme UF/MF membranes, have been around for a long time and are widely used in industry to remove germs and bulky molecules from wastewater. Because its pore size is substantially bigger than the dimensions of the dissolved metal ions and dye molecules, UF/MF membranes are seldom utilized directly to remove these contaminants. MOFs are particularly well-suited to this use because their porous structure greatly boosts their ability to extract metal ions and other tiny

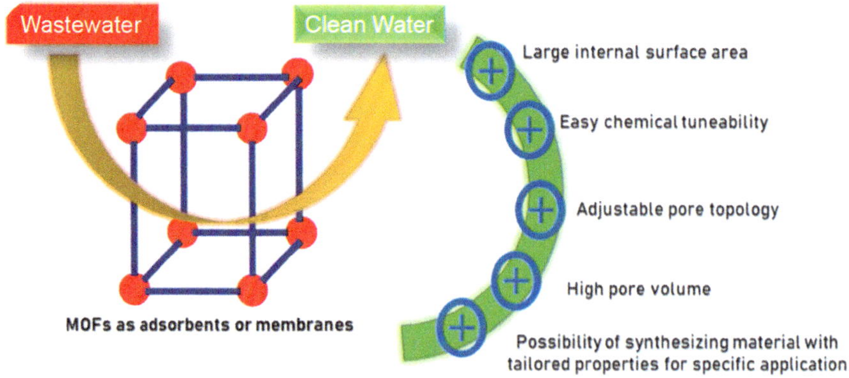

Figure 7.5 This example shows how MOFs may be used as membranes in the wastewater treatment process.

molecules from wastewater. Possible methods and results show that the MOF's enhanced membrane is an effective adsorbent for the detoxification of organic dyes and heavy metal ions [50–55]. Membrane filtration and the adsorbent properties of MOFs allowed for the continuous removal of metal ions from wastewater.

7.3.1 Forward Osmosis (FO) Membranes

These membranes are extensively utilized, notably for saltwater desalination, but to use them in the wastewater business and avoid forward osmosis (FO) constraints, they must have a stable, active, and thick layer on top [56]. Modified dense film-casting to generate thin-film porous asymmetric membranes (PMMs) coupled with magnetic water-instable MOFs ((magnetic) ZnO@MOF-5) improved FO process efficiency. Comparable studies found that FO performance increased PMM substrate mass (water/solute) transfer efficiency by controlling internal concentration. Despite their structural similarity to porous asymmetric membranes (PMMs), FO-TFN membranes are not widely used [57]. After several coating procedures were developed for facile and fast casting, such as those described in [58], UiO-66 in various noncovalent bonding forms improved polymer-interfacial contact during synthesis swapping.

Dai et al. conducted a study in which a thin-film nano composite (TFN) membrane with a suitable 2D MOF (copper 1,4-benzyl dicarboxylate) nano filter in a polyamide active layer was developed to enhance water flow and antifouling without decreasing FO selectivity [59]. The TFN membranes' reaction and posttreatment processes were similar to those of thin-film composite membranes, with the exception of the addition of a Cu-1,4-benzene dicarboxylate nanosheet to a trimethyl chloride/n-hexane solution at concentrations ranging from 0.03 to 0.15 wt/v% prior to interfacial polymerization. The parameters of the FO membrane process and the membrane characteristics determine the solute transport. The loading of UiO-66 affected water adsorption and permeance. In initialization with 20 mmol/L, the UiO66-TFN membranes beat virgin polymers in water permeance (0.52 L/m2hbar), phenol red retention (>99%), and Na_2SO_4 retention (94–96%) in FO processes [42]. Microorganisms' development and dispersion on the membrane surface diminish water flow and increase energy consumption during FO operations [60, 61]. Jun et al. and Firouzjaei et al. developed a TFN membrane with a graphene oxide (GOs)-Ag MOF to boost water permeability and anti-biofouling in FO procedures using Escherichia coli and sodium alginate (GOs-Ag-MOF). Due to its relatively high surface roughness, substantial negative surface charge, and low hydrophobicity, the GOs-Ag-MOF TFN membrane provided excellent fouling resistance through FO operations [41, 62].

7.3.2 Application and Effectiveness

Water, salt, water flow, and reverse solute flux are used to evaluate FO membranes. The low salt/water permeability ratio of FO membranes shows high selectivity. For the determination of FO salt loss, reverse solute flow is essential. A unique integrated thin-film PMM enhanced the flow of FO water. TFC is manufactured using porous substrates. The TFC material polyamide (PA) is produced by the interfacial polymerization of m-phenylenediamine (MPD) in water and trimesoyl chloride (TMC) in the organic phase. The substrate layer is less thick than the active layer, which resembles a sponge. TFC effectively rejects monovalent and divalent ions from FO, RO, and NF applications because of its small window size.

Hydration Technologies, Inc.'s TFC-like cellulose triacetate is a cellulose triacetate that resembles TFC (CTA). In a number of trials, hand-made TFC outperformed CTA in terms of water permeability, rejection, and membrane fouling

TFC-like cellulose triacetate from Hydration Technologies Inc. (CTA). [63, 64]. The water flow and rejection tests passed with flying colors. MOF nanoparticles. The combination of TFC with MOFs produces a nanocomposite thin-film material. Casting MOF in a monomer solution on a substrate's surface enhances its performance (TFN).

In a separate investigation, municipal wastewater-based Cu-1,4-benzene dicarboxylate nanosheet TFN membranes exhibited higher water flow and antifouling efficacy than virgin thin-film composite membranes [59]. In active-layer-draw and active-layer-feed solutions, thin-film PMMs displayed higher water flux than pure polyether sulfone membranes. Regardless of orientation mode, water flow increased from 0.5 to 2.0 M (NaCl) due to the rising osmotic pressure from FO membrane internal concentration polarization [56]. Cu-1,4-benzene dicarboxylate nanosheet and TFN membranes may improve the treatment of FO wastewater and desalination of seawater.

7.3.3 Reverse Osmosis (RO) Membranes

Numerous commercial RO membranes are optimized for high water flow and solute removal, since they are often used in water and wastewater treatment, as well as desalination. This was shown by adding a thin polyamide selective layer to a thin-film composite structure [61, 65]. In terms of MOF water flow and polyamide matrix compatibility, ZIF-8 may surpass TFN hydrophilic zeolite. Interfacial polymerization between m-phenylenediamine and trimethyl chloride introduced two MOFs (UiO-66 and MIL-125, 100 nm) into polysulfone thin-film composite membranes with various loadings (0–0.3%) [66]. HKUST-1 ($Cu_3(BCT_2)$) TFN RO membranes used the same strategy [67]. The active-layer thickness of the HKUST-1/RO membrane (29 nm) was significantly thinner than that of the commercial RO membrane (200 nm), which increased water flow and antifouling in comparison to the pure thin-film composite RO membrane with a polysulfone support layer without compromising $NaHCO_3$ rejection. Reversible accumulation-fragmentation chain transfer polymerization was used to synthesize an antibacterial multilayer membrane composed of phosphonium-conjugated GO-anchored Cu and trimesic acid-based MOF (pGO-Cu-MOF) [68]. pGO-Cu-MOF membranes generated considerable reactive oxygen species and decreased gram-positive and gram-negative bacteria by seven logs. An interlayer composed of MOF nanoparticles and a polysulfone layer that was thinner, less cross-linked, and hydrophobic than polyamide RO membranes enhanced RO support layer. On a polysulfone substrate with a thick selective polyamide layer, another research fabricated highly water-permeable TFN membranes with hydrophilic, stable MIL-101(Cr) for RO applications [69]. In order to accelerate membrane water flow, hydrostable MOF nanoparticles generated extremely permeable direct water channels in the polyamide layer.

MOF loading rates significantly influenced NaCl rejection and RO water flow in ZIF-8/TFN polyamide membranes [65]. Despite the fact that the 0.34 nm ZIF-8 pore can remove > 99% of hydrated Na+ and Cl from water, polyamide structural modification due to possible gaps between ZIF-8 and polyamide may have led to contradictory outcomes [65]. As MOF loading increased, aCl rejection decreased somewhat. Flux of water enhances concentration polarization and sodium chloride rejection. ZIF membranes generated greater NaCl water fluxes compared to RO membranes [67]. Fouling of the HKUST-1/RO membrane by bovine serum albumin After 8 hours of filtration, the pure composite thin-film ROS membrane saw a 40% quicker membrane flux fall than the HKUST-1/RO membrane, suggesting that the MOF/RO membrane demonstrated superior antifouling behavior against bovine serum albumin without impacting salt removal. The hydrophilicity and smoothness of the MOF/RO membrane increased, reducing organic molecule fouling.

In another study, pGO-Cu-MOF membranes with active layers, RO support, and interlayers exhibited significant antibacterial properties [57]. The multi-layered MOF membrane contained 2% to 5% more monovalent and divalent mixed salts (NaCl, $MgCl_2$, and $Ca(NO_3)_2$) than the pure RO membrane; however, the water flow was 15% lower owing to the resistance of the H_2O molecules. Initial concentrations had a significant impact on salt rejection. The positively charged active layer excludes cations through electrostatic repulsion, and the pGO-Cu-MOF interlayer exhibits superior steric exclusion compared to the pure RO membrane support layer [70].

7.3.4 Application and Effectiveness

For water flow and $NaHCO_3$ rejection, case studies compared HKUST-1/RO membranes to pure RO membranes with polysulfone support layers [67]. The acid HKUST-1 enhanced the water flow through the MOF-RO membrane by 33% (36.5L/m2h) and salt rejection by 96%. The virgin RO membrane rejected 94% of the $NaHCO_3$. The hydrophilicity and porosity of the polysulfone support layer restricted the diffusion of m-phenylene diamine into water, which explains these findings. Similar membranes eliminated 92% of phenol, although the NH2-MIL-88B RO membrane had a higher water flow rate (35.8 L/m2/h) than the TFC membrane (23.5L/m2/h) [71]. The structure, composition, and size of porous nanofillers affect the pore structure (size and porosity) of the polysulfone nanocomposite layer, which may explain performance differences (degree of polyamide crosslinking and thickness). In the polysulfone nanocomposite layer, the ideal quantity of porous nanofillers affects its porosity and pore size, hence enhancing RO processing. Thin-film nanocomposite (TFN) technology supported by a polysulfone layer, polyamide active layer, and non-woven support fabric is used in commercial RO membranes with MOFs improved technology developed by LG Chem NanoH2O [72] for the consumption of brackish water (Figure 7.6). At maximum flow, their seawater RO membrane rejects 99.89% of salt. They excel in wastewater reuse, industrial water, desalination of saltwater, ultrapure water, and worldwide utilities.

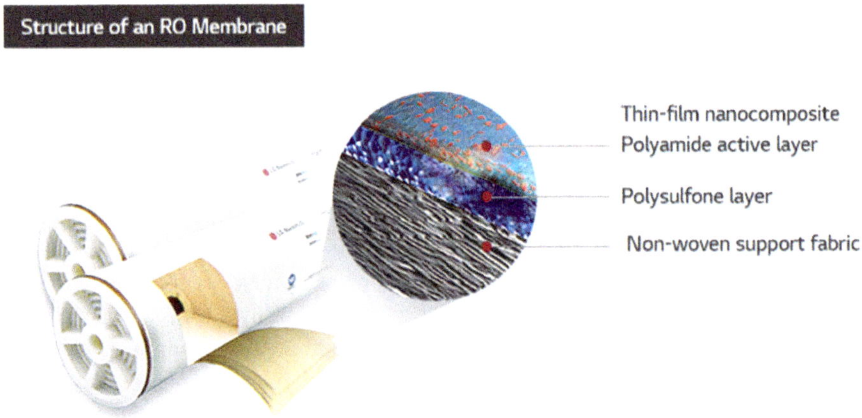

Figure 7.6 LG Chem's TFN-based RO membrane, depicted in a graphic. Adapted with permission from [72].

Polyamide TFN membranes with increased filler amounts have limited membrane selectivity owing to filler aggregation in the aqueous-organic phase [71]. One-step interfacial polymerization may also be used to build membranes, which is better for water flow than the two-step technique [73]. Additionally, MOF loading influences water permeance and NaCl rejection in highly stable TFN-MIL-101(Cr) membranes [74]. TFN-MIL-101(Cr) membranes with 0.025 w/v% MOFs outperformed thin-film composite membranes in terms of water permeance (1.5 L/m2 H). MIL-101(Cr) MOF concentrations gradually increased water permeance (3.0 L/m2hbar at 0.1 w/v%). The hydrophilic, porous MOF structure and minimal polyamide crosslinking increased water permeance because water molecules can travel in straight channels. MOF addition decreased NaCl rejection from 99.5% to 93.5%. Attributable to polyamide-aggregation interfacial defects and MOF nanoparticle aggregations' inner nonselective voids.

7.3.5 Nano Filter (NF) Membranes

Commercial nanofiltration membranes remove soluble inorganic and organic substances from water or organic liquids. Polydimethylsiloxane, polyamide, and polyimide are employed to manufacture commercial NF membranes because they are inexpensive, stable, and selective in organic solvents [75]. It is challenging to design high-flux, selective polymeric membranes for water filtration [76]. The membrane flux-selectivity trade-off was resolved by combining highly hydrophobic UiO-66 nanoparticles with polyamide membranes with a variety of functional surface morphologies. Co-electrospinning resulted in the formation of a Zn-based MOF-808 nanofibrous NF membrane with hydrophilic polyacrylonitrile nanofibers [77]. Cd2+ and Zn2+ were removed using MOF nanofibrous membranes that were reusable. Polyacrylonitrile containing MOF nanoparticles exhibited larger filaments, resulting in a rougher surface. Cu2+ and Co2+ were removed from a Zn-based

MOF (MOF-5) by phase inversion and immersion precipitation employing polymeric membranes (polyethersulfone, cellulose acetate, and polyvinylidene fluoride) [78, 79]. TFN membranes were produced by introducing stable MOF UiO-66 nanoparticles of 30, 100, and 500 nm into the selectively cross-linked polyamide layer on the polyethersulfone substrate. MOF TFN membranes used nanoparticle size and loading to remove Se and As. 0–1.25% UiO-66 nanoparticle loading. Nano cubes with hydrophilicity, a hollow shape, and a negative charge allowed for superior NF water permeance and salt rejection (NaCl and Na_2SO_4) [46]. To eliminate organic wastewater colors, stable MOF (HKUST-1)-reduced GO nanocomposite membranes containing polydopamine on cellulose acetate support layers were created. The vacuum-assisted fabrication of a ZIF-8@GO NF membrane with polyethyleneimine on a tubular ceramic substrate improved the organic solvent (methanol) NF performance for dye transfer [80].

In an iterative manner with w/v% nanoparticles (pore size and synthesized analogues = 0.6–1.0 nm), the polyamide/UiO-66 membrane increased water permeance by 85% (15.4 L/m2h bar) and solute rejection by 50%–99% by including nanoparticles (pore size and functionalized analogues = 0.6–1.0 nm) [81]. The polyamide/UiO-66 membrane increased water permeance by 85% (15.4 L/m2h bar) and solute rejection by 50%–99%. Pure polyamide composite thin-film membranes have a pure water permeability of 8.3 L/s. The thickness of the polyamide/UiO-66 membrane was 230 nm more than that of the pure polyamide membrane (205 nm). Similar pore geometry in thicker membranes decrease solvent permeability [82]. Due to the porosity of MOFs, the thicker membrane enhanced the permeation of organic solvents. In case studies, heavy metal removal was impacted by the porosity, pore size, hydrophobicity, water permeability, and antifouling of MOF-5-embedded polymeric membranes. Different polymers decreased the contact angle of MOF-polymer membranes, making them less hydrophobic than unmodified membranes [83, 84]. Improved pore size/porosity and decreased hydrophobicity significantly impacted the permeability and resistance of membranes. Hydraulic resistance and permeability were enhanced by polyethersulfone/MOF-5 and cellulose acetate/MOF-5. MOF-polymer membranes were more effective than fresh polymeric membranes in removing contaminants [78].

7.3.6 Application and Effectiveness

Polyacrylonitrile/MOF-808 nanofiltration membranes eliminate co-ions Cd2+ and Zn2+ [77]. While maintaining high water permeance (870 L/m2hbar), MOF membranes adsorb 225 and 287 mg/g of Cd2+ and Zn2+, respectively, while maintaining a water permeability of 870 L/m2hbar. Na+, Mg+, and Ca2+ inhibited Cd2+ removal in all MOF membranes by 20%. Na+, Mg2+, and Ca2+ competed for MOF membrane adsorption sites due to their dissimilar Pauling electronegativities [85]. Compared to thin-film composite membranes, ZIF-8-based hollow nanocube TFN membranes enhanced water permeance for Na_2SO_4 (19.4 L/m2h bar) and NaCl (14.5 L/m2h bar) [86]. There are three interpretations for these results: i. Hydrophilic nanocubes permit H_2O molecules to easily pass through the wet membrane surface and inner pores [87];

ii. The inner hollow space provides favorable flow channels and reduces mass transfer resistance by decreasing the diffusion space [88]; and iii. The produced boundary region between the polyamide polymer and the hydrophilic nanocubes provides flow channels [89]. Perhaps because of its increased hydrophilicity and negative surface charge, the ZIF-8-based hollow nanocube TFN membrane showed a substantially smaller water permeance reduction than the thin-film composite membrane for both humic acid and bovine serum albumin feed solutions [90].

In industrial applications, Direct Nanofiltration (dNF) membranes from Nxfiltration are lauded for their low-fouling hollow fiber structure and ability to remove organics and salt (hardness) from water in a single step. The strainer is enough pre-treatment. It claimed to have developed technology for creating nanoscale layers on a membrane support. Its methodology precisely controls the rejection and flow properties of the membrane used to treat problematic water. Implementation case studies show wastewater color, dye, pharmaceutical residue, microplastic, and endocrine disruptor removal rates. A specific study found that their (dNF 40 filtrations) membranes were utilized directly on wastewater at the wastewater treatment plant in Enschede, the Netherlands, including biological treatment and settling tanks. Pre-treatment for dNF40 filtration required a strainer with flux (20 L/m2h), >97% rejection of total organic carbon, and >80% rejection of a cocktail of micropollutants, predominantly pharmaceutical effluents [91].

7.3.7 Ultrafiltration (UF) Membranes

Ultrafiltration (UF) membranes are less expensive, more straightforward, and easier to maintain than FO, RO, and NF processes, despite membrane fouling affecting performance [92]. MOF-modified membranes are promising. Alumina hollow-fiber membranes impregnated with UiO-66 nanoparticles minimize humic acid and increase water flow [93]. Solvothermal sintering with differing alumina powder diameters generated nanoparticles of UiO-66 [58]. Immersion precipitation phase inversion was used to construct TMU-5 MOF-containing hydrophilic polyethersulfone UF membranes [94]. These membranes were tested using a milky oil-water solution. On the surfaces of trimesoyl chloride-polyvinylidene fluoride membranes, polyacrylic acid-immobilized ZIF-8 nanoparticles facilitated the in-situ polymerization of $C_3H_3NO_2$ monomer and N, N′-methylenebisacrylamide cross-linker with N, N′-methylenebisacrylamide [92]. GO nanosheet layers with porous modifiers that are hydrophilic nanoparticles of UiO-66 avoided stacking and conferred remarkable nanocomposite properties [95]. Nanocomposites of UiO-66@GO were used to create water-purifying and antifouling polyethersulfone membranes. Phase inversion generated a multitude of polyethersulfone membranes for dye removal using Cu terephthalate MOFs including GOs [96]. Polyethyleneimine, MOFs, laccase/polydopamine, and 3D modification developed a high-flux biocatalytic MOF membrane [44]. On the polyacrylonitrile support layer, polyethyleneimine polymers derived from water-stable MOFs (MIL-101-L, MIL-101-S, UiO25 67, UiO-66-NH2, and UiO66) were added.

Reusable, high-flow biocatalytic membranes eliminated contaminants across a broad pH range. Using acetone and thermally induced phase separation, 10%, 33%, and 67% MOF nanoparticles were disseminated in MIL-53(Fe)/polyvinylidene fluoride membranes [97]. Methylene blue was retained by the MIL-53(Fe)/polyvinylidene fluoride membrane by adsorption and catalytic oxidation. In a comparable study, [39] used phase inversion to produce superhydrophilic UiO-66-NH2 MOF membranes functionalized with poly(sulfobetaine methacrylate) in a polysulfone casting solution. Because raw UiO-66-NH2 particles scatter poorly in polysulfone casting solution, the manufacturing of phase-inversion membranes has difficulty uniformly dispersing raw MOF particles. To resolve the problem, poly(sulfobetaine methacrylate) brushes near MOF particles decreased agglomeration and boosted dispersion.

Similar studies developed polyacrylonitrile-co-maleic acid-polyethyleneimine-Ag-modified polyethersulfone membranes that are self-cleaning and antimicrobial [98]. Anti-biofouling MOF membranes include antibacterial Ag nanoparticles and negatively charged/hydrophilic amine functional groups. Antimicrobial Ag nanoparticles rupture the plasma membranes of bacteria, releasing negatively charged proteins. Electrostatic repulsion from the very hydrophilic and negatively charged -OH and -NH2 functional groups prevents the buildup of these proteins on the membrane surface, which explains why UF membranes need mechanical robustness [99].

Using surface functionalization-based etching and tannic acid, hydrophilic hollow ZIF-8/polysulfone membranes were made. They examined the mechanical strength of membranes. ZIF-8 particle addition promptly lowered polysulfone membrane elongation (about 15%–8%) with equivalent elasticity, corresponding with previous research demonstrating that adding different components to polysulfone membranes affected UF membrane strength [100]. 20% tannic acid on ZIF-8 nanoparticles increased the hollow ZIF-8/polysulfone membrane's elasticity and elongation (1.5 MPa). The homogenous distribution of ZIF-8 nanoparticles in the polysulfone membrane and the strong H-bonds between the nanoparticles' -OH groups and the polyethersulfone chain's -SOO groups likely improved the membrane's performance [101].

7.3.8 Application and Effectiveness

The water flow and humic acid elimination of UiO-66/alumina hollow-fiber membranes were examined [102]. MOF membranes have a pure water flow rate of 41,1 L/m2h, compared to 231 L/m2h for alumina hollow-fiber membranes, likely induced by UiO-66 nanoparticles covering the membrane. After 1,000 mg/L humic acid filtration, the pristine membrane's water flow decreased to 41.1 L/m2h due to fouling on the membrane surface and/or holes [103]. The electrostatic interaction between MOF particles and humic acid molecules decreased water flow to 25.7 L/m2h, or 98% more than the pure membrane. Polyethersulfone TMU-5 membrane was polluted with oil during testing. Milk powder impeded the flow of virgin polyethersulfone membranes. After cleaning, the pure membrane's water flow (contact angle: 67°, surface roughness: 44 nm) recovered poorly (flux recovery ratio = 25%), but the hydrophilicity and surface (98%) of the MOF membrane enhanced flux recovery. 5.2% surface roughness and 53° contact angle roughening at a pH of 5.5, polyacrylic acid/ZIF-8/polyvinylidene fluoride membranes removed Ni(II) (2 mg/L) and Na+ (15 000 mg/L) from synthetic wastewater [104]. Maximum Ni (II)/g adsorption capacity indicates that adsorption on the MOF membrane, electrostatic attraction, and hydrogen bonding remove Ni (II) (II).

Compared to other polyethersulfone membranes, UiO-66@GO polyethersulfone membranes transport 350% to 80% more water. The flux recovery ratio of UiO-66@GO was 89%, compared to 43% and 82% for polyethersulfone and GO/PES membranes, respectively: i. Ultrahigh MOF loading (67%) improve membrane hydrophilicity, facilitating water molecule tranport [105]; ii. During MOF membrane fabrication, numerous pores are created that serve as water transport channels, whereas the virgin polyvinylidene fluoride membrane showed no pores; iii. The MOF membrane's rough top surface increased the effective filtering area and membrane water flow compared to the pristine membrane [39].

This series pioneered the treatment and purification of wastewater using MOFs. It creates hydrogen for other implantation systems' energy solutions. It may offer clean water to one billion people. Another photocatalytic wastewater treatment system based on MOF separates water to remove contaminants. Rhodamine blue decomposed organic contaminants effectively. These photocatalytic, organic, and metal-manufactured base adsorption MOFs advance the use of metal-organic frameworks for wastewater treatment and other applications. According to research, MOF-polymer composites are comprised of inexpensive, environmentally benign materials. The compound eliminated mercury and lead from water samples. MOFs eliminated 1,6 times their own weight in mercury. Researchers frantically seeking low-cost, high-efficiency clean water solutions may benefit from MOF solutions.

Summary

1) Forward osmosis (FO), reverse osmosis (RO), ultrafiltration (UF), and nano filtration (NF) membranes may be classified. NF and RO molecules and ions need nano- and angstrom-scale holes. MOFs remove metal ions and other small molecules from wastewater more efficiently due to their porous nature. To increase water flow and antifouling, a thin-film composite membrane containing a 2D MOF (copper 1,4-benzenedicarboxylate) nano filter in a polyamide active layer was developed. UiO-66 loading influenced water adsorption and permeance.
2) TFC membranes outperform virgin polymers in water permeance (0.52 L/m2hbar), phenol red retention (99%), and Na_2SO_4 retention (94–96%) in FO processes. Cu-1,4-benzenedicarboxylate nano sheet TFN membranes employing municipal wastewater exhibited higher water flow and antifouling efficacy than virgin thin-film composite membranes [59]. Thin-film PMMs flowed more water than pure polyether sulfone membranes in active-layer-draw solution and active layer-feed solutions.
3) Osmotic pressure increased water flow from 0.5 to 2.0 M (NaCl), regardless of orientation mode. The enhanced FO membrane's internal concentration polarization modified a thin-film composite structure over a very thin polyamide

selective layer [61]. This showed effective water desalination of polyamide TFN RO membranes with hydrophobic zeolitic imidazolate framework-8 (ZIF-8).

4) MOF loading rates radically influenced RO water flow and ZIF-8/TFNpolyamide membrane NaCl rejection. ZIF membranes produced larger NaCl water flows than RO membranes. Multi-layered MOF membranes exhibited 2%–5% more monovalent and divalent mixed salts (NaCl, $MgCl_2$, and Ca $(NO_3)_2$) than the virgin RO membrane. The acid HKUST-1 increased MOF-RO membrane water flow by 33% (36.5L/m2h) and salt rejection by 96%. Virgin RO membrane rejected 94% $NaHCO_3$.
5) Porous nanofillers alter the polysulfone nanocomposite layer's pore structure and porosity. This may explain the membranes' performance differences (polyamide crosslinking and thickness). MOFs permeated 1.5 L/m2 H more water than thin-film composite membranes. One-step interfacial polymerization builds membranes with better water flow than two-step. Co-electrospinning developed a MOF (Zn-based MOF-808, Zr6O4(OH) 4(COOH) 6(BTC) 2) nanofibrous NF membrane with hydrophilic polyacrylonitrile nanofibers.

Acknowledgment

We appreciate the opportunity being provided by the book's editor. The authors are also grateful to all the authors and corporations that shared or permitted reviews of their work. Additionally, we appreciate the Department of Education's Bihar Bal Bhawan Kilkari's Science Section for providing the logistical assistance for this work.

References

1 Priyadarshini, M., Das, I., and Ghangrekar, M.M. (2019). Application of metal-organic framework in wastewater treatment and detection of pollutants: review. *The Journal of the Indian Chemical Society* 97: 507–512.

2 Organization of Economic Cooperation and Development (OECD) (2016). OECD Environmental Performance Reviews: chile 2016 [Internet]. OECD Environmental Performance Reviews. OECD. [cited 2022 Sep 6]. (OECD Environmental Performance Reviews). Available from: https://www.oecd-ilibrary.org/environment/oecd-environmental-performance-reviews-chile-2016_9789264252615-en.

3 United Nations Department of *Economic and Social Affairs:* Population Divisions (2017). World Population Prospects: the 2017 Revision. *Economic and Social Affairs* 1–53: 53.

4 Bedia, J., Muelas-Ramos, V., Peñas-Garzón, M. et al. (2019 January 7). A review on the synthesis and characterization of metal-organic frameworks for photocatalytic water purification. *Catalysts* 9 (1): 52.

5 Mahmood, T., Momin, S., Ali, R. et al. (2022). Technologies for removal of emerging contaminants from wastewater. Wastewater Treatment. [Working Title] [Internet]. IntechOpen [cited 2022 Sep 12] Available from: https://www.intechopen.com/online-first/81623.

6 Good practices for regulating wastewater treatment: legislations, policies and standards | UNEP – UN environment programme [internet]. [cited 2022 Sep 12]. Available from: https://www.unep.org/resources/report/good-practices-regulating-wastewater-treatment-legislations-policies-and-standards.

7 Taheran, M., Naghdi, M., Brar, S.K. et al. (2018). Emerging contaminants: here today, there tomorrow! *Environmental Nanotechnology, Monitoring and Management* 10: 122–126. Elsevier B.V.

8 Shree, R. (2011). *Causes of Water Pollution*. 44–55 p. Shikshamitra.

9 Khin, M.M., Nair, A.S., Babu, V.J. et al. (2012). A review on nanomaterials for environmental remediation. *Energy & Environmental Science* 5 (8): 8075.

10 Kramer, M., Schwarz, U., and Kaskel, S. (2006). Synthesis and properties of the metal-organic framework Mo3(BTC)2 (TUDMOF-1). *Journal of Materials Chemistry* 16 (23): 2245.

11 Ma, J., Shi, J., Ding, H. et al. (2017 March). Synthesis of cationic polyacrylamide by low-pressure UV initiation for turbidity water flocculation. *Chemical Engineering Journal* 312: 20–29.

12 Gianico, A., Bertanza, G., Braguglia, C.M. et al. (2015 September). Upgrading a wastewater treatment plant with thermophilic digestion of thermally pre-treated secondary sludge: techno-economic and environmental assessment. *Journal of Cleaner Production* 102: 353–361.

13 Yin, J., Zhu, G., and Deng, B. (2016 February). Graphene oxide (GO) enhanced polyamide (PA) thin-film nanocomposite (TFN) membrane for water purification. *Desalination* 379: 93–101.

14 Wang, L., Boutilier, M.S.H., Kidambi, P.R. et al. (2017 June 6). Fundamental transport mechanisms, fabrication and potential applications of nanoporous atomically thin membranes. *Nature Nanotechnology* 12 (6): 509–522.

15 Paul, E., Camacho, P., Sperandio, M., and Ginestet, P. (2006 July). Technical and economical evaluation of a thermal, and two oxidative techniques for the reduction of excess sludge production. *Process Safety and Environmental Protection* 84 (4): 247–252.

16 Tyagi, V.K. and Lo, S.L. (2011 September 12). Application of physico-chemical pretreatment methods to enhance the sludge disintegration and subsequent anaerobic digestion: an up to date review. *Reviews in Environmental Science and Biotechnology* 10 (3): 215–242.

17 Ramalingam, G., Pachaiappan, R., Kumar, P.S. et al. (2022 February). Hybrid metal-organic frameworks as an Exotic material for the photocatalytic degradation of pollutants present in wastewater: a review. *Chemosphere* 288: 132448.

18 Vinod, T. and Purnendu, B. (2009). Compendium of sewage treatment technologies. In: *National River Conservation Directorate Ministry of Environment & Forests Government of India*, 1–9.

19 Torad, N.L., Hu, M., Kamachi, Y. et al. (2013). Facile synthesis of nanoporous carbons with controlled particle sizes by direct carbonization of monodispersed ZIF-8 crystals. *Chemical Communications* 49 (25): 2521.

20 Li, Y. and Quan, X. (2017). Adsorption of sulfamethoxazole on nanoporous carbon derived from metal-organic frameworks. *Journal of Geoscience and Environment Protection* 05 (06): 1–8.

21 Dhakshinamoorthy, A., Li, Z., and Garcia, H. (2018). Catalysis and photocatalysis by metal-organic frameworks. *Chemical Society Reviews* 47 (22): 8134–8172.

22 Sharmin, E. and Zafar, F. (2016). Introductory chapter: metal-organic frameworks (MOFs). In: *Metal-Organic Frameworks*. InTech.

23 James, S.L. (2003). Metal-organic frameworks. *Chemical Society Reviews* 32 (5): 276.

24 Lu, W., Wei, Z., Gu, Z.Y. et al. (2014). Tuning the structure and function of metal-organic frameworks via linker design. *Chemical Society Reviews* 43 (16): 5561–5593.

25 Zhao, Y.P., Yang, H., Wang, F., and Du, Z.Y. (2014 September). A microporous manganese-based metal-organic framework for gas sorption and separation. *Journal of Molecular Structure* 1074: 19–21.

26 Abdi, J., Sisi, A.J., Hadipoor, M., and Khataee, A. (2022 February). State of the art on the ultrasonic-assisted removal of environmental pollutants using metal-organic frameworks. *Journal of Hazardous Materials* 424: 127558.

27 Huan, W., Xing, M., Cheng, C., and Li, J. (2019 January 22). Facile fabrication of magnetic metal-organic framework nanofibers for specific capture of phosphorylated peptides. *ACS Sustainable Chemistry & Engineering* 7 (2): 2245–2254.

28 Yang, D. and Gates, B.C. (2019 March 1). Catalysis by metal-organic frameworks: perspective and suggestions for future research. *ACS Catalysis* 9 (3): 1779–1798.

29 Gautam, S., Agrawal, H., Thakur, M. et al. (2020 June). Metal oxides and metal organic frameworks for the photocatalytic degradation: a review. *The Journal of Environmental Chemical Engineering* 8 (3): 103726.

30 Rasheed, T., Rizwan, K., Bilal, M., and Iqbal, H.M.N. (2020 March). Metal-organic framework-based engineered materials—fundamentals and applications. *Molecules* 25 (7): 1598.

31 Khan, N.A., Hasan, Z., and Jhung, S.H. (2013 January). Adsorptive removal of hazardous materials using metal-organic frameworks (MOFs): a review. *Journal of Hazardous Materials* 244–245: 444–456.

32 Tilgner, D., Friedrich, M., Verch, A. et al. A (2018 April). Metal-organic framework supported nonprecious metal photocatalyst for visible-light-driven wastewater treatment. *ChemPhotoChem* 2 (4): 349–352.

33 Panasonic (2013). Panasonic develops highly effective water purification system using new unique photocatalyst [Internet]. Headquarters News. [cited 2022 Sep 12]. Available from: https://news.panasonic.com/global/press/en130805-7.

34 Gu, Q., Ng, H.Y., Zhao, D., and Wang, J. (2020 April 1). Metal-organic frameworks (MOFs)-boosted filtration membrane technology for water sustainability. *APL Materials* [Internet] 8 (4): 040902. Available from: http://aip.scitation.org/doi/10.1063/5.0002905.

35 Le, T., Chen, X., Dong, H. et al. (2021 May 19). An evolving insight into metal organic framework-functionalized membranes for water and wastewater treatment and resource recovery. *Industrial & Engineering Chemistry Research* [Internet] 60 (19): 6869–6907. Available from: https://pubs.acs.org/doi/10.1021/acs.iecr.1c00543.

36 Lu, W., Wei, Z., Gu, Z.Y. et al. (2014). Tuning the structure and function of metal-organic frameworks via linker design [Internet]. *Chemical Society Reviews* The Royal Society of Chemistry. 43: 5561–5593. [cited 2022 Sep 12]. Available from: https://pubs.rsc.org/en/content/articlehtml/2014/cs/c4cs00003j.

37 Warsinger, D.M., Chakraborty, S., Tow, E.W. et al. (2018 June). A review of polymeric membranes and processes for potable water reuse. *Progress in Polymer Science* [Internet] 81: 209–237. Available from: https://linkinghub.elsevier.com/retrieve/pii/S0079670017300102.

38 Tajuddin, M.H.A., Jaafar, J., Hasbullah, H. et al. (2021 July 24)). Metal organic framework in membrane separation for wastewater treatment: potential and way forward. *The Arabian Journal for Science and Engineering* [Internet] 46 (7): 6109–6130. Available from: https://link.springer.com/10.1007/s13369-021-05509-7.

39 Sun, H., Tang, B., and Wu, P. (2017 June 28). Development of hybrid ultrafiltration membranes with improved water separation properties using modified superhydrophilic metal-organic framework nanoparticles. *ACS Applied Materials and Interfaces* [Internet] [cited 2022 Sep 12]. 9 (25): 21473–21484. Available from: https://pubs.acs.org/doi/abs/10.1021/acsami.7b05504.

40 Wibowo, A., Marsudi, M.A., Pramono, E. et al. (2021). Recent improvement strategies on metal-organic frameworks as adsorbent, catalyst, and membrane for wastewater treatment [Internet]. *Molecules. Multidisciplinary Digital Publishing Institute* 26: [cited 2022 Sep 12]. 5261. Available from: https://www.mdpi.com/1420-3049/26/17/5261/htm.

41 Jun, B.M., Al-Hamadani, Y.A.J., Son, A. et al. (2020). Applications of metal-organic framework based membranes in water purification: a review. *Separation and Purification Technology* 247: 116947. Elsevier.

42 Liu, T.Y., Yuan, H.G., Liu, Y.Y. et al. (2018 September 25). Metal-organic framework nanocomposite thin films with interfacial bindings and self-standing robustness for high water flux and enhanced ion selectivity. *ACS Nano* [Internet] [cited 2022 Sep 12]. 12 (9): 9253–9265. Available from: https://pubs.acs.org/doi/abs/10.1021/acsnano.8b03994.

43 Lee, J.Y., She, Q., Huo, F., and Tang, C.Y. (2015 October 5). Metal-organic framework-based porous matrix membranes for improving mass transfer in forward osmosis membranes. *Journal of Membrane Science* [Internet] [cited 2022 Sep 12]. 492 (492): 392–399. Available from: https://www.infona.pl//resource/bwmeta1.element.elsevier-69c6f198-585d-341e-9bb0-35e21f9b6b66.

44 Abdullah, N., Rahman, M.A., Dzarfan Othman, M.H. et al. (2018 October 1). Preparation, characterizations and performance evaluations of alumina hollow fiber membrane incorporated with UiO-66 particles for humic acid removal. *Journal of Membrane Science* [Internet] [cited 2022 Sep 12]. 563 (563): 162–174. Available from: https://www.infona.pl//resource/bwmeta1.element.elsevier-321244c7-9cd3-31ef-86a8-aa65e09fdd06.

45 Wu, M., Ye, H., Zhao, F., and Zeng, B. (2017 January 4). High-quality metal-organic framework ZIF-8 membrane supported on electrodeposited ZnO/2-methylimidazole nanocomposite: efficient adsorbent for the enrichment of acidic drugs. *Scientific Reports* [Internet] [cited 2022 Sep 12]. 7: Available from: https://pubmed.ncbi.nlm.nih.gov/28051129.

46 Liu, Y., Zhu, M., Chen, M. et al. (2019 March 1). A polydopamine-modified reduced graphene oxide (RGO)/MOFs nanocomposite with fast rejection capacity for organic dye. *Chemical Engineering Journal* 359: 47–57.

47 Zuo, J. and Chung, T.S. (2016 December 8). Metal-organic framework-functionalized alumina membranes for vacuum membrane distillation. *Water* (Switzerland) [Internet] [cited 2022 Sep 12]. 8 (12): 586. Available from: https://www.mdpi.com/2073-4441/8/12/586/htm.

48 Chen, H., Wu, S., Jia, X. et al. (2018 July 1). Atomic layer deposition fabricating of ceramic nanofiltration membranes for efficient separation of dyes from water. *AIChE Journal* [Internet] [cited 2022 Sep 12]. 64 (7): 2670–2678. Available from: https://www.researchgate.net/publication/322645762_Atomic_Layer_Deposition_Fabricating_of_Ceramic_Nanofiltration_Membranes_for_Efficient_Separation_of_Dyes_from_Water

49 Choudhury, R.R., Gohil, J.M., Mohanty, S., and Nayak, S.K. (2018). Antifouling, fouling release and antimicrobial materials for surface modification of reverse osmosis and nanofiltration membranes [Internet]. *Journal of Materials Chemistry A. The Royal Society of Chemistry* [cited 2022 Sep 12]. 6: 313–333. Available from: https://pubs.rsc.org/en/content/articlehtml/2018/ta/c7ta08627j.

50 Peng, Y., Huang, H., Zhang, Y. et al. (2018 January 15). A versatile MOF-based trap for heavy metal ion capture and dispersion. *Nature Communications* 2018 9:1 [Internet] [cited 2022 Sep 12]. 9 (1): 1–9. Available from: https://www.nature.com/articles/s41467-017-02600-2.

51 Zhou, X., Liu, W., Zhang, J. et al. (2017 October 18). Biogenic calcium carbonate with hierarchical organic-inorganic composite structure enhancing the removal of Pb(II) from wastewater. *ACS Applied Materials Interfaces* [Internet] [cited 2022 Sep 12]. 9 (41): 35785–35793. Available from: https://pubs.acs.org/doi/abs/10.1021/acsami.7b09304.

52 Wu, S., Ge, Y., Wang, Y. et al. (2018 August 3). Adsorption of Cr(VI) on nano Uio-66-NH 2 MOFs in water. *Environmental Technology* [Internet] 39 (15): 1937–1948. Available from: https://www.tandfonline.com/doi/full/10.1080/09593330.2017.1344732.

53 Jana, S., Saikia, A., Purkait, M.K., and Mohanty, K. (2011 May 15). Chitosan based ceramic ultrafiltration membrane: preparation, characterization and application to remove Hg(II) and As(III) using polymer enhanced ultrafiltration. *Chemical Engineering Journal* 170 (1): 209–219.

54 Wang, K., Tian, Z., and Yin, N. (2018 March 14). Significantly enhancing Cu(II) adsorption onto Zr-MOFs through novel cross-flow disturbance of ceramic membrane. *Industrial & Engineering Chemistry Research* [Internet] [cited 2022 Sep 12]. 57 (10): 3773–3780. Available from: https://pubs.acs.org/doi/abs/10.1021/acs.iecr.7b04850.

55 Yin, N., Wang, K., Wang, L., and Li, Z. (2016 December 15). Amino-functionalized MOFs combining ceramic membrane ultrafiltration for Pb (II) removal. *Chemical Engineering Journal* [Internet] [cited 2022 Sep 12]. 306: 619–628. Available from: https://linkinghub.elsevier.com/retrieve/pii/S1385894716310117.

56 Arjmandi, M., Peyravi, M., Pourafshari Chenar, M., and Jahanshahi, M. (2019 June 1). A new concept of MOF-based PMM by modification of conventional dense film casting method: significant impact on the performance of FO process. *Journal of Membrane Science* 579: 253–265.

57 Nik, O.G., Chen, X.Y., and Kaliaguine, S. (2012 September 15). Functionalized metal organic framework-polyimide mixed matrix membranes for CO2/CH4 separation. *Undefined* 413–414: 48–61.

58 Liu, X., Demir, N.K., Wu, Z., and Li, K. (2015 June 10). Highly water-stable zirconium metal-organic framework UiO-66 membranes supported on alumina hollow fibers for desalination. *Journal of the American Chemical Society* [Internet] [cited 2022 Sep 12]. 137 (22): 6999–7002. Available from: https://pubs.acs.org/doi/abs/10.1021/jacs.5b02276.

59 Dai, R., Zhang, X., Liu, M. et al. (2019 March 1). Porous metal organic framework CuBDC nanosheet incorporated thin-film nanocomposite membrane for high-performance forward osmosis. *Journal of Membrane Science* 573: 46–54.

60 Elder, A.C., Aleksandrov, A.B., Nair, S., and Orlando, T.M. (2017 October 3). Interactions on External MOF Surfaces: desorption of Water and Ethanol from CuBDC Nanosheets. *Langmuir* [Internet] [cited 2022 Sep 12]. 33 (39): 10153–10160. Available from: https://pubmed.ncbi.nlm.nih.gov/28877442.

61 Petersen, R.J. (1993). Composite reverse osmosis and nanofiltration membranes. *Journal of Membrane Science* 83: 81–150. Elsevier.

62 Firouzjaei, M.D., Shamsabadi, A.A., Aktij, S.A. et al. (2018 December 12). Exploiting synergetic effects of graphene oxide and a silver-based metal-organic framework to enhance antifouling and anti-biofouling properties of thin-film nanocomposite membranes. *ACS Applied Materials Interfaces* [Internet] [cited 2022 Sep 12]. 10 (49): 42967–42978. Available from: https://pubs.acs.org/doi/abs/10.1021/acsami.8b12714.

63 Yip, N.Y., Tiraferri, A., Phillip, W.A. et al. (2010 May 15). High performance thin-film composite forward osmosis membrane. *Environmental Science & Technology* [Internet] [cited 2022 Sep 12]. 44 (10): 3812–3818. Available from: https://pubs.acs.org/doi/abs/10.1021/es1002555.

64 Abu Tarboush, B.J., Arafat, H.A., Matsuura, T., and Rana, D. (2017 November 20). Recent advances in thin film composite (TFC) reverse osmosis and nanofiltration membranes for desalination. *Journal of Applied Membrane Science & Technology* [Internet] [cited 2022 Sep 12]. 10 (1): 41–50. Available from:https://amst.utm.my/index.php/amst/article/view/72.

65 Duan, J., Pan, Y., Pacheco, F. et al. (2015 February 5). High-performance polyamide thin-film-nanocomposite reverse osmosis membranes containing hydrophobic zeolitic imidazolate framework-8. *Journal of Membrane Science* 476: 303–310.

66 Kadhom, M., Hu, W., and Deng, B. (2017 June 14). Thin film nanocomposite membrane filled with metal-organic frameworks UiO-66 and MIL-125 nanoparticles for water desalination. *Membranes* (Basel) [Internet] [cited 2022 Sep 12]. 7 (2): 31. Available from: https://www.mdpi.com/2077-0375/7/2/31/htm.

67 Park, H.M., Jee, K.Y., and Lee, Y.T. (2017). Preparation and characterization of a thin-film composite reverse osmosis membrane using a polysulfone membrane including metal-organic frameworks. *Journal of Membrane Science* 541: 510–518.

68 Samantaray, P.K., Madras, G., and Bose, S. (2019 January 7). Water remediation aided by a graphene-oxide-anchored metal organic framework through pore- and charge-based sieving of ions. *ACS Sustainable Chemistry & Engineering* [Internet] [cited 2022 Sep 12]. 7 (1): 1580–1590. Available from: https://pubs.acs.org/doi/abs/10.1021/acssuschemeng.8b05354.

69 Ye, W., Liu, H., Lin, F. et al. (2019 October 10). High-flux nanofiltration membranes tailored by bio-inspired co-deposition of hydrophilic g-C3N4 nanosheets for enhanced selectivity towards organics and salts. *Environmental Science: Nano*

[Internet] [cited 2022 Sep 12]. 6 (10): 2958–2967. Available from: https://pubs.rsc.org/en/content/articlehtml/2019/en/c9en00692c.

70 Gutiérrez-Sevillano, J.J., Dubbeldam, D., Bellarosa, L. et al. (2013 October 10). Strategies to simultaneously enhance the hydrostability and the alcohol-water separation behavior of Cu-BTC. *Journal of Physical Chemistry C* 117 (40): 20706–20714.

71 Wang, L., Fang, M., Liu, J. et al. (2015 October 20). Layer-by-layer fabrication of high-performance polyamide/ZIF-8 nanocomposite membrane for nanofiltration applications. *ACS Applied Materials Interfaces* [Internet] [cited 2022 Sep 12]. 7 (43): 24082–24093. Available from: https://pubs.acs.org/doi/abs/10.1021/acsami.5b07128.

72 CHEM L. RO MembraneLG Chem [Internet]. [cited 2022 Sep 12]. Available from: https://www.lgchem.com/product/PD00000070.

73 Wang, L., Duan, S., Fang, M. et al. (2016 July 26). Surface modification route to prepare novel polyamide@NH2-MIL-88B nanocomposite membranes for water treatment. *RSC Advances* [Internet] [cited 2022 Sep 12]. 6 (75): 71250–71261. Available from: https://pubs.rsc.org/en/content/articlehtml/2016/ra/c6ra09080j.

74 Xu, Y., Gao, X., Wang, X. et al. (2016 November 1). Highly and stably water permeable thin film nanocomposite membranes doped with MIL-101 (Cr) nanoparticles for reverse osmosis application. *Materials* [Internet] [cited 2022 Sep 12]. 9 (11): Available from: /pmc/articles/PMC5457272.

75 Marchetti, P., Mechelhoff, M., and Livingston, A.G. (2015 December 2). Tunable-porosity membranes from discrete nanoparticles. *Scientific Reports* [Internet] [cited 2022 Sep 12]. 5 (1): 1–14. Available from: https://www.nature.com/articles/srep17353.

76 Cheng, X., Ding, S., Guo, J. et al. (2017). In-situ interfacial formation of TiO2/polypyrrole selective layer for improving the separation efficiency towards molecular separation. *Journal of Membrane Science* [Internet] [cited 2022 Sep 12]. 536: 19–27. Available from: https://www.cheric.org/research/tech/periodicals/view.php?seq=1558915.

77 Efome, J.E., Rana, D., Matsuura, T., and Lan, C.Q. (2018 June 6). Insight studies on metal-organic framework nanofibrous membrane adsorption and activation for heavy metal ions removal from aqueous solution. *ACS Applied Materials Interfaces* [Internet] [cited 2022 Sep 12]. 10 (22): 18619–18629. Available from: https://www.researchgate.net/publication/325158993_Insight_Studies_on_Metal-Organic_Framework_Nanofibrous_Membrane_Adsorption_and_Activation_for_Heavy_Metal_Ions_Removal_from_Aqueous_Solution.

78 Gnanasekaran, G., Balaguru, S., Arthanareeswaran, G., and Das, D.B. (2019 February 11). Removal of hazardous material from wastewater by using metal organic framework (MOF) embedded polymeric membranes. *Separation Science and Technology* Philadelphia) [Internet] [cited 2022 Sep 12]. 54 (3): 434–446. Available from: https://www.tandfonline.com/doi/abs/10.1080/01496395.2018.1508232.

79 He, Y., Tang, Y.P., Ma, D., and Chung, T.S. (2017). UiO-66 incorporated thin-film nanocomposite membranes for efficient selenium and arsenic removal. *Journal of Membrane Science* [Internet] [cited 2022 Sep 12]. 541: 262–270. Available from: https://academia.kaust.edu.sa/en/publications/uio-66-incorporated-thin-film-nanocomposite-membranes-for-efficie.

80 Yang, H., Wang, N., Wang, L. et al. (2018). Vacuum-assisted assembly of ZIF-8@GO composite membranes on ceramic tube with enhanced organic solvent nanofiltration performance. *Journal of Membrane Science* 545: 158–166.

81 Hu, Z., Peng, Y., Kang, Z. et al. (2015May 18). A modulated hydrothermal (MHT) approach for the facile synthesis of UiO-66-type MOFs. *Inorganic Chemistry* [Internet] [cited 2022 Sep 12]. 54 (10): 4862–4868. Available from: https://pubs.acs.org/doi/abs/10.1021/acs.inorgchem.5b00435.

82 Vandezande, P., Gevers, L.E.M., and Vankelecom, I.F.J. (2008 January 7). Solvent resistant nanofiltration: separating on a molecular level. *Chemical Society Reviews* [Internet] [cited 2022 Sep 12]. 37 (2): 365–405. Available from: https://pubs.rsc.org/en/content/articlehtml/2008/cs/b610848m.

83 Rodrigues Filho, G., Monteiro, D.S., Meireles, C.D.S. et al. (2008 July 4). Synthesis and characterization of cellulose acetate produced from recycled newspaper. *Carbohydrate Polymers* [Internet] [cited 2022 Sep 12]. 73 (1): 74–82. Available from: https://repositorio.unesp.br/handle/11449/39859.

84 Wu, G., Gan, S., Cui, L., and Xu, Y. (2008 August 30). Preparation and characterization of PES/TiO2 composite membranes. *Applied Surface Science* [Internet] [cited 2022 Sep 12]. 21 (254): 7080–7086. Available from: https://www.infona.pl//resource/bwmeta1.element.elsevier-16b41bd5-7343-3406-95d8-cba986bd99fb.

85 Bartolotti, L.J., Gadre, S.R., and Parr, R.G. (1980). Electronegativities of the elements from simple. xα theory. *Journal of the American Chemical Society* [Internet] [cited 2022 Sep 12]. 102 (9): 2945–2948. Available from: https://pubs.acs.org/doi/abs/10.1021/ja00529a013.

86 Liao, Z., Fang, X., Xie, J. et al. (2019 February 6). Hydrophilic hollow nanocube-functionalized thin film nanocomposite membrane with enhanced nanofiltration performance. *ACS Applied Materials Interfaces* [Internet] [cited 2022 Sep 12]. 11 (5): 5344–5352. Available from: https://pubs.acs.org/doi/abs/10.1021/acsami.8b19121.

87 Ji, Y.L., An, Q.F., Guo, Y.S. et al. (2016 March 8). Bio-inspired fabrication of high perm-selectivity and anti-fouling membranes based on zwitterionic polyelectrolyte nanoparticles. *Journal of Materials Chemistry A* [Internet] [cited 2022 Sep 12]. 4 (11): 4224–4231. Available from: https://pubs.rsc.org/en/content/articlehtml/2016/ta/c6ta00005c.

88 Hu, X., Wang, C., Li, J. et al. (2018 May 2). Metal-organic framework-derived hollow carbon nanocubes for fast solid-phase microextraction of polycyclic aromatic hydrocarbons. *ACS Applied Materials Interfaces* [Internet] [cited 2022 Sep 12]. 10 (17): 15051–15057. Available from: https://pubmed.ncbi.nlm.nih.gov/29648778.

89 Sun, H. and Wu, P. (2018 October 15). Tuning the functional groups of carbon quantum dots in thin film nanocomposite membranes for nanofiltration. *Journal of Membrane Science* 564: 394–403.

90 Rahimpour, A., Seyedpour, S.F., Aghapour Aktij, S. et al. (2018 May 1). Simultaneous improvement of antimicrobial, antifouling, and transport properties of forward osmosis membranes with immobilized highly-compatible polyrhodanine nanoparticles. *Environmental Science & Technology* [Internet] [cited 2022 Sep 12]. 52 (9): 5246–5258. Available from: https://pubs.acs.org/doi/abs/10.1021/acs.est.8b00804.

91 Nano – NX Filtration [Internet]. [cited 2022 Sep 12]. Available from: https://nxfiltration.com/products/nano/?utm_source=google&utm_medium=cpc&utm_campaign=search-nanofiltration&utm_content=nanofiltration&utm_term=nanofiltration&gclid=Cj0KCQjwjvaYBhDlARIsAO8PkE2fCcPqDmhdQdORAHaLcXe_HXQKwlCqHN9kJG5zuzpVd6gZtZEJkVwaAqaoEALw_w.

92 Marshall, A.D., Munro, P.A., and Trägårdh, G. (1993 March 1). The effect of protein fouling in microfiltration and ultrafiltration on permeate flux, protein retention and selectivity: a literature review. *Desalination* 91 (1): 65–108.

93 Gupta, K.M., Zhang, K., and Jiang, J. (2015 November 20). Water desalination through zeolitic imidazolate framework membranes: significant role of functional groups. *Langmuir* [Internet] [cited 2022 Sep 12]. 31 (48): 13230–13237. Available from: https://pubs.acs.org/doi/abs/10.1021/acs.langmuir.5b03593.

94 Gholami, F., Zinadini, S., Zinatizadeh, A.A., and Abbasi, A.R. (2018 April 3). TMU-5 metal-organic frameworks (MOFs) as a novel nanofiller for flux increment and fouling mitigation in PES ultrafiltration membrane. *Separation and Purification Technology* 194: 272–280.

95 Ma, J., Guo, X., Ying, Y. et al. (2017). Composite ultrafiltration membrane tailored by MOF@GO with highly improved water purification performance. *Chemical Engineering Journal* 313: 890–898.

96 Makhetha, T.A. and Moutloali, R.M. (2018 May 15). Antifouling properties of Cu(tpa)@GO/PES composite membranes and selective dye rejection. *Journal of Membrane Science* 554: 195–210.

97 Ren, Y., Li, T., Zhang, W. et al. (2019 March 5). MIL-PVDF blend ultrafiltration membranes with ultrahigh MOF loading for simultaneous adsorption and catalytic oxidation of methylene blue. *Journal of Hazardous Materials* [Internet] [cited 2022 Sep 12]. 365: 312–321. Available from: https://pubmed.ncbi.nlm.nih.gov/30447639.

98 Prince, J.A., Bhuvana, S., Anbharasi, V. et al. (2014 October 9). Self-cleaning metal organic framework (MOF) based ultra filtration membranes – a solution to bio-fouling in membrane separation processes. *Scientific Reports* 2014 4:1 [Internet] [cited 2022 Sep 12]. 4 (1): 1–9. Available from: https://www.nature.com/articles/srep06555.

99 Zhang, F., Zhang, W., Yu, Y. et al. (2013 April 1). Sol-gel preparation of PAA-g-PVDF/TiO2 nanocomposite hollow fiber membranes with extremely high water flux and improved antifouling property. *Journal of Membrane Science* [Internet] [cited 2022 Sep 12]. 432 (432): 25–32. Available from: https://www.infona.pl//resource/bwmeta1.element.elsevier-b3f5f837-de8d-3cf5-930d-2682d2ca76dd.

100 Xu, Z., Zhang, J., Shan, M. et al. (2014 May 15). Organosilane-functionalized graphene oxide for enhanced antifouling and mechanical properties of polyvinylidene fluoride ultrafiltration membranes. *Journal of Membrane Science* [Internet] [cited 2022 Sep 12]. 458 (458): 1–13. Available from: https://www.infona.pl//resource/bwmeta1.element.elsevier-53b4c8f8-9ffd-3c1c-83db-348e287f5b13.

101 Yang, Y., Zhang, H., Wang, P. et al. (2007 February 1). The influence of nano-sized TiO2 fillers on the morphologies and properties of PSF UF membrane. *Journal of Membrane Science* [Internet] [cited 2022 Sep 12]. 288 (1–2): 231–238. Available from: https://www.infona.pl//resource/bwmeta1.element.elsevier-abaa5a1d-87c4-345c-abfb-842ab05f8ebf.

102 Ruan, H., Guo, C., Yu, H. et al. (2016 November 23). Fabrication of a mil-53(al) nanocomposite membrane and potential application in desalination of dye solutions. *Industrial & Engineering Chemistry Research* [Internet] [cited 2022 Sep 12]. 55 (46): 12099–12110. Available from: https://pubs.acs.org/doi/abs/10.1021/acs.iecr.6b03201.

103 Yuan, W. and Zydney, A.L. (2000 December 1). Humic acid fouling during ultrafiltration. *Environmental Science & Technology* [Internet] [cited 2022 Sep 12]. 34 (23): 5043–5050. Available from: https://pubs.acs.org/doi/abs/10.1021/es0012366.

104 Li, T., Zhang, W., Zhai, S. et al. (2018 October). Efficient removal of nickel(II) from high salinity wastewater by a novel PAA/ZIF-8/PVDF hybrid ultrafiltration membrane. *Water Research* [Internet] 143: 87–98. Available from: https://linkinghub.elsevier.com/retrieve/pii/S0043135418304810.

105 Rahimpour, A., Jahanshahi, M., Khalili, S. et al. (2012 February1). Novel functionalized carbon nanotubes for improving the surface properties and performance of polyethersulfone (PES) membrane. *Desalination* 286: 99–107.

8

Prospects and Potentials of Microbial Applications on Heavy-Metal Removal from Wastewater

Dipankar Ghosh[1,*], *Shubhangi Chaudhary*[1], *and Snigdha Dhara*[1,2]

[1] *Microbial Engineering and Algal Biotechnology Laboratory, Department of Biosciences, JIS University, Kolkata, West Bengal, 700109, India*
[2] *Department of Biotechnology, Maulana Abul Kalam Azaj University of Technology, Nadia, West Bengal, 741249, India*
[*] *Corresponding author*

8.1 Introduction

The importance of water in the global economy cannot be overstated. Water covers the majority of Earth's surface (71%) and freshwater makes up only 3% of this [1]. Wastewater is water that has been affected by domestic, industrial, and commercial use. Heavy-metal-containing effluent is generated and dispersed into natural resources by human and industrial practices, rendering them unusable unreachable and harmful to human health and the ecology. To a greater extent, industrialization is to blame for environmental contamination, particularly in water, where rivers and lakes have become overrun with harmful compounds [2].

Heavy metals are a collection of substances that exist in nature but are unparalleled. Overconsumption and accumulation result from the continual release of heavy metallic constituents into water resources from industrial sectors(i.e., fertilizer, metallurgy, leather, aerospace, photography, mining, electroplating, pesticides, surface finishing, iron and steel, energy and fuel production, electrolysis, metal surface treatment, electro-osmosis, and appliance manufacturing) [3]. Heavier metals show higher including chromium (Cr), lead (Pb), zinc (Zn), arsenic (As), copper (Cu), nickel (Ni), cobalt (Co), cadmium (Cd), and mercury (Hg). Undesirable heavy-metal accumulation in biological systems leads to a number of illnesses and disorders that put human life in danger. Even in the parts per billion (ppb) range, they can cause health problems [4]. To remove heavy metallic residuals from effluents produced by versatile organizations that eventually take over and damage fresh water resources, biosorption has emerged as a potential alternative to traditional approaches.

Inhalation, percutaneous contact, and consumption have become the most common pathways of heavy-metal exposure. Every metal has its respective symptoms of toxicity indicators. Duration, dose, type, and chemical makeup of the heavy metal

Metal Organic Frameworks for Wastewater Contaminant Removal, First Edition. Edited by Arun Lal Srivastav, Lata Rani, Jyotsna Kaushal, and Tien Duc Pham.

all affect how much of a negative impact on health there will be. The effect could be toxic, mutagenic, neurotoxic, teratogenic, or carcinogenic in nature [5]. Multiple studies have shown that heavy metals cause cell damage and apoptosis by affecting cellular organelles and interacting with cellular constituents. These cause numerous organ abnormalities. Heavy-metal toxicity damages the body's fundamental systems and increases the risk of cancer development [6]. Metal ion pollution is extremely long-lasting, and the majority of them are non-biodegradable. Even circulatory, gastrointestinal, and nervous system disorders have been caused by the negative influences of heavy metals including Cr, Pb, Zn, As, Cu, Ni, Co, Cd, and Hg. These metals have an effect on other organs, leading to cancer, infertility, blindness, deafness, and brain impairment including death [7, 8].

8.2 Mainstream Avenues to Remediate Heavy Metals in Wastewater

Heavy metals are among the most significant contaminants that have an effect on freshwater reservoirs [9]. This is because industries dispose of significant amounts of wastewater that is contaminated with metals. Over the course of the last few years, wide divergences of conventional treatment methodologies have been applied in an effort to purge contaminated water of its heavy-metal content. Chemical precipitation, reverse osmosis, ion exchange, ultrafiltration, electrowinning, and phytoremediation are among the most utilized technologies, and these are discussed here in brief [10–12]:

- *Chemical precipitation*is the method that is utilized most frequently in the process of removing heavy metals from inorganic effluents. Precipitants have accelerated substantive efficaciousness to accelerated out dissolved metal ions, leading to the establishment of metal hydroxides, phosphates (insoluble solid particles), sulfides, and carbonates, which may have been easily segregated using accelerated methods of filtration and sedimentation.
- *Ion exchange.* The spontaneous exchange of ions between solid and liquid phases is the backbone of ion exchange. This technique can be broken down into its component parts for the accelerated exchange of ions within solid and liquid phases. Both cations and anions are exchanged with an electrolytic solution through an ion exchanger solid-resin bed and are expressed as chemically corresponding counter-ions of similar charge.
- *Membrane filtration.* In addition to metal-based ionic constituents, organic residues and suspended solid constituents can also be a good deal via the membrane filtering process. For the elimination of pollutants of various sizes, a membrane is a layer that is selective and makes contact between two homogeneous phases that have either a porous or non-porous structure [13].
- *Ultrafiltration (UF).*In this approach, diverse ranges of permeable membranes are used for separation with pore diameters ranging from 0.1 to 0.001 microns that allow water and solutes with low molecular weight to pass through while holding bulkier particles, macromolecules, and colloids. In ultrafiltration platforms,

a copolymer of malate and acrylic acids are generally implemented to segregate Ni (II), Zn (II), Cu (II), and Mn (II) from solutions that are aqueous in nature, achieving a removal efficiency of 98.8% by producing macromolecular complexes with the polymers that have been rejected by the membrane [14].

- *Microfiltration (MF)* is similar to ultrafiltration because it also works on the same principle. The significant difference between the two procedures is that MF extracts bigger solutes than UF refuses. The yeast-based bioaccumulation process (III) is a unique avenue that can eliminate metal ions from tap water having diverse ranges of metal ions including Cu (II), Cd (II), Pb (II), and Cr (II) through the involvement of cross-flow microfiltration (CFMF) [15].
- *Reverse osmosis (RO)* is used to remove heavy-metal effluents created by industry. In the RO process, the metal-enriched solution needs to transit through a semi-permeable membrane. The polyamide thin-film composite membrane TW30-1812–50 is used to segregate and separate Ni (II), Cu (II), and Zn (II) metal ions following the basic principle of reverse osmosis mechanisms [16].
- *Electrodialysis (ED)* is a revolutionary platform for liquid-hybrid membrane segregation technology that uses an electric potential or a concentration gradient to separate ionized species in a solution that flows through a membrane allowing ion exchange. An ED system has been used to extract arsenic, lead, manganese, and nitrate nitrogen from groundwater [17].
- *Photocatalysis* uses non-toxic semiconductors for the rapid and efficient obliteration of environmental contaminants. The stages of this approach include transfer, adsorption to the semiconductor surface, photocatalytic reactions at the surface, and eventually breakdown and removal of the contaminants at the interface region. Heavy metals in pharmaceutical waste have been photocatalytically degraded and retrieved using a selenium-doped ZnO nanocomposite semiconductor [18].

Techniques such as coagulation/flocculation [19], electrocoagulation [20], electroflotation [21], and electrodeposition [22] have been employed to separate heavy metals from polluted water resources in addition to these traditional methods. Incomplete metal removal, sludge production, high reagent and energy requirements, metal precipitate aggregation, and membrane fouling have all been limitations of the aforementioned approaches. Numerous biological methods for removing heavy metal from wastewater that fall under the categories of physical, chemical, or biological processes have already been covered. Regarding metal removal, physical and chemical methods are more expensive than biological treatments, but biological treatments lack effectiveness and rapidity. Physical and chemical treatment advantages and downsides are simplified in Table 8.1:

8.3 The Microbial Recycling Approach

To address the broad spectrum of pitfalls in conventional approaches to heavy-metal removal, microbial recycling needs to be the focus of future endeavors into toxic-metal recycling that is cost-effective. Biosorption/bioaccumulation procedures have recently been hailed as inventive, cost-effective, efficient, and environmentally

Table 8.1 Rewards and pitfalls of chemical avenues in heavy metals segregation [23–25].

Treatment	Advantages	Disadvantages
Chemical precipitation	Low cost of capital and a simple handling procedure.	Additional cost due to disposal of a large amount of sludge.
Ion exchange	Good removal of a wide range of heavy metals.	Adsorbents require regeneration or disposal.
Membrane filtration	Proper extraction of heavy metals.	Concentrated sludge production and expensive.
Oxidation	Rapid method for toxic-effluent extraction.	Increased energy costs and the formation of byproducts.
Adsorption	Flexibility and simplicity of design, ease of handling, and toxicity to harmful contaminants.	Adsorbents require regeneration.
Photocatalysis	Removes metals and organic pollutants simultaneously. Less dangerous byproducts.	Extended time period and restricted applications.
Coagulation/flocculation	Economically sustainable.	High sludge production and the formation of bulky particles.
Photochemical treatment	No sludge formation.	Emergence of a by-product.
Electrokinetic coagulation	Financially feasible.	Huge sludge releases.
Electrochemical treatment	Quick and efficient method for certain metal ions.	High energy costs and the formation of large particles.
Biological treatment	Achieves the removal of some metals	Technique yet to be established and commercialized.
Electrodialysis	Higher extraction selectivity.	Elevated expenses due to higher levels of energy consumption and fouling of membranes.

friendly methods for extracting heavy metals from wastewaters generated by diverse fields. Microbial metabolic cascades in biotic-viable microbial biosorbents are essential bioparts that can provide dynamic acceleration during bioaccumulation. The process requires the removal of a metal ion from the environment where the micro-organism is growing. Metal species accumulation has been best accomplished by organisms that can withstand large metal-ion loads [26]. It was discovered that when the concentration of the metal ion to be accumulated rises, the cell's morphology and physiology alter. Selecting micro-organisms that have been screened from contaminated environments can result in efficient bioaccumulation. Some additional biosorbents used for metal accumulation have been summarized in Table 8.2:

Table 8.2 Biosorbent efficiencies toward metal extraction [27–33].

Biosorbent Type	Metal Ion	Uptake capacity (mg/g)
Aspergillusniger	Pb (II)	172.32
Saccharomyces cerevisiae	Cr (III) & (VI)	11.1, 3.33
Penicillium simplicissimum	Cd (II), Zn (II) &Pb (II)	52.51, 65.63, 76.92
Pleurotus ostreatus	Cu (II), Zn (II), Ni (II)	8.07, 3.42, 20.14
E. coli	Ni (II)	6.90
Rhizobium spp.	Cd (II), Co (II)	135.32, 167.05
Spirogyra sp.	Pb (II)	140.01

8.4 General Overview of Heavy-Metal Pollution in Wastewater

Heavy metals have been defined as substances with a density more than 5 g/cm^3. This category includes a wide number of elements; however, the ones listed in Table 8.3 are the most common. These are the ones that matter in terms of the environment [34].

Heavy Metals in the Environment: Heavy metals are highly electronegative elements [35] with a specific gravity or atomic weight [36], and their ions are hazardous or poisonous in tiny concentrations [37, 38]. Because of the widespread use of heavy metals and their components in processes, non-biodegradable pollutants [39] that tend to accumulate in living creatures have become quite the serious environmental concern in recent years. Acid rain may cause heavy metals to migrate into the aquatic system from soils that have been eroded and broken down [40, 41]. Landfill leachate helps to detect different metal ionic constituents like iron, aluminum, arsenic, cadmium, copper, iron, manganese, mercury, nickel, silver, and zinc. However, landfill leachate produces difficulties during disassembly, which can induce secondary contamination [42].

- Heavy metals present in surface water: Industrial and municipal wastewater carry heavy metals into surface water. The leaching of chemical pollutants from landfills and the deposition of airborne particular matter can also affect surface water [43]. Water contamination can also be triggered by the leaching of pesticides, mineral fertilizers, and additional components from the soil. Heavy metals can be found in the environment in the following forms:
  - ionic state (the most toxic form to living organisms)
  - ions linked to a variety of ligand (complex compounds)
  - precipitated molecules
- Heavy metals present in soil: Industry and power engineering, air emissions, and landfilling are the main contributors of pollution caused by heavy metals in soil. Metals can be introduced to soil through fertilizer with sewage sludge and insecticides. Soil contamination happens as a result of transportation along route systems [44].

- Heavy metals present in the atmosphere: Heavy metals found in the atmosphere are mostly created by emissions of specific matter from industry, transportation, and power generation. The pollution of the atmosphere has a worldwide scope [45]. This pollution of the air has a detrimental impact on climate change, causes economic losses (primarily in agriculture and forestry), and increases deadly health hazards to human society and community.

A broad spectrum of heavy metals certainly has an adverse influence on aquatic flora and fauna and may offer a public health risk where contaminated organisms are used for food. When they exceed tolerance levels [46], they can cause poisoning, cancer, and brain damage. The authorities for environmental monitoring have set detrimental consequences for heavy-metal levels in drinking water because of their deleterious effects [47]. The hazardous impacts of some heavy metals have been summarized in Table 8.3:

Table 8.3 Sources, toxicity effects,andmaximum contaminant levels (MCL) for the most hazardous heavy metals [35, 41, 48, 49].

Heavy metal	Manufacturing industries	Entry route	Toxicity effect	MCL (mg/L)
Arsenic	Metal hardening, phosphate fertilizer, paints and textile, fungicides	Inhalation	Visceral cancers, liver andkidney damage, nausea andvomiting	0.052
Cadmium	Electroplating, pesticide fertilizer, welding, pigments	Ingestion	Renal disorders, lung damage, human carcinogenic	0.011
Mercury	Chlor-alkali, paper industry, scientific experiments, production of batteries	Inhalation andabsorption by skin	Rheumatoid arthritis, nervous system diseases, muscle coordination problems	0.000030 (vapor)
Chromium	Tanning rubber, electroplating, paints, photographic processes	Ingestion andabsorption by skin	Diarrhea, carcinogenic, vomiting, respiratory system damage	0.055
Lead	Battery production, automobile emission, paints, mining, pesticides	Ingestion	Kidney disease, nervous and circulatory system damage, loss of appetite	0.016
Copper	Electroplating, rayon, water pipes, fungicides	-	Insomnia, Wilson's disease, anemia, cystic fibrosis, schizophrenia, arthritis	0.252
Zinc	Galvanizing, the iron and steel industries	-	Lethargy, depression, hypertension, neurological diseases	0.84
Nickel	Electroplating, electrochemical processes, silver refineries	Inhalation	Chronic asthma, dermatitis, dizziness nausea and vomiting, encephalopathy	0.21

8.5 Techniques for Heavy-Metal Removal

Heavy metals have the utmost toxicity, causing substantial adverse effects on the environment, and wastewater purification from highly toxic substances has gained a lot of attention in recent years. As a result, effective methods for removing hazardous heavy metals (i.e. cadmium, chromium, lead, mercury, and arsenic) are necessary [50]. Some traditional methodologies for that goal are as follows:

i) **Chemical Precipitation**
One of the most extensively used methods for removing heavy metals from contaminated water and wastewater is chemical precipitation [51]. Its process is based on the inclusion of counter-ions to diminish the solubility of the ionic elements of contaminants dissolved in wastewater into solid particles [12, 52]. Setting the pH to basic conditions (pH 9–11) is the essential factor that enhances the removal of heavy metals by chemical precipitation. Lime and limestone are the most frequently used precipitant agents because they are widely available and affordable in the majority of countries [53]. Inorganic effluents having metal concentrations over 1000 mg/L can be efficiently treated with chemical precipitation. Utilizing chemical precipitation has additional benefits, such as its ease of use, minimal equipment requirements, and simple, secure operations. Sludge disposal's long-term environmental implications include retreatment, sluggish precipitation of metal, poor overall settling, and metal precipitate aggregation [54]. Heavy-metal treatment can take 2–4 hours in some circumstances, while other metal removals, such as mercury, cadmium, and lead, work slowly and inadequately [55]. Other types of this procedure are utilized, such as carbonate and sulfide precipitation, which are similar to hydroxide precipitation except for the pH range of the pretreated solution and the agents needed to achieve the contaminant ion precipitation [11].

ii) **Ion Exchange**
Ion exchange is a conventional chemical technology utilized in the removal of heavy metals from effluent that has proven to be effective in the industry. Ionexchanger is a substance that may exchange anions or cations with the materials around it [56]. As a conclusion, when ion exchangers contain cations, such as hydrogen and sodium ions, they are categorized as cation exchangers, while ion exchangers containing anions, such as hydroxyl and chloride, are classified as anion exchangers [57]. Due to the matrix's rapid contamination from organics and other effluent particles, this method has the drawback of being unable to handle concentrated metal solutions. Furthermore, the exchange of ions is a broad-spectrum process, which is extremely sensitive to the pH of the solution. Electrolytic recovery, also referred to as electrowinning, is one of the numerous techniques used to remove metals from processed water streams. A cathode plate and an insoluble anode are included in an aqueous metal-bearing solution before the solution is subjected to an electric current. One obvious downside is that corrosion might become a key limiting factor, necessitating frequent electrode replacements [58].

iii) **Adsorption**

Adsorption is another extremely effective physicochemical treatment for removing several types of contaminants from contaminated water and wastewater. The fundamental tenet of this approach is that mass ion transfer from the liquid phase to the surface of the solid phase is constrained by interactions with physical and/or chemical properties [11]. As a result, all adsorption methods rely on solid-liquid equilibrium and mass transfer, which depends on intraparticle or film formation diffusion, or both [59]. The adsorption process is divided into physical adsorption and chemical adsorption based on the types of intermolecular attractive interactions that allow adsorbents to precipitate from the solution into molecular scale pores. Van der Waals forces, or the attraction between the adsorbent and adsorbate, are what cause the former to occur. A chemical reaction occurs between the adsorbent and the adsorbate in chemical adsorption, also known as active adsorption [12]. Pollutant sorption onto solid sorbents involves three primary steps: (a) pollutant transport from the bulk solution to the sorbent surface; (b) adsorption on the particle surface; and (c) pollutant transit within the sorbent particle. The important parameters that influence the selection of the best adsorbent for treating inorganic wastewater seem to be technical applicability and cost-effectiveness [58].

iv) **Membrane Filtration**

Membrane filtration has attracted much interest for treating inorganic wastewater since it can remove suspended solids and organic molecules as well as inorganic pollutants like heavy metals. This is for a variety of distinct separation physicochemical methods, both continuous and batch [60, 61], that are relatively new compared to thermal treatments [62]. Membrane filtering procedures are categorized based on the size of particles that must be retained. Microfiltration (MIF), ultrafiltration (UF), nanofiltration (NF), reverse osmosis (RO), and electrodialysis (ED) have been some of the aquatic environmental treatment procedures utilized for membrane filtration [63–65]. They have excellent efficiency, are simple to operate, take up minimal space [66], and can be used to create pressure gradients, concentration gradients, or electrical potential gradients [67]. A general schematic diagram has been shown in Fig 5.1. Membranes have been formed of polymer, ceramic, or inorganic components, which consist of a lot of tiny pores and have a molecular weight ranging from 100 to 1000 Dalton. As a result, in the polymer matrix, the membrane should have a semipermeable and dense barrier layer.

v) **Microfiltration (MIF)**

Microfiltrarion is a technique for the removal of microbes from water used during medical applications [63]. The use of microfiltration (0.25–0.60 m) membrane-based sorbents featuring various functional groups to achieve high metal sorption under convective flow conditions is a unique approach [68]. With a membrane pore size ranging from 5 to 20 nm bigger than the dispersed metal ions, the microfiltration process (MF) operates at low transmembrane pressure. When the metal concentration ranges from 10 to 112 ppm at a pressure supply of (2–5 bar) and a pH value of (5–9.5), its metal-removal effectiveness surpasses

90%. The formation of sludge poses [9] a significant concern despite the low energy requirements and improved selectivity of the complexation-ultrafiltration and polymer-supported ultrafiltration (PSUF) approaches, which also involve selective binding [69, 70]. In an in-membrane bioreactor [71], microfiltration and ultrafiltration are both low-pressure membrane applications [67] that are defined by their capacity to filter out contaminants considering membrane pore sizes [61]. It entails applying intense pressure to a membrane that contains a solute on one side to force a solution across. On the other side of the membrane, the pure solvent will pass. The efficiency of reverse osmosis [72] deals with numerous factors including solute concentration, supplied pressure, water flux rate etc. Toxic metals [40] are removed from wastewater using the nanofiltration (NF) technique. It is trustworthy and uses a modest amount of energy. Reverse osmosis and nanofiltration technologies [66] have been used by many treatment systems to deal with one another sequentially. To remediate the harmful metals that wastewater contains,heavy-metal ions are removed via the electrochemical process known as electrodialysis (ED) [73], which also includes a membrane filtration method. The discharge of wastewater across a selectively charged membrane generates the electrical field's driving power. The polyelectrolyte membrane, which is categorized into two types—cation exchange and anion exchange—is inserted between the cathode and anode. Compared to reverse osmosis, this process is more efficient. As a result, it is typically used for the desalination of industrial effluents with TDS levels between 500 and 1500 ppm [62]. The voltage applied is the primary factor that affects its performance.

vi) **Photocatalysis**

The photocatalytic process in an aqueous suspension of semiconductors has received a lot of attention recently in the context of solar energy conversion. For the quick and effective eradication of environmental contaminants, this photocatalytic technique has been developed. When the semiconductor-electrolyte interface is illuminated with light energy greater than the semiconductor band gap, electron-hole pairs (e/h^+) are produced in the conduction band and valence band of the semiconductor, respectively [74, 75].

vii) **Electrochemical method**

This is an additional method of aquatic treatment that is capable of progressively removing different types of contaminants from wastewater, including harmful metal ions, organics, and inorganics [76–78]. Due to its ability to participate in ecological treatment, recycling, and monitoring in diverse ways, this method is more dependable and cost-effective when compared to other treatment techniques [20]. Additionally, electrochemical technologies are straightforward and simple to use. The wastewater treated by these methods is potable, colorless, and odorless with little sludge production and no secondary pollution development [51]. Additionally, they are smaller systems that only rely on the releasing ions that create "nascent" colloids in considerably more concentrated amounts for controlled and immediate reactions [79]. The ease of designing an electrochemical unit without solution dosing may demonstrate the ease of designing an electrochemical unit without solution dosing. Their use may still be constrained by a few flaws like the short lifespan of electrode materials and low current efficiency [80].

vii) **Electrocoagulation method**
Electricity and chemical coagulation [77], which are utilized to produce a clean electrochemical approach with a variety of applications for treating water and wastewater without the production of secondary pollutants [81], are included in the term "electrocoagulation" [82]. It is a quick and easy procedure as opposed to biological treatment methods, which cannot handle components with high levels of toxicity [78] and need a certain condition to produce coagulants. Because of this obstacle, and among the numerous wastewater treatment technologies, the electrocoagulation approach has the potential to resolve these issues for a number of reasons, including ecological compatibility, ingrained safety, energy efficiency, and economic efficiency [83]. Electrocoagulation has been derived from the theoretical aspects of the coagulation-flocculation approach through destabilizing negative colloidal particles by adding some metal salts (i.e. $Al_2(SO_4)_3$, $FeSO_4$, and $FeCl_3$) [66]. In addition, these colloidal particles will be surrounded by an electrical double layer comprised of positively charged cations [60, 84]. Additionally, a technique called electroflotation [85], which is paired with electrocoagulation for the purpose of removing contaminants, releases hydrogen to affirm natural buoyancy of the flocculated particles on the surface of wastewater [83, 86]. In a nutshell, the generation of the coagulant is the further advancement of this electrochemical method and depends on the supposition that "the pollutants contained in the wastewater are held in a solution by electrical charges" and happens in situ, is a component of the electro-oxidation process at the anode electrode [87], and is electro-oxidative. The interactions between the ions generated by oxidation of the artificial anode combine with the current provided to the electrocoagulation cell to compress the diffusion double layer surrounding the charged particles. The electrocoagulation process could be viewed as an electrocatalytic or auto-coagulating process that does not require any additional chemicals. Less sludge is formed when more chemicals are introduced compared to other traditional methods [51]. To remove a variety of pollutants, including heavy metals [20, 60], paint, detergents, TDS [88], COD, and TSS [89], among others, numerous researchers have employed the electrocoagulation approach in their research.

8.6 Microbial and Biological Approaches for Removing Heavy Metals from Wastewater

Over recent years, the massive growth in the human population and anthropogenic activities have significantly contributed to contamination caused by heavy metals, which is now being seen as a major threat to public health. Therefore, various strategies have been devised to provide effective remedial treatment to mitigate the effects of heavy-metal contamination, especially in water bodies [90]. Diverse ranges of chemical methodologies (i.e. oxidation, reduction, ion exchange, membrane filtration, and evaporation) are in practice, but as effective as these are, they are not cost-effective. As a result, micro-organisms such as bacteria, yeast, algae, and fungi have been sought after for use in the process to further promote an efficient, economically viable, and environmentally friendly alternative technology.

8.7 Biological Remediation Approaches for Heavy-Metal Removal

The conversion of harmful contaminants as heavy metals to less harmful substances and the removal or mitigation of concentrations of contaminants from the environment using a biological approach is termed "bioremediation"[91, 92]. It involves the use of plants, micro-organisms, associated derivatives, cells, tissues, and related characteristic properties to mitigate the negative impact on the surrounding environment. Heavy-metal biological remediation is a widely accepted and practical strategy because it is not only economical but also environmentally friendly [93]. The types of bioremedial approaches that are the most widely used based on the organism involved include:

i) **Mycoremediation:** This process of bioremediation involves the use of fungi or fungal-based technology to help degrade or remove contaminants from the surrounding environment. Fungi are effectively chosen for the purpose due to the presence of their unique and quite diverse features like metabolic capability, variation in morphology sound protein cogs, rapid growth under a range of conditions, development even under extreme conditions, low cost of cultivation, potent capability of absorption, and the presence of nonspecific intracellular and extracellular fungal-enzyme systems such as that of oxidases, catalases (CAT), chitinases, cellulases, cytochrome P450 mono-oxygenases, laccases, pectinases, ligninase, hemicellulase, lignocellulases, peroxidases, xylanases, and more that act synergistically to aid efficient decontamination. Fungi such as *Pleurotus ostreatus,Rhizophagus irregularis*, and *Funneliformis mosseae* [94] have been found to be quite effective for heavy-metal bioremediation in wastewater [95, 96].
ii) **Phytoremediation:** Phytoremediation, also termed "botanical bioremediation", involves the usage of plants to decontaminate or mitigate the contamination to help control its effects on the surrounding environment [97, 98]. The ability of plants to uptake and accumulate heavy metals varies significantly based on a plethora of factors including the mechanisms of ion uptake, genetics, morphology, physiology, and other characteristics [99]. The process of phytoremediation encompasses:
   - **Phytoextraction:** Phytoextraction is the process of harvesting and treating plant biomass that is rich in contaminants [91, 94, 100].
   - **Phytofiltration:** Phytofiltration has been termed "rhizofiltration", which involves the adsorption or precipitation of contaminants through plant roots or absorption into it, circumferential toward the root zone [91, 94].

- **Phytostabilization:** This process involves the use of plants and their parts to inhibit contaminant migration from the zone of contamination [91, 94].
- **Phytodegradation:** This process is described as the use of plants to degrade or transform contaminants into less toxic compounds [91, 94, 99].
- **Plant hyper accumulators:** Plant species have been participating in phytoremediation known to accumulate >100 mg/kg^{-1} of heavy metals (i.e., Cd, Cr, Co, or Pb) per unit biomass (dry weight). Plants also assimilate >1000 mg/kg^{-1} metal constituents (i.e. Ni, Cu, Se, As, or Al) and >10 000 mg/kg^{-1} metallic residues (i.e., Zn or Mn) in their unit biomass (dry weight) [98–102].

The aquatic macrophytes *Eichhornia crassipes* were studied to aid in the elimination of Pb from industrial effluents [97]. Water hyacinth has also been successfully used in wastewater treatment to effectively improve water quality; it accumulates trace elements such as Ag, Pb, Cd, etc., and is suitably efficient in the phytoremediation of wastewater contaminated with Cd, Cr, Cu, and Se [97, 99, 100, 102, 103].

iii) **Phycoremediation:** This refers to the usage of microalgae's ecological functions to aid in the treatment of heavy-metal contaminated wastewater [104]. Phycoremediation is sought after due to its efficacy, cost-effectiveness, and environmentally friendly approach. It involves both intracellular and extracellular processes. *Cladophora sp.* [105] has been found to be quite effective in phycoremediation [106].

iv) **Microbial remediation:** Microbial remediation has been inferred as the bioremediation of pollutants (including heavy metals) through the involvement of microbial physiological and metabolic efficiencies [83]. Microbial cellular factories are one pot complex enzymatic networksthat can transform and assimilate the toxic (oxidation) state of metals to a benign state to reduce environmental and human health hazards [98]. Secretions from metabolic activities in microbes can further be capable of dissolving heavy metals and particles containing heavy metals. Precipitation, biosorption, and transformation are the processes used by microbes to aid bioremediation [107].

v) **Zooremediation:** The cultivation and harvest of animals in order to remediate contaminants in aquatic systems is termed "zooremediation" [108]. Studies have found organisms such as *Crassostrea virginica* accumulating Cu, *Mytilus edulis* accumulating Pb, *Pinctada albina* accumulating Cd, and *Crassostrea virginica* accumulating Zn, proving to be useful for the purpose of bioremediation [109]. Much like Phytoremediation, zooremediation encompasses:
   - **Zooextraction:** This involves the harvest and treatment of animal biomass rich in the contaminant [108, 109].
   - **Zoostabilization:**This is the use of animals to inhibit the migration of contaminants from a particular zone. This includes the proper maintenance as well as supplementation of wild animal populations without the extraction of any animal biomass [108, 109].
   - **Zootransformation or zoodegradation:** This process involves the use of animals to degrade or transform the contaminants. It also involves the maintenance and supplementation of wild animal populations without the extraction of any animal biomass [108, 109].

- **Animal metal hyper accumulators:** In this process, animal species are known to accumulate metal ions (i.e., Cd, Cr, Co, and Pb) around >100 mg/kg^{-1} per unit animal biomass (dry weight); metallic residues (i.e., Ni, Cu, Se, As, and Al) >1000 mg/kg^{-1} per unit animal biomass (dry weight); metallic constituents, i.e., Zn and Mn >10 000 mg/kg^{-1} per unit animal biomass (dry weight). The current approach has been limited to industrial sustainability, as it involves animals (particularly invertebrates) for ethical reasons [108, 109].

Bioremediation is broadly classified into:

- **In-situ approaches:** In-Situ bioremediation techniques refer to those treatment methods that can be carried out at the site.In terms of heavy-metal bioremediation, it can be done using aquatic plants and/or microbial remediation methods. These systems can be commissioned in a short period, are easy to operate, and require less energy when compared to conventional treatment technologies [92, 107].
- **Ex-situ approaches:** Ex-situ bioremediation refers to the various biological processes involved in the remediation of a contaminated zone away from the initial zone of contamination. This process of bioremediation can be controlled and significantly enhance as per requirement [92, 107].

These approaches include the following techniques:

- **Bioaugmentation:** Bioaugmentation involves the introduction of indigenous micro-organisms to the contaminated zones to aid rapid microbial activity in the region [92, 110, 111].
- **Bioleaching:** Bioleaching is the process of solubilizing heavy metals from a contaminated area using specific, naturally occurring micro-organisms [96, 112–114].
- **Bioreactors:** Micro-organisms and/or biocatalysts require permissive micro-environments to carry out their maximal enzymatic reactivities for bioremediation. Bioreactors are the most potent remedy for this problem. Bioreactors create optimal conditions by mimicking an external environment where biochemical reactions occur to stimulate the influx and degradation of toxic contaminants. The bioreactors may indulge the extensive usage of enzymes, tissues, and micro-organisms for attaining a greater yield of bioremediation [92, 106].
- **Biosparging:** This approach involves the involvement of low-pressure air below the water table to acclivity the groundwater oxygen level to enhance the rate of microbial bioremediation of toxic contaminants [92].
- **Biostimulation:** Biostimulation involves the acceleration of bacterial growth to add to the process of bioremediation by mixing enriched nutrients to stimulate microbial activity, subsequently resulting in an increased rate of activity [92].
- **Bioventing:** Bioventing is a technique widely utilized as an in-situ mechanism for supplying air and nutrients to a zone of contamination to stimulate microbial activity. It requires low airflow and low oxygen rates to release contaminants into the environment through biodegradation.

Table 8.4 Micro-organisms and the method they are utilized for in bioremediation [95, 99, 111, 113, 115]:

Method	Organism	Metal ions
Bioaugmentation	*Cryptococcus humicolus*	Heavy metal ferricyanide
Bioleaching	*Pseudomonas aeruginosa*	Cd(II)
Bioreactors	*Aspergillus sp.*	Cr
Phytoremediation	*Eichhornia crassipis* (Water hyacinth)	Ar

The various micro-organisms used in these methods are summarized in Table 8.4.

8.8 Microbial Bioremediation Approaches

Microbial bioremediation of heavy-metal-polluted effluents are the most pivotal and effective environmentally friendly tools. In heavily contaminated environments, micro-organisms adeptly withstand heavy metals and their associated impacts with processes like oxidizing, binding, immobilizing, volatilizing, and/or transforming the heavy metals. There are diverse ranges of microbial communities that do exist in nature, and they can aid bioremediation with greater potency. *Methanothermobacter thermautotrophicus* can bioremediate heavy-metal ion Cr (VI) to Cr (III) in its hydroxide or oxide chemical state [110]. *Shewanella sp.* can carry out bacterial reduction and immobilization approaches to sequestrate heavy metals i.e., Cr(VI), Ar [103]. Different microbial communities like *Pseudomonas putida, Desulfovibrio desulfuricans*, and microalgae *Chlorella vulgaris* can carry out biosorption mechanisms to sequestrate heavy-metal ions including Cd(II), Ni(II), Cr(VI), and Pb [116–118]. However, processes that allow for the removal of heavy metals are:

- **Biosorption:** The micro-organisms are capable of taking up the heavy metal present in the zone of contamination by the process of biosorption [98]. It is evident at the binding sites found in the microbial cellular structure that it does not involve any energy. The extracellular polymeric substances have been of increasing interest for researchers since these have been found to be quite reactive with considerable effects on microbial characteristics including metal adsorption [113, 119, 120].
- **Bioaccumulation:** The intracellular collection of heavy metals is the mechanism by which the toxic metals are immobilized inside the cells of the micro-organisms by binding them to some proteins called metallothioneins (MTs).This method is termed "bioaccumulation"[104, 113, 119].
- **Biotransformation:** Biotransformation is essentially described as the process in which contaminants are converted into a non-toxic form by alteration of their structure [105, 113].

- **Biocatalysis:** Biocatalysis refers to a process that uses microbes or enzymes to speed up the transformation or degradation of a contaminant.
- **Bioleaching:** The process of solubilization of metals from solid substrates by the direct metabolism of leaching bacteria or extraction of the contaminant as products post metabolism is termed "bioleaching". The current studies have revealed that the process of mycoremediation using indigenous fungal isolates could be effective for leaching heavy metals from the industrial effluents contaminated by heavy metals, and it could be a promising technology for future use [95, 115].

Additionally, the existence of carboxyl, amino, phosphoryl, and sulfo-groups on the surface of microorganisms' cells functions as a naturally occurring potential site of ion exchanges with metal sinks. Additionally, through processes like proton exchanges or micro-precipitation of metals, extracellular polymeric materials on microbial cells surfaces also permit the attachment of heavy metals. Also, the biomass surfaces have a negative charge and bioremediation is accomplished by a variety of mechanisms, including redox reactions, adsorption, complexation, ion exchange, precipitation, and electrostatic attraction. Of these, the properties of biosorption and bioaccumulation are currently under use and detailed study. Further, we will also discuss the various ways in which these properties have been enhanced using engineering to better serve the deemed purpose of effective bioremediation.

8.9 Bioengineering Approaches on Microbes for Improving Heavy-Metal Removal from Wastewater

As discussed before, micro-organisms adapt to the heavy-metal contaminated environment due to thepresence or development of several properties. These properties, when studied in detail, can be utilized to effectively implement control over the negative impact on the environment. Several microbes and their properties of accumulation and biosorption are being studied so that they can be engineered to magnify the effect or perhaps prolong the efficacy. Some bioengineering approaches are as follows:

- **Ion import systems:** For effective improvement in the biological uptake of heavy metals, recent studies have focused on the mechanisms involved in uptake from periplasm into the cytoplasm of bacteria through the involvement of influx molecular pump proteins. These influx pump proteins belong to the three major transporter classes, i.e., channels, secondary carriers, and primary active transporters [97, 106].
- **Ion storage systems:** Upon uptake,heavy metals are stored in the cells.Efforts are now focused on improving the heavy-metal storage systems in microbes, especially regarding the metal-binding entities or metal-binding inclusion bodies inside the microbial cytoplasm to sequestrate and uptake heavy metals. This downplays toxic impacts through the establishment of oxidative stress. These metal-binding inclusion bodies are mostly proteinaceous in nature including peptide-producing enzymes and other polymers that can also effectively bind to heavy metals and aid sequestration [119, 121].

- **Metal-binding proteins:** Multiple ligand binding sites are being established through the involvement of heavy-metal-binding proteins. Multipotency of the ligand binding zones on heavy-metal binding proteins serve as binding pockets for the heavy-metal assimilation site in the bioaccumulation process. Moreover, heavy-metal based toxicants are being significantly accommodated inside small cysteine-rich binding pockets in metal-binding proteins. It causes toxic metal homeostasis. Moreover, it protects heavy-metal sequestering microbes against ubiquitous oxidative damage of DNA and other physiological mechanisms [104]. Catalytic motifs (i.e., high CxC and CC motifs) of these metal-binding proteins infer metal-thiolate linkages. Thus, metal-binding proteins, i.e., metallothioneins, have the capability to sequester zinc, cadmium, and copper, but these also bind to mercury, arsenic, nickel, and cobalt [119, 122].
- **Enzymatic production of metal-binding peptides and polymers:** Microbial cellular cytoplasm accumulates diverse ranges of polymers through potent enzymatic catalytic kinetics, which helps to assimilate heavy metals within microbial regimes [104]. Phytochelatin (a polymer of glutathione) isan ideal example of such a microbial polymer. These are commonly found in plants that possess heavy-metal resistance, and the associated efficacy can be improved by metabolic engineering to increase the pool of phytochelatin precursor compounds cysteine, γEC, and GSH by over-expressing the respective genes associated with their production [119, 121].
- **Ancestral sequence reconstruction:** Ancestral sequence reconstruction involves the research and study of ancient proteins that have a broader substrate range so that they can serve as templates to engraft new proteins (i.e., ABC transporter proteins) that have the desired phenotypic traits to support large-scale application. The current approach helps to fetch out ancestral amino acid-binding proteinsthat can bind to certain amino acids, i.e., L-arginine and L-glutamine within catalytic motifs of the ABC transporter super family. Moreover, this approach could possibly be a useful tool to identify riboswitches that could evolve into new heavy-metal transporters [119].
- **Application of genomics:** The use of genomics in bioremediation might make it possible to enhance the mechanisms involved in the process from the molecular level itself. This can be applied to form synthetic microbial communities, the characteristics and behaviors of which can be utilized to effectively combat pollution—especially pollution that is caused by heavy metals in water. It might include the use of biofilms and the use of genomics to enhance the properties of the film formed to perhaps aid the remediation[98, 115, 123].

8.10 Conclusion

Environmental toxicants in wastewater can undergo bioremediation techniques to reduce their amounts below threshold levels. Bioremediation technology is a practical, natural, environmentally benign approach and provides economic feasibility for large-scale implementation in the near future. Diverse ranges of microbial assemblages are key players in this process for removing heavy metals from wastewater. The removal of heavy metals from wastewater is essential for establishing different life forms and maintaining environmental eco-dynamics. The most potent aerobic bacterial communities for heavy-metal recycling from wastewater include *Pseudomonas* sp., *Alcaligenes* sp., *Sphingomonas* sp., *Rhodococcus* sp., *and Mycobacterium* sp. Hence, bioremediation of heavy metals from wastewater is an affordable and technologically feasible approach. However, certain pitfalls of bioremediation exist regarding its slowness and lower biocatalytic potential of microbial assemblages. Moreover, in a few instances, the bioremediation approach to heavy-metal removal from wastewater has generated some byproducts that couldpotentially become comparatively more toxic in nature than the original form of the heavy metals. Therefore, more extensive research on heavy-metal recycling from wastewater could be a future way forward to mitigate irregularity, incompleteness, and uncertainty through the identification of novel microbes, novel enzymatic activities, and novel metabolic networks of microbial communities to improve the potency of bioremediation of heavy metals from wastewater.

Acknowledgment

The authors show gratitude to JIS University Kolkata and JIS Group of Educational Initiatives for their support and encouragement. The authors would also like to thank Ms. Shrestha Debnath for her immense support in guiding and helping the graduate students.

References

1 Baroni, L., Cenci, L., Tettamanti, M., and Berati, M. (2007). Evaluating the environmental impact of various dietary patterns combined with different food production systems. *European Journal of Clinical Nutrition* 61 (2): 279–286.
2 Vieira, R.H. and Volesky, B. (2000). Biosorption: a solution to pollution? *International Microbiology* 3 (1): 17–24.
3 Das, N., Vimala, R., and Karthika, P. (2008). Biosorption of heavy metals–an overview. *Indian Journal of Biotechnology* 7: 159–169.
4 Monachese, M., Burton, J.P., and Reid, G. (2012). Bioremediation and tolerance of humans to heavy metals through microbial processes: a potential role for probiotics? *Applied and Environmental Microbiology* 78 (18): 6397–6404.
5 Verma, R. and Dwivedi, P. (2013). Heavy metal water pollution-A case study. *Recent Research in Science and Technology* 5 (5): 98–99.
6 Tchounwou, P.B., Yedjou, C.G., Patlolla, A.K., and Sutton, D.J. (2012). Heavy metal toxicity and the environment. *Molecular, Clinical and Environmental Toxicology* (101) 133–164.
7 Alluri, H.K., Ronda, S.R., Settalluri, V.S. et al. (2007). Biosorption: an eco-friendly alternative for heavy metal removal. *African Journal of Biotechnology* 6 (25): eISSN: 1684–5315.

8 Wuana, R.A. and Okieimen, F.E. (2011). Heavy metals in contaminated soils: a review of sources, chemistry, risks and best available strategies for remediation. *International Scholarly Research Notices* 2011: 20.

9 Joshi, N.C. (2017). Heavy metals, conventional methods for heavy metal removal, biosorption and the development of low cost adsorbent. *European Journal of Pharmaceutical and Medical Research* 4: 388–393.

10 Kanamarlapudi, S.L., Chintalpudi, V.K., and Muddada, S. (2018). Application of biosorption for removal of heavy metals from wastewater. *Biosorption* 18 (69): 70–116.

11 Behboody, G. (2020). Methods for removing heavy metals from water and wastewater: a review study. *Journal of Research in Environmental Health* 6 (2): 145–160.

12 Barakat, M.A. (2011). New trends in removing heavy metals from industrial wastewater. *Arabian Journal of Chemistry* 4 (4): 361–377.

13 Azimi, A., Azari, A., Rezakazemi, M., and Ansarpour, M. (2017). Removal of heavy metals from industrial wastewaters: a review. *ChemBioEng Reviews* 4 (1): 37–59.

14 Qiu, Y.R. and Mao, L.J. (2013). Removal of heavy metal ions from aqueous solution by ultrafiltration assisted with copolymer of maleic acid and acrylic acid. *Desalination* 329: 78–85.

15 Brady, D., Rose, P.D., and Duncan, J.R. (1994). The use of hollow fiber cross-flow microfiltration in bioaccumulation and continuous removal of heavy metals from solution by *Saccharomyces cerevisiae*. *Biotechnology and Bioengineering* 44 (11): 1362–1366.

16 Bakalár, T., Búgel, M., and Gajdošová, L. (2009). Heavy metal removal using reverse osmosis. *ActaMontanisticaSlovaca* 14 (3): 250.

17 Choi, S.Y., Keun-Young, P., Hee, J.K., and Kweon, J.H. (2015). Removal of heavy metal and nitrate nitrogen in polluted groundwater by electrodialysis process. *Journal of Water Resource and Hydraulic Engineering* 4: 412.

18 Shyni, L.S., Jagadish, K., Srikantaswamy, S., and Abhilash, M.R. (2016). Photocatalytic degradation and removal of heavy metals in pharmaceutical waste by selenium doped ZnOnano composite semiconductor. *Journal for Research* 2 (05).

19 Johnson, P.D., Girinathannair, P., Ohlinger, K.N. et al. (2008). Enhanced removal of heavy metals in primary treatment using coagulation and flocculation. *Water Environment Research* 80 (5): 472–479.

20 Un, U.T. and Ocal, S.E. (2015). Removal of heavy metals (Cd, Cu, Ni) by electrocoagulation. *International Journal of Environmental Science and Development* 6 (6): 425.

21 Merzouk, B., Gourich, B., Sekki, A. et al. (2009). Removal turbidity and separation of heavy metals using electrocoagulation–electroflotation technique: a case study. *Journal of Hazardous Materials* 164 (1): 215–222.

22 Addi, Y., Duverneuil, P., and Khouider, A.L. (2009). Electrodeposition of heavy metals (Cu; Ni; Zn and Cd) from industrial effluents. *ECS Transactions* 19 (32): 63.

23 AlJaberi, F.Y. and Mohammed, W.T. (2018). The most practical treatment methods for wastewaters: a systematic review. *Mesopotamia Environmental Journal* 1–28(Special Issue E).

24 Ahmaruzzaman, M. (2011). Industrial wastes as low-cost potential adsorbents for the treatment of wastewater laden with heavy metals. *Advances in Colloid and Interface Science* 166 (1–2): 36–59.

25 Mohammadi, T., Razmi, A., and Sadrzadeh, M. (2004August 15). Effect of operating parameters on Pb2+ separation from wastewater using electrodialysis. *Desalination* 167: 379–385.

26 Deng, X. and Wilson, D. (2001). Bioaccumulation of mercury from wastewater by genetically engineered *Escherichia coli*. *Applied Microbiology and Biotechnology* 56 (1): 276–279.

27 Iram, S., Shabbir, R., Zafar, H., and Javaid, M. (2015). Biosorption and bioaccumulation of copper and lead by heavy metal-resistant fungal isolates. *Arabian Journal for Science and Engineering* 40 (7): 1867–1873.

28 Ksheminska, H., Fedorovych, D., Babyak, L. et al. (2005). (III) and (VI) tolerance and bioaccumulation in yeast: a survey of cellular chromium content in selected strains of representative genera. *Process Biochemistry* 40 (5): 1565–1572.

29 Fan, T., Liu, Y., Feng, B. et al. (2008). Biosorption of cadmium (II), zinc (II) and lead (II) by *Penicilliumsimplicissimum*: isotherms, kinetics and thermodynamics. *Journal of Hazardous Materials* 160 (2–3): 655–661.

30 Javaid, A., Bajwa, R., Shafique, U., and Anwar, J. (2011). Removal of heavy metals by adsorption on *Pleurotusostreatus*. *Biomass and Bioenergy* 35 (5): 1675–1682.

31 Quintelas, C., Rocha, Z., Silva, B. et al. (2009). Removal of Cd (II), Cr (VI), Fe (III) and Ni (II) from aqueous solutions by an E. coli biofilm supported on kaolin. *Chemical Engineering Journal* 149 (1–3): 319–324.

32 Abd-Alla, M.H., Morsy, F.M., El-Enany, A.W., and Ohyama, T. (2012). Isolation and characterization of a heavy-metal-resistant isolate of Rhizobium *leguminosarumbv. viciae* potentially applicable for biosorption of Cd2+ and Co2+. *International Biodeterioration& Biodegradation* 67: 48–55.

33 Gupta, V.K. and Rastogi, A. (2008). Biosorption of lead from aqueous solutions by green algae Spirogyra species: kinetics and equilibrium studies. *Journal of Hazardous Materials* 152 (1): 407–414.

34 Babel, S. and Kurniawan, T.A. (2003). Low-cost adsorbents for heavy metals uptake from contaminated water: a review. *Journal of Hazardous Materials* 97 (1–3): 219–243.

35 Agarwal, S.K. (2009). *Heavy Metal Pollution*, 270. APH publishing.

36 Coelho, L.M., Rezende, H.C., Coelho, L.M. et al. (2015). Bioremediation of polluted waters using microorganisms. *Advances in Bioremediation of Wastewater and Polluted Soil* 10: 60770.

37 Sulaymon, A.H., Sharif, A.O., and Al-Shalchi, T.K. (2011). Effect of tubes bundle electrode on removal of cadmium from simulated wastewaters by electrodeposition. *Journal of Chemical Technology & Biotechnology* 86 (5): 651–657.

38 Pn, A.M. and Madhu, G. (2011). Removal of heavy metals from waste water using water hyacinth. *International Journal on Transportation and Urban Development* 1 (1): 48.

39 Sulaymon, A.H., Abdulmajeed, B.A., and Salman, A.B. (2017). Removal of cadmium from simulated wastewater by using stainless steel concentric tubes electrochemical reactor. *Desalination and Water Treatment* 68: 220–225.

40 Fu, F. and Wang, Q. (2011). Removal of heavy metal ions from wastewaters: a review. *Journal of Environmental Management* 92 (3): 407–418.

41 Siti, N., Mohd, H., Md, L.K., and Shamsul, I. (2013). Adsorption process of heavy metals by low-cost adsorbent: a review. *World Applied Sciences Journal* 28 (11): 1518–1530.

42 Liu, H.H. and Sang, S.X. (2010). Study on the law of heavy metal leaching in municipal solid waste landfill. *Environmental Monitoring and Assessment* 165 (1): 349–363.

43 Babák, L., Šupinova, P., Zichova, M. et al. (2012). Biosorption of Cu, Zn and Pb by thermophilic bacteria-effect of biomass concentration on biosorption capacity. *ActaUniversitatis Agriculturae etSilviculturaeMendelianaeBrun.* 60 (5): 9–18.

44 Ahalya, N., Ramachandra, T.V., and Kanamadi, R.D. (2003). Biosorption of heavy metals. *Research Journal of Chemistry and Environment* 7 (4): 71–79.

45 Zabochnicka-Świątek, M. and Krzywonos, M. (2014). Potentials of biosorption and bioaccumulation processes for heavy metal removal. *Polish Journal of Environmental Studies* 23 (2).

46 Sekhar, K.C., Kamala, C.T., Chary, N.S., and Anjaneyulu, Y. (2003). Removal of heavy metals using a plant biomass with reference to environmental control. *International Journal of Mineral Processing* 68 (1–4): 37–45.

47 Wang, D., Sun, W., Xu, Y. et al. (2004). Speciation stability of inorganic polymer flocculant–PACl. *Colloids and Surfaces A: Physicochemical and Engineering Aspects* 243 (1–3): 1–0.

48 Babel, S. and Kurniawan, T.A. (2005). *Various Treatment Technologies to Remove Arsenic and Mercury from Contaminated Groundwater: An Overview*, 1. Southeast Asian Water Environment.

49 Motsi, T., Rowson, N.A., and Simmons, M.J. (2009). Adsorption of heavy metals from acid mine drainage by natural zeolite. *International Journal of Mineral Processing* 92 (1–2): 42–48.

50 Abbas, A. (2015). Removal of heavy metals (Cr, Cd, Ni and Pb) using fresh water algae (Utriculariatenuissima, Utriculariatenuis&Zygogoniumericetorum) from contaminated water. *Journal of Biodiversity and Environmental Sciences (JBES)* 6 (5): 358–366.

51 Bazrafshan, E., Mohammadi, L., Ansari-Moghaddam, A., and Mahvi, A.H. (2015). Heavy metals removal from aqueous environments by electrocoagulation process–a systematic review. *Journal of Environmental Health Science and Engineering* 13 (1): 1–6.

52 Wang, L.K., Shammas, N.K., and Hung, Y.T. (eds.) (2009). *Biosolids Engineering and Management*, 7. Springer Science & Business Media.

53 Mirbagheri, S.A. and Hosseini, S.N. (2005). Pilot plant investigation on petrochemical wastewater treatmentfor the removal of copper and chromium with the objective of reuse. *Desalination* 171 (1): 85–93.

54 Aziz, H.A., Adlan, M.N., and Ariffin, K.S. (2008). Heavy metals (Cd, Pb, Zn, Ni, Cu and Cr (III)) removal from water in Malaysia: post treatment by high quality limestone. *Bioresource Technology* 99 (6): 1578–1583.

55 Peters, R.W., Ku, Y., and Bhattacharyya, D. (1985). Evaluation of recent treatment techniques for removal of heavy metals from industrial wastewaters. In: *AICHE Symposium Series*, 243, 165–203.

56 AlJaberi, F.Y. and Mohammed, W.T. (2018). Evaluation the effect of wastewater conductivity on the final stable voltage applied to electrocoagulation reactor. *Journal of Al-Nisour University College*, 6: 133–139.

57 Mohammed, A.A. and Mahmood, H.S. (2013). By ion exchange method on zeolite and purolite C105 resin. *Journal of Engineering* 19 (10).

58 Kurniawan, T.A., Chan, G.Y., Lo, W.H., and Babel, S. (2006). Physico–chemical treatment techniques for wastewater laden with heavy metals. *Chemical Engineering Journal* 118 (1–2): 83–98.

59 Tripathi, A. and Ranjan, M.R. (2015). Heavy metal removal from wastewater using low cost adsorbents. *Journal of Bioremediation & Biodegradation* 6 (6): 315.

60 Lekhlif, B., Oudrhiri, L., Zidane, F. et al. (2014). Study of the electrocoagulation of electroplating industry wastewaters charged by nickel (II) and chromium (VI). *Journal of Materials and Environmental Science* 5 (1): 111–120.

61 Teixeira, J.A., Vicente, A.A., and Press, C.R. (eds.) (2014). *Engineering Aspects of Food Biotechnology*. CRC Press. ISBN: 13: 978-1-4398-9546-7.

62 Majeed, B.A. and Zubaidy, M.A. (2016). Performance study of electrodialysis for treatment fuel washing wastewater. *Iraqi Journal of Chemical and Petroleum Engineering* 17 (4): 35–42.

63 World Health Organization, WHO (2004). *Guidelines for Drinking-water Quality*, 1. World Health Organization.

64 Ani, F.H., Fatlawi, A.H., and Taher, G.A. (2016). Removal lead and cadmium ions in industrial wastewater using graphene nano sheets. *Engineering and Technology Journal* 34 (9Part (A) Engineering).

65 Gupta, A., Joia, J., Sood, A. et al. (2016). Microbes as potential tool for remediation of heavy metals: a review. *Journal of Microbial and Biochemical Technology* 8 (4): 364–372.

66 Mamelkina, M.A., Cotillas, S., Lacasa, E. et al. (2017). Removal of sulfate from mining waters by electrocoagulation. *Separation and Purification Technology* 182: 87–93.

67 Tansel, B. (2008). New technologies for water and wastewater treatment: a survey of recent patents. *Recent Patents on Chemical Engineering* 1 (1): 17–26.

68 Bhattacharyya, D., Hestekin, J.A., Brushaber, P. et al. (1998). Novel poly-glutamic acid functionalized microfiltration membranes for sorption of heavy metals at high capacity. *Journal of Membrane Science* 141 (1): 121–135.

69 Harun, H., Hamid, H.A., Sunar, N.M. et al. (2018). Technologies used for heavy metals removal from wastewater. In: *Sustainable Environmental Technology*, Mr. Mohd Iskandar Faiz Amran, editor. 79–85. PenerbitUthm: Uthm. ISBN 9789672216445.

70 Kabdaşlı, I., Arslan, T., Ölmez-Hancı, T. et al. (2009). Complexing agent and heavy metal removals from metal plating effluent by electrocoagulation with stainless steel electrodes. *Journal of Hazardous Materials* 165 (1–3): 838–845.

71 Wang, W., Yang, Q., Zheng, S., and Wu, D. (2013). Anaerobic membrane bioreactor (AnMBR) for bamboo industry wastewater treatment. *Bioresource Technology* 149: 292–300.

72 Moura, R.C., Bertuol, D.A., Ferreira, C.A., and Amado, F.D. (2012). Study of chromium removal by the electrodialysis of tannery and metal-finishing effluents. *International Journal of Chemical Engineering* 2012.

73 Căpriţă, F.C. and Ene, A. (2020).Biosorption of heavy metals from the metallurgical industry wastewater by macroalgae. In: *AIP Conference Proceedings* 2218 (1): 030011. AIP Publishing LLC.

74 Herrmann, J.M. (1999). Heterogeneous photocatalysis: fundamentals and applications to the removal of various types of aqueous pollutants. *Catalysis Today* 53 (1): 115–129.

75 Wang, J.L. and Xu, L.J. (2012). Advanced oxidation processes for wastewater treatment: formation of hydroxyl radical and application. *Critical Reviews in Environmental Science and Technology* 42 (3): 251–325.

76 Lalmi, A., Bouhidel, K.E., Sahraoui, B., and el Houdaanfif, C. (2018). Removal of lead from polluted waters using ion exchange resin with Ca (NO3) 2 for elution. *Hydrometallurgy* 178: 287–293.

77 Fekete, É., Lengyel, B., and Cserfalvi, T. (2016). Electrocoagulation: an electrochemical process for water clarification. *Journal of Electrochemical Science and Engineering* 6 (1): 57–65.

78 Riyanto, A. and Hidayatillah, A. (2014). Electro-coagulation of detergent wastewater using aluminium wire netting electrode (AWNE). In *Proc. 2014 Research, Implementation and Education of Mathematics and Sciences Conf* 151–158.

79 Al Aji, B., Yavuz, Y., and Koparal, A.S. (2012). Electrocoagulation of heavy metals containing model wastewater using monopolar iron electrodes. *Separation and Purification Technology* 86: 248–254.

80 Walsh, F.C. (2001). Electrochemical technology for environmental treatment and clean energy conversion. *Pure and Applied Chemistry* 73 (12): 1819–1837.

81 Asadi, A., Emamjomeh, M.M., GHasemi, M., and MohammadianFazli, M. (2015). Efficiency of electrocoagulation process for lead removal from wastewater. *Journal of Inflammatory Diseases* 18 (6): 18–23.

82 Rani, K. and Elango, D. (2014). A quantitative comparison between electro coagulation and chemical dosing. *International Journal of Civil Engineering and Technology* 5(3): 245–251.

83 Rehman, A., Kimb, M., Reverberic, A., and Fabianoa, B. (2015). Operational parameter influence on heavy metal removal from metal plating wastewater by electrocoagulation process. *Transactions*43: ISSN 2283-9216.

84 Khaled, B., Wided, B., Béchir, H. et al. (2019). Investigation of electrocoagulation reactor design parameters effect on the removal of cadmium from synthetic and phosphate industrial wastewater. *Arabian Journal of Chemistry* 12 (8): 1848–1859.

85 Attia, H.G. (2013). Decolorization of direct blue dye by electrocoagulation process. *Journal of Engineering and Sustainable Development* 17 (1): 171–181.

86 Bayar, S., Yilmaz, A.E., Boncukcuoğlu, R. et al. (2013). Effects of operational parameters on cadmium removal from aqueous solutions by electrochemical coagulation. *Desalination and Water Treatment* 51 (13–15): 2635–2643.

87 Adeogun, A.I. and Balakrishnan, R.B. (2016). Electrocoagulation removal of anthraquinone dye Alizarin Red S from aqueous solution using aluminum electrodes: kinetics, isothermal and thermodynamics studies. *Journal of Electrochemical Science and Engineering* 6 (2): 199–213.

88 Abdul-Baqi, M.A. and Thamir, A. (2015). Removal of TDS from water using electrocoagulation. *Journal of Al-Rafidain Engineering* 23 (4): 85–97.

89 Ghanim, A.N. (2014). Optimization of pollutants removal from textile wastewater by electrocoagulation through RSM. *Journal of University of Babylon* 22 (2): 375–387.

90 Njoku, K.L., Akinyede, O.R., and Obidi, O.F. (2020). Microbial remediation of heavy metals contaminated media by *bacillus megaterium* and *Rhizopusstolonifer*. *Scientific African* 10: e00545. 90.

91 Wang, M., Chen, S., Jia, X., and Chen, L. (2021). Concept and types of bioremediation. In: *Handbook of Bioremediation: Physiological, Molecular and Biotechnological Interventions*, Mirza Hasanuzzaman and Majeti Narasimha Vara Prasad, editors. Academic Press & Elsevier, 3–8.

92 Sayqal, A. and Ahmed, O.B. (2021). Advances in heavy metal bioremediation: an overview. *Applied Bionics and Biomechanics* 2021: 8.

93 FauzulImron, M., Kurniawan, S.B., and Soegianto, A. (2019). Identification of mercury-resistant bacteria as a potential candidate for mercury bioremediation. 25: S66–S69.

94 Hassan, Z.U., Ali, S., Rizwan, M. et al. (2017). Role of bioremediation agents (bacteria, fungi, and algae) in alleviating heavy metal toxicity. In *Probiotics in Agroecosystem*, 517–537.

95 Assad, R., Rafiq, I., Bashir, I. et al. (2021). Mycoremediation of pollutants in aquatic environs. In: *Freshwater Pollution and Aquatic Ecosystems*, Rezwana Assad, Iflah Rafiq, Iqra Bashir, Irshad Ahmad Sofi, Zafar Ahmad Reshi, Irfan Rashid, editors; Apple Academic Press, 241–299.

96 Khan, I., Aftab, M., Shakir, S.U. et al. (2019). Mycoremediation of heavy metal (Cd and Cr)–polluted soil through indigenous metallotolerant fungal isolates. *Environmental Monitoring and Assessment* 191 (9): 1-1.

97 Khan, A.U., Khan, A.N., Waris, A. et al. (2022). Phytoremediation of pollutants from wastewater: a concise review. *Open Life Sciences* 17 (1): 488–496.

98 Dixit, R., Malaviya, D., Pandiyan, K. et al. (2015). Bioremediation of heavy metals from soil and aquatic environment: an overview of principles and criteria of fundamental processes. *Sustainability* 7 (2): 2189–2212.

99 Sabreena, H.S., Bhat, S.A., Kumar, V. et al. (2022). Phytoremediation of heavy metals: an indispensable contrivance in green remediation technology. *Plants* 11 (9): 1255.

100 Lone, M.I., He, Z.L., Stoffella, P.J., and Yang, X.E. (2008). Phytoremediation of heavy metal polluted soils and water: progresses and perspectives. *Journal of Zhejiang University Science B* 9 (3): 210–220.

101 Ashraf, M.A., Hussain, I., Rasheed, R. et al. (2017). Advances in microbe-assisted reclamation of heavy metal contaminated soils over the last decade: a review. *Journal of Environmental Management* 198: 132–143.

102 Kaur, S. and Roy, A. (2021). Bioremediation of heavy metals from wastewater using nanomaterials. *Environment, Development and Sustainability* 23 (7): 9617–40. 102.

103 Darma, A., Yang, J., Zandi, P. et al. (2022). Significance of *Shewanella*Species for the Phytoavailability and toxicity of arsenic – a review. *Biology* 11 (3): 472.

104 Danouche, M., El Ghachtouli, N., and El Arroussi, H. (2021). Phycoremediation mechanisms of heavy metals using living green microalgae: physicochemical and molecular approaches for enhancing selectivity and removal capacity. *Heliyon*7 (7): e07609.

105 Sargin, I., Arslan, G., and Kaya, M. (2016). Efficiency of chitosan-algal biomass composite microbeads at heavy metal removal. *Reactive and Functional Polymers* 98: 38–47.

106 Ahmad, S., Pandey, A., Pathak, V.V., and Tyagi, V.V. Kothari R. (2020). Phycoremediation: algae as eco-friendly tools for the removal of heavy metals from wastewaters. In: *Bioremediation of Industrial Waste for Environmental Safety*, Gaurav Saxena, Ram Naresh Bharagava, editors; Springer, 53–76.

107 Volarić, A., Svirčev, Z., Tamindžija, D., and Radnović, D. (2021). Microbial bioremediation of heavy metals. *HemijskaIndustrija* 75 (2): 103–115.

108 Kardousha, M.M. (2019). Status of mosquito diversity of the eastern part of arabian peninsula with focus on qatar. *Current Politics and Economics of the Middle East*Hauppauge 10 (1): 121–138.

109 Gifford, S., Dunstan, R.H., O'Connor, W. et al. (2007). Aquatic zooremediation: deploying animals to remediate contaminated aquatic environments. *Trends in Biotechnology* 25 (2): 60–65.

110 Yan, Y., Yan, M., Ravenni, G. et al. (2022). Novel bioaugmentation strategy boosted with biochar to alleviate ammonia toxicity in continuous biomethanation. *Bioresource Technology* 343: 126146.

111 Nzila, A., Razzak, S.A., and Zhu, J. (2016). Bioaugmentation: an emerging strategy of industrial wastewater treatment for reuse and discharge. *International Journal of Environmental Research and Public Health* 13 (9): 486.

112 Hassanien, W.A., Desouky, O.A., and Hussien, S.S. (2014). Bioleaching of some rare earth elements from Egyptian monazite using *aspergillusficuum* and *pseudomonas aeruginosa*. *Walailak Journal of Science and Technology (WJST)* 11 (9): 809–823.

113 Pande, V., Pandey, S.C., Sati, D. et al. (2022). Microbial interventions in bioremediation of heavy metal contaminants in agroecosystem. *Frontiers in Microbiology* 13: 824084-824084.

114 Jamil, N., Kumar, P., and Batool, R. (2019). Soil microenvironment for bioremediation and polymer production. *Environmental Chemistry* 420: ISBN: 978-1-119-59205-1.

115 MedfuTarekegn, M., ZewduSalilih, F., and Ishetu, A.I. (2020). Microbes used as a tool for bioremediation of heavy metal from the environment. *Cogent Food and Agriculture* 6 (1): 1783174.

116 John, R. and Rajan, A.P. (2022). Bioreactor level optimization of chromium(VI) reduction through *Pseudomonas putida APRRJVITS11* and sustainable remediation of pathogenic DNA in water. *Beni-Suef University Journal of Basic and Applied Sciences* 11 (1).

117 Joo, J.O., Choi, J.H., Kim, I.H. et al. (2015). Effective bioremediation of Cadmium (II), nickel (II), and chromium (VI) in a marine environment by using *Desulfovibriodesulfuricans*. *Biotechnology and Bioprocess Engineering* 20 (5): 937–941.

118 Dewi, E.R.S. and Nuravivah, R. (2018). Potential of microalgae chlorella vulgaris as bioremediation agents of heavy metal pb (lead) on culture media. In: *E3S Web of Conferences*, 31. EDP Sciences. 05010.

119 Diep, P., Mahadevan, R., and Yakunin, A.F. (2018). Heavy metal removal by bioaccumulation using genetically engineered microorganisms. *Frontiers in Bioengineering and Biotechnology. Frontiers Media S.A* 6: 157.

120 Hamdan, A.M., Abd-El-Mageed, H., and Ghanem, N. (2021). Biological treatment of hazardous heavy metals by streptomyces rochei ANH for sustainable water management in agriculture. *Scientific Reports* 11 (1).

121 Goswami, R.K., Agrawal, K., Shah, M.P., and Verma, P. (2021). Bioremediation of heavy metals from wastewater: a current perspective on microalgae-based future. *Letters in Applied Microbiology* 75 (4): 701–717.

122 Si, M. and Lang, J. (2018). The roles of metallothioneins in carcinogenesis. *Journal of Hematology and Oncology* 11 (1): 1–20.

123 Jaiswal, S. and Shukla, P. (2020). Alternative strategies for microbial remediation of pollutants via synthetic biology. *Frontiers in Microbiology* 11: 808.

9

Removal of Organic Contaminants from Aquatic Environments Using Metal-Organic Framework (MOF) Based Materials

Linkon Bharali and Siddhartha S. Dhar

Department of Chemistry, National Institute of Technology, Silchar, Cachar 788010, Assam, India

9.1 Introduction

Metal-organic frameworks (MOFs) are porous materials made up of inorganic metal atoms or clusters. Organic linker molecules encircle the ions. In brief, MOFs are crystalline hybrid materials. Due to their ability to be purposefully modified, MOFs are used in a variety of applications. Over the past few years, MOFs have attracted significant interest for catalysis, membranes, storage, drug delivery, separation, gas adsorption, etc., because of their permanent porosity, rich surface chemistry, and tunable pore sizes [1–5]. Although the construction, chemical alteration, and prospective applications of MOFs have all been discussed earlier by various researchers, there is still growing interest in the formation and potential applications of their composites. Because of the presence of a variety of functional groups on their frameworks, MOFs exhibit a higher ability to be chemically tailored compared to other porous materials, including active carbon, zeolites, mesoporous silica, etc. Different inorganic metals and organic ligands can be used in the manufacture of MOFs to accomplish particular applications [6]. Porous MOFs typically exhibit microporous characteristics (less than 2 nm), although the extent of the bi- or multipodal stiff organic linkers can be used to regulate pore diameters from several angstroms to several nanometers. The majority of research efforts over the past few years have been focused on the production of novel MOF structures and investigating their applications in different fields. Several MOFs are already commercially available [7–9]. In spite of having various efficient properties, MOFs have certain limitations. Sometimes, the full potential of MOFs cannot be utilized because of their poor chemical stability. Therefore, it is desirable to further improve the qualities and offer new functionality in order to meet the practical uses of MOFs. The combination of MOFs with various functional materials can be beneficial to both components, and utilizing combinations would also help to achieve more efficient use of the materials [10–12]. The MOF-based combined materials can be utilized in a variety of fields including environmental remediation.

Metal Organic Frameworks for Wastewater Contaminant Removal, First Edition. Edited by Arun Lal Srivastav, Lata Rani, Jyotsna Kaushal, and Tien Duc Pham.

MOFs are generally developed in a one-pot reaction by the self-assembly of metal atoms or ions and different ligands [13]. High temperatures and pressures during the reaction accelerate final product generation at the expense of single-crystal development. To produce crystalline materials, gentle, rate-controlled processes are desirable. The solvothermal/hydrothermal method is the most popular synthetic method. This technique involves dissolving ligands with metal ions or clusters in a solvent, encasing them in Teflon-lined autoclave, and heating them to a certain temperature [14]. High temperature and pressure are necessary for coordination reactions to occur. This technique has a number of benefits, including precise crystalline production, ease of use, and minimal energy usage [15]. The majority of well-known MOFs are manufactured using this approach, including University of Oslo (UiO), the zeolitic imidazolate framework (ZIF) series, the isoreticular metal-organic framework (IRMOF) series, and Materials of Institute Lavoisier (MIL) series [16].

The electrochemical approach and microwave/ultrasonic-aided methods are alternative synthetic procedures to prepare MOFs for solid-phase extraction. Utilizing field-assistance technology, microwave and ultrasonication approaches enable faster coordination reactions. Such processes reduce reaction times from days to minutes, unlike the solvothermal approach, which takes a longer time [17].

Another synthetic method is electrochemical synthesis. In an electrochemical cell, a metal plate serving as the anode is submerged in a solution containing the organic ligands to be synthesized electrochemically. With reaction periods of less than one hour, applying a voltage causes a layer of MOF to develop on the cathode [18]. Yan et al. reported the first utilization of MOFs in the analytical chemistry field. The group expanded their applications to sampling, high-performance liquid chromatography, solid-phase extraction, and gas chromatography [19].

Furthermore, various MOF-based materials were synthesized by different research groups. The enhanced efficiency of MOF composites containing active species, such as metal nanoparticles, nanorods, quantum dots, oxides, polymers, polyoxometalates, graphene, biomolecules, carbon nanotubes, and others, is still progressing. Such MOF-based composites have performed excellent activity that is beyond the capabilities of the individual components [20–27]. Although there has been significant development, certain applications of MOF-based composites still face difficulties in ensuring functioning between the host and guest species. Therefore, understanding the structural-functional link in MOF composites is essential for effectively and purposefully preparing certain MOF composites [28].

The synergistic relationship between MOFs and other functional materials has already been found in several literature reviews [9, 29–31]. The majority of them concentrated on developing MOF-based composites and utilizing them in catalysis or sensing applications [32–34]. Various synthetic techniques have so far been developed to produce MOF-based composites, and they can be categorized into the following: (i) formation of in-situ MOF cavities, (ii) modified surface encapsulation, (iii) step-by-step assembly of the heterostructure (sandwich-like structure), (iv) one-pot preparation method, and (v) self-sacrificed template method. All the synthetic processes require different environments and reagents.

Along with other significant applications, utilization of MOF-based composites in the area of environmental remediation has gained more focus in the last few years.

Recent studies have provided detailed descriptions of the environmental utilities of MOFs and MOF-based composites for the elimination of harmful pollutants, with a focus on the adsorption mechanism of different types of pollutants (such as heavy metals, organics, harmful gases, etc.) and the catalytic removal of various water (H_2O) pollutants utilizing photocatalysis, Fenton-like oxidation, reactions based on sulfate radicals, and electrocatalysis [35–41]. Various studies have found that water-stable metal-organic frameworks can be employed as efficient and recyclable substances for environmental cleanliness. The combination of MOFs with graphene oxide as well as carbon nanotubes are two examples of MOF-Cs that have been extensively described as adsorbents for separation processes (gas phase separation) and storage (for example H_2, CO_2, CH_4, etc.) [42–45], supercapacitors and batteries [46–48], and heterocatalysts and electrocatalysts to carry out organic synthesis and water oxidation reactions [49–52]. These composite materials have become viable options for environmental remediation applications that are economical as well as efficient pollution sensors for environmental control [53–58].

This chapter discusses various metal-organic-framework-based composites or materials and applicability in the field of environmental remediation. The chapter also describes different MOF-based catalysts that show excellent photocatalytic activities in the deduction of organic pollutants from water bodies.

9.2 MOF-Based Materials

Various metal-organic-framework-based composites have been reported to date. Such composites possess different characteristics and show different applications. Several MOF-based materials are discussed in Figure 9.1:

9.2.1 MOF–Metal Nanoparticle Materials

Metallic nanoparticles (MNP) have become a topic of great concern due to their extensive potential and real applications. Their distinctive physicochemical characteristics separate them from their bulk form [59–61]. Free MNPs are known to have greater surface energy, a tendency to aggregate, and fusion ability, which makes it difficult to store, process, and apply for longer durations. As a result, the remarkable qualities of metal nanoparticles are somewhat decreased. However, metal nanoclusters and nanoparticles can be effectively enclosed in systems with constrained void space to prevent agglomeration [62]. Porous metal-organic frameworks have persistent nanoscale voids or free channels and are thermally robust. Given their similarity to zeolites, MOFs can be utilized to support metal nanoparticles of specified sizes inside of the pores, evading the cause of aggregation and improving their usage in various applications like catalysis because MOFs provide significant confinement effects [63]. Metallic nanoparticle@MOF composites have been extensively used in heterogeneous catalysis, hydrogen storage, and sensing, based on the characteristics of the metal and framework supports, by utilizing the progress drawback of metal nanoparticles in MOF matrices, wide surface areas, and selectivity in size of metal-organic frameworks.

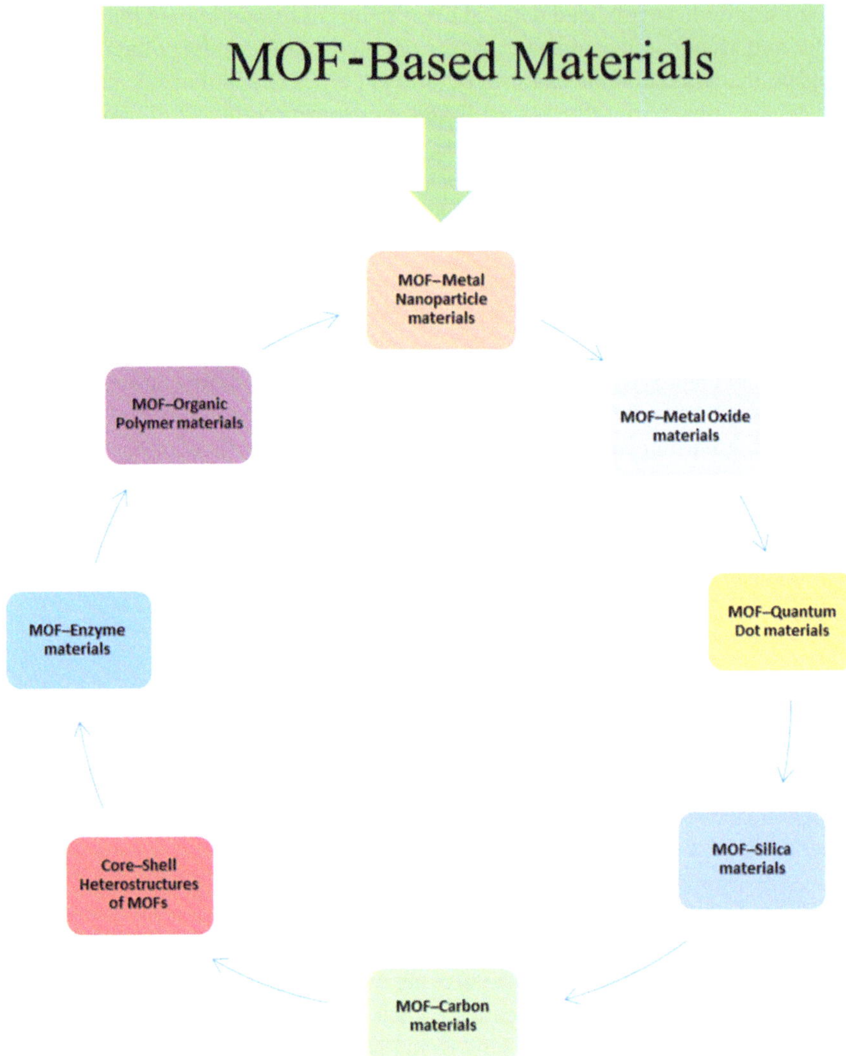

Figure 9.1 *Different MOF-based materials.*

9.2.2 MOF–MO Materials

Metal oxide (MO) nanomaterials (NP) with stretchable size, form, and crystalline quality can be applied to a variety of fields such as catalysis, electronics, solar energy harvesting, optics science, electrochemical energy conversion, and storage, etc. [64]. To improve efficacy, different functionalities can be introduced to metal oxides. It has been reported in the literature that metal oxides and MOFs, especially those with semiconducting or magnetic properties, may be incorporated into core—shell nanostructures. MO-NP@MOF nanocomposites can be developed in a similar way as MNP@MOF materials. One of the methods is the oxidative annealing of early loaded precursors that generates MO within the cavities of MOFs. Another method involves encapsulating already prepared, manufactured metal oxide nanoparticles into MOF matrices [65–67]. In the latter method, to enhance their affinity for MOFs and facilitate a structured crystal formation, the nanoparticles are often coated with appropriate surface functional groups such as RNH_2, RCOOH, etc. [68]. In addition to the core—shell MO-NP@MOF nanostructures, metal-organic frameworks can serve as patterns for the production of metal oxides with desired sizes and shapes [69, 70]. Fischer et al. reported syntheses of nanosized zinc oxide and TiO_2 from $Zn(C_2H_5)_2$ and $Ti(O\text{-}iPr)_4$, respectively, and introduced these into MOF-5. [71, 72] Furthermore, based on the iron oxide nanoparticles modified with an anionic polyelectrolyte and decorated with pyridine groups, Fe_3O_4@ZIF-8 and Fe_3O_4@HKUST-1 were effectively produced [73, 74].

9.2.3 MOF–Quantum Dot Materials

Quality based quantum scale semiconducting materials with distinct electrical and optical properties are simply known as quantum dots (QD). They have the benefits of low cost, greater molar absorption coefficients, photostability, and luminous quantum yields, depending on their size and optical characteristics. At present, there is high demand for the use of quantum dots in light emitting devices (LEDs) and solar devices [75, 76]. There are various methods for synthesizing composites with QD and metal-organic frameworks. Through the deposition of an nm-sized metal-organic framework shell in quantum dot@MOF composites, quantum dots can be maintained contrary to photochemical degradation while sustaining their important optical features. Such composites perform various applications, for example light harvesting, specific molecular sensing, and photochemical synthesis.

Maspoch et al. reported the first quantum dot@MOF. Their synthesis process involved a simple combination of fluorescent quantum dots with MOF precursors [77]. Again, through suitable surface functionalization, Buso et al. first demonstrated the inclusion of highly luminous multiple shell CdSe/ZnS/CdS/ quantum dots within cubic-pattern crystals of MOF-5 [78]. Furthermore, Biswal et al. reported encapsulation of graphene quantum dots in a ZIF-8 matrix without using any capping agents. The encapsulation resulted from graphene quantum dots with polar functional groups, such as carboxylic groups, hydroxyl groups, etc., adhering to the surfaces of the ZIF-8 nanocrystals [79].

9.2.4 MOF–Silica Materials

Due to their many uses in catalysis, drug release, separation, etc., silica nanoparticles have drawn a lot of attention. They offer novel material platforms for executing many nanoscale activities. With its stability, dielectric characteristics, porosity, and possibility for adding different functionalities, silica nanochemistry is the foundation for different advancements in this field [80–83]. There are two main types of MOF—silica composites that are common. Those are MOFs@SiO_2 and SiO_2@

MOFs composites. In the first type, for encouraging the development of microporous framework particles across the porous silica supports, silica shells act as surface coatings due to their mesoporous qualities. In the second type, silica nanoparticles are dispersed and incorporated into the pores and networks of MOFs; a MOF shell is also grown over a silica domain that has already been produced in MOF precursor solutions. The hydrophilic property of silica is highly influenced by silanol moieties on its surface, and this effect is inversely correlated with silica size. As a result, after the synthesis of nanocomposites with host materials, nanosized silica with a greater surface area might exhibit noticeable variations in the adsorption behavior [84]. Tetramethoxysilane was polycondensed using a sol-gel method in the nano passages or channels of highly permeable HKUST-1 and also CPL-5 ([Cu2(pzdc)2(dpe)], in which dpe = 1,2-di(4-pyridyl)-ethylene), having one- and 3 D channels, respectively. This produced subnanosized silica that was uniformly distributed throughout the host frameworks. Due to its nanoscale size and rigorous growth constraints, the resulting silica demonstrated a markedly reduced crystallization temperature. Significantly, the water-soluble silanol on the external boundary of the silica particles were enhanced, and as a result, the composites made of SiO_2 and metal-organic framework displayed extraordinary affinity for water-soluble molecules. This is due to the strong connection of the doped silica molecules with the adsorbates; the composites showed greater water adsorption compared to the original MOFs as well as specific adsorption of hydrophilic molecules. [85, 86] Furthermore, considering MOF@SiO_2 composites, nanoscale metal-organic frameworks have significant usability. For example, in the synthesis of aminofunctionalized nanoscale NH_2-MIL-101(Fe), due to the lower stability of the NH_2-MIL-101(Fe) particles in basic reaction circumstances, an alternate technique using sodium silicate where it acts as the silica source was used to coat the aminofunctionalized nanoscale NH_2-MIL-101(Fe) in the synthesis process [87]. By being aminofunctionalized, nanoscale MOFs were able to be loaded with a fluorophore (BODIPY) and a drug molecule used for anticancer treatment via covalent modifications. As a result, the MOF composites could be applied as a nano delivery carrier for imaging differentiating agents and anticancer remedies by slowly releasing the cargoes through nanoscale MOF degradation.

9.2.5 MOF–Carbon Materials

Carbon is a very appealing material for various applications in different fields since it has different allotropes (for example, graphite, fullerenes, diamond, etc.), microtextures with varying degrees of graphitization, it may exist as a zero-dimensional to three-dimensional material, and it can be found in different forms (fabric, powder, fiber, composite, foam, etc.) [88]. Nanocarbons, particularly carbon nanotubes (CNTs) and graphene, are receiving attention as the most viable participants for functional uses. They are desirable because of the grouping of the exceptional features of planer graphite with a greater surface region [89]. Until today, activated carbons, carbon monoliths, graphene oxide (GO, a derivative of graphene), and CNTs have all been used to synthesize a variety of MOF-nanocarbon composites that have been exclusively investigated for use in broad applications. From the literature, it was found that insertion of metal-organic frameworks to the meso and macropores of C (carbon) materials to synthesize hierarchically porous schemes can adjust the physical topographies and improve adsorbing qualities [90, 91].

De Oliveira et al. reported a composite material that contained MOFs inside the pores of activated carbon (AC), which was effective in removing aldicarb from rats. The synthesis process involved hydrothermal treatment. There was an in-situ addition of AC in varied proportions during preparation of Ln-succinate (Ln = Eu and Tb). To assess the adsorption of one of the most hazardous pesticides, called aldicarb, a biotic assessment of the Tb-MOF@activated carbon composite was carried out on rats. The average amounts of aldicarb adsorption that were seen in vitro over the course of ten minutes in an acidic medium were 46% for activated carbon and 41% for terbium-MOF. The standard values were considerably less than 77–78% for Tb-MOF@activated carbon composites comprising 50% and 40% of AC. That is undoubtedly a significant finding in poisoning cases, as the composite began to eliminate aldicarb from the stomach in an acid environment, greatly lowering the likelihood that it would enter the ileum and then the inner tissues. Research on the effects of aldicarb on injuries showed that action with a composite comprising 50% of AC completely preserved the reliability of tissues, demonstrating the efficient protection potential of the MOF composite [90]. Another MOF-carbon composite, MWCNT@MOF-5, was reported by Jiang et al. It was discovered that this highly porous interpenetrated MWCNT@MOF-5 composite increased hydrogen storing capacity [92].

9.2.6 Core–shell Structures of MOFs

The synthesis of core—shell structures by combining MOFs (MOF@MOF) is a favorable method for changing the porosity characteristics of the MOF and for introducing a new function without altering the distinctive properties of crystalline MOFs. Recently, two methods for synthesizing multifunctional core—shell structures have been proposed. A composite crystal might be developed by heteroepitaxially growing a shell metal-organic framework crystal on the external surface of different seeds. As a result, the segregation of the binary coordinating components with various metal centers or connecting linkers leads to various crystallographic areas [93]. The postproduction changes, which involve the specific reaction of the residue of an organic linker and the precise substitute of the framework metal atoms or ligands, is another technique for producing multifunctional frameworks with a core—shell structure. The inner core or the outer surface of the MOF crystals are the only places where these modifications to a MOF metal center or organic linker can occur. The epitaxially produced crystals with inner-shell structures were investigated by MacDonald et al., [94] Hosseini et al., [95, 96] Koh et al., [97] and Tang et al. [98]. Furukawa et al. reported the first instance of heteroepitaxial development of core—shell MOF single crystals [99]. Since then, much progress has been achieved in this field. Under solvothermal conditions, core—shell crystals with a zinc-MOF core and a copper-MOF shell were produced by employing the single crystals of a Zn-MOF, [$Zn_2(ndc)_2$(dabco)] (dabco = diazabicyclo-[2.2.2]octane), and seeds for the varied metallic epitaxial development of the iso reticular Copper-MOF, [$Cu_2(ndc)_2$(dabco)]. Along with the heterometallic epitaxial development of MOF-enlarged materials, variations in the bridging linker in the epitaxy have also been investigated. Using bdc and bdc-NH_2 linkers, Koh and co-works examined the generation of layered MOF heterostructures by seeding in a manner similar to the epitaxial development of the core—shell MOFs [97]. By dipping the seed crystals of MOF-5 (L = bdc) in a prepared solution of IRMOF-3 (L = bdc-NH_2), and vice versa, core—shell structures were developed. Due to the presence of orange IRMOF-3 and colorless MOF-5, the resulting composite crystals exhibited significant spatial color contrast. These structures maintained their porosity, and their surface area values ranged between IRMOF-3 and MOF-5 based on the product composition, which was governed by the ratio of the two particular linkers. On a porous alumina substrate, Yoo et al. reported possible generation of hybrid films by heteroepitaxially developing IRMOF-3 films on the MOF-5 seed crystalline surface.

9.2.7 MOF–Enzyme Materials

Enzymes are natural catalysts that have different properties such as high selectivity and reactivity as well as a specific nature under ordinary conditions, which contributes to environmentally friendly and sustainable processes in the chemical, medicinal, and nutrition sectors [100]. Because of their poor working stability, challenging recovery, and loss of activity under functioning settings, enzymes have unfortunately not been widely used in industrial applications. Enzymes that have been immobilized on a solid platform can be more stable, easier to separate and recover for recycling, and still retain functionality and selectivity [101, 102]. Silica materials with larger surface areas and greater mesoporosity have drawn lots of interest in the extensive investigation for suitable supports in enzyme immobilization [103–105]. Due to the poor affinities between enzymes and silica supports, leaching of the enzyme that is adsorbed is, however, frequently noticeable throughout the reaction phase. Porous MOFs may be a good choice to accommodate enzymes for catalytic applications due to their functionalizable porous structure and controllable, uniform pore diameters. However, the micropore measurement of the majority of MOFs prevents the access of larger enzyme molecues and may only allow for outer surface attachment with little enzyme load by an adsorption or covalent bonding mechanism [106–109]. Enzymatic catalytic applications are made possible by recent developments in mesoporous MOFs, and reports on the immobilization of enzymes on the inner side of the nanopores of metal-organic frameworks have been reported in various literatures [110–113].

9.2.8 MOF–Organic Polymer Materials

Organic polymers have different specific qualities that make them suitable for combining with other valuable materials to form composites. Some of those qualities are that it is simple to produce, light in weight, and it has decent thermal and chemical stability [114]. Particularly, restricted polymers offer unique and surprising characteristics that are different from those found in bulk form at nanoscale sizes [115]. There has been a lot of interest in polymer incorporation in crystal porous substrates with regulated and controllable nanochannel patterns. From several literatures, it has been found that In nano-reactors for polymerization, a variety of porous MOFs with channels of nanoparticles with varying sizes, forms, dimensions, and interfacial functional behaviors have been employed [116–118]. Particularly, MOFs with active sites on their pore surfaces can function as both nanomolds and polymer production catalysts [119–121]. The polymers that have been investigated so far range from polystyrene to polypyrroles. For instance, in the year 2005, the first polymerization of styrene was carried out in the nanochannels of $[M_2(bdc)_2(ted)]$, where M = Zn or Cu metal and ted = triethylenediamine. In later investigations, efforts in radical polymerization have been further focused on regulating the dimensions [119, 122], copolymer series [123], molecular weight [124], and tacticity [23, 125–127].

A fascinating case of host-guest polymerization within different frameworks has been reported recently. This generally identified that arranging polymer chains in a highly ordered crystalline structure is a challenging task. Uemura et al. have revealed a method for generating polymeric materials that display a crystalline arrangement that depends on ordered crosslinks. In this approach, divinyl cross linkers such as 2,5-divinylterephthalate (DVTP), were used as replacement ligands to formulate a MOF, $[Cu(DVTP)_x (bdc)_{1-x}(ted)_{0.5}]$ where x indicated the molar DVTP amount [128].

9.3 Environmental Effects of MOF-Based Materials

Water contamination is a severe issue as a result of inefficient pollution control. The contaminating elements in aquatic habitats have grown more complicated in comparison to earlier times. The safety of water is now threatened by some newly discovered contaminants. Therefore, there is a significant need to develop affordable technology for cleaning up water pollution [129]. The catalytic process can be utilized to break down or decompose the pollutants and turn them into safe degradation products [130]. To achieve controlled pore structure and a greater number of active sites, the goal is to plan and construct largely effective, non-hazardous, and chemically stable catalysts that are simple to generate in a variety of dimensions and forms [131, 132]. Several MOFs and MOF-based materials have been developed, and more are still being produced. Some widely used metal-organic frameworks are MOF-69C, MIL-100, MOF-5, MIL-101, MOF-74, ZIF-8, HKUST-1, ZIF-67, UiO-66, and MIL-53, etc. [133–138]. These materials offer significant benefits for both energy regeneration and environmental remediation.

Recent studies have demonstrated the wide range of potential applications for metal-organic-framework-derived carbons in the catalysis industry. They have outstanding activity in energy conversion and catalysis when employed as catalyst supports or catalysts to enhance the H_2 evolution reaction, oxygen evolution, and reduction reaction [139–141]. For example, it has showed excellent photocatalytic activity in the degradation of organic pollutants from water [142]. Because MOFs are so versatile, it should be feasible to strategically design both the active site and its environment with an unexpected level of accuracy, which would have significant benefits for catalysis. By directly synthesizing the desired scaffold or by postsynthetic modification, the catalytic activity can be performed at the organic [143] or inorganic [129] component. Various postsynthetic processes can be employed in this respect [144, 145].

In an attempt to deal with the energy crisis and environmental issues, photocatalysis has recently gained more attention [146]. Several materials have been investigated and used for photocatalysis applications. The following three basic phases make up the traditional photocatalytic process:

i) In order to produce charge-separated-excited states, photosensitizers first absorb solar radiation;
ii) Then, they make redox equivalents (h^+ and e^-) and move to the catalytic centers;
iii) Finally, at the reactive centers, their reactions with substrate molecules take place.

Consequently, an outstanding photocatalyst should have qualities such as strong sunlight absorption, a longer excited-state lifespan, a greater amount of charge-separated levels, and well charged movement.

A new class of photocatalytic material is thought to exist in some MOFs with characteristics resembling semiconductors. In contrast to conventional inorganic semiconductor materials, porous crystal MOF structures prevent the recombination of e^- and h^+ pairs and makes it easier to use charge carriers. As a result, such MOF materials show potential for photocatalysis as well as for the oxidation of water, carbon dioxide abatement, dye degradation, organic transformation, and hydrogen generation by water splitting reaction [147, 148]. Ramezanalizadeh et al. reported excellent photocatalytic activity of MOFs/$CuWO_4$ in the degradation of 4-nitrophenol and methylene blue dye. They also mentioned that the presence of metal-organic frameworks enhanced photocatalytic performance [149]. Furthermore, Zhou et al. synthesized Zr-based metal-organic frameworks that showed a greater adsorption for TC (tetracycline) [150]. By using the solvothermal approach, Azhar et al. developed HKUST-1 and UiO-66 and evaluated their potential to remove methylene blue (Mb) from wastewater. They can use their MOF formulations over a wide pH range. When compared to UiO-66, HKUST-1 had a better adsorption capacity [151]. For the treatment of Pb (II) in wastewater, Yin et al. combined ceramic membrane ultrafiltration with NH_2-functionalized MOFs (referred to as UiO-66-NH_2). The ceramic membrane could totally retain the MOF performance when certain parameters such as transmembrane pressure, reaction temperature, regular pore diameter, and cross-velocity were used at 0.15 MPa, 35°C, 50 nm, and 4.0 ms^{-1}, respectively. The performance of membranes was improved by the inclusion of MOFs, which altered their characteristics and added some new capabilities [152]. The coupling process helps magnetic separation of metal-organic frameworks from water bodies. By applying the solvothermal method, Xu et al. synthesized magnetically porous Ni@ MOFs-74 composite. In the elimination of rhodamine B (Rh B), the composite had greater effects in adsorption; the adsorption efficacies reached up to 4.1 mg/g^{-1} for ibuprofen and 177.8 mg/g^{-1} for rhodamine B dye [153]. A magnetic MOF, Cu-MOFs/ Fe_3O_4, was synthesized by Shi et al. utilizing the solvothermal process. Malacite green and Pb (II) adsorption studies were conducted. The two pollutants had adsorption efficiencies of 113.67 mg/g^{-1} for the dye and 219.00 mg/g for lead [154].

Furthermore, The functions of MOFs may be improved by combining them with graphene oxide (GO). Gaphene oxide is described as a soft, 2 D carbon nanomaterial made up of several functional groups that contain oxygen, such as carboxyl and epoxy groups. It has a sizable specific surface area, excellent electrical conductivity, and strong mechanical strength. The specific surface area of MOFs might be greatly increased by GO, which also exhibits a strong adsorption activity. According to Liang et al., the highest capacity of adsorption for Cr (VI) with GO@MIL-101 could reach 125 mg/g1, which was 20% higher than that attained with MIL-101 with an appropriate dose, while both the consistency of the crystalline composite and physical properties such as particle size were reduced with the increase of graphene oxide hosted in GO@MIL-101 [155].

Table 9.1 shows methods that have been applied in the removal of organic contaminants from water:

Toxic organic pollutants not only have a negative impact on the environment, water quality, air, and soil but they also have a negative impact on human health. Heavy metal pollutants in water are non-biodegradable and they accrue in the food chain, affecting various living organisms, including mankind. Environmental contamination caused by different toxic species such as dyes, antibiotics, pesticides, heavy metals, etc., lead to critical health issues in humans, which include kidney problems, chronic respiratory disease, cell damage, cancer, and neurological diseases [156]. Furthermore, aromatic nitro compounds and sulfur dioxide (SO_2) are considered to be toxic and harmful for human health [157]. Organic as well as pharmaceutical pollutants are regarded as emerging pollutants because they are often detected in water and have a high level of toxicity. To improve human health and

Table 9.1 Some methods applied for the removal of organic contaminants from water.

Sl No.	Pollutant	Removal Method	Reference
1	Heavy metals, for example: lead, cadmium, mercury	Adsorption	[129]
2	Atrazine	Catalytic ozonation	[132]
3	Levofloxacin	Photocatalytic degradation	[137]
4	Tetracycline	Luminescent quenching mechanism, adsorption	[150]
5	Trinitrotoluene	Bioremediation methods (phytoremediation)	[156]

environmental cleanliness, research groups have investigated various methods to remove these emerging pollutants.

Photodegradation experiments of organic dyes by MOF –Polymer compisites were carried out using UV-Vis spectrophotometer and curves obtained are shown in Figure 9.2.

MOFs have a lot of potential for removing contaminants from water. In this chapter, we go over various MOF preparation techniques. In these processes, the structural distributions, yields, energy requirements, purity, and efficacy of pollutant removal [159] of MOFs are intimately related to the choice and design of MOFs.

Figure 9.3 shows the preparation of copper-MOFs/Fe_3O_4 composite for the removal of lead, Pb (II) and malacite green (MG).

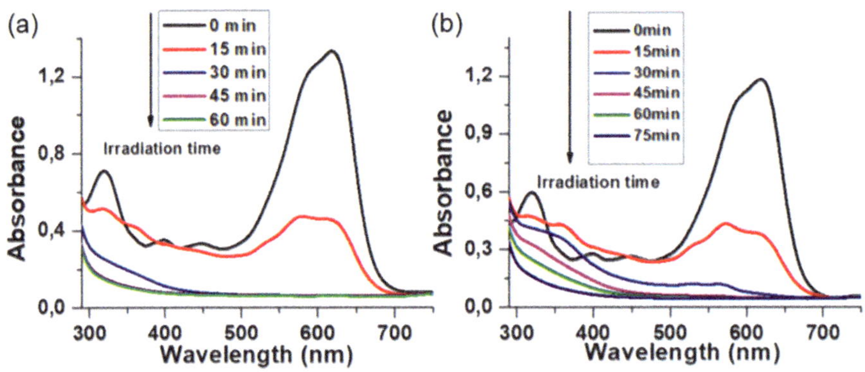

Figure 9.2 UV-Vis absorption curves showing photocatalytic degradation of Acid Black dye under UV lamp using MOF-based catalysts (A) 2% MIL-53/Polymer Composite and (B) 2% HKUST-1/Polymer composite Ref [158].

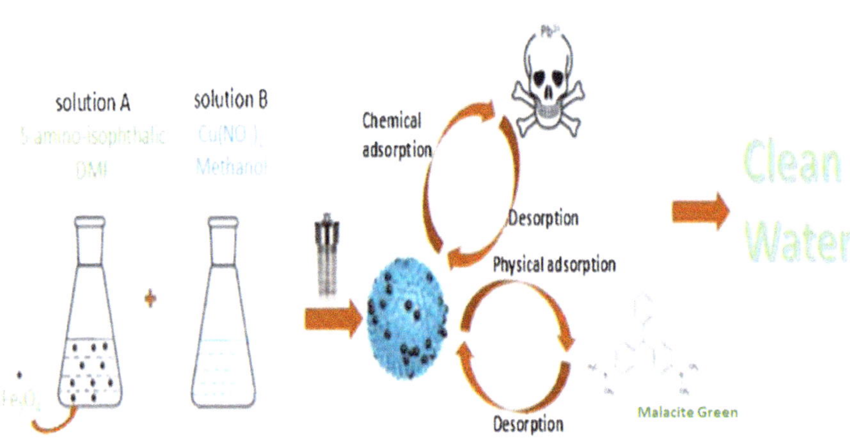

Figure 9.3 Schematic representation of preparation of copper-MOFs/Fe_3O_4 composite for the removal of lead, Pb (II) and malacite green (MG) Ref [154].

9.4 Conclusion

Metal-organic frameworks (MOFs), with simple variations of organic linkers with photocatalytically energetic groups like porphyrin and NH_2, can be employed as excellent photocatalysts. Furthermore, because of their high porosity, MOFs can serve as hosts for photoredox species such as valuable metals and semiconductor nanoparticles, etc., opening up new possibilities for photocatalytic and photodegradation applications. The typical uses of metal-organic frameworks in heterogeneous photocatalysis were discussed and summarized in this chapter, showing that MOFs are potential candidates for the process. Thus, different sections of the chapter discuss the various MOF-based materials. The chapter also describes the utility of MOF-based materials in the removal of organic contaminants from aquatic environs.

References

1 Eddaoudi, M., Kim, J., Rosi, N. et al. (2002 January 18). Systematic design of pore size and functionality in isoreticular MOFs and their application in methane storage. *Science* 295 (5554): 469–472.
2 Kreno, L.E., Hupp, J.T., and Van Duyne, R.P. (2010 October 1). Metal– organic framework thin film for enhanced localized surface plasmon resonance gas sensing. *Analytical Chemistry* 82 (19): 8042 6.
3 Bae, T.H., Lee, J.S., Qiu, W. et al. (2010 December 17). A high-performance gas-separation membrane containing submicrometer-sized metal-organic framework crystals. *Angewandte Chemie* 122 (51): 10059–10062.
4 Lee, J., Farha, O.K., Roberts, J. et al. (2009). Metal-organic framework materials as catalysts. *Chemical Society Reviews* 38 (5): 1450–1459.
5 Horcajada, P., Chalati, T., Serre, C. et al. (2010 February). Porous metal-organic-framework nanoscale carriers as a potential platform for drug delivery and imaging. *Nature Materials.* 9 (2): 172–178.
6 Yaghi, O.M., O'Keeffe, M., Ockwig, N.W. et al. (2003 June). Reticular synthesis and the design of new materials. *Nature* 423 (6941): 705–714.
7 Ikezoe, Y., Washino, G., Uemura, T. et al. (2012 December). Autonomous motors of a metal-organic framework powered by reorganization of self-assembled peptides at interfaces. *Nature Materials* 11 (12): 1081–1085.
8 Zhu, Q.L., Sheng, T.L., Fu, R.B. et al. (2011 March 14). Redox-responsive photochromism and fluorescence modulation of two 3D metal-organic hybrids derived from a triamine-based polycarboxylate ligand. *Chemistry–A European Journal* 17 (12): 3358–3362.
9 Zhu, Q.L. and Xu, Q. (2014). Metal-organic framework composites. *Chemical Society Reviews* 43 (16): 5468–5512.
10 Falcaro, P., Hill, A.J., Nairn, K.M. et al. (2011 March 15). A new method to position and functionalize metal-organic framework crystals. *Nature Communications* 2 (1): 1–8.

11 Khaletskaya, K., Reboul, J., Meilikhov, M. et al. (2013 July 31). Integration of porous coordination polymers and gold nanorods into core–shell mesoscopic composites toward light-induced molecular release. *Journal of the American Chemical Society* 135 (30): 10998–11005.

12 Ricco, R., Malfatti, L., Takahashi, M. et al. (2013). Applications of magnetic metal-organic framework composites. *Journal of Materials Chemistry A* 1 (42): 13033–13045.

13 Li, C., Qiu, W., Long, W. et al. (2014 November 1). Synthesis of porphyrin@ MOFs type catalysts through "one-pot" self-assembly. *Journal of Molecular Catalysis A: Chemical* 393: 166–170.

14 Qi, Y., Luo, F., Che, Y., and Zheng, J. (2008 February 6). Hydrothermal synthesis of metal– organic frameworks based on aromatic polycarboxylate and flexible bis (imidazole) ligands. *Crystal Growth and Design* 8 (2): 606–611.

15 Lei, G., Guan-Hua, L., and Ying, W. (2018 November 12). Hydrothermal synthesis, structure and properties of a 3D pillar-layered metal-organic framework based on amino-arenedisulfonate ligand. *Chinese Journal of Structural Chemistry* 37 (11): 1795–1804.

16 Li, H., Eddaoudi, M., O'Keeffe, M., and Yaghi, O.M. (1999 November). Design and synthesis of an exceptionally stable and highly porous metal-organic framework. *Nature* 402 (6759): 276–279.

17 Barton, H.F., Davis, A.K., Lee, D.T., and Parsons, G.N. (2018 June). Solvothermal synthesis of MIL-96 and UiO-66-NH2 on atomic layer deposited metal oxide coatings on fiber mats. *JoVE (Journal of Visualized Experiments)* 13 (136): e57734.

18 Campagnol, N., Van Assche, T., Boudewijns, T. et al. (2013). High pressure, high temperature electrochemical synthesis of metal-organic frameworks: films of MIL-100 (Fe) and HKUST-1 in different morphologies. *Journal of Materials Chemistry A* 1 (19): 5827–5830.

19 Zhou, Y.Y., Yan, X.P., Kim, K.N. et al. (2006 May 26). Exploration of coordination polymer as sorbent for flow injection solid-phase extraction on-line coupled with high-performance liquid chromatography for determination of polycyclic aromatic hydrocarbons in environmental materials. *Journal of Chromatography A* 1116 (1–2): 172–178.

20 Wei, Y., Han, S., Walker, D.A. et al. (2012 July 23). Nanoparticle core/shell architectures within MOF crystals synthesized by reaction diffusion. *Angewandte Chemie International Edition* 51 (30): 7435–7439.

21 Li, Z. and Zeng, H.C. (2013 May 14). Surface and bulk integrations of single-layered Au or Ag nanoparticles onto designated crystal planes {110} or {100} of ZIF-8. *Chemistry of Materials* 25 (9): 1761–1768.

22 Bradshaw, D., Garai, A., and Huo, J. (2012). Metal-organic framework growth at functional interfaces: thin films and composites for diverse applications. *Chemical Society Reviews* 41 (6): 2344–2381.

23 Uemura, T., Yanai, N., and Kitagawa, S. (2009). Polymerization reactions in porous coordination polymers. *Chemical Society Reviews* 38 (5): 1228–1236.

24 Esken, D., Turner, S., Wiktor, C. et al. (2011 October 19). GaN@ ZIF-8: selective formation of gallium nitride quantum dots inside a zinc methylimidazolate framework. *Journal of the American Chemical Society* 133 (41): 16370–16373.

25 Liu, Y., Zhang, W., Li, S. et al. (2014 January 28). Designable yolk–shell nanoparticle@ MOF petalous heterostructures. *Chemistry of Materials* 26 (2): 1119–1125.

26 Zhao, M., Deng, K., He, L. et al. (2014 February 5). Core–shell palladium nanoparticle@ metal-organic frameworks as multifunctional catalysts for cascade reactions. *Journal of the American Chemical Society* 136 (5): 1738–1741.

27 Li, L., Liu, X.L., Gao, M. et al. (2014). The adsorption on magnetic hybrid Fe 3 O 4/ HKUST-1/GO of methylene blue from water solution. *Journal of Materials Chemistry A* 2 (6): 1795–1801.

28 Lin, R., Yao, Y., Zulkifli, M.Y. et al. (2022). Binder-free mechanochemical metal-organic framework nanocrystal coatings. *Nanoscale* 14 (6): 2221–2229.

29 Yang, Q., Xu, Q., and Jiang, H.L. (2017). Metal-organic frameworks meet metal nanoparticles: synergistic effect for enhanced catalysis. *Chemical Society Reviews* 46 (15): 4774–4808.

30 Jiao, L., Wang, Y., Jiang, H.L., and Xu, Q. (2018 September). Metal-organic frameworks as platforms for catalytic applications. *Advanced Materials* 30 (37): 1703663.

31 Huang, Y.B., Liang, J., Wang, X.S., and Cao, R. (2017). Multifunctional metal-organic framework catalysts: synergistic catalysis and tandem reactions. *Chemical Society Reviews* 46 (1): 126–157.

32 Wu, C.D. and Zhao, M. (2017 April). Incorporation of molecular catalysts in metal-organic frameworks for highly efficient heterogeneous catalysis. *Advanced Materials* 29 (14): 1605446.

33 Liu, Y., Xuan, W., and Cui, Y. (2010 October 1). Engineering homochiral metal-organic frameworks for heterogeneous asymmetric catalysis and enantioselective separation. *Advanced Materials* 22 (37): 4112–4135.

34 Dhakshinamoorthy, A. and Garcia, H. (2012). Catalysis by metal nanoparticles embedded on metal-organic frameworks. *Chemical Society Reviews* 41 (15): 5262–5284.

35 Barea, E., Montoro, C., and Navarro, J.A. (2014). Toxic gas removal–metal-organic frameworks for the capture and degradation of toxic gases and vapours. *Chemical Society Reviews* 43 (16): 5419–5430.

36 Dias, E.M. and Petit, C. (2015). Towards the use of metal-organic frameworks for water reuse: a review of the recent advances in the field of organic pollutants removal and degradation and the next steps in the field. *Journal of Materials Chemistry A* 3: 22484–22506.

37 Feng, M., Zhang, P., Zhou, H.C., and Sharma, V.K. (2018 October 1). Water-stable metal-organic frameworks for aqueous removal of heavy metals and radionuclides: a review. *Chemosphere* 209: 783–800.

38 Jiang, D., Xu, P., Wang, H. et al. (2018 December 1). Strategies to improve metal organic frameworks photocatalyst's performance for degradation of organic pollutants. *Coordination Chemistry Reviews* 376: 449–466.

39 Khan, N.A., Hasan, Z., and Jhung, S.H. (2013 January 15). Adsorptive removal of hazardous materials using metal-organic frameworks (MOFs): a review. *Journal of Hazardous Materials* 244: 444–456.

40 Li, J., Wang, X., Zhao, G. et al. (2018). Metal-organic framework-based materials: superior adsorbents for the capture of toxic and radioactive metal ions. *Chemical Society Reviews* 47 (7): 2322–2356.

41 Liu, X., Zhou, Y., Zhang, J. et al. (2017 June 21). Iron containing metal-organic frameworks: structure, synthesis, and applications in environmental remediation. *ACS Applied Materials & Interfaces* 9 (24): 20255–20275.

42 Bian, Z., Xu, J., Zhang, S. et al. (2015 July 7). Interfacial growth of metal organic framework/graphite oxide composites through pickering emulsion and their CO2 capture performance in the presence of humidity. *Langmuir* 31 (26): 7410–7417.

43 Huang, A., Liu, Q., Wang, N. et al. (2014 October 22). Bicontinuous zeolitic imidazolate framework ZIF-8@ GO membrane with enhanced hydrogen selectivity. *Journal of the American Chemical Society* 136 (42): 14686–14689.

44 Kang, Z., Xue, M., Zhang, D. et al. (2015 August 1). Hybrid metal-organic framework nanomaterials with enhanced carbon dioxide and methane adsorption enthalpy by incorporation of carbon nanotubes. *Inorganic Chemistry Communications* 58: 79–83.

45 Zhou, X., Huang, W., Miao, J. et al. (2015 April 15). Enhanced separation performance of a novel composite material GrO@ MIL-101 for CO2/CH4 binary mixture. *Chemical Engineering Journal* 266: 339–344.

46 Bao, W., Zhang, Z., Zhou, C. et al. (2014 February 15). Multi-walled carbon nanotubes@ mesoporous carbon hybrid nanocomposites from carbonized multi-walled carbon nanotubes@ metal-organic framework for lithium sulfur battery. *Journal of Power Sources* 248: 570–576.

47 Wen, P., Gong, P., Sun, J. et al. (2015). Design and synthesis of Ni-MOF/CNT composites and rGO/carbon nitride composites for an asymmetric supercapacitor with high energy and power density. *Journal of Materials Chemistry A* 3 (26): 13874–13883.

48 Zhang, G., Hou, S., Zhang, H. et al. (2015 Apr). High-performance and ultra-stable lithium-ion batteries based on MOF-derived ZnO@ ZnO quantum Dots/C core–shell nanorod arrays on a carbon cloth anode. *Advanced Materials* 27 (14): 2400–2405.

49 Huang, X., Zheng, B., Liu, Z. et al. (2014 August 26). Coating two-dimensional nanomaterials with metal-organic frameworks. *ACS nano* 8 (8): 8695–8701.

50 Jahan, M., Liu, Z., and Loh, K.P. (2013 November 20). A Graphene oxide and copper-centered metal organic framework composite as a tri-functional catalyst for HER, OER, and ORR. *Advanced Functional Materials* 23 (43): 5363–5372.

51 Jia, H., Sun, Z., Jiang, D., and Du, P. (2015 July 14). Covalent cobalt porphyrin framework on multiwalled carbon nanotubes for efficient water oxidation at low overpotential. *Chemistry of Materials* 27 (13): 4586–4593.

52 Qiu, X., Wang, X., and Li, Y. (2015). Controlled growth of dense and ordered metal-organic framework nanoparticles on graphene oxide. *Chemical Communications* 51 (18): 3874–3877.

53 Liang, R., Shen, L., Jing, F. et al. (2015 May 13). Preparation of MIL-53 (Fe)-reduced graphene oxide nanocomposites by a simple self-assembly strategy for increasing interfacial contact: efficient visible-light photocatalysts. *ACS Applied Materials & Interfaces* 7 (18): 9507–9515.

54 Liu, N., Huang, W., Zhang, X. et al. (2018 February 1). Ultrathin graphene oxide encapsulated in uniform MIL-88A (Fe) for enhanced visible light-driven photodegradation of RhB. *Applied Catalysis B: Environmental* 221: 119–128.

55 Yang, C., You, X., Cheng, J. et al. (2017 January 1). A novel visible-light-driven In-based MOF/graphene oxide composite photocatalyst with enhanced photocatalytic activity toward the degradation of amoxicillin. *Applied Catalysis B: Environmental* 200: 673–680.

56 Yang, Y., Wang, W., Li, H. et al. (2017 June 15). NH2-MIL-53 (Al) nanocrystals anchored on the surface of RGO hollow spheres and its visible light degradation of methylene blue. *Materials Letters* 197: 17–20.

57 Wu, S.C., Yu, L.L., Xiao, F.F. et al. (2017 November 15). Synthesis of aluminum-based MOF/graphite oxide composite and enhanced removal of methyl orange. *Journal of Alloys and Compounds* 724: 625–632.

58 Wu, X., Cobbina, S.J., Mao, G. et al. (2016 May). A review of toxicity and mechanisms of individual and mixtures of heavy metals in the environment. *Environmental Science and Pollution Research* 23 (9): 8244–8259.

59 Astruc, D., Boisselier, E., and Ornelas, C. (2010 April 14). Dendrimers designed for functions: from physical, photophysical, and supramolecular properties to applications in sensing, catalysis, molecular electronics, photonics, and nanomedicine. *Chemical Reviews* 110 (4): 1857–1959.

60 Zhang, H., Jin, M., and Xia, Y. (2012). Enhancing the catalytic and electrocatalytic properties of Pt-based catalysts by forming bimetallic nanocrystals with Pd. *Chemical Society Reviews* 41 (24): 8035–8049.

61 You, H., Yang, S., Ding, B., and Yang, H. (2013). Synthesis of colloidal metal and metal alloy nanoparticles for electrochemical energy applications. *Chemical Society Reviews* 42 (7): 2880–2904.

62 White, R.J., Luque, R., Budarin, V.L. et al. (2009). Supported metal nanoparticles on porous materials. *Methods and Applications. Chemical Society Reviews.* 38 (2): 481–494.

63 Meilikhov, M., Yusenko, K., Esken, D. et al. (2010). Chromium (III) terephthalate metal organic framework (MIL-101): HF-free synthesis, structure, polyoxometalate composites, and catalytic properties. *European Journal of Inorganic Chemistry* 24: 3701–3714.

64 Vayssieres, L. (2004 January 1). On the design of advanced metal oxide nanomaterials. *International Journal of Nanotechnology* 1 (1–2): 1–41.

65 Hermes, S., Schröder, F., Amirjalayer, S. et al. (2006). Loading of porous metal-organic open frameworks with organometallic CVD precursors: inclusion compounds of the type [L n M] a@ MOF-5. *Journal of Materials Chemistry* 16 (25): 2464–2472.

66 Wang, W., Li, Y., Zhang, R. et al. (2011 May 15). Metal-organic framework as a host for synthesis of nanoscale Co3O4 as an active catalyst for CO oxidation. *Catalysis Communications* 12 (10): 875–879.

67 Lu, G., Li, S., Guo, Z. et al. (2012 Apr). Imparting functionality to a metal-organic framework material by controlled nanoparticle encapsulation. *Nature Chemistry* 4 (4): 310–316.
68 Ke, F., Qiu, L.G., Yuan, Y.P. et al. (2012). Fe 3 O 4@ MOF core–shell magnetic microspheres with a designable metal-organic framework shell. *Journal of Materials Chemistry* 22 (19): 9497–9500.
69 Hu, H., Wang, Z., Cao, L. et al. (2021 April). Metal-organic frameworks embedded in a liposome facilitate overall photocatalytic water splitting. *Nature Chemistry* 13 (4): 358–366.
70 Hall, A.S., Kondo, A., Maeda, K., and Mallouk, T.E. (2013 November 6). Microporous brookite-phase titania made by replication of a metal-organic framework. *Journal of the American Chemical Society* 135 (44): 16276–16279.
71 Müller, M., Hermes, S., and Kähler, K. (2008). MWE vandenBerg, M. *Muhler, RA Fischer Chemistry Material* 20: 4576–4587.
72 Müller, M., Zhang, X., Wang, Y., and Fischer, R.A. (2009). Nanometer-sized titania hosted inside MOF-5. *Chemical Communications* (1): 119–121.
73 Zhang, T., Zhang, X., Yan, X. et al. (2013 July 15). Synthesis of Fe3O4@ ZIF-8 magnetic core–shell microspheres and their potential application in a capillary microreactor. *Chemical Engineering Journal* 228: 398–404.
74 Bagheri, A., Taghizadeh, M., Behbahani, M. et al. (2012 September 15). Synthesis and characterization of magnetic metal-organic framework (MOF) as a novel sorbent, and its optimization by experimental design methodology for determination of palladium in environmental samples. *Talanta* 99: 132–139.
75 Nozik, A.J., Beard, M.C., Luther, J.M. et al. (2010 November 10;). Semiconductor quantum dots and quantum dot arrays and applications of multiple exciton generation to third-generation photovoltaic solar cells. *Chemical Reviews* 110 (11): 6873–6890.
76 Yang, Z., Chen, C.Y., Roy, P., and Chang, H.T. (2011). Quantum dot-sensitized solar cells incorporating nanomaterials. *Chemical Communications* 47 (34): 9561–9571.
77 Imaz, I., Hernando, J., Ruiz-Molina, D., and Maspoch, D. (2009 March 16). Metal-organic spheres as functional systems for guest encapsulation. *Angewandte Chemie International Edition* 48 (13): 2325–2329.
78 Buso, D., Jasieniak, J., Lay, M.D. et al. (2012 January 9). Highly luminescent metal-organic frameworks through quantum dot doping. *Small* 8 (1): 80–88.
79 Biswal, B.P., Shinde, D.B., Pillai, V.K., and Banerjee, R. (2013). Stabilization of graphene quantum dots (GQDs) by encapsulation inside zeolitic imidazolate framework nanocrystals for photoluminescence tuning. *Nanoscale* 5 (21): 10556–10561.
80 Hoffmann, F., Cornelius, M., Morell, J., and Fröba, M. (2006 May 12). Silica-based mesoporous organic–inorganic hybrid materials. *Angewandte Chemie International Edition* 45 (20): 3216–3251.
81 Slowing, I.I., Vivero-Escoto, J.L., Wu, C.W., and Lin, V.S. (2008). Mesoporous silica nanoparticles as controlled release drug 10.2217/nnm. 15.133 Nanomedicine (Lond.)(Epub ahead of print) future science group delivery and gene transfection carriers. *Advanced Drug Delivery Reviews* 60 (11): 1278–1288.

82 Huh, S., Chen, H.T., Wiench, J.W. et al. (2004 February 4). Controlling the selectivity of competitive nitroaldol condensation by using a bifunctionalized mesoporous silica nanosphere-based catalytic system. *Journal of the American Chemical Society* 126 (4): 1010–1011.

83 Slowing, I.I., Trewyn, B.G., Giri, S., and Lin, V.Y. (2007 May 21). Mesoporous silica nanoparticles for drug delivery and biosensing applications. *Advanced Functional Materials* 17 (8): 1225–1236.

84 Asay, D.B., Barnette, A.L., and Kim, S.H. (2009 February 12). Effects of surface chemistry on structure and thermodynamics of water layers at solid– vapor interfaces. *The Journal of Physical Chemistry C* 113 (6): 2128–2133.

85 Uemura, T., Hiramatsu, D., Yoshida, K. et al. (2008 July 23). Sol– gel synthesis of low-dimensional silica within coordination nanochannels. *Journal of the American Chemical Society* 130 (29): 9216–9217.

86 Uemura, T., Kadowaki, Y., Kim, C.R. et al. (2011 April 12). Incarceration of nanosized silica into porous coordination polymers: preparation, characterization, and adsorption property. *Chemistry of Materials* 23 (7): 1736–1741.

87 Taylor-Pashow, K.M., Della Rocca, J., Xie, Z. et al. (2009 October 14). Postsynthetic modifications of iron-carboxylate nanoscale metal– organic frameworks for imaging and drug delivery. *Journal of the American Chemical Society* 131 (40): 14261–14263.

88 Frackowiak, E. and Beguin, F. (2001 May 1). Carbon materials for the electrochemical storage of energy in capacitors. *Carbon* 39 (6): 937–950.

89 Vilatela, J.J. and Eder, D. (2012 March 12). Nanocarbon composites and hybrids in sustainability: a review. *ChemSusChem* 5 (3): 456–478.

90 de Oliveira, C.A., da Silva, F.F., Jimenez, G.C. et al. (2013). MOF@ activated carbon: a new material for adsorption of aldicarb in biological systems. *Chemical Communications* 49 (58): 6486–6488.

91 Qian, D., Lei, C., Hao, G.P. et al. (2012 November 28). Synthesis of hierarchical porous carbon monoliths with incorporated metal-organic frameworks for enhancing volumetric based CO2 capture capability. *ACS Applied Materials & Interfaces* 4 (11): 6125–6132.

92 Jiang, H., Feng, Y., Chen, M., and Wang, Y. (2013 August 21). Synthesis and hydrogen-storage performance of interpenetrated MOF-5/MWCNTs hybrid composite with high mesoporosity. *International Journal of Hydrogen Energy* 38 (25): 10950–10955.

93 Tsuruoka, T., Furukawa, S., Takashima, Y. et al. (2009 June 15). Nanoporous nanorods fabricated by coordination modulation and oriented attachment growth. *Angewandte Chemie* 121 (26): 4833–4837.

94 MacDonald, J.C., Dorrestein, P.C., Pilley, M.M. et al. (2000 November 29). Design of layered crystalline materials using coordination chemistry and hydrogen bonds. *Journal of the American Chemical Society* 122 (47): 11692–11702.

95 Dechambenoit, P., Ferlay, S., and Hosseini, M.W. (2005 November2). From tectons to composite crystals. *Crystal Growth & Design* 5 (6): 2310–2312.

96 Ferlay, S. and Hosseini, W. (2004). Crystalline molecular alloys. *Chemical Communications* (7): 788–789.

97 Koh, K., Wong-Foy, A.G., and Matzger, A.J. (2009). MOF@ MOF: microporous core–shell architectures. *Chemical Communications* (41): 6162–6164.
98 Noveron, J.C., Lah, M.S., Del Sesto, R.E. et al. (2002 June 12). Engineering the structure and magnetic properties of crystalline solids via the metal-directed self-assembly of a versatile molecular building unit. *Journal of the American Chemical Society* 124 (23): 6613–6625.
99 Furukawa, S., Hirai, K., Nakagawa, K. et al. (2009 February 23). Heterogeneously hybridized porous coordination polymer crystals: fabrication of heterometallic core–shell single crystals with an in-plane rotational epitaxial relationship. *Angewandte Chemie* 121 (10): 1798–1802.
100 Schmid, A., Dordick, J.S., Hauer, B. et al. (2001 January). Industrial biocatalysis today and tomorrow. *Nature* 409 (6817): 258–268.
101 Zhou, Z. and Hartmann, M. (2012 November). Recent progress in biocatalysis with enzymes immobilized on mesoporous hosts. *Topics in Catalysis* 55 (16): 1081–1100.
102 Zhou, Z. and Hartmann, M. (2013). Progress in enzyme immobilization in ordered mesoporous materials and related applications. *Chemical Society Reviews* 42 (9): 3894–3912.
103 Fried, D.I., Brieler, F.J., and Froeba, M. (2013 April). Designing inorganic porous materials for enzyme adsorption and applications in biocatalysis. *ChemCatChem* 5 (4): 862–884.
104 Hudson, S., Cooney, J., and Magner, E. (2008 October 27). Proteins in mesoporous silicates. *Angewandte Chemie International Edition* 47 (45): 8582–8594.
105 Hartmann, M. and Jung, D. (2010). Biocatalysis with enzymes immobilized on mesoporous hosts: the status quo and future trends. *Journal of Materials Chemistry* 20 (5): 844–857.
106 Liu, W.L., Lo, S.H., Singco, B. et al. (2013). Novel trypsin–FITC@ MOF bioreactor efficiently catalyzes protein digestion. *Journal of Materials Chemistry B* 1 (7): 928–932.
107 Shih, Y.H., Lo, S.H., Yang, N.S. et al. (2012 November). Trypsin-immobilized metal-organic framework as a biocatalyst in proteomics analysis. *ChemPlusChem* 77 (11): 982–986.
108 Jung, S., Kim, Y., Kim, S.J. et al. (2011 February 21). Bio-functionalization of metal-organic frameworks by covalent protein conjugation. *Chemical Communications* 47 (10): 2904–2906.
109 Qin, F.X., Jia, S.Y., Wang, F.F. et al. (2013). Hemin@ metal-organic framework with peroxidase-like activity and its application to glucose detection. *Catalysis Science & Technology* 3 (10): 2761–2768.
110 Lykourinou, V., Chen, Y., Wang, X.S. et al. (2011 July 13). Immobilization of MP-11 into a mesoporous metal-organic framework, MP-11@ mesoMOF: a new platform for enzymatic catalysis. *Journal of the American Chemical Society* 133 (27): 10382–10385.
111 Deng, H., Grunder, S., Cordova, K.E. et al. (2012 May 25). Large-pore apertures in a series of metal-organic frameworks. *science* 336 (6084): 1018–1023.

112 Xuan, W., Zhu, C., Liu, Y., and Cui, Y. (2012). Mesoporous metal-organic framework materials. *Chemical Society Reviews* 41 (5): 1677–1695.

113 Pisklak, T.J., Macías, M., Coutinho, D.H. et al. (2006 August). Hybrid materials for immobilization of MP-11 catalyst. *Topics in Catalysis* 38 (4): 269–278.

114 Zou, H., Wu, S., and Shen, J. (2008 September 10). Polymer/silica nanocomposites: preparation, characterization, properties, and applications. *Chemical Reviews* 108 (9): 3893–3957.

115 Uemura, T., Yanai, N., Watanabe, S. et al. (2010 October 5). Unveiling thermal transitions of polymers in subnanometre pores. *Nature Communications* 1 (1): 1–8.

116 Uemura, T., Kitagawa, K., Horike, S. et al. (2005). Radical polymerisation of styrene in porous coordination polymers. *Chemical Communications* (48): 5968–5970.

117 Uemura, T., Horike, S., Kitagawa, K. et al. (2008 May 28). Conformation and molecular dynamics of single polystyrene chain confined in coordination nanospace. *Journal of the American Chemical Society* 130 (21): 6781–6788.

118 Uemura, T., Hiramatsu, D., Kubota, Y. et al. (2007 June 25). Topotactic linear radical polymerization of divinylbenzenes in porous coordination polymers. *Angewandte Chemie International Edition* 46 (26): 4987–4990.

119 Yanai, N., Uemura, T., Ohba, M. et al. (2008 December 8). Fabrication of two-dimensional polymer arrays: template synthesis of polypyrrole between redox-active coordination nanoslits. *Angewandte Chemie International Edition* 47 (51): 9883–9886.

120 Uemura, T., Uchida, N., Higuchi, M., and Kitagawa, S. (2011 April 26). Effects of unsaturated metal sites on radical vinyl polymerization in coordination nanochannels. *Macromolecules* 44 (8): 2693–2697.

121 Uemura, T., Kitaura, R., Ohta, Y. et al. (2006 June 19). Nanochannel-promoted polymerization of substituted acetylenes in porous coordination polymers. *Angewandte Chemie International Edition* 45 (25): 4112–4116.

122 Uemura, T., Kadowaki, Y., Yanai, N., and Kitagawa, S. (2009 September 22). Template synthesis of porous polypyrrole in 3D coordination nanochannels. *Chemistry of Materials* 21 (18): 4096–4098.

123 Uemura, T., Ono, Y., and Kitagawa, S. (2008 June 5). Radical copolymerizations of vinyl monomers in a porous coordination polymer. *Chemistry Letters* 37 (6): 616–617.

124 Uemura, T., Ono, Y., Kitagawa, K., and Kitagawa, S. (2008 January 8). Radical polymerization of vinyl monomers in porous coordination polymers: nanochannel size effects on reactivity, molecular weight, and stereostructure. *Macromolecules* 41 (1): 87–94.

125 Uemura, T., Ono, Y., Hijikata, Y., and Kitagawa, S. (2010 April 7). Functionalization of coordination nanochannels for controlling tacticity in radical vinyl polymerization. *Journal of the American Chemical Society* 132 (13): 4917–4924.

126 Ishiwata, T., Furukawa, Y., Sugikawa, K. et al. (2013 April 10). Transformation of metal-organic framework to polymer gel by cross-linking the organic ligands preorganized in metal-organic framework. *Journal of the American Chemical Society* 135 (14): 5427–5432.

127 Uemura, T., Horike, S., and Kitagawa, S. (2006 July 17). Polymerization in coordination nanospaces. *Chemistry–An Asian Journal* 1 (1-2): 36–44.

128 Distefano, G., Suzuki, H., Tsujimoto, M. et al. (2013 April). Highly ordered alignment of a vinyl polymer by host-guest cross-polymerization. *Nature Chemistry* 5 (4): 335–341.

129 Bian, Y., Xiong, N., and Zhu, G. (2018 August 8). Technology for the remediation of water pollution: a review on the fabrication of metal organic frameworks. *Processes* 6 (8): 122.

130 Hao, M., Qiu, M., Yang, H. et al. (2021 March 15). Recent advances on preparation and environmental applications of MOF-derived carbons in catalysis. *Science of the Total Environment* 760: 143333.

131 Azhar, M.R., Vijay, P., Tadé, M.O. et al. (2018 April 1). Submicron sized water-stable metal organic framework (bio-MOF-11) for catalytic degradation of pharmaceuticals and personal care products. *Chemosphere* 196: 105–114.

132 Ye, G., Luo, P., Zhao, Y. et al. (2020 August 1). Three-dimensional Co/Ni bimetallic organic frameworks for high-efficient catalytic ozonation of atrazine: mechanism, effect parameters, and degradation pathways analysis. *Chemosphere* 253: 126767.

133 Xiong, Y., Chang, X., Qiao, X. et al. (2019 December 1). Co-MOF-74 derived Co3O4/graphene heterojunction nanoscrolls for ppb-level acetone detection. *Sensors and Actuators B: Chemical* 300: 127011.

134 Bhoria, N., Basina, G., Pokhrel, J. et al. (2020 July 15). Functionalization effects on HKUST-1 and HKUST-1/graphene oxide hybrid adsorbents for hydrogen sulfide removal. *Journal of Hazardous Materials* 394: 122565.

135 Joseph, L., Jun, B.M., Jang, M. et al. (2019 August 1). Removal of contaminants of emerging concern by metal-organic framework nanoadsorbents: a review. *Chemical Engineering Journal* 369: 928–946.

136 Molavi, H., Neshastehgar, M., Shojaei, A., and Ghashghaeinejad, H. (2020 May 1). Ultrafast and simultaneous removal of anionic and cationic dyes by nanodiamond/UiO-66 hybrid nanocomposite. *Chemosphere* 247: 125882.

137 Zhou, J., Liu, W., and Cai, W. (2019 December 15). The synergistic effect of Ag/AgCl@ ZIF-8 modified g-C3N4 composite and peroxymonosulfate for the enhanced visible-light photocatalytic degradation of levofloxacin. *Science of the Total Environment* 696: 133962.

138 Yu, D., Wu, M., Hu, Q. et al. (2019 April 5). Iron-based metal-organic frameworks as novel platforms for catalytic ozonation of organic pollutant: efficiency and mechanism. *Journal of Hazardous Materials* 367: 456–464.

139 Jin, H., Zhou, H., He, D. et al. (2019 August 5). MOF-derived 3D Fe-NS co-doped carbon matrix/nanotube nanocomposites with advanced oxygen reduction activity and stability in both acidic and alkaline media. *Applied Catalysis B: Environmental* 250: 143–149.

140 Li, K., Zhang, Y., Lin, Y.Z. et al. (2019 July 23). Versatile functional porous cobalt–nickel phosphide–carbon cocatalyst derived from a metal-organic framework for boosting the photocatalytic activity of graphitic carbon nitride. *ACS Applied Materials & Interfaces* 11 (32): 28918–28927.

141 Xie, A., Zhang, J., Tao, X. et al. (2019 November 20). Nickel-based MOF derived Ni@ NiO/N–C nanowires with core—shell structure for oxygen evolution reaction. *Electrochimica Acta* 324: 134814.

142 Zhao, S.N., Wang, G., Poelman, D., and Van Der Voort, P. (2018 November 12). Metal organic frameworks based materials for heterogeneous photocatalysis. *Molecules* 23 (11): 2947.

143 Hasegawa, S., Horike, S., Matsuda, R. et al. (2007 March 7). Three-dimensional porous coordination polymer functionalized with amide groups based on tridentate ligand: selective sorption and catalysis. *Journal of the American Chemical Society* 129 (9): 2607–2614.

144 Wu, C.D., Hu, A., Zhang, L., and Lin, W. (2005 June 29). A homochiral porous metal– organic framework for highly enantioselective heterogeneous asymmetric catalysis. *Journal of the American Chemical Society* 127 (25): 8940–8941.

145 Venkataraman, D., Gardner, G.B., Lee, S., and Moore, J.S. (1995 November). Zeolite-like behavior of a coordination network. *Journal of the American Chemical Society* 117 (46): 11600–11601.

146 Xiao, J.D., Li, D., and Jiang, H.L. (2018 August 8). Metal-organic frameworks for photocatalysis. *Scientia Sinica Chimica* 48 (9): 1058–1075.

147 Cui, Y., Zhang, J., He, H., and Qian, G. (2018). Photonic functional metal-organic frameworks. *Chemical Society Reviews* 47 (15): 5740–5785.

148 Xie, D.Y., Xing, H.B., Zhang, Z.G. et al. (2017). Porous hydrogen-bonded organometallic frameworks for adsorption separation of acetylene and carbon dioxide. *The Central Indiana Education Services Center* 68 (1): 154–162.

149 Ramezanalizadeh, H. and Manteghi, F. (2018 January 20). Synthesis of a novel MOF/CuWO4 heterostructure for efficient photocatalytic degradation and removal of water pollutants. *Journal of Cleaner Production* 172: 2655–2666.

150 Zhou, Y., Yang, Q., Zhang, D. et al. (2018 June 1). Detection and removal of antibiotic tetracycline in water with a highly stable luminescent MOF. *Sensors and Actuators B: Chemical* 262: 137–143.

151 Azhar, M.R., Abid, H.R., Sun, H. et al. (2017 March 15). One-pot synthesis of binary metal organic frameworks (HKUST-1 and UiO-66) for enhanced adsorptive removal of water contaminants. *Journal of Colloid and Interface Science* 490: 685–694.

152 Yin, N., Wang, K., Wang, L., and Amino-functionalized, L.Z. (2016 December 15). MOFs combining ceramic membrane ultrafiltration for Pb (II) removal. *Chemical Engineering Journal* 306: 619–628.

153 Xu, T., Hou, X., Liu, S., and Liu, B. (2018 March 15). One-step synthesis of magnetic and porous Ni@ MOF-74 (Ni) composite. *Microporous and Mesoporous Materials* 259: 178–183.

154 Shi, Z., Xu, C., Guan, H. et al. (2018 February 20). Magnetic metal organic frameworks (MOFs) composite for removal of lead and malachite green in wastewater. *Colloids and Surfaces A: Physicochemical and Engineering Aspects* 539: 382–390.

155 Bandosz, T.J. and Petit, C. (2011 February). MOF/graphite oxide hybrid materials: exploring the new concept of adsorbents and catalysts. *Adsorption* 17 (1): 5–16.

156 Bernstein, E.J., Schmidt-Lauber, C., and Kay, J. (2012 August 1). Nephrogenic systemic fibrosis: a systemic fibrosing disease resulting from gadolinium exposure. *Best Practice & Research Clinical Rheumatology* 26 (4): 489–503.

157 Galloway, J.N. (1995 December). Acid deposition: perspectives in time and space. *Water, Air, and Soil Pollution* 85 (1): 15–24.

158 Brahmi, C., Benltifa, M., Vaulot, C. et al. (2021 July 5). New hybrid MOF/polymer composites for the photodegradation of organic dyes. *European Polymer Journal* 154: 110560.

159 Kiiskila, J.D., Das, P., Sarkar, D., and Datta, R. (2015 March). Phytoremediation of explosive-contaminated soils. *Current Pollution Reports* 1 (1): 23–34.

10

Reformed Metal-Organic Frameworks (MOFs) for Abstraction of Water Contaminants – Heavy-Metal Ions

Prakash B. Rathod[1], Rahul A. Kalel[2], Mahendra Pratap Singh Tomar[1], Akshay Chandrakant Dhayagude[3], and Parshuram D. Maske[4]

[1]Department of Chemistry, Shri Sadguru Saibaba Science and Commerce College, Ashti, Maharashtra-442707
[2]Department of Chemistry, Padmabhushan Dr.Vasatraodada Patil Mahavidyalay, Tasgaon, Maharashtra-416312
[3]Department of Chemistry, MVP's K. K. Wagh Arts, Science and Commerce College, Pimpalgaon (B), Niphad, Nashik, Maharashtra 422209
[4]Department of Chemistry, S.G.V. & S.S.P'S, Art's Commerce and Science College, Onde, Maharashtra, 401605

10.1 Introduction

Groundwater reservoirs, which are essential to humanity and serve as the main supply of drinking water, are mostly contaminated by anthropogenic organic and inorganic pollutants. This contamination poses a concern to human health since it can poison both aquatic and terrestrial organisms. Monitoring and reducing possible pollution sources are vital. Examples include runoff from urban areas, sewage systems, railroads, highways, mining and hazardous waste disposal sites, landfills, and agricultural and industrial regions. Groundwater contamination can cause heavy metals to be dispersed throughout the environment via plant uptake.

The removal of heavy-metal ions from wastewater is crucial for a healthy environmental life on Earth. For the elimination of heavy metals from various wastewater resources, several methods have been used. There are three types of water treatment technologies: biotic action, phase-changing procedures, and sophisticated oxidation methods [1]. One of the most extensively used biotic methods is the use of aerobic- or anaerobic-triggered sludge. In the advanced oxidation method, oxidizing radicals were formed in situ to destroy the molecules of the impurities. The main shortcoming of this method is that the formation of oxidizing radicals is quite unselective. The adsorption of impure water molecules on porous materials is the basis of the phase segregation method. There are a number of techniques used to eliminate pollutants of wastewater such as reverse osmosis, extraction, distillation, nano-filtration, several phytotechnologies, etc. The reduction of environmental toxicity by cleaning wastewater is an interesting topic of research. Novel nanomaterials that

Metal Organic Frameworks for Wastewater Contaminant Removal, First Edition. Edited by Arun Lal Srivastav, Lata Rani, Jyotsna Kaushal, and Tien Duc Pham.

could be used as nano sorbates are currently being developed and their performance improved [2].

These methods of water purification could be categorized based on photocatalysis, adsorption, membranes, chemicals, and electrical processes [3]. Commercial methods of eradicating heavy metals are associated with limitations, as they are costly, less efficient, difficult to reuse due to fouling, and the production of sizable amounts of chemical sludge [4]. As a result, contamination incidents in developed countries with extensive documentation frequently go unrepaired.

MOFs are novel and fascinating materials due to their porous nature and they have gained interest in the last two decades. The first MOF was manufactured by Yaghi in 1995 [5]. After that, due to their characteristics, they were applied to environmental studies, energy safeguarding [6–9], sensing, molecular recognition, separation processes [10], drug delivery [11], non-linear optics, and luminescence and catalysis [12, 13]. This class of porous materials has quickly ascended to the top of the materials research hierarchy due to its exceptional ability to selectively adsorb significant amounts of guest species, easy chemical tuning, structural flexibility, vast surface area, tailorable pore size, and remarkable interior surface areas [14–16].

MOFs are composed of organic linkers that are arranged in three-dimensional lattices and metal ions/clusters as secondary building units (SBUs). MOFs are made up of 1-D chains, 2-D films, and 3-D frameworks. There are several benefits to employing MOFs as adsorbent materials. Metal centers and organic linkers combine to generate endless crystalline networks in MOFs [17]. Based on different organic functional groups of MOFs, these materials divide into different categories of hybrid material. The MOFs are crystalline compounds with highly flexible pore nature, dimensions, functional group, surface area, and stability [18, 19]. These characteristics enable them to be adaptable in the long-term removal of a variety of contaminants [20].

Several review articles have summarized the uses of MOFs for wastewater cleansing [15, 21–23]. However, the industrial solicitations of these MOFs have not been completely explored [23]. The several drawbacks of MOFs—such as their stability in aqueous mediums [24, 25]—limit their exploration on an industrial scale. Reviewing current techniques for employing MOFs to remove toxic impurities including heavy metals from wastewater streams is the major goal of this chapter. This is done by determining the most recent trends and knowledge gaps in the relevant scientific literature. This chapter summarizes all of the research given and provides a summary of the major pieces of information found in literature reports on metal-organic frameworks (MOF).

10.2 Metal-Organic Frameworks

Metal-organic frameworks are porous polymeric materials in which organic ligand molecules are connected with metal ions. Due to their easy synthesis, highly homogeny, and fine-tunable pore size they are a brand-new group of porous crystals. The pore size of these materials can be adjusted according to the need of analysis; hence, they are widely used in several fields of research such as environmental sciences, basic sciences, physical sciences, etc. with slight modifications in the methods of synthesis of this material with different organic moieties, it is possible to design and produce a new composite. Thus formed hybrid materials possess multifunctional character; therefore, they exhibit superior properties to other materials [26].

The components of the holes of framework materials are changeable. For ions and solvent molecules, the voids of materials are easily accessible. The structural integrity of such framework materials has been confirmed by powder X-ray diffraction analysis after their associations with several ions and solvent molecules. The specificity toward guest molecules by the MOFs through exchange study shows that the framework materials are highly selective. Further, these frameworks can act as catalysts for several transformations and selective adsorbents for the analyte and fluorescent sensors. As a result of their interaction with the guest molecule, these materials show changes in their luminescence and electrolytic and magnetic properties, and they act as efficient sensors. In MOFs containing paramagnetic transition metals or luminous lanthanides, this behavior has been established. There have been reports of catalytic behavior, and this field needs a lot more consideration [27].

It is possible to synthesize MOFs with large pore orifices with a minimum density of materials according to the isoreticular principle without altering its basic characteristic features. Hence, these materials are now demanded because of their ability to accommodate large-size guests with prominent selectivity. Further, in relation to the functionalization of these materials with metal and organic compounds, characteristic properties such as chemical and thermal stability remain constant. These qualities allow for a significant improvement of gas storage in MOFs and have enthused widespread research into their use in gas separation, bio-imaging, catalysis, etc. Recently, researchers have focused on the development of the synthesis of frameworks that can act as super crystals. Further, they possess the properties of nanomaterials [15].

10.3 Sorption Enrichment by Modification of MOFs

The adsorption capacities of MOFs were increased by their structural modification. An enormous investigation was carried out for structural modification of MOFs for succeeding in widening the sorption capabilities. The solvent molecule easily enters the interior channels of MOFs because of their porous nature. Both the inner and outer surfaces of the MOFs have been modified through the incorporation of solvent molecules. Modifications in the structures of MOFs, with the functional group modified, are shown in Table 10.1, which also denotes the selectivity of adsorption of the modified MOF toward toxic heavy-metal ions.

For the synthesis of such modified MOFs, two different synthetic strategies have been commonly applied. Firstly, the modification of the MOFs can be carried out before the construction of the MOF core structure, which is called the pre-synthetic method. On the other hand, MOFs can be modified after building their core structure. In the pre-synthetic modification approach, the reactions were carried out at

Table 10.1 Modifications of MOFs with various functional groups and the absorptivity toward toxic heavy-metal ions.

Sr. No.	MOF	Modifier	Contaminant adsorbed	Ref.
1	UiO-66	-2COOH	Cu^{2+}	[28]
2	UiO-66	-OH	Thorium ions	[29]
3	UiO-66	-NHC(S)NHMe	Pb^{2+}, Hg^{2+}, Cd^{2+}	[30]
4	MOF-5	-SH	Pb^{2+}, Hg^{2+}, Cd^{2+}	[31]
5	MOF-5	-Fe_3O_4	Pb^{2+}, Cd^{2+}	[30]
6	HKUST-1	-Na_2S	Hg^{2+}	[32]
7	MIL-68	-Na_2S	Pb^{2+}, Hg^{2+}, Cd^{2+}	[30]
8	MIL-53	-NH_2 (Na_2S)	Pb^{2+}, Hg^{2+}, Cd^{2+}	[30]

high temperatures. The applied high temperature may cause the breakdown of the functional group. Such possibility of the breakdown of the functional group in the postsynthetic method is negligible. Hence, postsynthetic modifications of MOFs are desired for their effectiveness as adsorbents in environmental remediation.

a) ***Pre-synthetic modification:*** During the adsorption of metal ions on MOFs, the efficiency of the MOF toward the metal ions was driven by the characteristic features of the metal ion and linker unit. Modification in the structure of the ligand has been achieved by the introduction of the target functional group in the MOFs. The added target group further enhances the substrate scope of different metal ions. The number of chemical functionalities was attached to the MOFs by the pre-synthetic modification approach. Moreover, as-synthesized MOFs show large adsorption capacities toward contaminants of water. In the pre-synthetic modification technique, the most prevalent functional group studied is -OH and -COOH. These functional groups are mostly embedded into different organic molecules as ligands.
b) ***Postsynthetic modification (PSM):*** MOFs modified by the postsynthetic strategy show increased structural stability and adsorption capacities. Therefore, PSM methods are more reliable compared to the pre-synthetic modification technique. In PSM, functional groups such as thiol (-SH), amino (-NH_2), quinine, and (RR'CN-NCRR') were added to pure MOFs. This functional group acted as organic hybrid materials for the coating of MOFs [30]. Further, for postsynthetic modifications, the coupling has been carried out by suitable reagents. The surface environments, pore characteristics, and metallic linkages of the modified MOFs enhanced the structure's chemical stability and efficiency of metal sorption. In postsynthetic alteration, there are certain detriments such as the modification requiring more time (1–5 days), emblematic reaction conditions, and the pore diameter-reliant production rate. For PSM, the most commonly used processes are liquid- and vapor-phase PSM. In between VP-PSM and LP-PSM, the VP-PSM offers an excellent yield, quick uniform distribution, and wide applicability. Hence, VP-PSM has drawn much attention from researchers.

10.4 Toxic-Metal Ion Adsorption by MOFs

In the last decade, the evolution of MOFs as promising materials for the detection and removal of heavy-metal ion contaminants for water purification have been reported. In this section of the chapter, we have discussed the sequential development of synthesis methods, structural features, and their physical and chemical properties. Large numbers of procedures have been developed by researchers to remove toxic-metal ions from environmental water samples using MOFs. The procedures developed so far are sorption, flocculation, precipitation, degradation, and coagulation. Among all these procedures, sorption is more advantageous than the others because the time required for the complete removal of toxic heavy-metal ions is very short.

Rick C. Schroden et al. synthesized macroporous MOFs by colloidal crystallization processes. In such a formed MOF, the thiol group as an organic hybrid connects to the titania and zirconia framework. The formed MOF possesses the capability of adsorption of toxic heavy-metal ions from the real water sample. For example, the hybrid macroporous MOF thus formed showed adsorption capacities for Hg^{2+} and Pb^{2+} of ~0.33–1.41 mmol/g and 0.27–1.24 mmol/g, respectively. Further, on reuse after acid washing, the hybrid macroporous MOF showed around two-thirds of their initial capacities toward the absorption of toxic metal ions [33]. Howarth et al. developed interesting and handy MOF-contained sponges for water cleaning. The MOF sponge is highly selective to oxyanions such as -SeO_3^{2-}, SeO_4^{2-}, AsO_4^{3-}, and PO_4^{3-}. Further, the MOFs are capable of removing the UO_2^{2+} as oxy cations. These oxy cations and anions are well-known contaminants of natural water resources [34].

10.4.1 MOFs for Mercury Adsorption

Ke et al. used postsynthetic modification to synthesize a MOF, SH-HKUST-1, in 2011. SH-HKUST-1 was synthesized by the solvothermal process. In HKUST-1, the thiol group was incorporated by treating the dithioglycol with HKUST-1. It was observed that the SH functional group added to the MOF in the position at which water molecules were already present. The number of thiol groups in the SH-HKUST-1 MOF was changed by increasing the mole ratio of dithioglycol. The octahedral structure of this MOF has been presented by SEM analysis. The thiol-functionalized MOF showed almost 100% adsorption of mercury from the water relative to non-functionalized MOF in two hours. Due to its large surface area and adsorption sites, this MOF shows high efficiency of adsorption [35]. Saleem et al. prepared UiO-66-NHC(S)NHMe as a modification of UiO-66-NH_2. Due to modification, ~25-fold enhancement in adsorption capacity was observed in 240 minutes. Further, the modified MOF has 99% adsorption efficiency while unmodified MOF shows only 4% [36]. 3D Co(II) MOF has been reported by Abbasi et al. as an effective mercury (II) adsorbent. At pH = 6, 3D Co(II) MOF showed ~70% of mercury adsorption in 100 minutes. The powder X-ray diffraction analysis demonstrated that, after use, the MOF structure had not suffered any appreciable degradation [37].

Liang et al. generated a different MOF, FJI-H12, in which the accessible -NCS group was enclosed in octahedral M_6L_4 cages of 2,4,6-tri(1-imidazolyl)-1,3,5-triazine linkers and Co(II) metal for the adsorption of mercury. For this MOF, mercury adsorption was controlled by the -NCS group and it followed both chemosorption and physisorption mechanisms. At a pH of 7, the highest absorption ability is 440 mg/g in up to one hour. Further, at different pH levels, the same adsorption rate was observed. The MOF can be recycled for next cycle simply by submerging it in KSCN solution for 24 hours [38]. A Zn-based MOF with a TATAB linker named PCN-100 was synthesized by Fang et al. in 2010. This MOF coordinates with the metal through chelation. After adsorption of Hg from DMF by the MOF, the ICP analysis of this framework showed that the mercury impregnation on the MOF was 1.38 mg [39]. In 2015, Chakraborty et al. formed AMOF-1—a new MOF containing Zn^{2+} ions and carboxylate as binding sites. The AMOF-1 has been tested for the mercury ion confiscation of contaminated water. In 18 hours, AMOF-1 showed a mercury ion adsorption efficiency of 94%. The time dependent adsorption rate study reveals that, in 36 hours, the adsorption rate rose up to 99%. The maximum adsorption capacity of the material is 78 mg/g. In AMOF-1, the ligand DMA reside in the rectangular channel of the MOF can act as an ion exchanger. This ion exchanger ligand is believed to be responsible for the adsorption process [40]. The other solvohydrothermally prepared MOF—Zn(hip)(L)•(DMF)(H_2O) by Luo et al. in 2015—acts as an efficient adsorbent for the removal of mercury from water. In the structure of this MOF, in addition to the inner wall of MOF, one hexagonal channel containing hydroxyl and acrylamide is present as an active site for adsorption. The extreme adsorption capability of 333 mg/g at a pH of 5 in less than an hour was achieved. The pH dependent study revealed that, on changing the pH at either side, adsorption efficiency falls. At very low mercury concentrations, the MOF shows upto 85% sorption capacity [41].

The removal of mercury ions (Hg^{2+}) from contaminated water was successfully presented by Halder et al. in 2017 using [Ni(3-bpd)$_2$(NCS)$_2$]n as a MOF. During the detection and adsorption of mercury ions using this MOF, the bonding between the free sulfur atom of the NCS ligand and the Hg^{2+} ion was believed to be responsible for the adsorption of the mercury ions. When exposed to many metal ions, the MOF's affinity for mercury was validated when it would only preferentially bind with Hg^{2+}. It is interesting to note that when mercury adsorption takes place, the color of the MOF switches from green to gray, making mercury ions in the solution visible to the naked eye. Additionally, a UV-visible absorption spectroscopy confirms that the MOF indicates selectivity toward mercury ions in the presence of other metal ions. This MOF showed ~94% removal of mercury ions from the water in two hours with a maximal capacity of adsorption of 713 mg/g [42]. The postsynthetically modified CrMIL-101-AS was prepared for the elimination of mercury (II) from the water sample by Liu et al. in 2014. In this MOF, the rigid –SH group in addition to the alkenyl function group was present. After six hours, ~99% of the Hg^{2+} ions in the water were removed by this MOF. Further, on decreasing the initial concentrations of Hg^{2+} by 100, this material showed 93% adsorption efficiency. Moreover, the material can be reused without losing its capabilities [43].

In 2013, Yee et al. reported that the Zr-DMBD is an efficient MOF to adsorb mercury (II) from water. The Zr-DMBD differs from UiO-66 in its functionalities with hard carboxylate and the free thiol group. The Zr-DMBD is made up of dimercapto-1,4-benzene dicarboxylic acid as a binding site. In 12 hours only, Zr-DMBD removed ~100% of mercury from the water. The luminescence of the material decreases by about ten times due to the adsorption of mercury from the water. Due to the quenching of its fluorescence, it can be used as a colorimetric sensor for the detection and separation of mercury from water [44]. In addition, the LMOF-263 MOF was developed by Rudd et al. in 2016. The LMOF-263 belongs to a class of isorecticular and fluorescent MOF. The toxic heavy-metal ion of the water selectively quenches the fluorescence of the MOF by up to 84%. For example, the Hg^{2+} ion quenches the fluorescence by 84% and Pb^{2+} by 64% in the presence of other metal ions. Hence, this MOF was considered a highly sensitive and selective material for the detection and exclusion of heavy-metal ions. The highest capacity of adsorption, 380 mg/g, was achieved in only half an hour. The interaction between the sulfonate functionality of the MOF and the heavy-metal ion containing acid in the water solution is believed to be the foundation of the adsorption mechanism [45].

The ZIF-90-SH is another MOF synthesized by postsynthetical modification in a one-step method by Bhattacharjee et al. in 2015. Further, their capacity to remove mercury ions from the water has been reported. The powder XRD analysis of ZIF-90-SH represented that the structural features of the original ZIF-90 MOF were retained after incorporation of a thiol functional group. The highest mercury adsorption capacity of this ZIF-90-SH MOF was 22 mg/g. The distinct observation about this MOF is that the adsorption efficiency decreases when the concentration of mercury ion in the contaminated water is increased [46]. In 2013, Sohrabi et al. demonstrated that the MOF HKUST-1 modified with a thiol functional group enclosing silica nanocomposite can be used for the removal of toxic mercury from water by sorption. The silica nanocomposite was composed of thiol immobilized silica nanoparticles $SH@SiO_2$ and $Cu(BTC)_2$. The nanocomposite-modified MOF attained equilibrium in only 60 minutes at pH = 6 and the utmost mercury adsorption capacity was 210 mg/g. For this material, the active sites of the MOF protonate at an acidic pH, and the material is selective to mercury ions over the other heavy-metal ion contaminants of water [47].

The postsynthetically modified MOF MIL-101-Thymine, containing –SH as a functional group, was synthesized by Luo et al. The MIL-101-Thymine was applied for the removal of mercury ions from water. During the sorption, the two thymine units of the MOF interacted with the Hg^{2+} ions. The system attained equilibrium in only 200 minutes at pH = 6. A maximum adsorption capacity of 52 mg/g was observed for sorption of mercury ion contaminants in water. MIL-101-Thymine has been highly selective to mercury ions over other metal ions in water. Further, the effect of change in pH on the adsorption capacity of MIL-101-Thymine was studied. The pH dependence studies revealed that maximum adsorption capacity is only at the optimum pH of 6, and it decreases on changing the pH on either side [48].

In 2016, Mon et al. demonstrated that bio-mimic MOFs are relatively stable for the removal of mercury ions. The bio-mimic MOF contains thioalkyl chains and has a honeycomb structure with hexagonal tunnels. Due to the presence of thioalkyl chains, they show exceptionally high mercury ion adsorption capacity. For the mercury ions, bio-mimic MOFs show an adsorption capacity of 900 mg/g in water and water–methanol mixture. In addition, they are effective at adsorbing methyl mercury from water. The formed MOF can be reused after washing with dimethyl sulfide because in the presence of DMS the adsorption process is reversed [49]. Xiong et al., in 2017, studied the sorption of hydrothermally synthesized MOF-74-Zn for the amputation of Hg^{2+} ions from water. The detailed investigation revealed that physical sorption was significant in the adsorption mechanism. At pH = 6, equilibrium was achieved and the maximum adsorption capacity was 63 mg/g. On decreasing the concentration of Hg^{2+} contaminant, the adsorption capability increased upto 72% [50].

10.4.1.1 Magnetic MOFs for Mercury Adsorption

Huang at al. (2015) introduced a magnetic MOF called Fe_3O_4@SiO_2@HKUST-1. This MOF has been tested for mercury ion sorption from aquatic conditions. For the synthesis of such a magnetic MOF, the postsynthetic modification approach was applied and the material HKUST-1 was functionalized with bismuthol (Bi-I). The added functional group Bi-I acted as an active site of the adsorption. It was discovered that, relative to non-magnetic MOF, the magnetic MOF had high adsorption capacity in a short time period. The magnetic MOF Fe_3O_4@SiO_2@HKUST-1 has 99% adsorption capacity in ten minutes. The MOF is highly selective to mercury in the presence of other toxic heavy-metal ions such as Pb^{2+} and Cr^{2+}. At pH = 3, the MOF has a constant adsorption value in the range of 2–9 [51].

10.4.2 MOFs for Lead Adsorption

The Zr-based amino-containing MOF called UiO-66-NHC(s)NHMe was used for the adsorption of lead ions from water solutions by Saleem et al. in 2015. This material showed the highest capacity of adsorption of Pb^{2+} as being 232 mg/g [36]. Abbasi et al. demonstrated that the 3D Co(II) framework can remove lead contaminants in a water medium. The system attains equilibrium in ~100 minutes, and the effect of pH on the sorption value of the MOF with a lead concentration in water has been nicely presented [37]. Further, the silica and thiol-linked framework material, i.e., HSmSi@MOF-5, has been investigated for the exclusion of Pb ions from water as demonstrated by Zhang et al. in 2016. For this material, the highest adsorption capacity was 312 mg/g with an equilibration time of only 30 minutes. In an acidic pH, the protons compete with the metal for adsorption, and in a basic pH, the metal ions got precipitate out in the solution itself. Hence the pH of 6 has been selected as the ideal pH for the sorption of lead by this MOF [52].

Tahmasebi et al. exposed another mechanochemically synthesized azine-functionalized TMU-5 MOF for the adsorption of Pb(II) in 2014. According to reports, the highest adsorption capability was achieved in only 15 minutes at ~251 mg/g for the TMU-5. At low pH, the active sites of the MOF protonate; hence, the pH of 10 was selected as the optimum [53]. Chakraborty et al. presented an anionic framework containing Zn^{2+} and tetracarboxylate called AMOF-1 for the elimination of Pb^{2+} ions from water in 2016. The highest lead ion adsorption value was 71 mg/g after 24 hours of equilibrium [40]. Rahimi et al. used a Cu-terephthalate MOF for lead-ion adsorption. The Cu-terephthalate MOF shows the highest ability of adsorption at ~80 mg/g at a pH of 7 with an equilibrium time of 120 minutes [54]. Additionally, Zou et al. (2013) described an effective technique for creating HKUST-1. The functionalization of HKUST-1 has been carried out by using polyoxometalate and the formed MOF was represented as HKUST-1-MW@$H_3PW_{12}O_{40}$. These formed MOF's show a greater selectivity and adsorptivity for lead(II) ions within aqueous systems. A different framework that can remove lead (II) was described by Jamali et al. in 2016. The rod-like structure of Dy(BTC)(H_2O)$(DMF)_{1.1}$ contains lanthanide. The extreme adsorption capability of this MOF was 98 mg/g of Pb^{2+} in ten minutes at a pH of 7. Further, the MOF can be reused in up to five successive cycles with retention of an adsorption capacity of ~94–98% [55].

In 2011, Qin et al. presented a highly efficient Pb^{2+} ion adsorbent. The substance MnO_2-MOF is created by oxidizing $MnSO_4$ with $KmnO_4$. This MOF shows an exceptionally high adsorption capacity of 917 mg/g in one hour. Because of proton release during adsorption, the pH of the starting solution, which was 6 before adsorption, decreased to 5. The hydroxyl group of the MOF and Pb(II) metal ions were associated with the inner-sphere complexation [56]. In addition, MOF-5 exhibits a maximal adsorption ability of 659 mg/g at 45°C, according to research by Rivera et al. in 2016. MOF-5 contains both the acidic and basic sites; hence, on changing the pH from 5 on either side, the adsorption value increases [57].

10.4.2.1 Magnetic MOFs for Lead Removal

Magnetic MOF composites for the sorption of lead (II) containing iron oxide nanoparticles were synthesized by Ricco et al. in 2015. The formed MOF presents as MIL-53(Al@100aBDC). For this MOF, the maximal uptake capacity of Pb^{2+} was ~492 mg/g with an equilibration time of six hours. Further, the effect of the amount of amino group of a BDC unit on the adsorption of lead has been investigated. The experimental results demonstrated that the adsorption of lead by MOFs enhances drastically when the percentage of the amino group is greater than 50 [58].

10.4.3 MOFs for Cadmium Adsorption

The -SO_3H functionalized framework $Cu_3(BTC)_2$-SO_3H was effectively used for the sorption of Cd (II) by Wang et al. The parent material was functionalized by PSM with sulfonic acid. The ideal pH for effective adsorption was 6, as below this pH, protons blocked the active center of the material, and above pH = 6, cadmium ions precipitated as hydroxide salts. The -SO_3H group of materials offers chelating coordination with Cd(II) [59]. AMOF-1 was created in 2015 and was defined as a potential material for the adsorption of Cd(II). Due to the free tetracarboxylate and zinc metal, the uppermost adsorption value was 41 mg/g after an equilibration of 24 hours. AMOF is nearly the highest selective toward Cd(II) ions of contaminated water [40]. Fang et al. published PCN-100, another framework material for the removal of Cd(II). The PCN-100 consists of $Zn_4O(CO_2)_6$ SBUs and TATAB linking sites. Through the linkers used to enclose the metal ions, it was discovered that the MOF displayed a chelating coordination mode. ICP analysis was used to track cadmium(II) adsorption, and the results showed a 1.6 mg Cd(II) absorption per unit [39].

In 2015, Saleem et al. presented that the MOF UiO-66- NHC(S)NHMe is further proficient for removing Cd (II) ions with a supreme ability of 49 mg/g [36]. The 3D Co(II) MOF, as another framework for Cd(II) removal, was provided by Abbasi et al. The ideal pH for the adsorption of Cd (II) is ~6. In the acidic medium, the active sites of MOFs are unavailable for binding with metal ions. The nano form of the MOF shows better adsorption capabilities than MOF. The PXRD study was evident in the structural constancy of the MOF after adsorption [37]. Zhang et al. presented a different methodology for removing Cd (II) from water in 2016. The PSM approach for the structural modification of Zn-containing MOF-5 has been used. The material attained an equilibrium of sorption in 30 minutes, and the estimated capability of adsorption was 98 mg/g larger than MOF-5. Further, the azine-functionalized TMU-5 MOF was demonstrated by Tahmasebi et al. for effective Cd (II) sorption in 2014. The greatest adsorption rate in this case was 43 mg/g in an equilibrium period of 15 minutes [53].

The additional stable polyoxometalate-modified form (HKUST1-MW@ $H_3PW_{12}O_{40}$) of the HKUST-1-MW MOF was tested by Zou et al. in 2013. The increased stability resulted from the Keggin polyoxometalate filling the MOF pores. Microwave irradiation was used to create the MOF, and during that process, polyoxometalate was added. The Cd^{2+} ions removed by chemical adsorption had a capacity of ~32 mg/g in 80 minutes of equilibration. For this MOF, up to the saturation limit, the sorption capacity increases further on an increase in the temperature of the system [60].

10.4.3.1 Magnetic MOFs for Cadmium Adsorption

The magnetic Cu-terephthalate MOF was synthesized by Rahimi et al. (2015) and was tested for its capacity to remove Cd^{2+} ions from water. This MOF contained the number of carboxylate groups that enhance the adsorption ability of the MOF. In the alkaline medium, the surface of the MOF became anionic, which offered a strong electrostatic force of attraction with the cationic metal-ion contaminant of the aqueous medium; hence, efficient adsorption of metal ions on the surface of the MOF took place. The maximal Cd(II) ion elimination capability was ~100 mg/g [54].

10.4.4 MOFs for Chromium Removal

The MOF UiO-66-NHC(S)NHMe was synthesized by Saleem et al. via postsynthetic modification of UiO-66 for potential application in the removal of chromium(III) from water. This material possesses -NH_2 groups in its structure; hence, they are water stable and alkaline in nature. Relative to unmodified MOF UiO-66, the modified MOF showed ~25-fold increases in metal-ion adsorption capacity [36]. Tahmasebi et al. synthesized three MOFs, namely TMU-4, TMU-5, and TMU-6, with imine and azine functional groups using the mechanochemical method for Cr(III) metal-ion adsorption from water. The ability of TMU-5 to remove metal ions was the subject of the investigation. The MOF displayed a three-dimensional, highly organized network of interconnected pores covered in azine groups. For this MOF, the uppermost adsorption aptitude was 123 mg/g for Cr(III) and it increased with pHs up to 10 [53].

In 2016, a composite of chitosan-UiO-66 MOF was synthesized by grinding and microwave irradiation for the exclusion of ~94 mg/g of Cr(VI) from water. The adsorption value was much higher than that of chitosan as an adsorbent. The increased capacity of Cr(VI) adsorption was due to the strong electrostatic force of attraction between the metal ions of the water and -O/-NH_2 groups of MOF. The presence of hydroxyl groups in the system blocked the active sites of the framework; therefore, the basic medium diminished the adsorption capacity [61]. The copper-benzenetricarboxylate-based MOF, Cu-BTC, showed Cr(VI) ion adsorption in the neutral environment with the highest adsorption capacity of 48 mg/g [62].

In 2016, Li et al. reported on the silver-triazolato MOF, named -1-NO3-, for the adsorption of Cr(VI) in the form of $Cr_2O_7^{2-}$ by the ion-exchange mechanism. Chromium ascetically adsorbed and achieved equilibrium after four hours. Further, the recyclability of this MOF was tested, and a reusability study showed that this MOF was effective for up to four repeated adsorption-release cycles. Additionally, this material has proven a discernment to $Cr_2O_7^{2-}$ adsorption over other anions [63]. In 2015, Zheng and coworkers discovered that the Zr-based MOF known as ZJU-100 was a cationic material for the removal of Cr(VI). The material was synthesized by the postsynthetic modification approach. In the PSM, -CH_3 groups were added on the MOF-867. Only ten minutes had passed when the adsorption equilibrium was established, resulting in a 96% decrease in the solution's $Cr_2O_7^{2-}$ concentration (50 ppm initial concentration). It was discovered that ZJU101's initial adsorption capacity was 324 times greater than that of MOF-867. Additionally, ZJU-101 demonstrated superior $Cr_2O_7^{2-}$ adsorption selectivity than other interfering anions [64]. Li et al. investigated the cobalt-based zeolitic imidazolate framework ZIF-67 for its Cr(VI) adsorption capabilities in 2015. A rapid adsorption equilibrium was attained, and the capacity of adsorption gradually increased on increasing their contact time with Cr(VI) contained in water. The adsorption equilibrium was attained in three hours with the highest capability of sorption being in the range of pH 7 to 9. Further, on decreasing the pH of the system, the adsorption ability of the MOF increased. Ion exchange between -OH groups of frameworks and $Cr_2O_7^{2-}$ of contaminated water is the mechanism of the adsorption [65].

Furthermore, Cr(V) was removed from the water by using isonicotinate-N-oxide containing MOF TUM-30 by Aboutorabi et al. The TMU-30 with rhombic 1D tunnels possessed the ability to efficiently remove the Cr(VI) with the highest adsorption value of 145 mg/g and equilibrium attained in only ten minutes. It worked very well in the broad range of pH = 2 to pH = 9, as the variation in pH would not change the structural features of the MOF. This MOF had high selectivity to Cr(VI) than other ions [66]. Finally, Rapti et al. synthesized an amino-functionalized, alginic acid containing UiO-66-based MOF on the reflux of starting materials in 2016. The name for this material is given as MOR-1-HA. During the adsorption of heavy-metal ions, the ion exchange between Cl^- ions of MOF and $Cr_2O_7^{2-}$ took place. For MOR-1HA, it was observed that the alginic acid coating was responsible for the decrease in adsorption capacity relative to its unmodified analog, MOR-1. The highest Cr(VI) ion adsorption capacity for MOR-1-HA was ~280 mg/g. The effect of change in pH on the ability of the MOF to adsorb the Cr(V) was negligible. MOR-1-HA showed high selectivity toward $Cr_2O_7^{2-}$ in the presence of other anions such as SO_4^{2-} [67]. Recently, the use of MOFs for the separation of anions from water has been considered a more standard method than photocatalytic reduction [68].

10.4.4.1 Magnetic MOFs for Chromium Removal

The magnetic microspheres of Fe_3O_4@MIL-100Fe as MOFs for the removal of Cr(IV) from an aqueous medium were presented by Yang et al. The Fe_3O_4@MIL-100Fe MOF shell on which the composite was built was grown in place (by hydrothermal reaction) using Fe_3O_4 crystal seeds. The framework's maximal capacity of adsorption was 18 mg/g at an ideal pH of 2. The electrostatic interfaces of anion $HCrO_4^-$ of the contaminant and cationic Fe_3O^+ site of the framework was promoted in an acidic medium. Hence, on increasing the pH, the adsorption capacity decreased. The microspheres had a quick uptake of metal ions, and after two hours, adsorption equilibrium was attained [69].

10.4.5 MOFs for Arsenic Removal

The other extremely poisonous metalloid obtainable in polluted water is arsenic (As). In water, arsenic is present in two forms, namely arsenite (AsO_3^{3-}) and arsenate (AsO_4^{3-}). The toxic arsenic enters the environment from orchards and glass and electronic industrial wastewater. Moreover, arsenic enters the human blood stream through erosion, and it is dangerous to health. Hence, nowadays, researchers make a lot of effort for the detection and removal of As for environmental remediation [70]. For the removal of arsenic from the water, Jie Li et al. synthesized a MOF called [MIL-53(AL)] and tested their capacity toward the removal of arsenate from wastewater. Further, the effect of change in MOF concentration and arsenate removal capacity has also been investigated. It has been observed that, at a minimum equilibrium concentration of 10 g/L, the highest arsenic exclusion ability was ~105.6 mg/g and the lowest capability was 15.4 mg/g. Moreover, other ions present in the water such as PO_4^{3-} did not considerably interfere with the exclusion capability of [MIL-53(AL)] [71].

The equivalent framework prepared by A. Vu is called MIL-53(Fe), and it has an As(V) adsorption capability of 21.27 mg/g from an aqueous medium [72]. For both materials, an association between oxyanions of arsenic and the trivalent metal spot was responsible for arsenate adsorption. Studies on the effect of the metal site on adsorption revealed that the Fe metal site provided more efficient adsorption than the Al and Cr sites of the framework. MIL-100IJFe was discovered to have the maximum adsorption capacity since water is considerably easy to coordinate to the Fe node than others. The bond flexibility in various such framework derivatives was determined [73].

Jie Li and coworkers have investigated the Zn-based MOF ZIF-8's capacity to adsorb trace amounts of arsenate in an aqueous medium. This material displays the utmost capability of sorption toward As(V) than any other reported in the literature

[74]. This material provided the highest efficiency and capability of adsorption of a trace amount of arsenate due to its characteristic properties, such as rate, sorption capability, recyclability, etc. [74]. A carboxylic acid as a linker and an Fe node containing framework called Fe-BTC was further used for the removal of As(V) by adsorption in 2012. The Fe-BTC was synthesized by autogenic stress method. It showed ~96% efficiency in the removal of As(V) from contaminated water at pH 4. The material is ~37 fold, ~7 times more effective in the elimination of As(V) than solid iron and Fe_2O_3 nanomaterial [75].

The sources of the As(III) and As(V) are sodium arsenite and sodium arsenate as dodecahydrate salt, respectively. The framework of ZIF-8 exhibited reasonable abilities for the elimination of 49 mg/g of As(V) and 60 mg/g of As(III) at a pH of 7. The actual MOF was synthesized at room temperature by a slight variation in the procedure. The size of As is larger than the pores of the framework ZIF8; hence, the As sorption took place at the outer surface of the material. Diffusion between the framework and the contaminant arsenic metal controlled the rate of sorption [76]. In 2015, Li et al. recommended the material MOF-808 as an adsorbing material for As(V). The material produced by microwave irradiation has a relatively small size, ~150–200 nm than that formed by the solvothermal method. The maximal exclusion of As(V) achieved was 95% in 30 minutes. As(V) sorption by MOF-808 was due to contact between the Zr spot of the material and As(V) [77]. In a medium of pH = 2, the outmost As(V) deletion capacity was 303 mg/g using a UiO-66 framework by Wang et al. The MOF showed encouraging adsorption capabilities over the range of pH = 1 to 10. It was also discovered that the presence of other anions in the solution had little effect on the uptake of arsenic ions. The hydroxyl group of the zirconium node or ligand BDC of the framework coordinated with the As(V) and was responsible for the sorption confirmed by FT-IR and powder XRD. One Zr_6 cluster was found to be capable of adsorbing seven different species of arsenic in equilibrium [77]. Arsenic metal in both the oxidation state III and V were removed from the sample by its sorption with thiol-adapted UiO-66 in 2016. After six hours of equilibration time, this material could remove 40 mg/g of As(II) and 10 mg/g of As(V), respectively [78].

10.4.6 MOFs for Heavy Metals Phosphate Removal

Hydrous zirconia is well known for its fascination with phosphate and phosphonic-acid-containing material. This is due to the strong interaction between the surfaces of the Zr-OH group of MOFs and the oxygen atoms of the phosphate and phosphonic-acid-containing compounds. In the environment, water P-containing materials are also dangerous to the water. Hence, Gu et al. developed a Zr-based MOF known as UiO-67, which was efficiently useful for the removal of P- containing impurities from the water sample [79]. Further, the comparison between MOF UiO-66 and UiO-66-NH_2 has been investigated to test their efficiency for the adsorptive removal of a phosphate anion. In between UiO-66 and UiO-66-NH_2, the -NH_2 group interacted with the phosphate anions by hydrogen bonding in addition to electrostatic interactions. In the neutral medium (i.e., pH = 7), UiO-66-NH_2 showed a moderate increase in adsorption capacity of ~28 mg/g due to the added effect of electrostatic and hydrogen bonding interactions with the phosphate anions. The UiO-66 MOFs could be easily regenerated and reused for the removal of phosphate anions from the water sample. Therefore, they can act as effective and selective adsorbents for phosphate removal in the cleaning of water [80].

10.4.7 MOFs for Nickel Adsorption

In 2016, Wang et al. developed the chitosan-MOF by the postsynthetic method. During synthesis, the MOF was reacted with the chitosan by using microwave irradiations. The chitosan acted as an active site for the removal of Ni^{2+} from the water. In the chitosan, the $-NH_2$ were present, which acted as a Lewis base, and the heavy-metal ion Ni^{2+} from the water acted as a Lewis acid. The Lewis acid–base interaction between the MOF and the Ni^{2+} ion of water played a crucial role in the process of adsorption and removal of Ni contaminants from the water. The highest adsorption competency achieved by this chitosan-modified MOF was 60 mg/g after eight hours at 200 ^{0}C with a medium pH of ~5 [61].

10.4.8 MOFs for Selenium Adsorption

Another element occurring in drinking water is selenium (Se), which is a naturally occurring element. Industrial and natural activities such as flue-gas desulfurization, mining and petroleum refinery discharges, agricultural runoff, and erosion of natural deposits add Se into the drinking water. The Se is present in the water as its anions -selenite (SeO_3^{2-}) and selenate (SeO_4^{2-}). Several Zr-based MOFs were used to remove these anions from drinking water; therefore, a number of Zr-based MOFs were tested for the removal of selenite and selenate from the water. In between those tested MOFs, the MOF named NU-1000 showed the maximum adsorption capacity and fastest uptake rate for Se oxyanions. The fast uptake rate and high adsorption capacity are due to two factors; one of them is that ~30 orifices of the MOF facilitate fast diffusion and adsorption. Another factor is the labile hydroxyl and water ligands on the Zr6 node of NU-1000 [81].

10.4.9 MOFs for Uranium Adsorption

The extraction of uranium from water using MOFs was first reported by Carboni et al. in 2013 [82]. The Zr-MOF, UiO-68, has intrinsic water stability and has the ability to adsorb and diffuse the big size particles; hence, they were selected as the adsorbent for the extraction of uranium [83]. The functional group, N-diphenylphosphorylurea, is well known for the removal of lanthanides and actinides from water [84]; therefore, a Zr-MOF called UiO-68 was modified by incorporating phosphorylurea through phenyldicarboxylate linkage. The two phosphorylurea-modified MOFs formed UiO-68-PIJO$(IJOEt)_2$ and UiO-68-PIJO$(IJOH)_2$. The resultant modified derivetives showed an increased affinity to the uranyl cation $IJUO_2^{2+}$. They show enhanced adsorption capacities of 217 mg/g and 188 mg/g to uranyl of water and sea water, respectively.

In addition, the MOFs HKUST-168 and Tb-MOF-7669 were examined for uranium adsorption and exclusion from water [85, 86]. It is a well-known fact that carboxylates have a great affinity to uranyl dication through the coordination of oxygen of carboxylate with uranyl. Hence, benzenetricorboxylate linkers were used for the adsorption of uranyl dication by these MOFs. Further, the benzenetricarboxylate showed charge dipole interactions with uranyl. To understand the effect of the surface charge of the MOF on the adsorption of uranyl, the adsorption was carried out at different pHs. On increasing the pH from 2 to 6, the surface of the MOF became strongly negative. As the strength of the negative charge on the surface of the MOF increased, the electrostatic interaction between uranyl dications and the MOF increased; hence, the adsorption of uranyl on the surface of MOF boosts.

10.5 Future Perspective

Enormous amounts of research have been carried out on the use of MOFs due to their characteristic properties such as high flexibility, economical production, and great proficiency. The capabilities of MOFs to use on a bulk scale for industrial wastewater treatment plants are striking and encouraging. MOFs have a characteristic feature of adjustable structural and electrical properties. Hence, they can act as an efficient photodegradable catalyst and adsorbent that can be extra credible. The chemical properties and prospective applications of these substances offer challenges for the development and superiority of novel synthetic MOFs. The potential of MOFs in wastewater treatment is encouraging while additional research will be essential—specifically regarding the scaling up of their application to fulfill the requirements of the bulk samples of the world. In the pertained chapter, we have précised the usages of MOFs in foundations of adsorption and photodegradation for EC remediation in wastewater. Recently, studies of the applications of these materials have received much attention from researchers. The pore size of the MOF should be modified to accommodate the type of pollutants in the water. To increase their interactions with toxic heavy metals, the characteristic properties of MOFs such as their acidity, basicity, H-bonding ability, and electrostatic character can be enhanced by their functionalization. The composites of these MOFs showing large photocatalytic activity can be produced by hybridizing such MOFs with organic linkers, metals, and other inorganic semiconducting materials. The formed functionalized hybrid MOFs will possess relatively minimum electron-hole recombination and will enhance the rate of photoexcitation of MOFs. Finally, nowadays, MOFs are the most demanded material in environmental remediation. Due to their astonishing and favorable characteristic properties in EC cleanup, MOFs are particularly tantalizing materials for further analyses in a variety of conceivable approaches [25].

10.6 Future Scope

MOFs have considerable potential for adsorbing and eliminating heavy-metal ions from water. Further, MOFs show significant selectivity toward toxic metal ions. The extensive investigation of MOFs for the removal of heavy-metal ions as a contaminant of water has a major shortcoming due to its poor water stability. In addition to water stability, the chemical stability of MOFs is a major concern where specific data is required [87, 88]. The present overview realizes that, up to now, a single MOF has not been reported by researchers that can act as a gold-standard adsorbent of heavy-metal ions. Hence, there is a still future scope for researching the development of such MOFs. Further developments of MOFs containing ionic liquids [89] and fluorescent probes[90, 91] can also be possible. As discussed earlier, biocompatible MOFs were used for drug delivery, and they can also be used for drug sequestration from DNA [92, 93]. The design and development of new MOFs and investigation of the mechanism of adsorption with heavy-metal ions has a great future scope. With the advance of reusable MOFs and their composites, one can apply them as adsorbents in the continuous filtration of water. Moreover, the potable, use-and-throw-type filters of MOFs containing adsorbents will be available. The engineered versions of MOFs also have a bright future scope. Finally, due to belonging to a class of porous material, MOFs can be used as adsorbents for the detection and removal of heavy-metal contaminants of real water samples such as lake water, river water, seawater, boar-well water, etc. The increased understanding of the characteristic properties of MOFs as adsorbents will help the new design and development of filters for environmental remediation.

10.7 Conclusions

The present chapter provides a summary of findings concerning the abstraction of toxic metals from the environmental water mimicking contaminated water and real water samples of natural water utilizing modified MOFs. In this chapter, we have discussed the different methods used for the functionalization of MOFs with their active sites. For most of the frameworks, groups such as amino, carboxylates, and thiol functionalization at the active sites of the MOF will be more beneficial for its application in environmental water remediation. In addition, for each metal, the adsorption mechanism on functionalized MOFs has been discussed. The efficiency of materials relative to their adsorption value, equilibrium time, and suitable pH for the successive removal of metal-ion contaminants for each framework has been mentioned precisely. Finally, the future perspective in the field of development for more suitable, efficient, and compatible materials for environmental remediation has been discussed.

References

1 Cheremisinoff, P.N. (2019). *Handbook of Water and Wastewater Treatment Technology [Internet]*. Routledge. Available from: https://www.taylorfrancis.com/books/9781351441896.

2 El-Qanni, A. (2017). Development of sustainable nanosorbcats based technology for hydrocarbons and organic pollutants recovery from industrial wastewater.

3 Qasem, N.A.A., Mohammed, R.H., and Lawal, D.U. (2021 December 8). Removal of heavy metal ions from wastewater: a comprehensive and critical review. *Npj Clean Water* [Internet] 4 (1): 36. Available from: https://www.nature.com/articles/s41545-021-00127-0.

4 Shannon, M.A., Bohn, P.W., Elimelech, M. et al. (2008 March 20). Science and technology for water purification in the coming decades. *Nature* [Internet] 452 (7185): 301–310. Available from: https://www.nature.com/articles/nature06599.

5 Yaghi, O.M., Li, G., and Li, H. (1995 December). Selective binding and removal of guests in a microporous metal-organic framework. *Nature* [Internet].378 (6558): 703–706. Available from: http://www.nature.com/articles/378703a0.

6 Hasan, Z., Jeon, J., and Jhung, S.H. (2012 March). Adsorptive removal of naproxen and clofibric acid from water using metal-organic frameworks. *Journal of Hazardous Materials* [Internet] 209–210: 151–157. Available from: https://linkinghub.elsevier.com/retrieve/pii/S0304389412000210.

7 Martinez-Costa, J.I., Leyva-Ramos, R., and Padilla-Ortega, E. (2018 December 1). Sorption of diclofenac from aqueous solution on an organobentonite and adsorption of cadmium on organobentonite saturated with diclofenac. *Clays and Clay Minerals* [Internet] 66 (6): 515–528. Available from: https://link.springer.com/10.1346/CCMN.2018.064119.

8 Trickett, C.A., Helal, A., Al-Maythalony, B.A. et al. (2017 August 25). The chemistry of metal-organic frameworks for CO_2 capture, regeneration and conversion. *Nature Reviews Materials* [Internet] 2 (8): 17045. Available from: http://www.nature.com/articles/natrevmats201745.

9 Kreno, L.E., Leong, K., Farha, O.K. et al. (2012 February 8). Metal-organic framework materials as chemical sensors. *Chemical Reviews* [Internet] 112 (2): 1105–1125. Available from: https://pubs.acs.org/doi/10.1021/cr200324t.

10 Yaghi, O.M., Kalmutzki, M.J., and Diercks, C.S. (2019). Liquid- and gas-phase separation in MOFs. In: *Introduction to Reticular Chemistry [Internet]*. 365–393. Wiley. Available from: https://onlinelibrary.wiley.com/doi/10.1002/9783527821099.index

11 Rojas, S., Arenas-Vivo, A., and Horcajada, P. (2019 June). Metal-organic frameworks: a novel platform for combined advanced therapies. *Coordination Chemistry Reviews* [Internet] 388: 202–226. Available from: https://linkinghub.elsevier.com/retrieve/pii/S0010854518305599.

12 Dhakshinamoorthy, A., Asiri, A.M., and Garcia, H. (2019 October 20). 2D metal-organic frameworks as multifunctional materials in heterogeneous catalysis and electro/photocatalysis. *Advanced Materials* [Internet] 31 (41): 1900617. Available from: https://onlinelibrary.wiley.com/doi/abs/10.1002/adma.201900617

13 Yang, D. and Gates, B.C. (2019 March 1). Catalysis by metal organic frameworks: perspective and suggestions for future research. *ACS Catalysis* [Internet] 9 (3): 1779–1798. Available from: https://pubs.acs.org/doi/10.1021/acscatal.8b04515.

14 Farha, O.K., Eryazici, I., Jeong, N.C. et al. (2012 September 12). Metal-organic framework materials with ultrahigh surface areas: is the sky the limit? *Journal of the American Chemical Society* [Internet] 134 (36): 15016–15021. Available from: https://pubs.acs.org/doi/10.1021/ja3055639.

15 Furukawa, H., Cordova, K.E., O'Keeffe, M., and Yaghi, O.M. (2013 August 30). The chemistry and applications of metal-organic frameworks. *Science* (80-) [Internet] 341 (6149): Available from: https://www.science.org/doi/10.1126/science.1230444

16 Wang, C., Liu, X., Keser Demir, N. et al. (2016). Applications of water stable metal-organic frameworks. *Chemical Society Reviews* [Internet] 45 (18): 5107–5134. Available from: http://xlink.rsc.org/?DOI=C6CS00362A.

17 Butova, V.V., Soldatov, M.A., Guda, A.A. et al. (2016 March 31). Metal-organic frameworks: structure, properties, methods of synthesis and characterization. *Russian Chemical Reviews* [Internet] 85 (3): 280–307. Available from: http://stacks.iop.org/0036-021X/85/i=3/a=280?key=crossref.4839b386ef90605011a70c97dd55e408.

18 Eddaoudi, M., Li, H., and Yaghi, O.M. (2000 February 1). Highly porous and stable metal–organic frameworks: structure design and sorption properties. *Journal of the American Chemical Society* [Internet] 122 (7): 1391–1397. Available from: https://pubs.acs.org/doi/10.1021/ja9933386.

19 Han, T., Li, C., Guo, X. et al. (2016 December). In-situ synthesis of SiO2@MOF composites for high-efficiency removal of aniline from aqueous solution. *Applications of Surface Science* [Internet] 390: 506–512. Available from: https://linkinghub.elsevier.com/retrieve/pii/S0169433216317548.

20 Hasan, Z. and Jhung, S.H. (2015 February). Removal of hazardous organics from water using metal-organic frameworks (MOFs): plausible mechanisms for selective adsorptions. *Journal of Hazardous Materials* [Internet] 283: 329–339. Available from: https://linkinghub.elsevier.com/retrieve/pii/S0304389414007912.

21 Dhakshinamoorthy, A., Li, Z., and Garcia, H. (2018). Catalysis and photocatalysis by metal organic frameworks. *Chemical Society Reviews* [Internet] 47 (22): 8134–8172. Available from: http://xlink.rsc.org/?DOI=C8CS00256H.

22 Kumar, P., Bansal, V., Kim, K.-H., and Kwon, E.E. (2018 June). Metal-organic frameworks (MOFs) as futuristic options for wastewater treatment. *Journal of Industrial and Engineering Chemistry* [Internet] 62: 130–145. Available from: https://www.sciencedirect.com/science/article/abs/pii/S1226086X1730713X

23 Bedia, J., Muelas-Ramos, V., Peñas-Garzón, M. et al. (2019 January 7). A review on the synthesis and characterization of metal organic frameworks for photocatalytic water purification. *Catalysts* [Internet] 9 (1): 52.Available from: https://www.mdpi.com/2073-4344/9/1/52.

24 Rego, R.M., Kuriya, G., Kurkuri, M.D., and Kigga, M. (2021 February). MOF based engineered materials in water remediation: recent trends. *Journal of Hazardous Materials* [Internet] 403: 123605. Available from: https://linkinghub.elsevier.com/retrieve/pii/S0304389420315910.

25 Russo, V., Hmoudah, M., Broccoli, F. et al. (2020 October 23). Applications of metal organic frameworks in wastewater treatment: a review on adsorption and photodegradation. *Frontiers of Chemical Engineering* [Internet] 2: Available from: https://www.frontiersin.org/articles/10.3389/fceng.2020.581487/full.

26 Zhu, Q.-L. and Xu, Q. (2014 March 18). Metal-organic framework composites. *Chemical Society Reviews* [Internet] 43 (16): 5468–5512. Available from: http://xlink.rsc.org/?DOI=C3CS60472A.

27 Rowsell, J.L.C. and Yaghi, O.M. (2004 August). Metal-organic frameworks: a new class of porous materials. *Microporous and Mesoporous Materials* [Internet] 73 (1–2): 3–14. Available from: https://linkinghub.elsevier.com/retrieve/pii/S1387181104001295.

28 Zhao, X., Liu, D., Huang, H., and Zhong, C. (2016 April). Highly selective and sensitive metal-organic framework fluorescent probe for Cu2+ through rational design of binding sites. *Microporous and Mesoporous Materials* [Internet] 224: 149–154. Available from: https://linkinghub.elsevier.com/retrieve/pii/S1387181115006496.

29 Moghaddam, Z.S., Kaykhaii, M., Khajeh, M., and Oveisi, A.R. (2018 April). Synthesis of UiO-66-OH zirconium metal-organic framework and its application for selective extraction and trace determination of thorium in water samples by spectrophotometry. *Spectrochimica Acta, Part A: Molecular and Biomolecular Spectroscopy* 194: 76–82.

30 Wen, J., Fang, Y., and Zeng, G. (2018 June). Progress and prospect of adsorptive removal of heavy metal ions from aqueous solution using metal-organic frameworks: a review of studies from the last decade. *Chemosphere* 201: 627–643.

31 Wu, Y., Xu, G., Liu, W. et al. (2015 July). Postsynthetic modification of copper terephthalate metal-organic frameworks and their new application in preparation of samples containing heavy metal ions. *Microporous and Mesoporous Materials* 210: 110–115.

32 Kobielska, P.A., Howarth, A.J., Farha, O.K., and Nayak, S. (2018 March). Metal-organic frameworks for heavy metal removal from water. *Coordination Chemistry Reviews* 358: 92–107.

33 Schroden, R.C., Al-Daous, M., Sokolov, S. et al. (2002). Hybrid macroporous materials for heavy metal ion adsorption. *Journal of Materials Chemistry* [Internet] 12 (11): 3261–3267. Available from: http://xlink.rsc.org/?DOI=B204065B.

34 Howarth, A.J., Liu, Y., Hupp, J.T., and Farha, O.K. (2015). Metal-organic frameworks for applications in remediation of oxyanion/cation-contaminated water. *CrystEngComm* [Internet] 17 (38): 7245–7253. Available from: http://xlink.rsc.org/?DOI=C5CE01428J.

35 Ke, F., Qiu, L.-G., Yuan, Y.-P. et al. (2011 November). Thiol-functionalization of metal-organic framework by a facile coordination-based postsynthetic strategy and enhanced removal of Hg2+ from water. *Journal of Hazardous Materials* [Internet] 196: 36–43. Available from: https://linkinghub.elsevier.com/retrieve/pii/S0304389411010879.

36 Saleem, H., Rafique, U., and Davies, R.P. (2016 February). Investigations on post-synthetically modified UiO-66-NH 2 for the adsorptive removal of heavy metal ions from aqueous solution. *Microporous and Mesoporous Materials* [Internet] 221: 238–244. Available from: https://linkinghub.elsevier.com/retrieve/pii/S1387181115005326.

37 Abbasi, A., Moradpour, T., and Van Hecke, K. (2015 May). A new 3D cobalt (II) metal-organic framework nanostructure for heavy metal adsorption. *Inorganica Chimica Acta* [Internet] 430: 261–267. Available from: https://linkinghub.elsevier.com/retrieve/pii/S002016931500153X.

38 Liang, L., Chen, Q., Jiang, F. et al. (2016). In-situ large-scale construction of sulfur-functionalized metal-organic framework and its efficient removal of Hg(ii) from water. *Journal of Materials Chemistry A* [Internet] 4 (40): 15370–15374. Available from: http://xlink.rsc.org/?DOI=C6TA04927C.

39 Fang, Q.-R., Yuan, D.-Q., Sculley, J. et al. (2010 December 20). Functional mesoporous metal–organic frameworks for the capture of heavy metal ions and size-selective catalysis. *Inorganic Chemistry* [Internet] 49 (24): 11637–11642. Available from: https://pubs.acs.org/doi/10.1021/ic101935f.

40 Chakraborty, A., Bhattacharyya, S., Hazra, A. et al. (2016). Post-synthetic metalation in an anionic MOF for efficient catalytic activity and removal of heavy metal ions from aqueous solution. *Chemical Communications* [Internet] 52 (13): 2831–2834. Available from: http://xlink.rsc.org/?DOI=C5CC09814A.

41 Luo, F., Chen, J.L., Dang, L.L. et al. (2015). High-performance Hg 2+ removal from ultra-low-concentration aqueous solution using both acylamide- and hydroxyl-functionalized metal-organic framework. *Journal of Materials Chemistry A* [Internet] 3 (18): 9616–9620. Available from: http://xlink.rsc.org/?DOI=C5TA01669J.

42 Halder, S., Mondal, J., Ortega-Castro, J. et al. (2017). MOF for selective detection and removal of Hg 2+ in aqueous medium: a facile strategy. *Dalton Transactions* [Internet] 46 (6): 1943–1950. Available from: http://xlink.rsc.org/?DOI=C6DT04722J.

43 Liu, T., Che, J., Hu, Y. et al. (2014 October 20). Alkenyl/Thiol-derived metal-organic frameworks (MOFs) by means of postsynthetic modification for effective mercury adsorption. *Chemistry—A European Journal* [Internet] 20 (43): 14090–14095. Available from: https://chemistry-europe.onlinelibrary.wiley.com/doi/abs/10.1002/chem.201403382

44 Yee, -K.-K., Reimer, N., Liu, J. et al. (2013 May 29). Effective mercury sorption by thiol-laced metal-organic frameworks: in strong acid and the vapor phase. *Journal of the American Chemical Society* [Internet] 135 (21): 7795–7798. Available from: https://pubs.acs.org/doi/10.1021/ja400212k.

45 Rudd, N.D., Wang, H., Fuentes-Fernandez, E.M.A. et al. (2016 November 9). Highly efficient luminescent metal-organic framework for the simultaneous detection and removal of heavy metals from water. *ACS Applied Materials and Interfaces* [Internet] 8 (44): 30294–30303. Available from: https://pubs.acs.org/doi/10.1021/acsami.6b10890.

46 Bhattacharjee, S., Lee, Y.-R., and Ahn, W.-S. (2015). Post-synthesis functionalization of a zeolitic imidazolate structure ZIF-90: a study on removal of Hg(II) from water and epoxidation of alkenes. *CrystEngComm* [Internet] 17 (12): 2575–2582. Available from: http://xlink.rsc.org/?DOI=C4CE02555E.

47 Sohrabi, M.R. (2014 February 6). Preconcentration of mercury(II) using a thiol-functionalized metal-organic framework nanocomposite as a sorbent. *Microchimica Acta* [Internet] 181 (3–4): 435–444. Available from: http://link.springer.com/10.1007/s00604-013-1133-1.

48 Luo, X., Shen, T., Ding, L. et al. (2016 April). Novel thymine-functionalized MIL-101 prepared by post-synthesis and enhanced removal of Hg 2+ from water. *Journal of Hazardous Materials* [Internet] 306: 313–322. Available from: https://linkinghub.elsevier.com/retrieve/pii/S0304389415302946.

49 Mon, M., Lloret, F., Ferrando-Soria, J. et al. (2016 September 5). Selective and efficient removal of mercury from aqueous media with the highly flexible arms of a

BioMOF. *Angewandte Chemie International Edition* [Internet] 55 (37): 11167–11172. Available from: https://onlinelibrary.wiley.com/doi/abs/10.1002/anie.201606015

50 Xiong, Y.Y., Li, J.Q., Le, G.L. et al. (2017 February). Using MOF-74 for Hg2+ removal from ultra-low concentration aqueous solution.*Journal of Solid State Chemistry* [Internet] 246: 16–22. Available from: https://linkinghub.elsevier.com/retrieve/pii/S0022459616304169.

51 Huang, L., He, M., Chen, B., and Hu, B. (2015). A designable magnetic MOF composite and facile coordination-based post-synthetic strategy for the enhanced removal of Hg 2+ from water. *Journal of Materials Chemistry A* [Internet] 3 (21): 11587–11595. Available from: http://xlink.rsc.org/?DOI=C5TA01484K.

52 Zhang, J., Xiong, Z., Li, C., and Wu, C. (2016 September). Exploring a thiol-functionalized MOF for elimination of lead and cadmium from aqueous solution. *Journal of Molecular Liquids* [Internet] 221: 43–50. Available from: https://linkinghub.elsevier.com/retrieve/pii/S0167732216305219.

53 Tahmasebi, E., Masoomi, M.Y., Yamini, Y., and Morsali, A. (2015 January 20). Application of mechanosynthesized azine-decorated zinc(II) metal-organic frameworks for highly efficient removal and extraction of some heavy-metal ions from aqueous samples: a comparative study. *Inorganic Chemistry* [Internet] 54 (2): 425–433. Available from: https://pubs.acs.org/doi/10.1021/ic5015384.

54 Rahimi, E. and Mohaghegh, N. (2016 March 12). Removal of toxic metal ions from sungun acid rock drainage using mordenite zeolite, graphene nanosheets, and a novel metal-organic framework. *Mine Water and the Environment* [Internet] 35 (1): 18–28. Available from: http://link.springer.com/10.1007/s10230-015-0327-7.

55 Jamali, A., Tehrani, A.A., Shemirani, F., and Morsali, A. (2016). Lanthanide metal-organic frameworks as selective microporous materials for adsorption of heavy metal ions. *Dalton Transactions* [Internet] 45 (22): 9193–9200. Available from: http://xlink.rsc.org/?DOI=C6DT00782A.

56 Qin, Q., Wang, Q., Fu, D., and Ma, J. (2011 August). An efficient approach for Pb(II) and Cd(II) removal using manganese dioxide formed in situ. *Chemical Engineering Journal* [Internet] 172 (1): 68–74. Available from: https://linkinghub.elsevier.com/retrieve/pii/S1385894711006279.

57 Rivera, J.M., Rincón, S., Ben Youssef, C., and Zepeda, A. (2016). Highly efficient adsorption of aqueous Pb(II) with mesoporous metal-organic framework-5: an equilibrium and kinetic study. *Journal of Nanomaterials* [Internet] 2016: 1–9. Available from: https://www.hindawi.com/journals/jnm/2016/8095737.

58 Ricco, R., Konstas, K., Styles, M.J. et al. (2015). Lead(II) uptake by aluminium based magnetic framework composites (MFCs) in water. *Journal of Materials Chemistry A* [Internet] 3 (39): 19822–19831. Available from: http://xlink.rsc.org/?DOI=C5TA04154F.

59 Wang, Y., Ye, G., Chen, H. et al. (2015). Functionalized metal-organic framework as a new platform for efficient and selective removal of cadmium(ii) from aqueous solution. *Journal of Materials Chemistry A* [Internet] 3 (29): 15292–15298. Available from: http://xlink.rsc.org/?DOI=C5TA03201F.

60 Zou, F., Yu, R., Li, R., and Li, W. (2013 August 26). Microwave-assisted synthesis of HKUST-1 and functionalized HKUST-1-@H 3 PW 12 O 40 : selective adsorption of heavy metal ions in water analyzed with synchrotron radiation. *ChemPhysChem* [Internet] 14 (12): 2825–2832. Available from: https://chemistry-europe.onlinelibrary.wiley.com/doi/abs/10.1002/cphc.201300215

61 Wang, K., Tao, X., Xu, J., and Yin, N. (2016 December 5). Novel chitosan-MOF composite adsorbent for the removal of heavy metal ions. *Chemistry Letters* [Internet] 45 (12): 1365–1368. Available from: http://www.journal.csj.jp/doi/10.1246/cl.160718.

62 Maleki, A., Hayati, B., Naghizadeh, M., and Joo, S.W. (2015 August). Adsorption of hexavalent chromium by metal organic frameworks from aqueous solution. *Journal of Industrial and Engineering Chemistry* [Internet] 28: 211–216. Available from: https://www.sciencedirect.com/science/article/abs/pii/S1226086X15000660

63 Li, -L.-L., Feng, X.-Q., Han, R.-P. et al. (2017 January). Cr(VI) removal via anion exchange on a silver-triazolate MOF. *Journal of Hazardous Materials* [Internet] 321: 622–628. Available from: https://linkinghub.elsevier.com/retrieve/pii/S0304389416308482.

64 Zhang, Q., Yu, J., Cai, J. et al. (2015). A porous Zr-cluster-based cationic metal-organic framework for highly efficient Cr 2 O 7 2− removal from water. *Chemical Communications* [Internet] 51 (79): 14732–14734. Available from: http://xlink.rsc.org/?DOI=C5CC05927E.

65 Li, X., Gao, X., Ai, L., and Jiang, J. (2015 August). Mechanistic insight into the interaction and adsorption of Cr(VI) with zeolitic imidazolate framework-67 microcrystals from aqueous solution. *Chemical Engineering Journal* [Internet] 274: 238–246. Available from: https://linkinghub.elsevier.com/retrieve/pii/S1385894715004647.

66 Aboutorabi, L., Morsali, A., Tahmasebi, E., and Büyükgüngor, O. (2016 June 6). Metal-organic framework based on isonicotinate N -oxide for fast and highly efficient aqueous phase Cr(VI) adsorption. *Inorganic Chemistry* [Internet] 55 (11): 5507–5513. Available from: https://pubs.acs.org/doi/10.1021/acs.inorgchem.6b00522.

67 Rapti, S., Pournara, A., Sarma, D. et al. (2016). Rapid, green and inexpensive synthesis of high quality UiO-66 amino-functionalized materials with exceptional capability for removal of hexavalent chromium from industrial waste. *Inorganic Chemistry Frontiers* [Internet] 3 (5): 635–644. Available from: http://xlink.rsc.org/?DOI=C5QI00303B.

68 Wang, C.-C., Du, X.-D., Li, J. et al. (2016 September). Photocatalytic Cr(VI) reduction in metal-organic frameworks: a mini-review. *Applied Catalysis, B: Environmental* [Internet] 193: 198–216. Available from: https://linkinghub.elsevier.com/retrieve/pii/S092633731630296X.

69 Yang, Q., Zhao, Q., Ren, S. et al. (2016 December). Fabrication of core-shell Fe3O4@MIL-100(Fe) magnetic microspheres for the removal of Cr(VI) in aqueous solution. *Journal of Solid State Chemistry* [Internet] 244: 25–30. Available from: https://linkinghub.elsevier.com/retrieve/pii/S0022459616303541.

70 Choong, T.S.Y., Chuah, T.G., Robiah, Y. et al. (2007 November). Arsenic toxicity, health hazards and removal techniques from water: an overview. *Desalination* [Internet] 217 (1–3): 139–166. Available from: https://linkinghub.elsevier.com/retrieve/pii/S0011916407004791.

71 Li, J., Wu, Y., Li, Z. et al. (2014 Octoberl). Characteristics of arsenate removal from water by metal-organic frameworks (MOFs). *Water Science and Technology* [Internet] 70 (8): 1391–1397. Available from: https://iwaponline.com/wst/article/70/8/1391/18226/Characteristics-of-arsenate-removal-from-water-by.

72 Vu, T.A., Le, G.H., Dao, C.D. et al. (2015). Arsenic removal from aqueous solutions by adsorption using novel MIL-53(Fe) as a highly efficient adsorbent. *RSC Advances* [Internet] 5 (7): 5261–5268. Available from: http://xlink.rsc.org/?DOI=C4RA12326C.

73 Jun, J.W., Tong, M., Jung, B.K. et al. (2015 January 2). Effect of central metal ions of analogous metal-organic frameworks on adsorption of organoarsenic compounds from water: plausible mechanism of adsorption and water purification. *Chemistry—A European Journal* [Internet] 21 (1): 347–354. Available from: https://chemistry-europe.onlinelibrary.wiley.com/doi/abs/10.1002/chem.201404658

74 Li, J., Wu, Y., Li, Z. et al. (2014 November 26). Zeolitic imidazolate framework-8 with high efficiency in trace arsenate adsorption and removal from water. *The Journal of Physical Chemistry C* [Internet] 118 (47): 27382–27387. Available from: https://pubs.acs.org/doi/10.1021/jp508381m.

75 Zhu, B.-J., Yu, X.-Y., Jia, Y. et al. (2012 Apr 19). Iron and 1,3,5-benzenetricarboxylic metal-organic coordination polymers prepared by solvothermal method and their application in efficient As (v) removal from aqueous solutions. *The Journal of Physical Chemistry C* [Internet] 116 (15): 8601–8607. Available from: https://pubs.acs.org/doi/10.1021/jp212514a.

76 Jian, M., Liu, B., Zhang, G. et al. (2014 January). Adsorptive removal of arsenic from aqueous solution by zeolitic imidazolate framework-8 (ZIF-8) nanoparticles. *Colloids Surfaces A Physicochem Engineering Aspects* [Internct] 465: 67–76. Available from: https://linkinghub.elsevier.com/retrieve/pii/S0927775714007936.

77 Li, Z.-Q., Yang, J.-C., Sui, K.-W., and Yin, N. (2015 December). Facile synthesis of metal-organic framework MOF-808 for arsenic removal. *Materials Letters* [Internet] 160: 412–414. Available from: https://www.sciencedirect.com/science/article/abs/pii/S0167577X15303621

78 Audu, C.O., Nguyen, H.G.T., Chang, C.-Y. et al. (2016). The dual capture of As V and As III by UiO-66 and analogues. *Chemical Science* [Internet] 7 (10): 6492–6498. Available from: http://xlink.rsc.org/?DOI=C6SC00490C.

79 Zhu, X., Li, B., Yang, J. et al. (2015 January 14). Effective adsorption and enhanced removal of organophosphorus pesticides from aqueous solution by Zr-Based MOFs of UiO-67. *ACS Applied Materials and Interfaces* [Internet] 7 (1): 223–231. Available from: https://pubs.acs.org/doi/10.1021/am5059074.

80 Lin, K.-Y.A., Chen, S.-Y., and Jochems, A.P. (2015 June). Zirconium-based metal organic frameworks: highly selective adsorbents for removal of phosphate from water and urine. *Materials Chemistry and Physics* [Internet] 160: 168–176. Available from: https://linkinghub.elsevier.com/retrieve/pii/S0254058415300225.

81 Howarth, A.J., Katz, M.J., Wang, T.C. et al. (2015 June 17). High efficiency adsorption and removal of selenate and selenite from water using metal-organic frameworks. *Journal of the American Chemical Society* [Internet] 137 (23): 7488–7494. Available from: https://pubs.acs.org/doi/10.1021/jacs.5b03904.

82 Carboni, M., Abney, C.W., Liu, S., and Lin, W. (2013). Highly porous and stable metal-organic frameworks for uranium extraction. *Chemical Science* [Internet] 4 (6): 2396. Available from: http://xlink.rsc.org/?DOI=c3sc50230a.
83 Schaate, A., Roy, P., Godt, A. et al. (2011 June 6). Modulated synthesis of Zr-based metal-organic frameworks: from nano to single crystals. *Chemistry—A European Journal* [Internet] 17 (24): 6643–6651. Available from: https://chemistry-europe.onlinelibrary.wiley.com/doi/10.1002/chem.201003211
84 Matrosov, E.I., Goryunov, E.I., Baulina, T.V. et al. (2010 May 20). First complexes of N-diphenylphosphorylureas with actinides and lanthanides: synthesis and structure. *Doklady Chemistry* [Internet] 432 (1): 136–139. Available from: http://link.springer.com/10.1134/S0012500810050058.
85 Feng, Y., Jiang, H., Li, S. et al. (2013 August). Metal-organic frameworks HKUST-1 for liquid-phase adsorption of uranium. *Colloids Surfaces A Physicochem Eng Asp* [Internet] 431: 87–92. Available from: https://linkinghub.elsevier.com/retrieve/pii/S0927775713003178.
86 Yang, W., Bai, Z.-Q., Shi, W.-Q. et al. (2013). MOF-76: from a luminescent probe to highly efficient U VI sorption material. *Chemical Communications* [Internet] 49 (88): 10415–10417. Available from: http://xlink.rsc.org/?DOI=C3CC44983A.
87 Burtch, N.C., Jasuja, H., and Walton, K.S. (2014 October 22). Water stability and adsorption in metal-organic frameworks. *Chemical Reviews* [Internet] 114 (20): 10575–10612. Available from: https://pubs.acs.org/doi/10.1021/cr5002589.
88 Howarth, A.J., Liu, Y., Li, P. et al. (2016 March 9). Chemical, thermal and mechanical stabilities of metal-organic frameworks. *Nature Reviews Materials* [Internet] 1 (3): 15018. Available from: http://www.nature.com/articles/natrevmats201518.
89 Kalel, R.A. and Gaikwad, D.S. (2022). Cooperative catalysis: condensation-aromatization for synthesis of 2-(4-nitrophenyl)-1H-benzimidazole by silica immobilized Brønsted-Lewis acidic ionic liquid (Si-BLAIL). *Journal of the Indian Chemical Society* 9 (7): 100550.
90 Kalel, R.A. (2022). Mini-review on a polymers film detector for chloroform vapour: julolidine as fluorescent molecular rotors (JCFMRs). Chemical Papers.
91 Kalel, R., Mora, A.K., Ghosh, R. et al. (2016 September 22). Interaction of a julolidine-based neutral ultrafast molecular rotor with natural DNA: spectroscopic and molecular docking studies. *Journal of Physical Chemistry B* [Internet] 120 (37): 9843–9853. Available from: https://pubs.acs.org/doi/10.1021/acs.jpcb.6b04811.
92 Kalel, R., Mora, A.K., Patro, B.S. et al. (2017). Synergistic enhancement in the drug sequestration power and reduction in the cytotoxicity of surfactants. *Physical Chemistry Chemical Physics* [Internet] 19 (37): 25446–25455. Available from: http://xlink.rsc.org/?DOI=C7CP05042A.
93 Mora, A.K., Basu, A., Kalel, R., and Nath, S. (2019). Polymer-assisted drug sequestration from plasma protein by a surfactant with curtailed denaturing capacity. *Physical Chemistry Chemical Physics* [Internet] 21 (13): 7127–7136. Available from: http://xlink.rsc.org/?DOI=C8CP03576H.

11

Application of Algal-Polysaccharide Metal-Organic Frameworks in Wastewater Treatment

Dharitri Borah[1], Jayashree Rout[2], and Thajuddin Nooruddin[3]

[1] Department of Environmental Science, Arunachal University of Studies, Namsai-792103, Arunachal Pradesh, India
[2] Department of Ecology and Environmental Science, Assam University, Silchar-788011, Assam, India
[3] Department of Microbiology, Bharathidasan University, Tiruchirappalli-620024, Tamil Nadu, India

11.1 Introduction

The freshwater resource accounts for a very small amount on Earth: approximately 3%. It has been documented that about two-thirds of the population of the world suffers from water shortages for at least a month across their lifetime [1]. To lessen the environmental impression, the most effective techniques require accountability to energy efficiency, economic viability of resources, materials, stopping production of deadly secondary pollutants, life cycle analysis with recyclability, and reuse [2].

11.1.1 Water Pollutants and Sources

Organic pollutants are concomitant to oil crude/petroleum products (diesel, gasoline, kerosene etc.); chlorinated solvents (freons and perchloroethylene); organic solvents (methyl ketone and alcohols); disinfection by-products (chloroform); food-processing waste; perchlorates (explosives, fireworks, and rocket fuels); trihalomethanes (dibromochloromethane and bromodichloromethane); etc. Inorganic pollutants are metals and their compounds, inorganic fertilizers, acidity, etc. [2] Different industrial sectors like base-metal and iron-ore mining, cement manufacturing, coalmining and production, electricity generation, foundries, iron/steel/lead smelting, meat processing/rendering, oil/gas, pesticide/petrochemical production, petroleum refining, phosphorus (P)/nitrogen (N) fertilizer plants, pulp and paper mills, tanning and leather finishing, and textile manufacturing are accountable for creating pollutants like toxic metal sludge, arsenic, hot water, solvents, lead, cadmium, biological oxygen demand, oil, pesticides and toxic intermediates, hydrocarbons, nutricnts, toxic dyes, etc. [3] The fertilizer, pesticide, and electrochemical industries are the principal sources of metal pollutants [4]. However, natural sources like soil erosion, weathering of rocks and minerals, and volcanic eruptions

Metal Organic Frameworks for Wastewater Contaminant Removal, First Edition. Edited by Arun Lal Srivastav, Lata Rani, Jyotsna Kaushal, and Tien Duc Pham.

also contribute to heavy-metal pollution [5]. The urbanized river water presented more spatial temporal heterogeneity with high aggravation of nutrients and metals, which was affected by industrial activities, hydrodynamic conditions, and land use mostly in summer [6].

11.1.2 Common Wastewater Treatment Techniques

Depending on economic stability, different countries reuse either treated or untreated wastewater (1.5–6.6%) for at least irrigation [7]. Primary treatment includes screening, which removes suspended particles (50–60%) and reduces biological oxygen demand (BOD) (20–30%). Secondary treatment is biological treatment, reducing BOD/suspended solids by up to 85%. Finally, in tertiary treatment, 99% of pollutants (bacteria from secondary treatment, heavy metals) are removed. Common wastewater treatment technology includes advanced oxidations (UV/O_3, UV/H_2O_2, UV, O_3, fenton/photo processes, sonochemical processes), phase-changing technologies (adsorption using activated carbon, carbon nanotubes, biochar, and clay minerals), biological processes (aerobic or anaerobic digestion), etc. [8, 9] Aerobic processes are more pertinent to wastewaters with biodegradable chemical oxygen demand (COD) <1000 mg/L, while anaerobic processes are useful for biodegradable COD >4000 mg/L [10]. Different biofilm reactors like aerobic membrane bioreactors (MBR), anaerobic-aerobic granular biofilm bioreactors, anaerobic-aerobic fixed film bioreactors (FFB), and integrated anaerobic/aerobic fluidized bed reactors are accomplished for removing pollution load [11]. Chemical precipitation pertaining to heavy-metal separation includes hydroxides and sulfides [12]. The chemical and membrane methods result in large formations of sludge requiring posttreatment [13].

11.1.3 Metal-Organic Frameworks for Wastewater Treatment

Metal-organic frameworks (MOFs) are presented by their porous crystalline constituents, which are arranged by assembling metal ions and the bridging of organic ligands held by strong coordination bonds. The porous organic-inorganic hybrid structures are highly ordered and present surface areas up to 7000 m^2/g [14]. Although considered highly crystalline, non-crystalline MOFs (amorphous/liquid/glasses) also have emerged [15]. MOFs are excellent selective adsorbents pertaining to disparate functional groups (C–O–C, -OH, -COOH etc.) added or modified to the MOFs [16]. MOFs have a wide range of applications like drug delivery, gas storage, photo-based reactions, water purification, optics, contrast agents, chemical sensors, dust removal, polymerization, luminescence, catalysis, magnetism, etc. [17–20] MOFs are applied for bioremediation pertaining to their photocatalytic properties [21]. In wastewater/effluents, MOFs adsorb heavy metals (Cu^{2+} and Pb^{2+}, etc.), tetracycline, dyes (methylene blue, malachite green, and rhodamine B), osmosis (forward/reverse), filtration (ultra/nano), produced water purification (organic/inorganic pollutants), aliphatic hydrocarbon, oil, radioactive-material, phosphate, algae-cyanobacteria bloom, pesticide, etc., removal [6, 9, 20–30]. Different MOFs viz. Ce-MOF, Co(II)-MOF, and

Fe-MOF are the sensors of ochratoxin A, MnO^{4-}, and chlorogenic acid respectively [27]. Shapable MOFs are available in the form of spheres, membranes, chains, and higher dimensional structures. [14] MOFs applied to water purification are viz. zeolite imidazolate frameworks (ZIF-8), Materials of Institute Lavoisier (MIL) [MIL-n are trivalent metal-based porous carboxylates such as chromium(III), vanadium(III) and iron(III) and continue to the p-elements such as aluminum(III) and gallium(III)]; University of Oslo (UiO-66) [zirconium–carboxylate]; C300, copper-based copper benzene-1,3,5-tricarboxylate, etc. [16] Application of MOFs has benefits like the use of safe and compatible building blocks, they are energy efficient, require the use of less solvent, have a continuous production potential, and can be designed for specific functions [31]. Different MOFs act as photocatalysts and are useful for the complete removal of pollutants [9].

11.1.4 Polysaccharide-Metal-organic Frameworks (Ps-MOFs)

In MOF adsorbents, the organic part is mostly from non-renewable origins (petroleum), limiting applicability pertaining to toxicity and non-biodegradability [18]. In nature, polysaccharides (Ps) are extracted from biological origins like plants, animals (crustaceans), and microbes (yeasts, fungi, and bacteria) including algae and cyanobacteria. The Ps as organic cross-linkers present biocompatibility, low toxicity, and selectivity with high stability, keeping the MOFs intact [32]. The application of Ps as such in wastewater treatment is not very useful pertaining to hydrophilicity, low mechanical strength, regeneration, or reusability [33]. The Ps porous starch-chitosan-UiO-66-COOH composite adsorbed sulphanilamide [34]. The chitosan-coated Fe_3O_4@Cd-MOF was an effective adsorbent (~103.09 mg/g) as compared to amoxicillin [35]. Magnetic Fe_3O_4/ZIF-67@AmCs (aminated chitosan) beads are an efficient adsorbent (~119 mg/g) of the heavy metal Cr(VI) [36]. However, chitin and chitosan are of animal origin (sponges, molluscs, and crustaceans) and mass production is limited. Cellulose-derived MOFs are efficient adsorbents (for dyes, metals, drugs, and pesticides) and are catalysts (oxidation, reduction, and degradation) for separate organic pollutants [21]. Algal cellulose can be harvested all year round, requiring no productive land. Algae can be cultivated in wastewater for phycoremediation (bioremediation by algae and cyanobacteria), and biomass production can be maximized by optimization. The cellulose microfibrils obtained from algae are thicker (up to 30 nm) than woods (width, 5 nm) with a distinct orientation [37]. Also, the crystallinity index (>60%) is higher than with plants [38]. Algae and cyanobacteria are rich in polysaccharides like cellulose, hemicelluloses, alginates, chitin-/chitosan-like molecules, fucans, carrageenans, etc. [39] However, only alginate-based MOFs are widely applied in wastewater treatment [19, 40, 41]. Therefore, it is worthwhile studying the algal polysaccharides (AlPs) from different groups of algae with potential applications in MOF production. The synthesis and characterizations of available and other plausible AlPs-MOFs for wastewater treatment are discussed in this chapter. Additionally, the regeneration, recycling, and durability of AlPs-MOFs are discussed contingent on scalability, economic viability, and ecological sustainability.

11.2 Polysaccharides in Algae/cyanobacteria (AlPs)

11.2.1 Polysaccharides in Cyanophyceae

Cellulose, a long-chain polymer of D-glucose, is among the most abundant polysaccharides in nature. Interestingly, the cyanobacterium *Synechococcus* sp. was found to possess cellulose [38]. Under optimized cultivation conditions, cellulose production (crystallinity index: 65%) improved up to ~14% (dry weight) in 12 days for strain CM12. Cellulose was also detected in *Anabaena* sp., *Oscillatoria* sp., *Gloeocapsa* sp., *Phormidium autumnale, Scytonema hofmanni, Nostoc muscorum*, etc. [42] Also, cellulose production can be improved by expressing cellulose synthase genes of bacterium *Acetobacter xylinum* in cyanobacteria like *Anabaena* sp [43]. Cellulose was augmented two-fold in *Thermosynechococcus vulcanus* on reducing the temperature (31°C from 57°C) within a day [44]. The extracellular polysaccharides (EPSs) from cyanobacteria can be extracted by water, EDTA, or NaOH solvents (Figure 11.1a, 11.1b). Interestingly, cyanobacteria spend most of their energy (70%) in EPSs [45]. It is plausible to improve EPSs with stresses, optimization of nutrients, light (duration/intensity), temperature, etc. EPSs are often sulfated and acidic appertaining to uronic acids, detectable by the carboxylic group. The EPS of *Cyanothece epiphytica* was predominantly composed of glucose (42.49%). The uronic acid content was 5.3% with a hydrophobicity of ~69% on treating with ozone for ten seconds. [46] The EPS was thermally stable (265°C). The species *Nostoc muscorum* yielded 126.73 μg/mL of EPS within 12 days. The EPS was rich in uronic acid (24.61%), which was affirmed by the -COOH functional group [47]. *Nostoc microscopicum* produced 0.90g/L of EPS in 44 days [48]. Salt stress (NaCl) further improved the EPS (88μg/mL) in *Lyngbya stagnina* in six days [49]. Under continuous light (9.8 W/m^2), *Limnothrix redekei* (PUPCCC 116) released 614 μg /mL of EPSs by 21 days [50]. Therefore, cyanobacterial PS is a plausible resource for the organic ligands in AlPs-MOF production (Table 11.1).

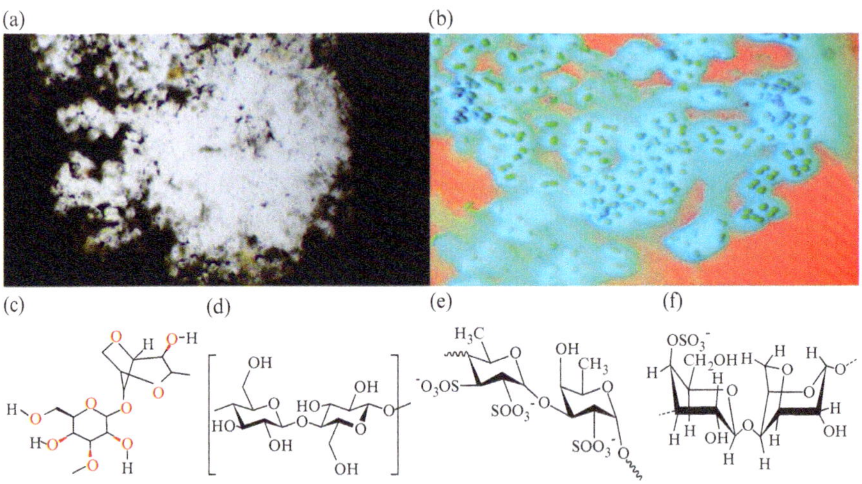

Figure 11.1 Polysaccharide in cyanobacteria/algae. The extracellular polysaccharide of cyanobacteria (a-b), chemical structure of agar (c), cellulose (d), fucoidan (e), and K-carrageenan (f).

Table 11.1 Characteristics of different PSs from algae/cyanobacteria.

Class	Species	PS	Content	Functional group	Crystallinity/Degree of crystallinity	References
Cyanophyceae	*Cyanothece* sp. CCY 0110	EPS	1.8 g/L	Stretching of C–H (2934 cm^{-1}), bending C–H (1423 cm^{-1}), – OH/ – NH (3424 cm^{-1})	XRD peak at $2\theta = 5.83°$ (15.1 A°), 10.13o (8.7 A°), 19.23° (4.6 A°), 21.04° (4.2 A°) and 24.70° (3.6 A°) corresponding to *d*-spacing	[67]
Cyanophyceae	*Limnothrix redekei* PUPCCC 116	EPS	313 µg EPS/ mg protein/ day	N–H stretching (3779 cm^{-1}), O–H stretching (3405 cm^{-1}), stretching of C–H (2928 cm^{-1}), COOH (1657 cm^{-1}), CH3 Bending (1384 cm^{-1}), C–N stretching (1270 cm^{-1}), C–O stretching (1052 cm^{-1})	-	[50]
Cyanophyceae	*Microcystis aeruginosa*	EPS		Phosphorylated compounds bands (1044 cm^{-1}), polysaccharides ring vibration (1164 cm^{-1}), COO– (1410 cm^{-1}; 1618 cm^{-1}), amides C=O (1660 cm^{-1}), RCOOR, C=O (1742 cm^{-1}), C–H at 2936 cm^{-1}, O–H at 3350–3470 cm^{-1}		[68]
Cyanophyceae	*Nostoc muscorum*	EPS	126.73 µg/ mL	-OH (3317.10 cm^{-1}), C–H (2928 cm^{-1}), -CH_3COOH (1606.95 cm^{-1}), C=O (1416.27 cm^{-1}), C–O vibration (1027.04 cm^{-1})	XRD presented broad diffraction peak for amorphous nature of 2θ in the range of 20–30°	[47]

(Continued)

Table 11.1 (Continued)

Class	Species	PS	Content	Functional group	Crystallinity/Degree of crystallinity	References
Cyanophyceae	*Oscillatoria* sp.	EPS	700 mg/g of carbohydrate in EPS	O–H (3400–3448 cm^{-1}), asymmetrical C–H stretching (2924 cm^{-1}), C–H stretching (symmetrical) 2854 cm^{-1}, C–H bending (1400–1380 cm^{-1}), C–O Stretching (1040–1074 cm^{-1}), carboxylate (1636 cm^{-1})		[69]
Chlorophyceae	*Scenedesmus* sp. SB1	EPS	Increase in 1.8-fold (48 mg/L)	OH (3445.34 cm^{-1}), CH (2866.82 cm^{-1}), C=O (1645.76 cm^{-1}), S=O (1375.69 cm^{-1}), C–H (750.29 cm^{-1})	XRD peak at $2\theta = 30.57°$ (2.92 A°), 34.74° (2.58 A°), 34.95° (2.56 A°), 35.72° (2.51 A°), 38.43° (2.34 A°), 40.35° (2.23 A°), 41.69° (2.16 A°), 41.98° (2.15 A°), 47.04° (1.92 A°), 48.75° (1.86 A°), 56.72° (1.62 A°), and 59.02° (1.56 A°) with respective interplanar *d*-spacing.	[70]
Chlorophyceae	*Ulva lactuca*	Cellulose	2.2%	- O–H (3500–3200 cm^{-1}), C–H (3000–2800 cm^{-1}), H–C–H scissor vibration (1420 cm^{-1}), H–C–H tip Vibration (1315 cm^{-1})	XRD peak at $2\theta = 16°$, 22°, and 34°	[71]
Phaeophyceae	*Ascophyllum nodosum*	Cellulose	460.9 mg/g	-OH (3300 cm^{-1}), skeletal vibrations C–O–C (1110 cm^{-1}), stretching C–O (1060 cm^{-1})	26%	[72]

	Ecklonia cava	Cellulose	233.9 mg/g	-OH (3300 cm^{-1}), skeletal vibrations C–O–C (1110 cm^{-1}), stretching C–O (1060 cm^{-1})	23%	[72]
	Laminaria saccharina	Cellulose	513.3 mg/g	-OH (3300 cm^{-1}), skeletal vibrations C–O–C (1110 cm^{-1}), stretching C–O (1060 cm^{-1})	55%	[72]
	Undaria pinnatifida		220.6 mg/g	-OH (3300 cm^{-1}), skeletal vibrations C–O–C (1110 cm^{-1}), stretching C–O (1060 cm^{-1})	22%	[72]
Rhodophyceae	*Gelidium sesquipedale*	Cellulose nanocrystal	Increase in ~14%	-OH (3350–3250 cm^{-1}), C-O-C asymmetric stretching (1157 cm^{-1}); β-glucosidic ether linkages (C–O-C): 1110 and 899 cm^{-1}	Predominance of cellulose I, XRD peak at 2θ =14.8° (110), 16.4° (11–0), 22.4° (002); crystallinity index: 39–87%	[73]

EPS: Extracellular polysaccharides

11.2.2 Polysaccharides in Chlorophyceae

The group chlorophyceae presents morphologically diverse members from unicellular to multicellular forms. Mostly, the predominant PS is cellulose (~70%). The native cellulose is cellulose I (Iα and Iβ allomorphs), which is the foremost cell wall constituent of *Chaetomorpha*, *Cladophora*, *Microdictyon* and *Rhizoclonium* of Cladophorales [37, 51]. A few species of genera like *Boergesenia*, *Siphonocladus*, *Dictyosphaeria*, and *Valonia* (Siphonocladales) are rich in cellulose I. Cellulose Iα is the dominant microfibril in algae. In a few forms of algae, the major component of the cell wall is cellulose with a high degree of crystallinity (Figure 11.1d). *Cladophora* sp. accumulates high cellulose with a crystallinity of 95% appertaining to the thick microfibrils [37]. The upraised crystallinity avoids absorbing more moisture from the environment. For *Cladophora* sp., cellulose density is ~1.64 g/cm^3. An elevated cellulose crystallinity was ascertained in *Apjohnia* sp., *Chaetomorpha* sp., *Rhizoclonium* sp., *Dictyosphaeria* sp., etc. [52] *Bryopsis maxima* (siphonales) contains n-glucan (10%) and (1–3) ~/I–D-xylan (90%). ^{1}H NMR confirmed the ubiquity of cellulose [53]. Also, chlorophyceae produces diverse sulfated PS. The members of chlorophycean walls are composed of cellulose pectin complexes and hydroxyproline-rich glycol proteins [54].

The fibrillar cell walls pertaining to Ulvophyceae and Charophyceae are composed of diverse PS and proteoglycans. The cell walls bear a resemblance to cellulosic plants in *Oedogonium* sp., microalgae *Chlorella*, *Trebouxia*, etc., which possess cellulose. *Chlorella* presents a PS of fibrillar layers, composed of cellulose, hemicellulose, and pectin. The trilaminar cell wall in *Scenedesmus quadricauda* is constituted by cellulose and pectin [55]. A bilaminar cell wall (pectin, glycoprotein, and cellulose) was presented by *Oedogonium bharuchae*. The bilayer structure of the microalgae *Nannochloropsis gaditana* was primarily composed of cellulose (~75%) [39]. A few potential microalgaes for cellulose production are *Coelastrella* sp., *Staurastrum* sp., *Chlorococcum* sp., *Chaetosphaeridium* sp., etc. [55] The macroalgae *Ulva* species presents two types of polysaccharides: cellulose and Ulvans [54]. Pertinent to the absence of lignin, the cellulose extracted is purer than the plants. However, algal cellulose is composed of other sugars like xylose rather than pure ß-1,4 glucan [56]. A yield of ~20% crude cellulose (α-cellulose-18.5%) was derived from *Chaetomorpha aerea* [57]. *Caulerpa imbricata* of the order bryopsidales presented β-cellulose-3.1%.

11.2.3 Polysaccharides in Rhodophyceae

The group rhodophyceae (red algae) contains diverse PSs. Agar and carrageenan are derived from marine forms and are available commercially (Figure 11.1c, 11.1f). The major composition is the sulfated galactans. Floridean starch is the storage carbohydrate. The structural PSs are xylans, mannans, and cellulose. In rhodophyceae, cellulose remains in a moderate amount (<10%). The cell wall (secondary) of coralline red algae is formed by ~22% of cellulose, while primary walls present only ~8% [58]. Cellulose was ascertained in the order Nemaliales, Gelidiales,

Gracilarilariales, Cryptonemiales, Corallinales, Gigartinales, Rhodymeniales, Ceramilaes, etc. [57] Crude cellulose was ascertained in the species of *Galaxaura oblongata* (1.8%), *Pterocladia heteroplatos* (8.7%), *Gracilaria fergusonii* (4.3%), *Grateloupia lithophila* (3%), *Acanthophora dendroides* (3.3%), etc. Cellulose extracted from *Gracilaria dura*, *G. longa*, and *Hypnea musciformis* resembled that of higher plants [59, 60]. Interestingly, the regenerated cellulose, cellulose II, was isolated as of *Bangia fuscopurpurea* during the conchocelis phase [61]. Carrageenan contains up to 40% ester sulfate content. The gel is firm in κ-carrageenan. Carrageenan is credibly modified to provide more thermal stability and hydrophobicity and to provide a barrier for water vapor [62]. Agar, a PS, is extracted from red algae like *Gelidium*, *Ceramium*, *Gracilaria*, *Acantkopeltis*, etc. Agar/agarose are excellent gelling agents with greater strength. The sulfated agar extracted from macroalga *Ahnfeltia plicata* (24.53% yield) presented gel firmness of 1152.50 g/cm^2 [63].

11.2.4 Polysaccharides in Phaeophyceae

The PSs in phaeophyceae (brown algae) are sulfated anionic fucoidans predominated by fucose (Figure 11.1e). Pertinent to uronic acid in PSs, heavy metals like Pb^{2+}, Zn^{2+}, Ni^{2+}, Cd^{2+}, Co^{2+}, etc., can be separated from wastewater [19]. Alginates are ascertained in phaeophyceae, composed of uronic acid. The presence of α-L-guluronic and β-D-mannuronic acid blocks are accountable for the characteristics of the PS. Industrially, alginates are related to food, textile industries, pharmaceuticals, wastewater treatment, etc. [64] Alginates are primarily the salts of Na/Ca. The macroalga viz. *Laminaria* spp., *Durvillea antartica*, *Ecklonia maxima*, and *Macrocystis pyrifera* are commercial producers of alginates [41]. In phaeophyceae, the fibrillar skeleton in the cell wall is the cellulose surrounded by the matrix of alginic acid. The residues after the extraction of alginic acid is algic cellulose, which is similar to cellulose I. Cellulose is detected in the order Dictyotales, Fucales, and Scytosiphonales [57]. The species were *Dictyota divaricata* (5.2%), *Padina pavonica* (6.1%), *Spatoglossum asperum* (2.2%), *Stoechospermum marginatum* (4.5%), *Pocockiella variegata* (9.5%), *Hormophysa triquerta* (~8%), and *Hydroclathrus clathratus* (~10%). Alginates have been characterized from *Fucus vesiculosus* (18.3%), *Sargassum muticum* (25.6%), *Bifurcaria bifurcata*, *Carpodesmia tamariscifolia* (~17%), *Laminaria ochroleuca* (27.5%), *Durvillaea antarctica* (37–52%), etc. [65, 66] Alginate was the sole PS in *Sargassum turbinarioides* (molecular weight: $5.528 \times 105\,g\,mol^{-1}$), collected from Nosy Be (Madagascar), Indian ocean [41].

11.3 Synthesis of Algal Polysaccharide MOFs (ALPs-MOFs)

MOFs are synthesized by solvothermal, electrochemical, microwave-assisted, mechanochemical, sonochemical, direct precipitation methods, etc. [9] The solvothermal method uses thermal energy (353–453 K) for a duration of 48–96 hours. The electrochemical method uses electrical energy (273–303 K) and requires comparatively less time for synthesis (10–30 minutes). Although external energy augmentation for slow evaporation is not required, it takes a long time (seven days to seven months) to develop MOFs [74]. In large scale production, linkers are mixed with a metal source solution for precipitation together with crystallization. This process is followed by filtration, drying, processing, and shaping, while the solvents can be recycled [31].

11.3.1 Alginate-MOFs

To be more appealing, MOFs should be shapable (beads, fibers, and membranes). MOF-alginate composites were derived with zeolitic-imidazolate framework-8 (ZIF-8-), Hong Kong University of Science and Technology (HKUST)-1, ZIF-67, Material Institute de Lavoisier-100 (MIL-100(Fe)) [14]. Bead-shaped MOFs were emanated by adding an aqueous solution (3 mL) of sodium alginate (SA) (10 mg/mL) to 15 mL of copper nitrate trihydrate (30 mg/mL). The composition was left overnight and washed (water/ethanol). The fiber-shaped hydrogels were derived by injecting SA by a syringe. For developing the membrane, alginate gel was casted on a slide, and dropped in Cu^{2+} solution (Figure 11.2). Also, in-situ development of rough, wrinkled MOF beads (ZIF-8 in alginate) were ascertained with a large number of pores [40]. The MOF removed ofloxacin with an efficacy of ~98%. Since powdered MOFs are not suitable for wastewater treatment, fabrication with alginates increases its applicability. Dyes are among the great contaminants in wastewater. Malachite green could be removed by Ca-Alginate MOF. The Ca-Alginate MOF was synthesized pertaining to ion exchange in $CaCl_2$ solution with the calcination temperature 900 °C [75].

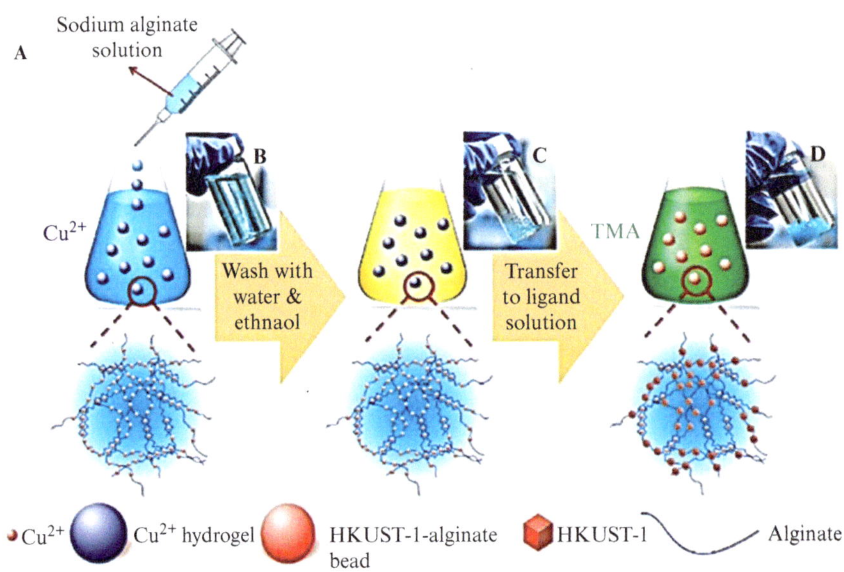

Figure 11.2 Schematic presentation of alginate-MOFs[14]. Re-printed with permission from Copyright © 2016, American Chemical Society.

Phosphonate MOF-incorporated alginate hydrogels were prepared. The alginate hydrogels were derived with Cu/Al [19]. Alginate solution (2.5%) was dropped into each $CuSO_4/AlCl_3$ solution (2.5%) separately. The magnetic stirring was conducted overnight by a syringe pump adjusted to 100 μL/min. The hydrogels (2.48 mg/mL) were finally transferred to ligand–water N, N'-piperazine-bis(methylenephosphonic acid) solution at pH 7.5. The composite MOF-alginate beads (UiO-66@ABs) were prepared by a dripping method [28]. Different volumes of UiO-66 nanoparticles (0.05 g mL^{-1}) were added to SA (5% in water). The composite solution was added to a specially designed capillary tube with a syringe pump while dropping into an aqueous solution of $Ba(CH_3COO)_2$. The composite was incubated for 24 hours for complete cross-linking. The cationic dyes (malachite blue) and heavy metals were removed by fabrication of the MOF-alginate composite beads [20].

The adsorption of contaminants in wastewater can be augmented by the presence of functional groups. The UiO-66/GOCOOH was prepared in a composition of UiO-66 (0.5 g) together with GOCOOH (10–25% wt) in distilled water (5 mL). A homogenous composite was obtained by sonication for two hours. The composite of UiO-66/GOCOOH was added bit by bit to the aqueous solution of SA (0.5g/5 mL) and stirred continuously. Finally, the composition was mixed with $CaCl_2$ (300 mL). The beads were washed with water and fabricated in alginate solution without UiO-66/GOCOOH. The capsules from the biochar alginate can be developed for wastewater treatment (Pb^{2+}) with a recyclability of ten times [76]. The capsules are the mixture of biochar (5%), $CaCl_2$ (1%), and xanthan gum (0.15%) in sodium alginate (0.1%) with Tween 20 (0.1%). The capsule membrane formed as soon as biochar and $CaCl_2$ were added to the alginate. The MOF-polymer composites were generated from calcium alginate, gelatin, and HKUST-1. The MOF composite was useful in adsorbing dyes like rhodamine B, methyl violet, malachite green, auramine O, methylene blue, etc. The suspension was vigorously stirred and dropwise added to a $CaCl_2$ (0.1 M) solution at a constant rate. The beads were washed by distilled water after 24 hours of incubation in the $CaCl_2$ [77]. Also, the aerogel MOFs potentially remove antibiotics from water [78]. SA (2g/100 mL) was dropped into 100 mL distilled water. The solution was dropped into a mixture of $CuCl_2.2H_2O$ (0.05M) and $FeCl_36H_2O$ (0.05M). The alginate-Fe-Cu hydrogel was washed and added to ethanol with 1,3,5-benzenetricarboxylic acid (18 hours). To separate tetracycline, MIL-125(Ti)/MIL-53(Fe)/CNT@Alg beads were fabricated [36]. A 50 mL alginate in water (5%) was prepared by vigorous stirring. An amount of 5 g MIL125(Ti)/MIL53(Fe) and carbon nanotube (CNT) were mixed (5–20%) with distilled water (50 mL), and sonicated (two hours). The solution was added to alginate dropwise and mixed for one hour. Finally, the composite was added to $CaCl_2$ solution (2%) by a syringe under slow stirring for one hour. The beads (273.77 m^2/g) were washed and dried. The alginate-MOF-hydrogel preparation involved several phases. However, smooth-surfaced 3D alginate-MOF hydrogel was prepared in a single step [79].

SA was dissolved in water to form a homogenous solution (2% w/v). Separately, $Co(NO_3)_2$ (40 mg/mL) and 2-methylimidazolate (30 mg/mL) were added to methanol

(100 mL). The composition was incubated and stirred for 24 hours. The alginate-MOF hydrogels were washed (ethanol) and dried at 60°C. Likewise, beads of ZIF8/Ps-MOF hydrogel based on alginate/carrageenan were prepared [80]. The hydrogel performed excellent stress (1.14 MPa), elastic modulus (1.70 MPa), adsorption capacity (2887 mg/g), and water stability (seven days). The MagFePC/SA hydrogel was derived by combining magnetic Fe (MagFe)/porous carbon(PC), SA, agarose, and polyacrylic acid [81]. The hydrogel was an excellent adsorbent of norfloxacin. The hydrogels presented excellent toughness, Fenton activity, more pores, stability, and reusability. Zr-MOF-Alg composites performed methylene blue adsorption (4.361 mg/g) in wide pH (pH1–3) and salinity ranges (up to 1 M NaCl) [82]. The composite was prepared by trimesic acid (450 mg) in water (15 mL). Zr-Alg hydrogels were added, incubated (48 hours) at 50 °C, and then washed with water.

11.3.2 Cellulose-MOFs

Cellulose is an opulent polymer in nature. These are water insoluble, with polymerization of ~15,000 glucose units. Among all the allomorphs, cellulose I with a monoclinic crystal structure presents more thermal stability, which is required for MOF development [21]. In general, microalgae/cyanobacteria are lignin free. *Gelidium elegans* is composed of lignin. Pure cellulose from red algae is extracted by alkali treatment, bleaching together with acid hydrolysis [73]. With a top-down approach, cellulose can be extracted in several forms like fibers (micro/nano) or crystals with ordered structures and hydrogen bonding. Cellulose can be used in different applications pertaining to its degree of crystallinity, polymerization, oxidation, crystalline phase composition, mechanical properties, functional composition, etc. [72, 83, 84] In this regard, cellulose pertinent to macroalgae/seaweed is considered to be capable of withstanding thermal and chemical treatments pertaining to intensified crystallinity leading to more inertness [85].

Apart from different uses like the production of antibacterial substances, energy storage, catalysis, drug delivery, MOFs combined with cellulose has a great potential in bioremediation [86]. Cellulose aerogels are emanated by mixing MOFs with cellulose nanocrystals. Under the process of biomineralization, MOFs are grown directly in cellulose nanocrystals [18]. The adsorption potentiality of UiO-66-NH_2@CA, a cellulose aerogel, was ~89 mg/g [87]. The adsorption propensity of the metals is improved by groups like sulfate, xanthate, carboxylate, etc. The MOFs based on cellulose are hybrids, developed by applying in-situ and ex-situ approaches [21]. The in-situ method was opted for in the construction of ZIF-8@CNF@cellulose foams (Figure 11.3) [22]. In one of the in-situ approaches, MOF metals or the linkers are entrenched into cellulose prior to processing into foams or aerogels, followed by freeze-drying together with casting. The addition of metals is favorable pertaining to the formation of interactions of metals and cellulose [21].

Ex-situ synthesis involves the initial synthesis of MOFs by conventional methods, the addition of MOFs into cellulose, and finally processing it into the required forms. To allow MOFs to grow on the cellulosic surface, nucleation sites (carboxyl/ester/thiol) must be created by chemical processes like oxidation, esterification, or carboxymethylation. Once the process is over, the cellulosic surface is loaded over

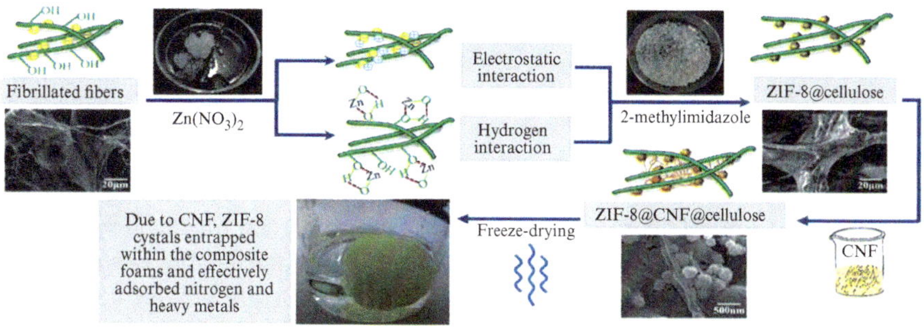

Figure 11.3 Schematic illustration of the preparation of ZIF-8@CNF@cellulose green foam[22]. Re-printed with permission from Copyright © 2018 Elsevier Ltd.

by MOFs with the interaction of metal ions; carboxyl and thiol group together with the amino and thiol group [86].

CNF@Al-MIL-53 aerogels were synthesized from the cellulose of the green alga *Cladophora* sp [88]. The cellulose from *Cladophora* was oxidized with TEMPO, and carboxyls were introduced to the surface of cellulose nanofibrils (CNF). Ionic exchange with Al^{3+} formed an intermediate complex (CNF–COO–Al^{3+}). The fabrication included ion exchange, interfacial synthesis, freeze-drying, and cross-linking. In brief, the aqueous suspension of CNF@Al-MIL-53 nanofibers in different concentrations (0.2–2 mg/L) was treated with an ultrasonicator and sealed in a vessel for air removal. The vessel was enclosed in liquid nitrogen for one hour, and freeze-dried for 48 hours to obtain the aerogels. Separately, terephthalic acid was added to double the molar strengths of NaOH in the water. Aerogels were appended in Na_2BDC, while $Al(NO_3)_3.9H_2O$ H_2O solution was added dropwise. The composition was incubated for ten hours, washed, and freeze-dried.

11.3.3 Agar-MOFs

The adsorption potential of ZIF was enhanced by combining nanosheets of C_3N_4 to form small (14.95 nm) and highly dispersed MOFs integrated with agar aerogels. These hybrid aerogels presented greater adsorption ability with photocatalytic activity [89]. To synthesize the agar hybrid aerogel, the cloudy agar suspension (0.4%) was heated to a clear solution. To the solution, ZIF-8/C_3N_4 heteronanosheets were added in different concentrations (Figure 11.4). The composite was sonicated for two hours and shifted to 6-wells (9.6 cm^2) and 24-wells (2 cm^2) plates. The plates

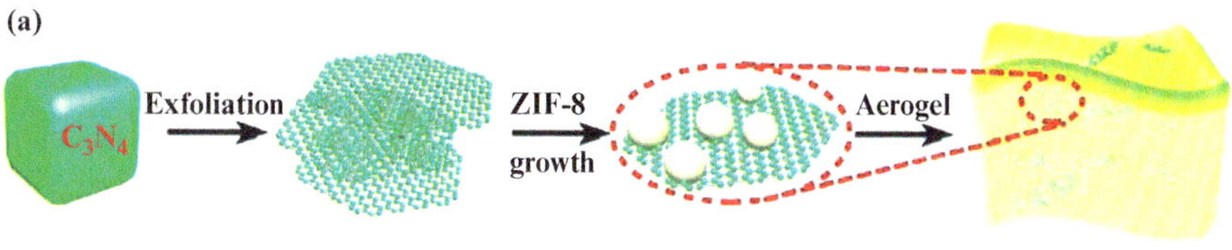

Figure 11.4 Schematic illustration of the fabrication of ZIF-8/C_3N_4 heterostructure [89]. Re-printed with permission from Copyright © 2017, American Chemical Society.

with hybrid aerogels were left undisturbed for two hours, frozen for two hours at -80°C, and freeze-dried.

11.4 Characterization of AlP-MOFs

The physicochemical characterization of MOFs is important for practical applications in treating wastewater (Table 11.2, Figure 11.1). Characteristics like pore size/shape, surface area, spatial distribution, pore network structure, etc., are determined by small-angle X-ray scatterings (SAXS) and wide-angle X-ray scatterings (WAXS). While SAXS are useful for the characterization of nano/micro structured materials (1-200 nm) and in-situ crystal growth/nucleation, WAXS are applicable for large-sized crystalline polymer structures [90]. Additionally, pore size/volume, surface area, and other textural properties can be characterized by nitrogen adsorption and desorption (196°C). The principle works due to nitrogen adsorption on the MOF's surface. The isotherm occurs through adsorption and is linked to the boiling temperature of the liquid nitrogen.

The morphological features of MOFs are determined by scanning electron microscopy (SEM) or transmission electron microscopy (TEM). Although sensitive to high-energy bearing electrons, TEM provides the structural information of MOFs. Damage pertinent to radiation is lessened by cryo-TEM. A smooth surface indicating continuous nucleation was ascertained for cellulose nanofibril@Al-MIL-53 MOFs [88]. It also revealed a pore diameter of ~10μm with interconnected cellular networks. SEM (2D) indicated quasi-spherical particles for MIL125(Ti)/MIL-53(Fe)/CNT@Alg [36]. The alginate microbeads were elongated in shape. SEM requires pre-treatment processes like coating with Au. This limitation is overcome by introducing field emission scanning electron microscopy (FESEM). FESEM is equipped with a field emission gun with extreme focus efficacy and high resolution. SEM/energy-dispersive X-ray (EDX) spectroscopy ascertain the elemental distribution in the MOFs. In HKUST-1-/ZIF-67-alginate composites, C, O, and Cu were found to be uniformly distributed [14]. However, Na/N were absent in HKUST-1-alginate.

Inductively coupled plasma optical emission spectroscopy (ICP-OES) is applied to ascertain elements in traces. However, this process requires prior treatment (dissolution) using H_2SO_4, HNO_3, H_2O_2, HF, etc. To surpass this issue, wavelength dispersive X-ray fluorescence (WDXRF) is directly applied to the MOFs. The elements are excited together, each producing detectable characteristic emissions [74]. Nano-computed tomography (nano-CT), an advanced technique, is pertinent to ascertain the 3D morphology of MOFs using microscopes (X-ray imaging) [19]. The MOF samples are rotated on a stage with an interval of 1° over 180°. CT is reconstructed with angular projection images with the provided software.

MOFs are required to be stable in wastewater. The stability test is performed by the addition of MOFs in water at a known pH. The MOFs are filtered, washed, dried, and reweighed after a fixed time. However, the stability must be affirmed by XRD and N_2 adsorption isotherms.

Table 11.2 Characteristics of polysaccharide (PS) based MOFs.

Ps	MOFs	Functional group	Thermal Stability	Crystallinity	References
Cellulose	UiO-66-NH_2@CA	-	**Cellulose aerogel (CA)**-T_0: 230.5 °C T_{max}: 332.2 °C T_f: 469.1 °C; **UiO-66@CA**- Increases thermal stability, T_0 (by 26.2 °C), T_{max} (by 20.7 °C), T_f (by 60.7 °C); **UiO-66-NH_2@CA**-Further increases thermal stability, T_0 (by 26.7 °C), T_{max} (by 24.6 °C), T_f (by 62.1 °C)	XRD peaks for UiO-66; UiO-66-NH_2; $2\theta = 7.6°, 8.7°, 15.6°, 17.5°$	[87]
Cellulose	ZIF-8@CA [cellulose aerogels/ zeolitic-imidazolate framework hybrids]	**Characteristic cellulose aerogel band at:** 896 cm^{-1}/ 1545 cm^{-1}/ 1645 cm^{-1}/ 2888 cm^{-1} **Characteristic ZIF-8 band at** 679 cm^{-1}, 1310 cm^{-1}, 1430 cm^{-1}	Decomposition temperature range: 120–400 °C; at 550 °C: residues of 10–30 wt%	XRD peaks were indistinguishable for ZIF-8@CA and only ZIF-8 at (011)/ (002)/(112)/(222); Typical diffraction peak for cellulose at (020)	[94]
Cellulose	MCNC@Zn-BTC [magnetic cellulose nanocrystal/MOF composite]	O–H (3415 cm^{-1}), C–H (2902 cm^{-1}), C–C bond vibrations (1160; 1063 cm^{-1}), Fe–O (581 cm^{-1})	-	For Fe3O4, XRD 2θ: = 30.2°/35.5°/43.1°/53.6°/57.1°/ 62.6. For CNC, 2θ =15.7° (101), characteristic for type I cellulose	[93]

(Continued)

Table 11.2 (Continued)

Ps	MOFs	Functional group	Thermal Stability	Crystallinity	References
Cellulose	ZIF-8@CNF@ cellulose foam (CNF, cellulose nanofiber)	C–H (1500– 1300 cm^{-1}), C–O (1300–900 cm^{-1}), -OH stretching and blending vibrations (3200–3600 cm^{-1} and 1627 cm^{-1}), C=N stretching (1584 cm^{-1}), 440 cm^{-1} (440 cm^{-1})	-	For pure cellulose foam and cellulose foam, $2\theta = 22.6°$ and weak peaks at 14.9°/16.5°; cellulose peak at $2\theta = 7.3°$ (011); ZIF-8 peak $2\theta = 10.4°$ (002), 12.7° (112), 14.7° (022), 16.4° (013), and 18.0° (222)	[22]
Alginate	Beads of alginate polymer-MIL (MIL-alg)	In MIL/MIL-alg, presence of C–H (out-of plane bending vibrations pertaing to benzene) absent in alginate	-	For MIL-alg beads, wide peak at 3–5°; 10–11°, the peaks are absent in alginates	[77]
Alginate	MOF-functionalized aerogel (MAlgs)/ MOF-derived carbon/alginate aerogel (CMAlgs)	In carbonized CMAlgs aerogel, C–H(2951 cm^{-1}), C–O-C stretching vibration (930 cm^{-1})	-	For CMAlg-Fe-Cu aerogel, XRD peak at $2\theta = 18.3°$ (111), 30.1° (220), 35.4° (311), 43.0° (400), 56.9° (511), and 62.5° (440).	[78]
Alginate	MIL-125(Ti)/ MIL-53(Fe)/CNT (carbon nanotube)@Alg (alginate) composite microbeads	Fe–O (650 cm^{-1}), Ti–O (735 cm^{-1}), C–H (1101 cm^{-1}), C–O (1291 cm^{-1}), carboxyl (1385 cm^{-1}, 1581 cm^{-1}), vibration C–H (799 cm^{-1}), stretching vibration-C–H (2916 cm^{-1}), C–O stretching (1019 cm^{-1}), OH (3250 cm^{-1})	-	In XRD, CNT peak $2\theta = 25.81°$ (100), 43.28° (101); For MIL-125(Ti)/MIL-53(Fe), $2\theta = 9.4°$; for alginate composite microbeads, all peaks with low intensity	[36]

Alginate+ chitosan	MIP-202/CA (chitosan-alginate) beads composite	NH2 (3380 cm^{-1}, 3495 cm^{-1}); MIP200 (600–1650 cm^{-1})	Stable up to 260 °C, at 100 °C, 12% loss in weight, 60% weight lost at 700 °C	XRD peak for MIP-202, $2\theta = 8.3°$ (111) 9.9° (200), 14° (222), 20° (420), and 21.5° (440); broad peaks for bead composite were observed with low intensity peaks	[95]
Alginate	Composite sheet matrix-alginate immobilized nano zirconium tungestovanadate	Band for zirconium tungestovanadate (600–750 cm^{-1}), – COOH (1630 cm^{-1}- 1419 cm^{-1}), – OH (3500 cm^{-1}), C–N stretching (1172 cm^{-1}), C–C aromatic stretching (1373 cm^{-1})	-	$2\theta = 17.4341$, 43.0067, 62.351, 66.0471, and 78.8197° with small shifting	[96]

The polymorphy and chemical structure of the MOFs are ascertained by Raman spectroscopy. In the Raman test, CMAlg aerogel (carbon/alginate aerogel) presented two resonance peaks (1350;1580 cm^{-1}), referred as peak D (graphitized sp^2 carbon) and peak G (disordered sp^3 carbon). The defects pertaining to carbon were discerned by the I_D/I_G intensity of the peaks [78]. The characterization for CMAlg-Fe (0.84), CMAlg-Cu (0.82), and CMAlg-Fe-Cu (0.83) aerogels confirmed that Fe created more defects and provided more adsorption sites.

Fourier-transform infrared spectroscopy (FTIR) presents the chemical compositions by determining the functional groups present in the MOFs in high resolutions covering a wide range of the infrared spectrum. Additionally, attenuated total reflectance (ATR)-FTIR is advantageous for quick applicability on thick samples. In FTIR, the adsorption appertaining to a heavy metal in wastewater can be analyzed. The bands of the FTIR spectrum were found to be intensified and shifted from 3342 (-OH), 2928 (-CH), 1613 (antisymmetric COO), 1417 (symmetric COO), 1244 (antisymmetric stretching P=O) to 3351/2929/1617/1419/1257 cm^{-1} after adsorption of Li^+ [19]. This indicated the carboxylic/phosphonate groups were effective Li^+ adsorbents. In alginates, the spectra pertinent to FTIR/FT-Raman confirmed the presence of functional groups of O-H, S=O, C–O–C/C-OH stretching nodes [91]. In carrageenan, -OH was detected in 3000–3600 cm^{-1}. The peaks for O=S=O, glycosidic bond, ether group were detected at 1241 cm^{-1}, 1069 cm^{-1}, and 922 cm^{-1}, respectively [92]. The FTIR spectrum of agar ascertained the occurrence of -OH (3200–3400 cm^{-1}), C–H stretching/bending vibration (3000–2800/1400–1200 cm^{-1}), and C–O stretching vibration (1100 cm^{-1}) [63].

The crystallinity of MOFs is characterized by X-ray diffraction (XRD) analysis. The structural damage or collapse in the pores, polymorphic forms, crystallinity decrease, and lattice parameters are ascertained by the broadening of peaks. The crystallinity for UiO-66 was indicated by sharp diffraction peaks at $2\theta=7.4$ and 8.7 at the crystalline plane of (111) and (200), respectively [28]. However, the alginate beads were characterized by broad peaks corresponding $2\theta=13.6$, 25.9, 39.8 with low crystallinity. The characteristic UiO peak was observed in UiO-66@ABs (alginate organic linkers), which was not distinctive when UiO concentration (0.2 mL) decreased.

The stability in disparate temperatures is measured by thermogravimetric analysis (TGA). TGA is performed usually with a ramp of temperatures of 10–20 °C/min (25–500/800 °C) in the N_2 atmosphere.

The magnetic properties are analyzed by vibrating sample magnetometry analysis. The Zn-BTC coated magnetic cellulose nanocrystals lost their magnetization from a level of 61.69 to 41.42 emu/g [93]. An external magnetic field separated the composite within 30 seconds.

11.5 Adsorption Mechanism of AlPs-MOFs

The adsorption process involves different mechanisms like electrostatic interaction, hydrogen bonding, acid-base/π-π/hydrophobic interactions, and selective adsorption (pore/size) [9]. The interaction between the pollutant and the absorbent is either physical adsorption or chemical adsorption. Often, the pollutants are held in the adsorbent pores by Van der Waals forces. Hence, physical adsorption is conducive in terms of regeneration (Figure 11.5) [17]. However, true chemical bonding occurs in chemical adsorption. Electrostatic interaction is the most frequent mechanism. In this mechanism, charge variation is influenced by pH. In AlP-MOFs, functional groups are held by the hydrogen bonding. The hydrophobic interaction is useful for water-insoluble pollutants. It is plausable to hasten pollutant removal by AlPs by enhancing the interactions between adsorbents and pollutants. Also, the raising of accessibility to the active sites augments adsorption.

In phosphonate-MOF-alginate, Li^+ was adsorbed by the synergistic effects of phosphonate and carboxyl groups. The phosphonate group in the MOF was more selective to Li^+ adsorption by the effect of size confinement. As normalized conductivity increases, Li^+ adsorption was inhibited by Alg(Al). The fabricated pMOF@Alg(Al) adsorbed more Mg^{2+} as normalized conductivity decreased (0.7912 to 0.7520). Li^+ easily binded to the negatively charged carboxylic group pertaining to the small radius. However, a high repulsion force of Al^{3+} greatly affected Li^+ due to its small size (Figure 11.6) [19]. The Cr(VI) adsorbed (82%) by MOFs@ABs (alginate beads) was dependent upon several factors like pH, agitation speed, temperature, chromium concentration, bead dosages, etc. [28] Adsorption decreased as pH dropped, which could be attributed to increased H^+ hindering adsorption at the active sites of the MOFs. Also, an unstable interaction may occur as pH drops. Only $Cr_2O_7{}^{-2}$ occured in the solution, and the sites for two positive sites were occupied by $Cr_2O_7{}^{-2}$. In the structure, one corner was shared by two

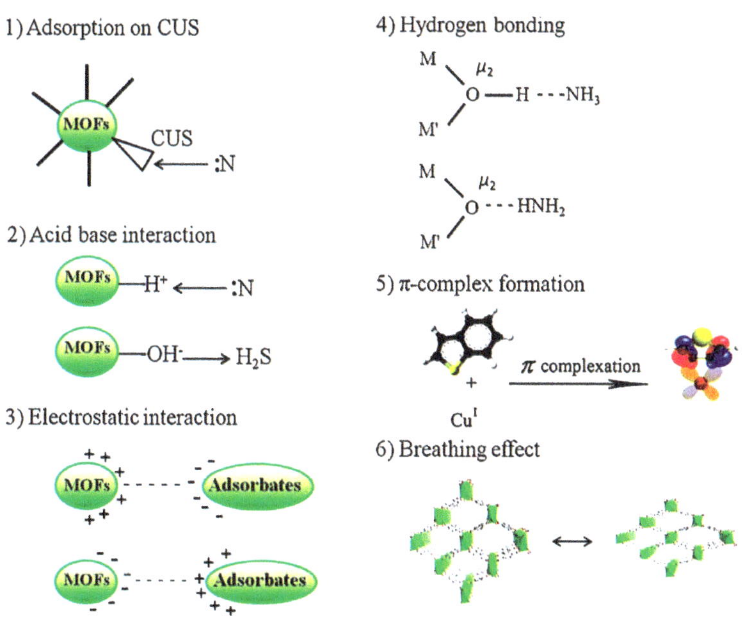

Figure 11.5 Adsorption mechanism of hazardous materials over MOFs [17]. Re-printed with permission from Copyright © 2012 Elsevier.

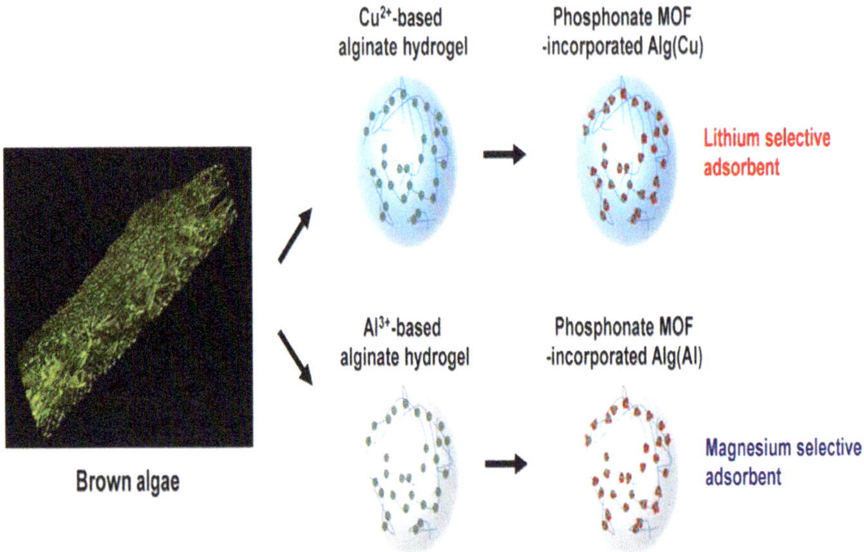

Figure 11.6 Fabrication of Cu^{2+} and Al^{3+}-based alginate hydrogels [19]. Re-printed with permission from Copyright © 2018 Elsevier.

tetrahedra, and adsorption by UiO-66 (0.8 mL)@Abs was lowered. Agitation speed hastened the mass transfer due to the Cr(VI) reaching the active sites of the adsorbents. Initially, the agitation speed (0 to 100rpm) increased adsorption via the endothermic process. Raising the temperature hastened diffusion rates by UiO-66 (0.8 mL) @AB (alginate MOF) across the boundary layer. Although alginate is negatively charged, the positive charge of MOFs turned MOFs@ABs into positive. Therefore, adsorption occurred by electrostatic attraction between the Cr(VI) and the positively charged MOFs. An increase in the pollutant dose (methylene blue and Cu^{2+}) also increased adsorption by the UiO-66/GOCOOH@SA (alginate MOF) [20]. An increase in pH (pH3[76.30]–11[98.76 mg/g]) hastened adsorption on GOCOOH@SA beads. This may pertain to the ample negative charge of the MOFs strengthening the electrostatic interaction. The surface charge of the adsorbents were also controlled by the pH. Often, an increase in bead dose decreases the adsorption due to aggregation. Adsorption of methylene blue dye onto the surface of UiO-66/GOCOOH@SA was attributed to Lewis acid–based, electrostatic, n–π/ π–π interactions. However, adsorption efficiency decreased with a raise in pH (>5). It might be due to decreased mobility of the pollutant Pb^{2+} as exchangeable forms decreased. This pertained to the contact probability between adsorbent and pollutant [76]. Also, the adsorption pertaining to antibiotic ciprofloxacin was augmented by the coordination effect, π-π, hydrogen bonding, and electrostatic attractions [80]. The adsorption by ZIF-8/Ps hydrogel was pseudo-second order. Adsorption is contingent on the geometrical attributes between the MOFs and the dye [97]. Minimum BBR-250 adsorption on MOFs occurred when the pore sizes were small for the dye. Bigger pore sizes presented higher adsorption (73%).

Chemical structures, relative to the amount of shaping agents, are important in increasing adsorption [77]. Tetracycline removal was 43.08 to 86.33% with an increase in dose from 0.01 to 0.08 g. Adsorption is reliant on contact time, functional groups, and surface area. Interestingly, often the adsorption is a combined effect contingent on the hydrophobic interaction and π-π interaction together with H-bond. Active sites increased with adsorbent amount [36]. Adsorption (98.98 mg/g) by ZIF-9/SA/PVA (polyvinyl) was at its maximum at pH 5. The mechanism was pseudo-second order and followed Freundlich isotherm [98]. The hydrogel performed excellent stability and recyclability together with mechanical strength.

Pesticides greatly pollute an aquatic ecosystem. The fabricated ZIF-8-on-Zn2@SA performed excellently at low limits of detection (0.14×10^{-6} M) pertaining to exposed excessive nitrogen sites and chemical stability with greater recovery (>98%) [99]. The $La(OH)_3$@SA/PAM (polyacrylamide) hydrogels were excellent adsorbents pertaining to crystal violet (993.29 mg/g), methylene blue (1610.34 mg/g) and malachite green (3000.08 mg/g). The comprehensive removal was observed under ultraviolet radiation. Adsorption was augmented by the H-bond together with electrostatic interactions [100].

11.6 Regeneration of AlPs-MOFs

Regeneration of MOFs is important for real applications in wastewater treatment (Figure 11.7). The fabricated alginate hydrogels were regenerated by adsorption/desorption experiments without acid. The hydrogels were washed (with ethanol three times and de-ionized water four times) [19]. Recovery without acid treatment is useful, since it reduces the processing cost and environmental impact. Regeneration studies on UiO-66@ABs were conducted initially with a chrome solution (5 mL; 50 mg/L). After eight hours of contact, UiO-66@ABs with adsorbed Cr(VI) were dried and treated with different eluents like deionized water, base (NaOH, 0.1 M), acid (HCl -0.1, 0.5, 1, and 2M), and salts (NaH_2PO_4, 0.1 M/$CaCl_2$, 0.1 M). The experiment was

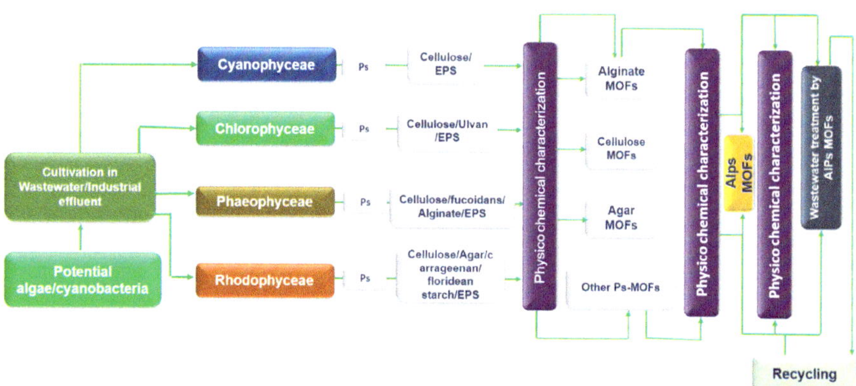

Figure 11.7 Production of AlPs-MOFs from algae/cyanobacteria.

conducted at 25 °C, 500 rpm for 24 hours [28]. The 100 mL solution was added to acidic diphenylcarbazide (DPC) (900 mL)at regular intervals to form a complex [Cr(VI)DPC]. Desorption was at its maximum for HCl at its highest strength (2M). An increase in proton concentration augmented electrostatic attraction with Cr(VI) ions. Salts and bases were not useful in desorptions. The fabricated UiO-66/GOCOOH@SA composite beads were tested for reusability/durability for methylene blue (MB) and Cu^{2+} [20]. The adsorbents were soaked in the eluents ethanol and acid (0.1 M) for regeneration by constant stirring for two hours. The separation of MB (12%) and Cu^{2+} (7%) decreased after the fifth cycle. The microbeads of MIL-125(Ti)/MIL-53(Fe)/CNT@Alg were regenerated (six cycles) with NaOH (0.01 M) solution with constant stirring for one hour [36]. The efficiency pertaining to microbeads reduced from 65.10 (42.02 mg/g) to 53.50% (35.22 mg/g). The alginate-fabricated ZIF-8 hydrogel separated ofloxacin via the processes of solvent exchange and microchannel adsorption up to ~98% (adsorption capacity, ~161 mg/g). The hydrogel continued its adsorption ability (~86%) beyond the fouth cycle [40]. The regeneration of MAlgs/CMAlgs (carbon/alginate) aerogels were accomplished in dimethylformamide and methanol [78]. Tetracycline adsorption by the aerogels was lost 8–22% after the sixth cycle due to the destruction of the binding positions of the adsorbents. The experiment was conducted with tetracycline (30 mg/L) at a ratio of 0.8 mg/mL. The regeneration of ZIF-8/C_3N_4-Agar aerogel was regenerated by photocatalytic activity (visible light) for 50 minutes [89]. The hybrid aerogel was reused (five times) as water was squeezed out. The regeneration process involved simultaneous adsorption pertaining to organic pollutants together with photocatalytic degradation. This degradation progression continued to provide active positions for further adsorption. Cellulose-nanocrystal (CNC)/MOF composite was regenerated by mild acid treatment (HCl) for five cycles [93]. The composite (20 mg) was appended to the acid (20 mL) and shaken for four hours. Once the adsorbed pollutants Pb (II) were removed, the composite was separated by magnetic field, washed, and freeze-dried.

11.7 Conclusion and Future Prospects

Algae and cyanobacteria are potential resources to make PSs useful in MOF production. The PS can be extracted from the algae/cyanobacteria cultivated in wastewater, industrial effluents, and seawater in closed oropen photobioreactors (raceways). The metals should be considerably uniform in the polymeric matrix with good adhesion properties increasing efficiency [16]. Also, for efficient wastewater treatment, the AlPs must be characterized in at least three phases: presynthesis, postsynthesis, and recycling of the MOFs. Geometry, charge, functional groups, subunits with bioactive sites, thermal stability, etc., are the important characteristics to be considered [18]. The EPs from algae/cyanobacteria are suitable for AlP-MOF production by the naturally occurring functional groups, hydrophilicity/hydrophobicity, and thermal stability [46, 47, 101].

The challenges pertaining to PS-MOFs are low symmetry of organic ligands, synthesis of ordered crystalline structures, maintenance of porosity and active opening sites, stability in wastewater, etc. To overcome the limitations, different pre-synthetic and postsynthetic modifications and eliminations have been opted. Modifications like functionalization for electrostatic interactions, acid-base interactions, π-π interactions, and hydrogen bonding should be considered [9]. However, industrial applicability demands technology improvement, strategic scalability, and regeneration with a life cycle assessment. AlP-MOFs are stable, recyclable, sustainable, eco-friendly, and cost-effective materials for wastewater treatment.

References

1 Mekonnen, M.M. and Hoekstra, A.Y. (2016). Four billion people facing severe water scarcity. *Science Advances* 2: e1500323.

2 Sharma, R.K., Gulati, S., and Puri, A. (2014). Green chemistry solutions to water pollution. In: *Water Reclamation and Sustainability* (ed. S. Ahuja), 57–75. Elsevier.

3 Kjellstrom, T., Lodh, M., McMichael, T. et al. (2006). Air and water pollution: burden and strategies for control. In: *Disease Control Priorities in Developing Countries* (ed. D.T. Jamison, J.G. Breman, A.R. Measham et al.), 817–832. Oxford University Press.

4 Koc-Jurczyk, J. (2013). Microbial removal of heavy metals from wastewater. *Inż Ekologiczna* 34: 166–172.

5 Burakov, A.E., Galunin, E.V., Burakova, I.V. et al. (2018). Adsorption of heavy metals on conventional and nanostructured materials for wastewater treatment purposes: a review. *Ecotoxicology and Environmental Safety* 148: 702–712.

6 Li, Y., Xu, Z., and Wang, W.X. (2022). Effective flocculation of harmful algae *Microcystis aeruginosa* by nanoscale metal-organic framework NH-MIL-101(Cr). *The Chemical Engineering Journal* 433: 134584.

7 Kesari, K.K., Soni, R., Jamal, Q.M.S. et al. (2021). Wastewater treatment and reuse: a review of its applications and health implications. *Water, Air, & Soil Pollution* 232: 208.

8 Rodriguez-Narvaez, O.M., Peralta-Hernandez, J.M., Goonetilleke, A., and Bandala, E.R. (2017). Treatment technologies for emerging contaminants in water: a review. *The Chemical Engineering Journal* 323: 361–380.

9 Russo, V., Hmoudah, M., Broccoli, F. et al. (2020). Applications of metal organic frameworks in wastewater treatment: a review on adsorption and photodegradation. *Frontiers of Chemical Science and Engineering* 2: 581487.

10 Chan, Y.J., Chong, M.F., Law, C.L., and Hassell, D.G. (2009). A review on anaerobic-aerobic treatment of industrial and municipal wastewater. *The Chemical Engineering Journal* 155: 1–18.

11 Seow, T.W., Lim, C.K., Norb, M.H.M. et al. (2016). Review on wastewater treatment technologies. *International Journal of Applied Environmental* 11: 111–126.

12 Pohl, A. (2020). Removal of heavy metal ions from water and wastewaters by sulfur-containing precipitation agents. *Water, Air, & Soil Pollution* 231: 503.

13 Qasem, N.A.A., Mohammed, R.H., and Dahiru, U. (2021). Lawal removal of heavy metal ions from wastewater: a comprehensive and critical review. *NPJ Clean Water* 4: 36.
14 Zhu, H., Zhang, Q., and Zhu, S. (2016). Alginate hydrogel: a shapeable and versatile platform for in-situ preparation of MOF-polymer composites. *ACS Applied Materials* 8: 17395–17401.
15 Fonseca, J., Gong, T., Jiao, L., and Jiang, H.L. (2021). Metal-organic frameworks (MOFs) beyond crystallinity: amorphous MOFs, MOF liquids and MOF glasses. *Journal of Materials Chemistry A* 9: 10562–10611.
16 Elrasheedy, A., Nady, N., Bassyouni, M., and El-Shazly, A. (2019). Metal organic framework based polymer mixed matrix membranes: review on applications in water purification. *Membranes* 9: 88.
17 Khan, N.A., Hasan, Z., and Jhung, S.H. (2013). Adsorptive removal of hazardous materials using metal-organic frameworks (MOFs): a review. *The Journal of Hazardous Materials* 244– 245: 444–456.
18 Nadar, S.S., Vaidya, L., Maurya, S., and Rathod, V.K. (2019). Polysaccharide based metal organic frameworks (polysaccharide–MOF): a review. *Coordination Chemistry Reviews* 396: 1–21.
19 Park, S.H., Kim, K., Lim, J.H. et al. (2019). Selective lithium and magnesium adsorption by phosphonate metal-organic framework-incorporated alginate hydrogel inspired from lithium adsorption characteristics of brown algae. *Separation and Purification Technology* 212: 611–618.
20 Eltaweil, A.S., Omer, A.M., El-Aqapa, H.G. et al. (2021). Chitosan based adsorbents for the removal of phosphate and nitrate: a critical review. *Carbohydrate Polymers* 274: 118671.
21 Abdelhamid, H.N. and Mathew, A.P. (2022). Cellulose–metal organic frameworks (CelloMOFs) hybridmaterials and their multifaceted applications: a review. *Coordination Chemistry Reviews* 451: 214263.
22 Ma, S., Zhang, M., Nie, J. et al. (2019). Lightweight and porous cellulose-based foams with high loadings of zeolitic imidazolate frameworks-8 for adsorption applications. *Carbohydrate Polymers* 208: 328–335.
23 Ahsan, M.A., Jabbari, V., Imam, M.A. et al. (2020). Nanoscale nickel metal organic framework decorated over graphene oxide and carbon nanotubes for water remediation. *Science of the Total Environment* 698: 134214.
24 Jun, B.M., Al-Hamadani, Y.A.J., Son, A. et al. (2020). Applications of metal-organic framework-based membranes in water purification: a review. *Separation and Purification Technology* 247: 116947.
25 Mahmoodi, N.M., Oveisi, M., Taghizadeh, A., and Taghizadeh, M. (2020). Synthesis of pearl necklace-like ZIF-8@chitosan/PVA nanofiber with synergistic effect for recycling aqueous dye removal. *Carbohydrate Polymers* 227: 115364.
26 Zhao, R., Ma, T., Zhao, S. et al. (2020). Uniform and stable immobilization of metal-organic frameworks into chitosan matrix for enhanced tetracycline removal from water. *The Chemical Engineering Journal* 382: 122893.
27 Priyadarshini, M., Das, I., and Ghangrekar, M.M. (2020). Application of metal organic framework in wastewater treatment and detection of pollutants: review. *The Journal of the Indian Chemical Society* 97: 507–512.

28 Daradmare, S., Xia, M., Le, V.N. et al. (2021). Metal-organic frameworks/alginate composite beads as effective adsorbents for the removal of hexavalent chromium from aqueous solution. *Chemosphere* 270: 129487.

29 Gul Zaman, H., Baloo, L., Pendyala, R. et al. (2021). Produced water treatment with conventional adsorbents and MOF as an alternative: a review. *Materials* 14: 7607.

30 Vargas, K.G.K. and Qi, Z. Quezada-NovoaV. (2021). Iron (Fe) metal-organic frameworks: a new class of superior and sustainable phosphate adsorbents. *The Journal of Environmental Chemical Engineering* 9: 106849.

31 Julien, P.A., Mottillo, C., and Friščić, T. (2017). Metal-organic frameworks meet scalable and sustainable synthesis. *Green Chemistry* 19: 2729.

32 Musarurwa, H. and Tavengwa, N.T. (2022). Application of polysaccharide-based metal organic framework membranes in separation science. *Carbohydrate Polymers* 275: 118743.

33 Qi, X., Tong, X., Pan, W. et al. (2021). Recent advances in polysaccharide-based adsorbents for wastewater treatment. *Journal of Cleaner Production* 315: 128221.

34 Jia, X., Zhang, B., Chen, C. et al. (2021). Immobilization of chitosan grafted carboxylic Zr-MOF to porous starch for sulfanilamide adsorption. *Carbohydrate Polymers* 253: 117305.

35 Pooresmaeil, M. and Namazi, H. (2021). Chitosan coated Fe_3O_4@Cd-MOF microspheres as an effective adsorbent for the removal of the amoxicillin from aqueous solution. *International Journal of Biological Macromolecules* 191: 108–117.

36 Omer, A.M., Abd El-Monaem, E.M., El-Subruiti, G.M. et al. (2021). Fabrication of easy separable and reusable MIL-125(Ti)/MIL-53(Fe)binary MOF/CNT/Alginate composite microbeads for tetracycline removal from water bodies. *Scientific Reports* 11: 23818.

37 Mihranyan, A. (2011). Cellulose from cladophorales green algae: from environmental problem to high-tech composite materials. *Journal of Applied Polymer Science* 119: 2449–2460.

38 Zhao, C., Li, Z., Li, T. et al. (2015). High-yield production of extracellular type-I cellulose by the cyanobacterium *synechococcus* sp. PCC 7002. *Cell Discovery* 1: 15004.

39 Scholz, M.J., Weiss, T.L., Jinkerson, R.E. et al. (2014). Ultrastructure and composition of the *nannochloropsis gaditana* cell wall. *Eukaryotic Cell* 13: 1450–1464.

40 Yu, R. and Wu, Z. (2022). The adsorption property of in-situ synthesis of MOF in alginate gel for ofloxacin in the wastewater. *Environmental Technology* 2: 1–12.

41 Fenoradosoa, T.A., Ali, G., Delattre, C. et al. (2010). Extraction and characterization of an alginate from the brown seaweed *Sargassum turbinarioides* Grunow. *The Journal of Applied Phycology* 22: 131–137.

42 Nobles, D.R., Romanovicz, D.K., and Brown, R.M., Jr. (2001). Cellulose in cyanobacteria. Origin of vascular plant cellulose synthase? *Plant Physiology* 127: 529–542.

43 Su, H.Y., Lee, T.M., Huang, Y.L. et al. (2011). Increased cellulose production by heterologous expression of cellulose synthase genes in a filamentous heterocystous cyanobacterium with a modification in photosynthesis performance and growth ability. *Botanical Studies* 52: 265–275.

44 Kawano, Y., Saotome, T., Ochiai, Y. et al. (2011). Cellulose accumulation and a cellulose synthase gene are responsible for cell aggregation in the cyanobacterium *thermosynechococcus vulcanus* RKN. *Plant & Cell Physiology* 52: 957–966.

45 Bhatnagar, M. and Bhatnagar, A. (2019). Diversity of polysaccharides in cyanobacteria. In *Microbial Diversity in Ecosystem Sustainability and Biotechnological Applications* (ed. T. Satyanarayana, B. Johri, and S. Das), 447–496. Singapore: Springer.

46 Borah, D., Nainamalai, S., Gopalakrishnan, S. et al. (2018). Biolubricant potential of exopolysaccharides from the cyanobacterium *Cyanothece epiphytica*. *Applied Microbiology and Biotechnology* 102: 3635–3647.

47 Borah, D., Rethinam, G., Gopalakrishnan, S. et al. (2020). Ozone enhanced production of potentially useful exopolymers from the cyanobacterium *Nostoc muscorum*. *Polymer Testing* 84: 106385.

48 Ramachandran, L., Marappa, N., Sethumadhavan, K. et al. (2020). Glycoprotein prompted plausible bactericidal and antibiofilm outturn of extracellular polymers from *nostoc microscopicum*. *Biotechnology and Applied Biochemistry* 191: 284–298.

49 Jindal, N., Singh, D.P., and Khattar, J.I.S. (2013). Optimization, characterization, and flow properties of exopolysaccharides produced by the cyanobacterium *Lyngbya stagnina*. *Journal of Basic Microbiology* 53: 1–11.

50 Khattar, J.I.S., Singh, D.P., Jindal, N. et al. (2010). Isolation and characterization of exopolysaccharides produced by the cyanobacterium *Limnothrix redekei* PUPCCC. *Biotechnology and Applied Biochemistry* 162: 1327–1338.

51 Nicolai, E. and Preston, R.D. (1952). Cell-wall studies in the Chlorophyceae. I. A general survey of submicroscopic structure in filamentous species. *Proceedings of the Royal Society London B* 140: 244–274.

52 Richmond, P.A. (1990). Occurrence and functions of native cellulose. In: *Biosynth Biodegrad Cellul* (ed. C.H. Haigler), 5–23. CRC Press.

53 Maeda, M., Fukushi-Fujikura, Y., and Otsuru, O. (1990). Cellulose in the cell wall of the siphonous green alga, *Bryopsis maxima*. *Carbohydrate Research* 207: 91–99.

54 Domozych, D.S., Ciancia, M., Fangel, J.U. et al. (2012). The cellwalls of green algae: a journey through evolution and diversity. *Frontiers in Plant Science* 3: 1–7.

55 Zanchetta, E., Damergi, E., Patel, B. et al. (2021). Algal cellulose, production and potential use in plastics: challenges and opportunities. *Algal Research* 56: 102288.

56 Baldan, B., Andolfo, P., Navazio, L. et al. (2001). Cellulose in algal cell wall: an "in situ" localization. *The European Journal of Histochemistry* 45: 51–56.

57 Siddhanta, A.K., Kumara, S., Mehta, G.K. et al. (2013). Cellulose contents of some abundant Indian seaweed species. *Natural Product Communications* 8: 497–500.

58 Martone, P.T., Janot, K., Fujita, M. et al. (2019). Cellulose-rich secondary walls in wave-swept red macroalgae fortify flexible tissues. *Planta* 250: 1867–1879.

59 Usov, A.I. (2011). Polysaccharides of the red algae. *Advances in Carbohydrate Chemistry and Biochemistry* 65: 115–217.

60 Toffanin, R., Knutsen, S.H., Bertocchi, C. et al. (1994). Detection of cellulose in the cell wall of some red algae by 13C NMR spectroscopy. *Carbohydrate Research* 262: 167–171.

61 Gretz, M.R., Aronson, J.M., and Sommerfeld, M.R. (1980). Cellulose in the cell walls of the Bangiophyceae (Rhodophyta). *Science* 207: 779–780.
62 Tye, Y.Y., Abdul Khalil, H.P.S., Cheng, Y.K. et al. (2018). Preparation and characterization of modified and unmodified carrageenan-based films. *The Wood and Biofiber International Conference (WOBIC 2017) IOP Publishing IOP Conf. Series: Materials Science and Engineering* 368: 012020.
63 Zhang, Y., Fu, X., Duan, D. et al. (2019). Preparation and characterization of agar, agarose, and agaropectin from the red alga *Ahnfeltia plicata*. *Journal of Oceanology and Limnology* 37: 815–824.
64 Hecht, H. and Srebnik, S. (2016). Structural characterization of sodium alginate and calcium alginate. *Biomacromolecules* 17: 2160–2167.
65 Rioux, L.E., Turgeon, S.L., and Beaulieu, M. (2007). Characterization of polysaccharides extracted from brown seaweeds. *Carbohydrate Polymers* 69: 530–537.
66 Belattmania, Z., Kaidi, S., and Atouani, S.E. (2020). Isolation and FTIR-ATR and ^{1}H NMR characterization of alginates from the main alginophyte species of the atlantic coast of Morocco. *Molecules* 25: 4335.
67 Mota, R., Guimarães, R., Büttel, Z. et al. (2013). Production and characterization of extracellular carbohydrate polymer from *Cyanothece* sp. *CCY 0110*. *Carbohydrate Polymer* 92: 1408–1415.
68 Kuang, X., Shao, J., Chen, A. et al. (2016). Effects of bloom-forming cyanobacterial extracellular polymeric substances on the adsorption of cadmium onto kaolinite: behaviors and possible mechanisms. *SpringerPlus* 5: 542.
69 Parikh, A. and Madamwar, D. (2006). Partial characterization of extracellular polysaccharides from cyanobacteria. *Bioresource Technology* 97: 1822–1827.
70 Angelaalincy, M., Senthilkumar, N., Karpagam, R. et al. (2017). Enhanced extracellular polysaccharide production and self sustainable electricity generation for PAMFCs by *Scenedesmus* sp. *SB1*. *ACS Omega*. 2: 3754–3765.
71 Wahlström, N., Edlund, U., Pavia, H. et al. (2020). Cellulose from the green macroalgae *Ulva lactuca*: isolation, characterization, optotracing, and production of cellulose nanofibrils. *Cellulose* 27: 3707–3725.
72 Bogolitsyn, K., Parshina, A., and Aleshina, L. (2020). Structural features of brown algae cellulose. *Cellulose* 27: 9787–9800.
73 El Achaby, M., Kassab, Z., Aboulkas, A. et al. (2018). Reuse of red algae waste for the production of cellulose nanocrystals and its application in polymer nanocomposites. *International Journal of Biological Macromolecules* 106: 681–691.
74 Bedia, J., Muelas-Ramos, V., Peñas-Garzón, M. et al. (2019). A review on the synthesis and characterization of metal organic frameworks for photocatalytic water purification. *Catalysts* 9 (52): 1–43.
75 Mahreni, M., Ramadhan, R.R., Pramadhana, M.F. et al. (2022). Synthesis of metal organic framework (MOF) based Ca-Alginate for adsorption of malachite green dye. *Polymer Bulletin* doi: 10.1007/s00289-022-04086-5.
76 Do, X.H. and Lee, B.K. (2013). Removal of Pb^{2+} using a biochar alginate capsule in aqueous solution and capsule regeneration. *The Journal of Environmental Management* 131: 375e382.

77 Tanimoto, Y. and Noro, S. (2021). Influence of carbohydrate polymer shaping on organic dye adsorption by a metal-organic framework in water. *RSC Advances* 11: 23707.

78 Kong, Y., Han, K., Zhuang, Y., and Shi, B. (2022). Facile synthesis of MOFs-templated carbon aerogels with enhanced tetracycline adsorption performance. *Water* 14: 504.

79 Yuan, Z., Kong, Y., Wang, X., and Shi, B. (2019). Novel one step preparation of 3D alginate based MOFs hydrogel for water treatment. *New Journal of Chemistry* 43: 7202–7208.

80 Chai, Y., Zhang, Y., Wang, L. et al. (2022). In situ one-pot construction of MOF/hydrogel composite beads with enhanced wastewater treatment performance. *Separation and Purification Technology* 295: 121225.

81 Liu, D., Gu, W., Zhou, W. et al. (2022). Magnetic Fe/carbon/sodium alginate hydrogels for efficient degradation of norfloxacin in simulated wastewater. *Journal of Cleaner Production* 369: 133239.

82 Klein, S.E., Sosa, J.D., Castonguay, A.C. et al. (2020). Green synthesis of Zr-based metal-organic framework hydrogel composites and their enhanced adsorptive properties. *Inorganic Chemistry Frontiers* 7: 4813–4821.

83 El-Sakhawy, M. and Hassan, M.L. (2007). Physical and mechanical properties of microcrystalline cellulose prepared from agricultural residues. *Carbohydrate Polymers* 67: 1–10.

84 Kazharska, M., Ding, Y., Arif, M. et al. (2019). Cellulose nanocrystals derived from *Enteromorpha prolifera* and their use in developing bionanocomposite films with water soluble polysaccharides extracted from *E. prolifera*. *International Journal of Biological Macromolecules* 134: 390–396.

85 Bar-Shai, N., Sharabani-Yosef, O., and Zollmann, M. (2021). Seaweed cellulose scaffolds derived from green macroalgae for tissue engineering. *Scientific Reports* 11: 11843.

86 Liu, X., Xiao, Y., Zhang, Z. et al. (2021). Recent progress in metal organic frameworks@cellulose hybrids and their applications. *Chinese Journal of Chemistry* 39: 3462–3480.

87 Lei, C., Gao, J., Ren, W. et al. (2019). Fabrication of metal-organic frameworks@cellulose aerogels composite materials for removal of heavy metal ions in water. *Carbohydrate Polymers* 205: 35–41.

88 Zhou, S., Apostolopoulou-Kalkavoura, V., da Costa, M.V.T. et al. (2020). Elastic aerogels of cellulose nanofibers@Metal–organic frameworks for thermal insulation and fire retardancy. *Nano-Micro Letters* 12: 9.

89 Zhang, W., Shi, S., Zhu, W. et al. (2017). Agar aerogel containing small-sized zeolitic imidazolate framework loaded carbon nitride: a solar-triggered regenerable decontaminant for convenient and enhanced water purification. *ACS Sustainable Chemistry & Engineering* 5: 9347–9354.

90 Goesten, M.G., Stavitski, E., Juan-Alcañiz, J. et al. (2013). Small-angle X-ray scattering documents the growth of metal-organic frameworks. *Catalysis Today* 205: 120–127.

91 Pereira, L., Sousa, A., Coelho, H. et al. (2003). Use of FTIR, FT-Raman and ^{13}C-NMR spectroscopy for identification of some seaweed phycocolloids. *Biomolecular Engineering* 20: 223–228.

92 Mahmood, W.A.K., Khan, M.M.R., and Yee, T.C. (2014). Effects of reaction temperature on the synthesis and thermal properties of carrageenan ester. *Journal of Physical Science* 25: 123–138.

93 Wang, N., Ouyang, X.K., Yang, L.Y., and Omer, A.M. (2017). Fabrication of a magnetic cellulose nanocrystal/metal-organic framework composite for removal of Pb(II) from water. *ACS Sustainable Chemistry & Engineering* 5: 10447–10458.

94 Bo, S., Ren, W., Lei, C. et al. (2018). Flexible and porous cellulose aerogels/zeolitic imidazolate framework (ZIF-8) hybrids for adsorption removal of Cr(IV) from water. *Journal of Solid State Chemistry* 262: 135–141.

95 Diab, K.E., Salama, E., Hassan, H.S. et al. (2021). Bio-zirconium metal-organic framework regenerable bio-beads for the effective removal of organophosphates from polluted water. *Polymers* 13: 3869.

96 Elkady, M.F., Hassan, H.S., and El-Sayed, E.M. (2015). Basic violet decolourization using alginate immobilized nanozirconium tungestovanadate matrix as cation exchanger. *Journal of Chemistry, (2015)* 1–10. DOI: https://doi.org/10.1155/2015/385741

97 Au, V.K.M. (2020). Recent advances in the use of metal-organic frameworks for dye adsorption. *Frontiers in Chemistry* 8: 708.

98 Zhang, G., Chen, H., Yang, G. et al. (2022). Preparation of in situ ZIF-9 grown on sodium alginate/polyvinyl alcohol hydrogels for enhancing Cu (II) adsorption from aqueous solutions. *Journal of Inorganic and Organometallic Polymers* doi: 10.1007/s10904-022-02463-1.

99 Jia, W., Fan, R., Zhang, J. et al. (2022). Smart MOF-on-MOF hydrogel as a simple rod-shaped core for visual detection and effective removal of pesticides. *Small* 18: 2201510.

100 Qian, D., Bai, L., Wang, Y. et al. (2019). A bifunctional alginate-based composite hydrogel with synergistic pollutant adsorption and photocatalytic degradation performance. *Industrial & Engineering Chemistry Research* 58: 13133–13144.

101 Borah, D., Gopalakrishnan, S., and Nooruddin, T. (2021). Carbohydrate biolubricants from algae and cyanobacteria. *The Journal of Polymers and the Environment* 29: 3444–3458.

12

Ecological Risk Assessment of Heavy Metal Pollution in Water Resources

Swati Singh[1,*] *and K. V. Suresh Babu*[2]

[1]*CSIR-National Botanical Research Institute, Lucknow - 226001, India*
[2]*University of Cape Town, Cape Town, South Africa*
* *Corresponding author*

12.1 Introduction

Heavy metal pollution in the global environment is a serious problem due to its potential toxicity and accumulation in natural habitats [1]. These arise from many natural and man-made sources. In ecosystems, metal contamination may result directly from atmospheric precipitation, environmental weathering, or the release of domestic, municipal, agricultural, or industrial waste products [2, 3]. The term heavy metal refers to any metallic component that has a comparatively high density of more than 5 g/cm3 [4]. Many industrial, agricultural, medical, domestic, and technical tenders have led to their extensive release in the surroundings, raising worries over their possible effects on human health and the environment [5]. Certain metals are required in very small amounts for metabolic function in the human body while others cause several serious diseases. These metallic elements are measured systemic toxins that are recognized as causing damage to multiple organs, even at a modest intensity of exposure. The heavy metal toxicity is dependent on numerous aspects comprising the exposure route, dose, and chemical composition, as well as the age, genetics, and nourishing position of a contact person [6]. Heavy metals have naturally occurring origins and originate in the Earth's crust. Most environmental contamination and human exposure results from anthropogenic actions such as coal-mining activity, industrial production, and various other agricultural uses of metal-containing composites [7]. Environmental pollution can also occur through atmospheric deposition, metal corrosion, and leaching of various metals from water resources into soil and groundwater [8]. Natural events such as volcanic eruptions and soil weathering also contribute considerably to heavy metal pollution. Industrialized sources consist of metal handling in factories, the burning of coal in thermal power plants, petroleum combustion, nuclear power stations, and wood preservation and its processing [9].

Metal Organic Frameworks for Wastewater Contaminant Removal, First Edition. Edited by Arun Lal Srivastav, Lata Rani, Jyotsna Kaushal, and Tien Duc Pham.

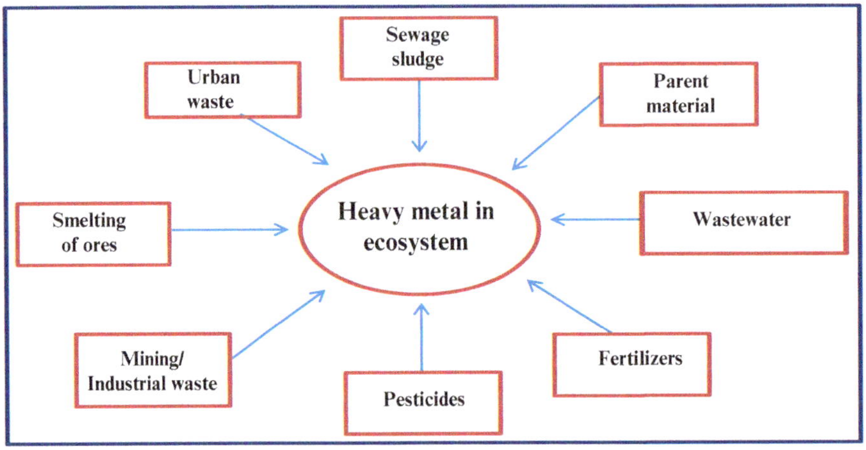

Figure 12.1 Heavy metals system by natural and anthropogenic activities.

Heavy metals come into the aqueous system by natural and anthropogenic activities (Figure 12.1). Throughout the past two decades, heavy metals have resulted from various anthropogenic events that have contributed to the damage of natural sources and the environment [10]. Water pollution is a major environmental and socio-economic problem due to these rudiments. Several water quality management approaches have been executed to protect water ecosystems from pollution [11; 12]. The heavy metals appraised in this paper embrace several important metals and metaloids of natural and biological toxicity, such as lead (Pb), zinc (Zn), iron (Fe), copper (Cu), cadmium (Cd), mercury (Hg), arsenic (As), chromium (Cr), manganese (Mn), and nickel (Ni) [13; 14]. To recognize the effect of heavy metal pollution on water, this chapter will look at its primary sources, sites of impact, and assessment systems.

12.2 Natural and Anthropogenic Sources of Heavy Metals in the Environment

Non-essential heavy metals were generated in the water system through an extensive range of procedures and routes by several sources [10, 15]. The natural sources embrace dry and wet deposits of atmospheric salts, water, soil, and rock exchanges. Anthropogenic sources include the speedy expansion of industrialization [16, 17].

The existence of heavy metals in water from natural sources depends on the native geology, hydrology, and different geochemical features of the aquifer [18]. The primary source of the constituents contaminating the water body is the permanent deposit of rock such as limestone, dolomite, shale, and sandstone. The interface of water with igneous rocks such as granite, gabbro, nepheline syenite, basalt, andesite, and ultramafic also underlies some of the major components [19]. Typical natural resources or elements that exaggerate the level of elements upon dissolution are magnetite, hematite, goethite, siderite (Fe); calcite, cuprite, malachite, azurite (Cu); chromite (Cr); kaolinite, montmorillonite, arsenic trioxide, orpiment, arsenopyrite

(As); calamine, smithsonite (Zn); and pyrolusite, rhodochrosite (Mn) [20]. As also present in the sulfide-containing mineral deposits, particularly those linked with gold minerals and hydrous iron oxide ore [21, 22]. Some minor elements like Cd, Co, Mn are found in the Earth's crust along with other minerals [23]. In addition, Ni, Pb, and Hg are dumped in the water system by the dry or wet fall of atmospheric aerosols designed by wind-blown dust, volcanic releases, and forest fires [24, 25].

The rapid speed of urbanization and industrialization reduces the carrying capacity of water rapidly [26]. The absorption levels of Hg in aquatic systems have mostly increased due to agriculture and human activities such as plowing and logging, domestic sewage release, atmospheric deposition from solid waste ignition, coal and oil combustion, pyrometallurgical processes, and excavating deeds [27]. Surface runoff from snow or rain transports Hg-polluted soil to neighboring water systems [28]. Industrial development, social change, and rapid race for expansion suddenly reduce the carrying capacity of water. Industrial developments that are accountable for contaminating water with Hg comprise construction of chlor alkalis, fluorescent lamps, batteries, thermometers, and electronic switches. The chemical industry has been one of the major intentionally polluting sources of mercury in the domain [29, 30]. An anthropogenic source of Ni is from corrosive metal pipes and containers [31]. Lead in the water body comes from the precipitation of mixtures such as paint, petrol, and aerosols formed from high-temperature manufacturing procedures such as coal smelting, combustion, and production.

Cd arrives into water bodies through industrialized expulsion and stimulated cylinder analysis [32]. Cadmium metal is also found in phosphate fertilizers and acts as one of the major substances of contaminating proxy in water bodies [33]. Cu is commonly present in drinking water from copper pipes, industrial waste, and additives intended to control algae growth [34]. Fe and Mn in water bodies comes from sewages, landfill leachate, sewage, and acid-mining drainage [35]. Anthropogenic causes of Cr contain the industrial wastewater discharge from various types of industries, for example, metallurgy (alloys), refractory (chrome), chemical, tanning, and others. Aquatic ecosystem sources include non-ferrous mining, combustion of fossil fuels and wastes, mineral extraction, pesticides, and poultry and swine feed additives [36]. Other sources include municipal and industrial waste (Kerr & Craw, 2021), wood stabilizers, and sweltering of arsenious gold ores. Zn in the water system likely arises due to ore-mining events. Various heavy metals in the marine system are added through acid mine drainage (AMD) at high levels, which is one of the most perceptible environmental pressures of mining productivity [37]. AMD results from the corrosion of sulfide-bearing minerals open to enduring circumstances, resultant in acidic pH and high levels of dissolved metals (e.g., As, Cd, Cu, Zn) and ions (e.g., sulfates) [38].

12.3 Impacts of Heavy Metal Pollution

Heavy metals move in the human body through drinking water attained from numerous causes such as reservoir, rivers, wells, ponds, etc. The presence of metals in drinking water in excess of the suggested limits approved by numerous national and international organizations (Table 12.1) has been reported to be a health hazard

Table 12.1 Drinking water standards.

S.No	Heavy metals	WHO (μg/L)	USEPA (μg/l)	EU,1998 (μg/l)	BIS (ISO: 10500,2012) (μg/l)
1	Mn	400	50	50	100
2	Ni	70	100	20	20
3	As	10	10	10	10
4	Cr	50	100	50	50
5	Hg	6	2	1	1
6	Cu	2000	1300	2000	50
7	Cd	3	5	5	3
8	Zn	NGL**	5000	NM*	5000
9	Pb	10	15	10	10
10	Fe	NGL**	300	200	300

NM* NM stands for Not Mentioned NGL stands for No Guideline.
NGL** No Guideline.
World Health Organization (WHO); United States Environment Protection Agency (USEPA); Europen Standards (EN); Bureau of Indian Standards (BIS).

(Figure 12.2). Heavy metal pollution is colorless and odorless, so it is challenging to notice. It does not obviously harm the environment in the short term [39]. Nevertheless, when it surpasses environmental easiness, or when various environmental circumstances change, heavy metals in soil can become active and cause serious ecological damage [40]. So heavy metal pollution is generally referred to as a chemical time bomb [41, 42]. If atmospheric air and water bodies are polluted, the pollution problem can certainly be reversed by weakening and self-purification after closing the causes of pollution [43]. However, it is quite difficult to practice dilution or self-purification methods to remove heavy metal contamination and improve soils [44]. Some soils polluted with heavy metals may take a hundred or two hundred years to improve or recover. Consequently, heavy metal contamination requires a moderately high cost of treatment and a relatively long remediation cycle [45, 46]. In previous years, soil pollution was mostly caused by a particular heavy metal. However, in current instances, more cases have been found to be instigated by a variety of heavy metals [47, 48]. Complex adulteration instigated by different types of heavy metals will always increase the individual contamination by heavy metals.

Ni and Hg are hazardous and are a reason for DNA (deoxyribonuclic acid) damage. Ni also causes allergies, hair loss, anemia, and other toxicity in the system [49]. Pb, one of the most common heavy metals with normally anticipated limits, is a metabolic poison and enzyme inhibitor [50]. It can also harm nerve connections and causes blood-related disorders. In addition, the biochemical properties of lead interfere with heme synthesis, which leads to hemorrhagic destruction [51]. Fe and Mn at low concentrations is desirable for enzyme action [52], but at very high concentration, it assembles in muscles and the liver and mainly disturbs the central nervous structure [53]. Cr, known as a dangerous toxic agent, can cause dermatitis and ulcers on human

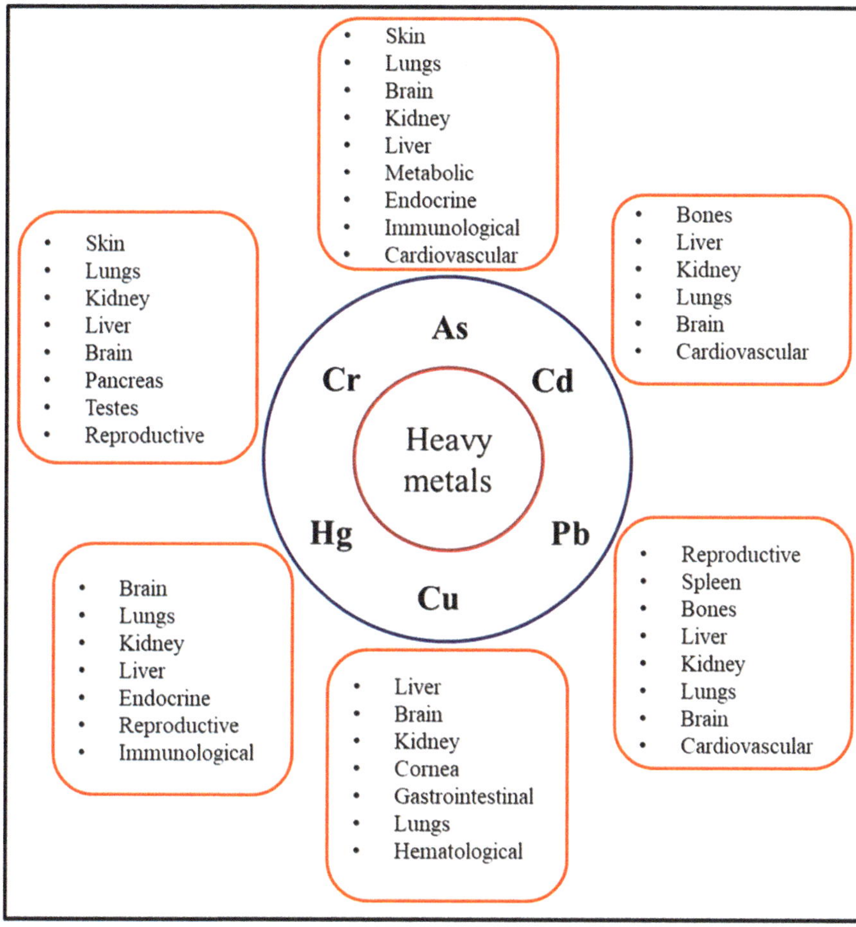

Figure 12.2 Toxic effects of heavy metals.

skin. Prolonged exposure to it can lead to liver, kidney, circulatory, and nervous-tissue damage. Higher concentrations can cause respiratory complications, hormonal changes, skin lesions, hyperpigmentation, and chronic renal failure [54]. Zn is required in very low concentrations to act as a catalyst in the enzyme activity of the living system, but it accumulates in the liver [55]. Chronic health effects of Zn include birth deficiencies, nervous system disorders, organ damage, cancer, and immune-system damage [56]. Cd, categorized as a toxic trace element, accumulates with age, especially in the kidneys. It is known to cause cancerous disease and cardiac illnesses. Industrially polluted consumption water causes kidney disease. Long-term contact can change calcium in the bones [57] and damage the kidneys [58]. Cd metallothionein (a protein that binds to additional essential metals to render them unavailable) may interfere with the body's capability to regulate Zn and Cu concentrations, which causes an increase in urinary zinc [59]. Long-term exposure to Cu or high concentrations can cause nervous system disorders and chronic diseases such as liver and kidney failure. Elevated levels of Cu in drinking water can also be the reason for various other diseases like vomiting, abdominal pain, nausea, diarrhea, and anemia [60].

12.4 Water Quality Assessment Using Pollution Indices

Several organizations and researchers are currently placing emphasis on water quality assessment to lessen the influence of pollutants on social wellbeing and the surroundings. The hydrology-based study shows the development of several methods to identify the origin of source and overall admittance to water quality [61]. Widespread literature surveys evidence that statistical techniques (factorial analysis) and heavy metal pollution indices (geoaccumulation index) are the most expedient and operative methods of water quality assessment. Various environmental indices, namely, the geoaccumulation index, contamination factor, and pollution loading index, are widely used in heavy metal quantification [62] (Table 12.2). The average interior crustal values are generally used as background values for all index calculations [63].

Table 12.2 Heavy metal pollution index used in environmental pollution monitoring.

S.N	Pollution Index used	Study area	Area	References
1	Heavy metal pollution index (HPI), heavy metal evaluation index (HEI), degree of contamination (mCd), enrichment factor (EF), Muller's geoaccumulation index (Igeo), and the anthropogenic enrichment assessment (IAP)	Greek	Water	[68]
2	Pollution index of groundwater (PIG), ecological risk index (ERI), and hierarchical cluster analysis (HCA)	Ojoto, suburban southeast Nigeria	Ground water	[69]
3	Water pollution index (WPI)	West Bengal, India	Water	[70]
4	HPI	Bangladesh	Water	[71]
5	Water quality index (WQI) and synthetic pollution index (SPI)	Pakistan	Water	[72]
6	Igeo	Marinduque, Philippines	Soil	[73]
7	Igeo	Indonesia	Water	[74]
8	EF and Igeo	River Hornad, Slovakia	Water	[75]
9	Igeo, EF, and human health risk index	Swat river, Pakistan	Water	[76]
10	EF, contamination factor (CF), and Igeo	Iran	Water	[77]
11	EF, CF, and Igeo	Rybnik water reservoir, Poland	Water	[78]
12	EF and CF	Tamaki Estuary, Auckland, New Zealand	Water	[79]

Table 12.2 (Continued)

S.N	Pollution Index used	Study area	Area	References
13	Igeo	Dongting Lake, Central China	Water	[80]
14	CF, EF, and Igeo	Thermaikos Gulf, N. Greece	Water	[81]
15	Igeo and CF	Poland	Water	[82]
16	Igeo	Liwa Oasis (UAE)	Soil	[83]
17	EF, CF, Igeo, and pollution load index (PLI)	Western fringes of the Niger Delta	Forcados River and adjoining soils	[84]
18	Igeo, CF, contamination degree (CD), PLI, EF, and the potential ecological risk index (PERI)	Bangladesh	Water	[85]
19	EF, CF, Igeo, and PERI	Turkey	Water	[86]
20	CF, Igeo, EF, CD, pollution index, and modified pollution index	Southwestern Nigeria	Soil	[87]
21	Igeo, potential ecological risk	High Moulouya, Morocco	Water	[88]
22	Igeo, EF, and PERI	Brazil	Water	[89]
23	Igeo, CF, ecological risk index (ERI), and PERI	Poland	Soil	[90]
24	Igeo, CF, and degree of contamination (DC)	Bizerte Lagoon, Northern Tunisia	Soil	[91]
25	Igeo, contamination degree (Cdeg), CF, modified contamination degree (mCdeg), Nemerow pollution index (PI), PLI, and PERI	Punjab, India	Soil	[92]

12.4.1 Heavy Metal Pollution Index (HPI)

The HPI is used to evaluate overall water quality. The following model (Equations 12.1 and 12.2) calculates the index as follows:

$$Qi = \sum_{i=1}^{n} \frac{\left[\boldsymbol{Mi}(-)\boldsymbol{Ii}\right]}{\boldsymbol{Si} - \boldsymbol{Ii}} \times 100 \tag{12.1}$$

Mi: Measured parameter value. Ii: Ideal parameter or uppermost desirable value. Si: Standard or acceptable parameter value. The (–) sign signifies statistical variance of the two standard values discounting the algebraic sign [64].

$$\boldsymbol{Heavy\ Pollution\ Index(HPI)} = \sum_{i=1}^{N} \frac{\boldsymbol{Wi} * \boldsymbol{Qi}}{\boldsymbol{Wi}} \tag{12.2}$$

Where Qi is a sub index calculation, Wi is the parameter weight assigned to the ith value [65].

Samples are weighed based on significance parameters that are dispersed between 0 (zero) and 1 (one) [66]. The value for each element is also considered as inversely proportional to the standard. Water quality is classified on the basis of the heavy metal pollution index as: low heavy metal pollution (HPI < 100), heavy metal pollution risk at threshold (HPI = 100), and high heavy metal pollution (HPI > 100) [67].

12.4.2 Statistical Technique

Factor analysis (FA) is a multivariate arithmetic system, which subtracts and categorizes a huge number of metals and in turn analyzes the heavy metals origin source in the water ecosystem. The complex leading steps in factor analysis are the extraction of features from a large data set and the selection of revolving methods [93]. The purpose of abstraction is to decrease a huge number of metals into features. The most common technique to extract factors is principal component analysis (PCA) [94]. The principal component method comprises the group of eigenvalues and eigenvectors (loading factor or weighting) from a four-sided matrix (either covariance or correlation) composed of the dataset [95]. The eigenvector with the highest eigenvalue is called the principal component. Statistically, they are uncorrelated or orthogonal to those acquired by reproducing the original correlated heavy metals with eigenvectors (loading factors) [96]. The following calculation demonstrates the component generated via PCA:

$$\mathbf{Y}_{(i*j)} = \mathbf{a}_{i1}\mathbf{x}_{1j} + \mathbf{a}_{i2}\mathbf{x}_{2j} + \mathbf{a}_{i3}\mathbf{x}_{3j} + \text{-------} + \mathbf{a}_{im}\mathbf{x}_{mj}$$

Y = heavy metal score; a = loading vector; x = measured value (heavy metal); I = heavy metal number; J = sample number; and m = total number (heavy metal).

The first principal component (PC1) accountable for the highest potential fraction of the total number of variances in the dataset and the second principal component (PC2) explains the maximum number of the remaining variance and so on [97]. In principal component analysis, only a small number of PCs with an eigenvalue greater than 1 (Kaiser criterion) are retained. The number of principal components is less than or equal to the number of the original variables (metals) [98]. The rotational method is applied to easily explain the source of origin of the respective metals [99]. In several studies, it has been found that the literature has cited orthogonal varimax rotations for data analysis because the uncorrelated factors are more certainly interpretable. This method is furthermost extensively recognized in the field of hydrology to classify the source of origin of the several heavy metal pollutants. The result of factor analysis can categorically recognize the correlated metals that have a single source of origin in the study area. The loading plot produced in this process helps in the conception of software produced results. Researchers with previous related knowledge of the study region will effortlessly explain the natural or anthropogenic events that are accountable for heavy metal pollution in aquatic system.

12.5 MOFs for Heavy Metal Contaminant Removal from Water

Heavy metal contaminated water is a worldwide environmental concern. Heavy metals are highly toxic water pollutants, which have the potential to cause various serious health problems for both humans and wildlife due to their bioaccumulation and non-degradable properties [100]. The charge and effectiveness of heavy metal removal from water bodies remains a major challenge for effective water purification. Several methods have been envisaged to ensure that the targets set by international regulations are achieved. One of the most efficient methods is removal through the adsorption method. In recent times, the class of MOFs porous polymeric materials has been identified, consisting of metal ions linked together by organic bridging ligands, and molecular coordination is a new development at the interface between chemistry and materials science [101]. This method is effective primarily because of its ease, as it does not require high temperatures, and it allows for the removal of multiple substances at a time.

MOFs have attracted increased attention as promising resources for aqueous-phase adsorption elimination of emerging pollutants (ECs). Properties such as large adsorption capability, high surface area, porosity, structure, and recycling capacity give MOFs an advantage over conformist adsorbents [102]. The disadvantaged constancy of MOFs in water is the main competition for their real-world environmental use. The performance of MOFs and their sensitivities to the pollutants targeted for elimination can be modulated by practical selection of the organic linker and metal ion.

The metal-organic structures are known as evolving adsorbent materials. Several characteristics of MOFs fit them among competitors suitable for adsorption. The unique features of MOFs such as huge surface area, versatile functionality, porosity, and high thermal strength give support to the adsorption process [103]. In particular, huge surface area and high porosity facilitate the accessibility of adsorption sites and the diffusion of pollutants across the framework. MOFs can be arranged with ultra-high surface area and porosity, with pores well organized due to their crystalline nature [104].

The pores' shape and size can be adjusted by suitable assortment of the organic linker and their connectivity with the metal nodes. Furthermore, in-situ structural and postsynthetic alterations are useful in MOFs to achieve unique material features without disturbing the underlying network topology [105]. The prospect of modifying the structural properties of MOFs by careful arrangement of metal and organic linkers is one of the major benefits of MOFs compared to their organic and inorganic counterparts. Therefore, MOFs have the potential to be targeted with increased target specificity. Ultimately, its pore size, morphology, and chemical properties control the interactions between MOFs and adsorbates. As an outcome of their simple synthesis and large surface area, MOFs are more active and progressive than conformist adsorbents such as silica, zeolites, and carbonaceous solids. Permanent pores existing in the crystal network deliver space for guest molecules, and the diffusion of foreign molecules within the pores depends on the structural features of the MOF. Due to their unique structural properties and a wide range of compositions, MOFs have proven useful for many efficient adsorption processes [106]. The MOFs' properties have already been proficiently exploited, especially in applications for gas adsorption [107]. In view of the exceptional physicochemical properties of MOFs as confirmed by their larger adsorption capacity, water-stable MOFs can be functional for the active management of water and wastewater [108].

12.6 Conclusion

In conclusion, this review paper summarizes the middling values of various heavy metals for water study, which are based on the guidelines of organizations like WHO and USEPA. Apart from this, suggestions have been presented for the use and removal of various indicators. The causes of heavy metals in the water ecosystem are natural and anthropogenic. The weathering of mineral-rich rocks in natural sources and atmospheric salt precipitation occur as a result of natural progress such as forest fires, volcanic eruptions, etc. Major anthropogenic sources consist of wastewater discharge and industrial waste (e.g., sludge). Agriculture and domestic events also exaggerate the level of heavy metals in the surrounding water system. Combustion of fossil fuels, ignition of municipal as well as industrial waste, and aerosols generated by vehicles and industrial emissions—released as dry and wet deposition—also pollute the aquifer. The concentration of heavy metals in drinking water in excess of the limits approved by numerous national and international organizations causes various acute and serious diseases. These can be non-fatal such as muscle and physical weakness, or they can be fatal to an extent, for example brain, nervous system disorders, and even major diseases like cancer. Water quality checks are essential for the shelter of human health and the environment. The first step is to access the overall quality of the water and then recognize the source of the pollutants to reduce the pollution levels. The heavy metal pollution index is a well-documented system for examining water conditions related to heavy metals. Factor analysis proved to be an active technique to recognize the source of origin of heavy metals polluting the water body. The subsequent application of both methods characterizes the actual condition of the water body being considered and supports the preparation of a management plan to reduce the level of pollution.

References

1 Ayangbenro, A.S. and Babalola, O.O. (2017). A new strategy for heavy metal polluted environments: a review of microbial biosorbents. *International Journal of Environmental Research and Public Health* 14 (1): 94.

2 Madhav, S., Ahamad, A., Singh, A. K., Kushawaha, J., Chauhan, J. S., Sharma, S., and Singh, P. (2020). Water pollutants: sources and impact on the environment and human health. *Sensors in Water Pollutants Monitoring: Role of Material*, 43–62.

3 Liu, J. and Han, G. (2021). Tracing riverine sulfate source in an agricultural watershed: constraints from stable isotopes. *Environmental Pollution* 288: 117740.

4 Duffus, J.H. (2002). "Heavy metals" a meaningless term?) (IUPAC TECHNICAL REPORT. *Pure and Applied Chemistry* 74 (5): 793–807.
5 Anand, U., Reddy, B., Singh, V.K. et al. (2021). Potential environmental and human health risks caused by antibiotic-resistant bacteria (ARB), antibiotic resistance genes (ARGs) and emerging contaminants (ECs) from municipal solid waste (MSW) landfill. *Antibiotics* 10 (4): 374.
6 Sisodia, N.S. (2020). Effect of heavy metal toxicity on environment & health. *International Journal for Research in Applied Sciences and Biotechnology* 7 (6): 248–254.
7 Medfu Tarekegn, M., Zewdu Salilih, F., and Ishetu, A.I. (2020). Microbes used as a tool for bioremediation of heavy metal from the environment. *Cogent Food & Agriculture* 6 (1): 1783174.
8 Izydorczyk, G., Mikula, K., Skrzypczak, D. et al. (2021). Potential environmental pollution from copper metallurgy and methods of management. *Environmental Research* 197: 111050.
9 Weldeslassie, T., Naz, H., Singh, B., & Oves, M. (2018). Chemical contaminants for soil, air and aquatic ecosystem. In: *Modern Age Environmental Problems and Their Remediation*, 1–22. Cham: Springer.
10 Akhtar, N., Syakir Ishak, M.I., Bhawani, S.A., and Umar, K. (2021). Various natural and anthropogenic factors responsible for water quality degradation: a review. *Water* 13 (19): 2660.
11 Patinvoh, R.J. and Taherzadeh, M.J. (2019). Challenges of biogas implementation in developing countries. *Current Opinion in Environmental Science & Health* 12: 30–37.
12 Ukhurebor, K.E. & Adetunji, C.O. (2021). Relevance of biosensor in climate smart organic agriculture and their role in environmental sustainability: what has been done and what we need to do? In: *Biosensors in Agriculture: Recent Trends and Future Perspectives*, 115–136. Cham: Springer.
13 de Souza Miranda, R., Boechat, C.L., Bomfim, M.R. et al. (2022). Phytoremediation: a sustainable green approach for environmental cleanup. In: *Phytoremediation Technology for the Removal of Heavy Metals and Other Contaminants from Soil and Water*, 49–75. Elsevier.
14 Zafeiraki, E., Kasiotis, K.M., Nisianakis, P. et al. (2022). Occurrence and human health risk assessment of mineral elements and pesticides residues in bee pollen. *Food and Chemical Toxicology* 161: 112826.
15 Singh, R., Ahirwar, N.K., Tiwari, J., and Pathak, J. (2018). Review on sources and effect of heavy metal in soil: its bioremediation. *International Journal of Research in Applied, Natural and Social Sciences* 2018: 1–22.
16 Balogun, M.A., Anumah, A.O., Adegoke, K.A., and Maxakato, N.W. (2022). Environmental health impacts and controlling measures of anthropogenic activities on groundwater quality in Southwestern Nigeria. *Environmental Monitoring and Assessment* 194 (5): 1–36.
17 Zak, D., Hupfer, M., Cabezas, A. et al. (2021). Sulphate in freshwater ecosystems: a review of sources, biogeochemical cycles, ecotoxicological effects and bioremediation. *Earth-Science Reviews* 212: 103446.
18 Paul, R., Brindha, K., Gowrisankar, G. et al. (2019). Identification of hydrogeochemical processes controlling groundwater quality in Tripura, Northeast

India using evaluation indices, GIS, and multivariate statistical methods. *Environmental Earth Sciences* 78 (15): 1–16.

19 Rao, N.C., Giri, R.K., Sharma, A., and Pandey, A. (2020). Lamprophyres from the Indian shield: a review of their occurrence, petrology, tectonomagmatic significance and relationship with the Kimberlites and related rocks. *Episodes Journal of International Geoscience* 43 (1): 231–248.

20 Saha, P. and Paul, B. (2020). Water quality assessment techniques. In: *Sustainable Agriculture Reviews*, 40. 179–216. Cham: Springer.

21 Duran, C.J., Dubé-Loubert, H., Pagé, P. et al. (2019). Applications of trace element chemistry of pyrite and chalcopyrite in glacial sediments to mineral exploration targeting: example from the churchill province, northern Quebec, Canada. *Journal of Geochemical Exploration* 196: 105–130.

22 Martins, N.P., Srivastava, S., Simão, F.V. et al. (2021). Exploring the potential for utilization of medium and highly sulfidic mine tailings in construction materials: a review. *Sustainability* 13 (21): 12150.

23 Moiseenko, T.I., Dinu, M.I., Gashkina, N.A., and Kremleva, T.A. (2019). Aquatic environment and anthropogenic factor effects on distribution of trace elements in surface waters of European Russia and Western Siberia. *Environmental Research Letters* 14 (6): 065010.

24 Singh, S. (2022). Forest Fire Emissions: A contribution to global Climate change. *Frontiers in Forests and Global Change*, 244.

25 Singh, S., Suresh Babu, K.V. (2021). Forest Fire Susceptibility Mapping for Uttarakhand State by Using Geospatial Techniques. In: Rai, P.K., Singh, P., Mishra, V.N. (eds) *Recent Technologies for Disaster Management and Risk Reduction. Earth and Environmental Sciences Library*. Springer, Cham. https://doi.org/10.1007/978-3-030-76116-5_11.

26 Jia, Z., Cai, Y., Chen, Y., and Zeng, W. (2018). Regionalization of water environmental carrying capacity for supporting the sustainable water resources management and development in China. *Resources, Conservation and Recycling* 134: 282–293.

27 Okereafor, U., Makhatha, M., Mekuto, L. et al. (2020). Toxic metal implications on agricultural soils, plants, animals, aquatic life and human health. *International Journal of Environmental Research and Public Health* 17 (7): 2204.

28 Gustaytis, M.A., Myagkaya, I.N., and Chumbaev, A.S. (2018). Hg in snow cover and snowmelt waters in high-sulfide tailing regions (Ursk tailing dump site, Kemerovo region, Russia). *Chemosphere* 202: 446–459.

29 Gollakota, A.R., Gautam, S., and Shu, C.M. (2020). Inconsistencies of e-waste management in developing nations–Facts and plausible solutions. *Journal of Environmental Management* 261: 110234.

30 Shah, M.P., (Ed.) (2021). *Biological Treatment of Industrial Wastewater*. 5. Royal Society of Chemistry.

31 Thakur, A. and Kumar, A. (2021). Sustainable inhibitors for corrosion mitigation in aggressive corrosive media: a comprehensive study. *Journal of Bio-and Tribo-Corrosion* 7 (2): 1–48.

32 Faizan, M., Hayat, S., and Pichtel, J. (2020). Effects of zinc oxide nanoparticles on crop plants: a perspective analysis. In: *Sustainable Agriculture Reviews*, 41. 83–99. Cham: Springer.

33 Zamparas, M., Kapsalis, V.C., Kanteraki, A.E. et al. (2019). Novel composite materials as P-adsorption agents and their potential application as fertilizers. *The Global NEST Journal* 21 (1): 48–57.

34 Chubaka, C.E., Whiley, H., Edwards, J.W., and Ross, K.E. (2018). Lead, zinc, copper, and cadmium content of water from South Australian rainwater tanks. *International Journal of Environmental Research and Public Health* 15 (7): 1551.

35 Iqbal, M., Abbas, M., Nisar, J. et al. (2019). Bioassays based on higher plants as excellent dosimeters for ecotoxicity monitoring: a review. *Chemistry International* 5 (1): 1–80.

36 Singh, S., Naik, T.S.K., Chauhan, V. et al. (2022). Ecological effects, remediation, distribution, and sensing techniques of chromium. *Chemosphere* 307: 135804.

37 Valenzuela-Diaz, M.J., Navarrete-Calvo, A., Caraballo, M.A. et al. (2020). Hydrogeochemical and environmental water quality standards in the overlap between high mountainous natural protected areas and copper mining activities (Mapocho river upper basin, Santiago, Chile). *Journal of Hydrology* 588: 125063.

38 Nguegang, B., Masindi, V., Msagati, T.A.M., and Tekere, M. (2021). The treatment of acid mine drainage using vertically flowing wetland: insights into the fate of chemical species. *Minerals* 11 (5): 477.

39 Jaishankar, M., Tseten, T., Anbalagan, N. et al. (2014). Toxicity, mechanism and health effects of some heavy metals. *Interdisciplinary Toxicology* 7 (2): 60.

40 Vardhan, K.H., Kumar, P.S., and Panda, R.C. (2019). A review on heavy metal pollution, toxicity and remedial measures: current trends and future perspectives. *Journal of Molecular Liquids* 290: 111197.

41 Queiroz, H.M., Nóbrega, G.N., Ferreira, T.O. et al. (2018). The Samarco mine tailing disaster: a possible time-bomb for heavy metals contamination? *Science of the Total Environment* 637: 498–506.

42 Wang, Y., Wang, G., Sun, M. et al. (2022). Environmental risk assessment of the potential "Chemical Time Bomb" of ion-adsorption type rare earth elements in urban areas. *Science of the Total Environment* 822: 153305.

43 Brown, K.P., Gerber, A., Bedulina, D., and Timofeyev, M.A. (2021). Human impact and ecosystemic health at Lake Baikal. *Wiley Interdisciplinary Reviews: Water* 8 (4): e1528.

44 Junjun, G. (2021). RETRACTED ARTICLE: characteristics of heavy metal pollutants in groundwater based on fuzzy decision making and the effect of aerobic exercise on teenagers. *Arabian Journal of Geosciences* 14 (17): 1–11.

45 Wang, L., Rinklebe, J., Tack, F.M., and Hou, D. (2021). A review of green remediation strategies for heavy metal contaminated soil. *Soil Use and Management* 37 (4): 936–963.

46 Xu, D.M., Fu, R.B., Liu, H.Q., and Guo, X.P. (2021). Current knowledge from heavy metal pollution in Chinese smelter contaminated soils, health risk implications and associated remediation progress in recent decades: a critical review. *Journal of Cleaner Production* 286: 124989.

47 Jin, M., Yuan, H., Liu, B. et al. (2020). Review of the distribution and detection methods of heavy metals in the environment. *Analytical Methods* 12 (48): 5747–5766.

48 Peng, H., Chen, Y., Weng, L. et al. (2019). Comparisons of heavy metal input inventory in agricultural soils in North and South China: a review. *Science of the Total Environment* 660: 776–786.

49 Obasi, P.N. and Akudinobi, B.B. (2020). Potential health risk and levels of heavy metals in water resources of lead–zinc mining communities of Abakaliki, southeast Nigeria. *Applied Water Science* 10 (7): 1–23.

50 Nagajyoti, P.C., Lee, K.D., and Sreekanth, T.V.M. (2010). Heavy metals, occurrence and toxicity for plants: a review. *Environmental Chemistry Letters* 8 (3): 199–216.

51 Islam, M.T. (2017). Oxidative stress and mitochondrial dysfunction-linked neurodegenerative disorders. *Neurological Research* 39 (1): 73–82.

52 Hermes-Lima, M. (2004). Oxygen in biology and biochemistry: role of free radicals. *Functional Metabolism: Regulation and Adaptation* 1: 319–366.

53 Carocci, A., Catalano, A., Lauria, G., Sinicropi, M. S., and Genchi, G. (2016). 16 A Review on Mercury Toxicity in Food. *Food Toxicology*, 315.

54 Barrera-Díaz, C.E., Lugo-Lugo, V., and Bilyeu, B. (2012). A review of chemical, electrochemical and biological methods for aqueous Cr (VI) reduction. *Journal of Hazardous Materials* 223: 1–12.

55 Mohanna, C. and Nys, Y. (1999). Effect of dietary zinc content and sources on the growth, body zinc deposition and retention, zinc excretion and immune response in chickens. *British Poultry Science* 40 (1): 108–114.

56 Chasapis, C.T., Loutsidou, A.C., Spiliopoulou, C.A., and Stefanidou, M.E. (2012). Zinc and human health: an update. *Archives of Toxicology* 86 (4): 521–534.

57 Ou, Y., Bloom, M.S., Nie, Z. et al. (2017). Associations between toxic and essential trace elements in maternal blood and fetal congenital heart defects. *Environment International* 106: 127–134.

58 Na Nagara, V., Sarkar, D., Luo, Q. et al. (2022). Health risk assessment of exposure to trace elements from drinking black and green tea marketed in three countries. *Biological Trace Element Research* 200 (6): 2970–2982.

59 Dallinger, R., Berger, B., Hunziger, P., and Kgi, J.H. (1997). Metallothionein in snail Cd and Cu metabolism. *Nature* 388 (6639): 237–238.

60 Jin, T., Lu, J., and Nordberg, M. (1998). Toxicokinetics and biochemistry of cadmium with special emphasis on the role of metallothionein. *Neurotoxicology* 19 (4–5): 529–535.

61 Xue, J., Wang, Q., & Zhang, M. (2022). A review of non-point source water pollution modeling for the urban–rural transitional areas of China: research status and prospect. *Science of the Total Environment* 154146.

62 Panghal, V., Singh, A., Kumar, R. et al. (2021). Soil heavy metals contamination and ecological risk assessment in Rohtak urban area, Haryana (India). *Environmental Earth Sciences* 80 (21): 1–20.

63 Ahamad, M.I., Song, J., Sun, H. et al. (2020). Contamination level, ecological risk, and source identification of heavy metals in the hyporheic zone of the Weihe River, China. *International Journal of Environmental Research and Public Health* 17 (3): 1070.

64 Poonam, T., Tanushree, B., and Sukalyan, C. (2013). Water quality indices-important tools for water quality assessment: a review. *International Journal of Advances in Chemistry* 1 (1): 15–28.

65 Tyagi, S., Sharma, B., Singh, P., and Dobhal, R. (2013). Water quality assessment in terms of water quality index. *American Journal of Water Resources* 1 (3): 34–38.

66 Prasanna, M.V., Praveena, S.M., Chidambaram, S. et al. (2012). Evaluation of water quality pollution indices for heavy metal contamination monitoring: a case study from Curtin Lake, Miri City, East Malaysia. *Environmental Earth Sciences* 67 (7): 1987–2001.

67 Boateng, T.K., Opoku, F., Acquaah, S.O., and Akoto, O. (2015). Pollution evaluation, sources and risk assessment of heavy metals in hand-dug wells from Ejisu-Juaben Municipality, Ghana. *Environmental Systems Research* 4 (1): 1–12.

68 Karaouzas, I., Kapetanaki, N., Mentzafou, A. et al. (2021). Heavy metal contamination status in Greek surface waters: a review with application and evaluation of pollution indices. *Chemosphere* 263: 128192.

69 Egbueri, J.C. (2020). Groundwater quality assessment using pollution index of groundwater (PIG), ecological risk index (ERI) and hierarchical cluster analysis (HCA): a case study. *Groundwater for Sustainable Development* 10: 100292.

70 Hossain, M. and Patra, P.K. (2020). Water pollution index–A new integrated approach to rank water quality. *Ecological Indicators* 117: 106668.

71 Haque, M., Niloy, N.M., Nayna, O.K. et al. (2020). Variability of water quality and metal pollution index in the Ganges River, Bangladesh. *Environmental Science and Pollution Research* 27 (34): 42582–42599.

72 Solangi, G.S., Siyal, A.A., Babar, M.M., and Siyal, P. (2019). Groundwater quality evaluation using the water quality index (WQI), the synthetic pollution index (SPI), and geospatial tools: a case study of Sujawal district, Pakistan. *Human and Ecological Risk Assessment: An International Journal, 26*(6), 1529-1549.

73 Monjardin, C.E.F., Senoro, D.B., Magbanlac, J.J.M. et al. (2022). Geo-accumulation index of manganese in soils due to flooding in boac and mogpog rivers, marinduque, philippines with mining disaster exposure. *Applied Sciences* 12 (7): 3527.

74 Wardani, N.K., Prartono, T., and Sulistiono, S. (2020). Sediments quality based on geo-accumulation index in heavy metals (Pb, Cu, and Cd) of cengkok coastal waters, Banten bay. *Jurnal Pendidikan IPA Indonesia* 9 (4): 574–582.

75 Singovszka, E. and Balintova, M. (2019). Enrichment Factor and geo-accumulation index of trace metals in sediments in the river Hornad, Slovakia. In *IOP Conference Series: Earth and Environmental Science* (Vol. 222, No. 1, p. 012023). IOP Publishing.

76 Liu, M., Xu, Y., Nawab, J. et al. (2020). Contamination features, geo-accumulation, enrichments and human health risks of toxic heavy metal (loids) from fish consumption collected along Swat river, Pakistan. *Environmental Technology & Innovation* 17: 100554.

77 Seifi, M., Mahvi, A.H., Hashemi, S.Y. et al. (2019). Spatial distribution, enrichment and geo-accumulation of heavy metals in surface sediments near urban and industrial areas in the Persian Gulf. *Desalination and Water Treatment* 158: 130–139.

78 Loska, K., Cebula, J., Pelczar, J. et al. (1997). Use of enrichment, and contamination factors together with geoaccumulation indexes to evaluate the content of Cd, Cu, and Ni in the Rybnik water reservoir in Poland. *Water, Air, and Soil Pollution* 93 (1): 347–365.

79 Abrahim, G.M.S. and Parker, R.J. (2008). Assessment of heavy metal enrichment factors and the degree of contamination in marine sediments from Tamaki Estuary, Auckland, New Zealand. *Environmental Monitoring and Assessment* 136 (1): 227–238.

80 Yao, Z. (2008). Comparison between BCR sequential extraction and geo-accumulation method to evaluate metal mobility in sediments of Dongting Lake, Central China. *Chinese Journal of Oceanology and Limnology* 26 (1): 14–22.

81 Christophoridis, C., Dedepsidis, D., and Fytianos, K. (2009). Occurrence and distribution of selected heavy metals in the surface sediments of Thermaikos Gulf, N. Greece. assessment using pollution indicators. *Journal of Hazardous Materials* 168 (2–3): 1082–1091.

82 Diatta, J.B., Chudzinska, E., and Wirth, S. (2008). Assessment of heavy metal contamination of soils impacted by a zinc smelter activity. *Journal of Elementology* 13 (1).

83 Al-Taani, A.A., Nazzal, Y., Howari, F.M. et al. (2021). Contamination assessment of heavy metals in agricultural soil, in the Liwa area (UAE). *Toxics* 9 (3): 53.

84 Abija, F.A. and Abam, T.K.S. (2018). Application of geoaccumulation and pollution load indices in the assessment of heavy metal contamination in sediments of the forcados river and adjoining soils, Western Niger Delta. *Journal of Geoscience and Environmental Research (JOGER)* 1 (1): 35–51.

85 Kabir, M., Islam, M., Hoq, M., and Tusher, T.R. (2020). Appraisal of heavy metal contamination in sediments of the Shitalakhya River in Bangladesh using pollution indices, geo-spatial, and multivariate statistical analysis. *Arabian Journal of Geosciences* 13 (21): 1–13.

86 Ustaoğlu, F., Tepe, Y., and Aydin, H. (2020). Heavy metals in sediments of two nearby streams from Southeastern Black Sea coast: contamination and ecological risk assessment. *Environmental Forensics* 21 (2): 145–156.

87 Kolawole, T.O., Olatunji, A.S., Jimoh, M.T., and Fajemila, O.T. (2018). Heavy metal contamination and ecological risk assessment in soils and sediments of an industrial area in Southwestern Nigeria. *Journal of Health and Pollution* 8 (19).

88 Bouzekri, S., El Hachimi, M.L., Touach, N. et al. (2019). The study of metal (As, Cd, Pb, Zn and Cu) contamination in superficial stream sediments around of Zaida mine (High Moulouya-Morocco). *Journal of African Earth Sciences* 154: 49–58.

89 da Silva Junior, J.B., de Carvalho, V.S., Sousa, D.S. et al. (2022). A risk assessment by metal contamination in a river used for public water supply. *Marine Pollution Bulletin* 179: 113730.

90 Rutkowski, P., Diatta, J., Konatowska, M. et al. (2020). Geochemical referencing of natural forest contamination in Poland. *Forests* 11 (2): 157.

91 Saidi, I., Said, O.B., Abdelmalek, J.B. et al. (2018 November). Assessment of Heavy metal contamination in the sediment of the Bizerte Lagoon in Northern Tunisia. In: *Conference of the Arabian Journal of Geosciences* (pp. 41–44). Cham: Springer.

92 Kaur, J., Bhat, S.A., Singh, N. et al. (2022). Assessment of the heavy metal contamination of roadside soils alongside Buddha Nullah, Ludhiana,(Punjab) India. *International Journal of Environmental Research and Public Health* 19 (3): 1596.

93 Winderbaum, L., Ciobanu, C.L., Cook, N.J. et al. (2012). Multivariate analysis of an LA-ICP-MS trace element dataset for pyrite. *Mathematical Geosciences* 44 (7): 823–842.

94 Yidana, S.M., Ophori, D., and Banoeng-Yakubo, B. (2008). A multivariate statistical analysis of surface water chemistry data—The Ankobra Basin, Ghana. *Journal of Environmental Management* 86 (1): 80–87.

95 Locurto, C., Fortin, E., and Sullivan, R. (2003). The structure of individual differences in heterogeneous stock mice across problem types and motivational systems. Genes. *Brain and Behavior* 2 (1): 40–55.

96 Korre, A. (1999). Statistical and spatial assessment of soil heavy metal contamination in areas of poorly recorded, complex sources of pollution. *Stochastic Environmental Research and Risk Assessment* 13 (4): 260–287.

97 Isaak, D.J., Luce, C.H., Chandler, G.L. et al. (2018). Principal components of thermal regimes in mountain river networks. *Hydrology and Earth System Sciences* 22 (12): 6225–6240.

98 Ringnér, M. (2008). What is principal component analysis? *Nature Biotechnology* 26 (3): 303–304.

99 Abdi, H. and Williams, L.J. (2010). Principal component analysis. *Wiley Interdisciplinary Reviews: Computational Statistics* 2 (4): 433–459.

100 Okoro, H.K., Fatoki, O.S., Adekola, F.A. et al. (2012). A review of sequential extraction procedures for heavy metals speciation in soil and sediments.

101 Horike, S., Nagarkar, S.S., Ogawa, T., and Kitagawa, S. (2020). A new dimension for coordination polymers and metal-organic frameworks: towards functional glasses and liquids. *Angewandte Chemie International Edition* 59 (17): 6652–6664.

102 Dhaka, S., Kumar, R., Deep, A. et al. (2019). Metal-organic frameworks (MOFs) for the removal of emerging contaminants from aquatic environments. *Coordination Chemistry Reviews* 380: 330–352.

103 Sun, C.Y., Qin, C., Wang, X.L., and Su, Z.M. (2013). Metal-organic frameworks as potential drug delivery systems. *Expert Opinion on Drug Delivery* 10 (1): 89–101.

104 Jiang, G., Osman, S., Senthil, R.A. et al. (2022). Hierarchically porous carbon derived from magnesium-based metal-organic frameworks as advanced active material for supercapacitor. *Journal of Energy Storage* 49: 104071.

105 Yu, Q., Li, Z., Cao, Q. et al. (2020). Advances in luminescent metal-organic framework sensors based on post-synthetic modification. *TrAC Trends in Analytical Chemistry* 129: 115939.

106 Safaei, M., Foroughi, M.M., Ebrahimpoor, N. et al. (2019). A review on metal-organic frameworks: synthesis and applications. *TrAC Trends in Analytical Chemistry* 118: 401–425.

107 Khan, S., Guan, Q., Liu, Q. et al. (2021). Synthesis, modifications and applications of MILs metal-organic frameworks for environmental remediation: the cutting-edge review. *Science of the Total Environment*, 810, 152279.

108 Mrinalini, M. and Prasanthkumar, S. (2019). Recent advances on stimuli-responsive smart materials and their applications. *ChemPlusChem* 84 (8): 1103–1121.

13

Organic Contaminants in Aquatic Environments: Sources and Impact Assessment

*Shipa Rani Dey, Priyanka Devi, and Prasann Kumar**

Department of Agronomy, School of Agriculture, Lovely Professional University, Punjab, India, 144411
** Corresspsonding author*

13.1 Introduction

As part of wastewater pollution, many organic pollutants include dyes, humic substances, petroleum, surfactants, phenolic compounds, pesticides, and pharmaceuticals. There is a possibility that harmful chemicals can be produced during the process of disinfecting water when there are organic contaminants present in it. In most cases, farm wastewater consists of decay products from organic matter and pharmaceuticals like antibiotics, which lead to the formation of humic substances (humic acid, fulvic acid, humin) that need to be separated from the wastewater before it is released into the environment. As a result of the abundance of organic pollutants, a variety of industrial, domestic, and agricultural applications are provided.

There is no doubt that organic pollutants will make their way into wastewater even if they are only used in the home or at the workplace. Trace contaminants in wastewater treatment processes do not effectively remove them from the effluent, and this may result in their retention in the discharge of sewage into aquatic environments [1–3].

There are several ways in which organic materials used for agricultural purposes are carried away from the farm instead of being disposed of in a surface water body. For instance, some organic materials might be disposed of through runoff into a surface water body, and others might be disposed of through infiltration into a groundwater body. Drinking water can be derived directly from the waters that receive trace pollutants, thereby making it a possible source of pollution in the area. It is also possible to utilize these waters indirectly by using them to recharge a water supply. There are many different qualities to be found in a freshwater environment, which can be categorized as both surface water and groundwater systems combined.

Metal Organic Frameworks for Wastewater Contaminant Removal, First Edition. Edited by Arun Lal Srivastav, Lata Rani, Jyotsna Kaushal, and Tien Duc Pham.

The presence of newly discovered pollutants in aquatic systems is one of the problems that is of concern on a global scale. In addition to the chemicals themselves, emerging pollutants include a wide range of substances, both synthetic and natural, that are not regulated or are regulated in a manner that exposes living beings to the possibility of adverse health effects (along with their transformation products and metabolites) [4]. There are several examples of emerging pollutants in the environment: nanomaterials, surfactants, biocides, plasticizers, potentially toxic elements [5–7], pharmaceuticals as well as personal care products [8], illicit drugs (Ntshani and Tavengwa), flame retardants [9], and pesticides [10]. Because of this, it has become apparent that the world is continuing to produce synthetic chemicals, although there is a growing awareness of the harmful effects these chemicals have on both humans as well as the environment.

There is no doubt that the use of a variety of chemical compounds in consumer goods and industrial processes has remained widespread in recent years, and some of these compounds are classified as EOCs as a result of their continual release into the environment. Although EOCs have been detected in a variety of aquatic environments through a variety of routes, such as effluents from wastewater treatment plants and livestock activities, the majority of EOCs have been detected in aquatic environments [11, 12]. If these EOCs are continually released into the environment, there is a possibility that ecosystem degradation will occur. The other major concern is that they may accumulate in non-target species and cause adverse effects on those species by bioaccumulating inside them. It has been suggested that biomagnification through the food system could also pose a danger to humans, according to other potential concerns.

Even though there is an increasing number of anthropogenic organic compounds being found in continental waters, only a small fraction of these compounds have proven to be of interest to researchers studying marine pollution to date. To consider chemical substances as potential marine contaminants that need to be targeted, several characteristics must be taken into account. To be considered hazardous chemicals, they either have to fall into the category of chemicals with a high volume of production or they have to be produced as by-products in substantial quantities, or else they have to be continuously released into the marine environment. Aside from this, these substances must also be able to withstand biotic and abiotic processes of degradation, as well as have a high probability of being bioaccumulated, which makes them a potential threat to either marine ecosystems or human health [13].

13.2 The Various Forms and Causes of Chemical Pollutants

Among the different kinds of environments that can introduce contaminants into water, two distinct categories are worth mentioning. It is critical to distinguish between two different types of pollution: point pollution and nonpoint pollution [14]. There are many types of pollution, but the type that is referred to as "point pollution" can be distinguished by the fact that the sources of the chemicals that are polluting the environment can be easily identified. Factories, processing plants, and other similar establishments are examples of point pollution.

There is a significant amount of oil that is released into the ocean during the process of flushing tanks. Additionally, several other factors contribute to the contamination of water. As a result of the reaction between sulfur dioxide (SO_2) in the air and water vapor in the atmosphere, sulfuric acid is produced, which is then precipitated to the ground as a result of the reaction. As a result of runoff, this acid will find its way into water bodies, where it will cause a great number of animals to be harmed, either because they live in the water bodies or because they drink from the water bodies.

Petroleum products: Crude oil is one of several petroleum by-products that are used to make synthetic compounds. In turn, these compounds may serve a wide range of functions, from lubrication to energy generation to plastics manufacturing and beyond. Raw materials for these compounds come from crude oil processing. Unfortunately, because of the improper storage and use of these chemicals, they make their way into water bodies and decrease the overall quality of the water as a result. The vast majority of products derived from oils are toxic to animals when consumed in sufficient quantities. There is a group of hazardous substances known as polychlorinated biphenyls, which are also known as PCBs, which pollute water sources.

Pesticides: In terms of chemicals, herbicides, insecticides, and fungicides are examples of chemical agents that are used to control weeds, insects, and other organisms (for the control of fungi). The majority of these chemicals are biodegradable, meaning that they can quickly transform into harmless or less harmful forms over a short period. However, some of these chemicals are not biodegradable, and they remain toxic for a long period. The chemicals that have been used to treat plants with these chemicals are not biodegradable. They can be absorbed into the brains and tissues of animals that consume plants that have been exposed to these chemicals. In addition to chlordane, dichlorodiphenyltrichloroethane (DDT) is also contained in these chemicals. If an animal that has been tainted with chemicals is consumed by another species, then the chemicals are transmitted up the food chain when the other species consumes that animal. As a result of this process, we refer to it as "biomagnification.". Overabundant glyphosate, a component of the herbicide known as glyphosate, is highly toxic to fish due to its potential to kill beneficial insects that prey on pest insects as well as poison them. It should be noted that glyphosate is also lethal to beneficial insects that are susceptible to it.

Nitrates: A lack of regulation of pesticides and fertilizers in farming areas contributes to the contamination of drinking water sources with nitrates from pesticides and fertilizers used in farming areas. The fact that the NO_3^- ion has a negative charge makes it impossible for soils to adsorb it.

Heavy metalloids: It is well known that several metals are harmful to organisms, including lead, mercury, copper, and selenium. It has been established that these chemicals pollute waterways when they reach them from their mines, mills, factories, businesses, cars, and their respective mothers containing these toxins [15]. Heavy metals can be absorbed by animals through the consumption of plants and animals that have previously ingested those metals.

Hazardous waste: Hazardous waste can be classified into four different classes or categories depending on its characteristics: toxic, reactive, corrosive, and flammable wastes. Typically, these kinds of materials can reach water bodies without being treated or stored properly because there are not enough facilities available to treat or store them. Wastes have the potential to enter the water system in a variety of different ways, depending on the route they take. If animals scavenge for food from dead or living organisms that have been contaminated by these wastes, they are more likely to become victims of these wastes than animals that eat dead or living organisms that have been contaminated by these wastes. As a result of the inherent characteristics of an animal, it may also be toxic when consumed by other animals besides humans as a result of its inherent characteristics.

Excess organic matter: There is a possibility that excessive amounts of fertilizer, organic matter, and other nutrients can pollute water if they are applied in large amounts. The decomposition process that occurs when these organisms pass away in the water can cause a prevalence of high decomposition processes, which can lead to eutrophication phenomena and severe depletion of oxygen at a level that is lethal for the organisms. As a consequence of this situation, algae and aquatic plants grow at a rapid rate. It has been reported that many oxygen-dependent organisms, such as fish and other animals, have died as a result of the process of eutrophication. There can be contamination of water that is used to encourage plant growth that may result in contamination of the water. Due to landfills, the landscape surrounding the area could deteriorate, and pollutants could leak into the groundwater as a result of contamination from landfills.

Sediment: In agricultural fields, mining areas, and roads, a large number of soil particles are washed into water bodies due to runoff. The lack of measures aimed at conserving soil water leads to an absence of adequate vegetation cover. The pollution that makes its way to the ocean is capable of killing marine life and disrupting aquatic ecosystems when it reaches the ocean. Despite the enormous area they cover, the oceans are extremely susceptible to the effects of pollution. In addition to causing health issues for people, polluted water is directly or indirectly responsible for the death of an unfathomably large number of animals (Figure 13.1).

13.3 Increasing Contaminant Occurrence in Aquatic Systems

There is a strong increase in the amount of newly discovered pollutants released into the environment with the rapid urbanization and industrialization of the world, as well as the general improvement in the quality of people's lives as a result of these factors. As a result of waste from municipal, household, industrial, and hospital facilities as well as leachate from landfills, urban stormwater, and agricultural runoff, newly discovered pollutants are often introduced into aquatic systems by wastewater from these sources [16]. In addition to these pharmaceuticals, anti-retroviral drugs are also commonly found in these environments [17]. A lot of pesticides and veterinary drugs have been found in runoff from agricultural land, as well as occasionally in urban stormwater, where they can be found in abundant [18]. In turn, it has been found that the source of the effluent and the runoff often have an impact on the composition of

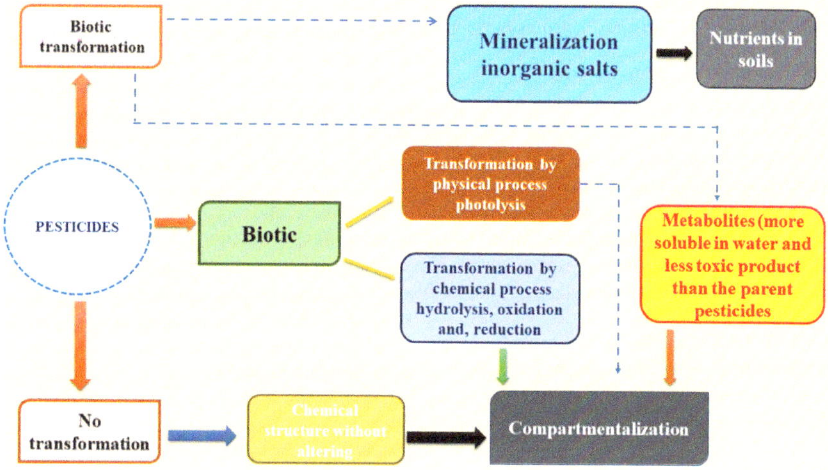

Figure 13.1 Chemical contaminants.

the contaminants in them. There are many different applications in which newly discovered pollutants can be used, and as a result, they are everywhere in the aquatic systems of the world due to their wide variety of applications.

The number of pharmaceuticals on the market is more than 5000 right now, and 631 of them have been identified by scientists as being present in the freshwater systems of several countries [19, 20–23]. Therefore, the high number of emerging pollutants detected in aquatic systems in Africa and Latin America could be attributed more to a lack of advanced mass spectrometry and chromatographic equipment rather than new pollutants less frequently released into the environment [24]. (Table 13.1).

Table 13.1 Examples of emerging pollutants frequently detected in aquatic systems.

Contaminants	Environmental concern	Main sources
Pharmaceutical products	• Potentially acute and chronic harmful effects on aquatic life are a major issue. • Antibiotics promote bacteria and gene growth that are resistant to antibiotics.	Domestic wastewater, aquaculture, hospital effluent, livestock breeding
Personal care products	• Although PCPs have the potential to cause a decline in the populations of aquatic wildlife, it is extremely unlikely that they will cause acute toxicity. • Pose chronic toxicity to aquatic organisms	Domestic wastewater, landfills
Pesticides	• Toxic to wildlife, marrow cells, and earthworms; can cause cancer; has been linked to childhood leukemia and Parkinson's disease; carcinogenic.	Agricultural and urban runoff

(Continued)

Table 13.1 (Continued)

Contaminants	Environmental concern	Main sources
Disinfection products	• A significant number of the by-products produced by disinfection are either mutagenic, genotoxic, cytotoxic, carcinogenic, or teratogenic.	Domestic wastewater
Fire retardants	• Accumulate mainly in the serum, liver, and kidneys, and have the potential to have a negative effect on reproductive, developmental, and other outcomes across the body.	Industrial and domestic effluent, firefighting discharges
Hydrocarbons	• High toxicity, mutagenicity, and carcinogenicity.	Combustion in the industry as well as in biomass and fossil fuels
Food additives	• According to research based on ecotoxicological research, sweeteners have very low toxicity in relation to the environment even when used in large amounts.	Domestic wastewater

13.4 Identifying Potential Points of Entry for New Pollutants into Aquatic Systems

Pollutant release mechanisms and the physicochemical properties of emerging pollutants both play a significant part in the path that all these contaminants take before reaching aquatic habitats. By contrast, nonpoint-source pollution refers to a process by which pollutants enter the environment from a variety of different sources as opposed to one single source [25]. Frequently, environmental regulators will determine the maximum concentrations that can be tolerated in effluents from industrial and municipal sources. The list of contaminants that are commonly regulated does not, however, include substances that have been discovered in the recent past. These pollutants are discharged into the environment as a result of combustion, industrial flue gas, and pesticide spraying [26].

According to the findings of a study conducted in France, the accumulation of microplastics after a rainfall event increased by a factor of five [27]. It is important to remember that the molecular weight and density as well as volatility of a pollutant all play a role in determining how far it will travel after it is released into the atmosphere. Compounds that have both a high level of volatility and a long half-life are capable of being transported over long distances because they both have a high level of volatility.

Earlier studies conducted in France and China indicate that microplastic particles are capable of traveling up to a maximum distance of 95 kilometers when transported by air, and this has been shown by a previous study that was conducted in France and China.

Microplastics and lead are two of the most common pollutants found in the environment. An in-depth study in Lake Michigan showed that PCBs that were

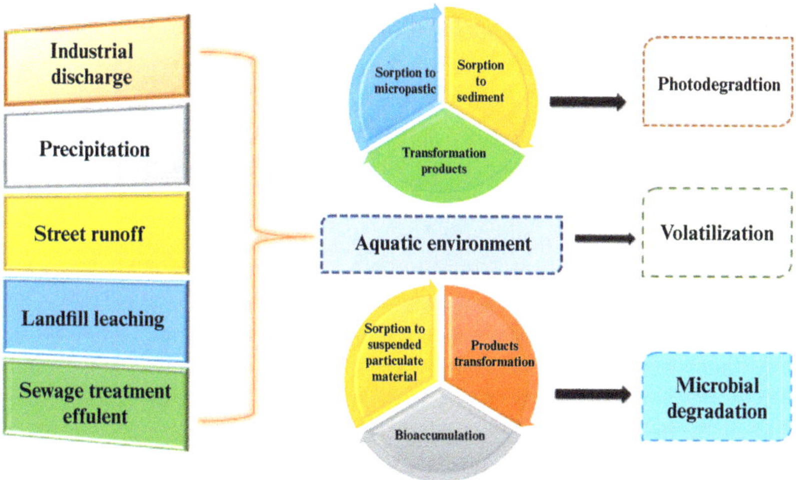

Figure 13.2 Industrial discharge and aquatic environments.

particle-bound were deposited less with each transit by aircraft of 40 kilometers compared with PCBs that were not particle-bound [28]. There is a high probability that polybrominated Diphenyl Ethers (PBDEs) and PCBs are gravitationally sedimented rather than being transported through the air as a result of their relatively heavy molecular weight.

Figure 13.2 shows different sources of pollutants and their discharge into aquatic environments:

13.5 Groups of Trace Pollutants and ECs

13.5.1 Polybrominated Diphenyl Ethers (PBDEs)

There are many consumer products as well as commercial items that use PBDEs as flame retardants, including computers, textiles, televisions, furniture, pipes, and other building materials [29, 30]. It is not surprising that these are essential chemicals that are widely used in consumer and commercial products, as PBDEs are used as flame retardants in these cases. As a result, brominated flame retardants have been known to linger in the environment for long periods of time [31], accumulate in living organisms [32], and be capable of causing endocrine disruption [31]. However, PBDEs are very poorly soluble in water, and they can only be detected at a detection level of picograms per liter, and detecting higher levels of their presence in freshwater is not possible [33, 34].

Pesticides: There are three types of pesticides: insecticides, fungicides, and herbicides, but there are also subcategories of pesticides depending on the type of pest they are meant to control [35–37]. Besides these, there are other types of pesticides that fall into these categories. It has been shown that both organochlorines and organophosphates have been capable of causing adverse effects on humans and other organisms in the past [38, 39].

Chloroacetanilides: The most common form of herbicide that is used to keep grassy and broadleaf weeds under control is pre-emergence herbicides like chloroacetanilides. In accordance with a recent study by [40, 41], metolachlor has been found in groundwater and surface water in the United States more frequently than any other pesticide. It was found in both groundwater and surface water in all 50 states. In the United States, both surface water and groundwater have been found to contain traces of alachlor, particularly in agricultural regions that produce maize [42]. Since 1994 when acetochlor was substituted for alachlor, there has been a downward trend in both the concentrations of alachlor in the environment as well as the frequency with which it is detected [43].

Organochlorines: Almost all of these compounds have at least one atom of chlorine that is bound to another atom via a covalent bond in some way. As an insecticide, these are often used in agricultural settings for many reasons; they are known for their high toxicity, they are hydrophobic, their solubility is extremely low, and Nhan et al., 2001 reported that they have a high environmental persistence [39].

Organophosphates: Due to their neurotoxic properties, these compounds are frequently used in the pest control industry. It is possible that these pesticides might be more hazardous (strong cholinesterase inhibitors) than organochlorines or carbamates, although this is not guaranteed. They are considered to be less consistent and less bioaccumulative than the majority of other pesticide classes [44, 37, 45, 46].

Pyrethroids: As a result of the hydrophobic properties of these compounds, which are capable of being persistent in the environment, these compounds are frequently used as insecticides. It has been found recently that pyrethroids, which are less toxic compared to other pesticides (especially organochlorines), are being used to replace pesticides with a higher toxicity level to reduce pesticide use. Among the most commonly used pyrethroids are bifenthrin, cypermethrin, and esfenvalerate, and esfenvalerate has even been found in surface runoff water [47].

13.6 Pharmaceuticals and Personal Care Products (PPCPs)

Antimicrobials: It is important to understand that antimicrobials are substances that are capable of killing microorganisms such as bacteria, fungi, or viruses, as well as inhibiting their reproduction and spread. In comparison to other antibiotics such as erythromycin, oleandomycin, tylosin, salinomycin, and tiamulin, roxithromycin has a longer half-life in soil than these other antibiotics [48]. Several studies have demonstrated that the antiseptic triclosan, which is widely used, is hydrophobic in nature and has a low water solubility, making it a particularly effective antiseptic [49, 50]. The presence of triclosan in wastewater and surface waters has been demonstrated despite these characteristics [49, 41]. In addition to these antiseptics, other commonly used antiseptics are present in the influent and effluent of WWTPs. According to research, there are some antiseptics in this category including biphenylol and chloroprene.

Synthetic hormones: An animal's endocrine glands are responsible for releasing hormones that occur naturally in the body, such as insulin as well as thyroxine, which are responsible for controlling every aspect of an animal's life, including its behavior, growth, and reproduction [51]. A phytohormone is an equivalent compound that is naturally produced by plants to control the differentiation and growth of various plant tissues. Isoflavones and lignans are examples of phytohormones that may be found in plants.

Other PPCPs: It is relevant to point out that several chemicals are commonly used that are not classified as PPCPs, including fragrance (acetophenone), insect repellent (N,N-diethylmetatoluamide, or DEET), and stimulant (caffeine). A variety of these organisms have been found in freshwater environments in Europe as well as the United States [40, 41, 52].

Occurrence of micropollutants in water: Currently, a great deal of research is being carried out on the micropollutants that are found in surface water. There are many micropollutants released into rivers, lakes, and reservoirs as a result of untreated discharges from sewage treatment plants. These micropollutants can either get deposited in sediments as a result of the hydrological impact that they cause or they may get carried to other locations by the movement of water that they create. As a result of the chemical and biological breakdown of these composites, their waste products are discharged into surface water as a result of the chemical and biological breakdown. Nevertheless, very few micropollutants can remain in surface waters for long periods and build up over time [53]. The presence of micropollutants in natural waters is influenced both by their physicochemical properties as well as their bioavailability as micropollutants [54]. To determine the fate of micropollutants, it is important to take into consideration their physicochemical properties, their interaction with the environment, their transit and retention, as well as the process of transformation and accumulation that they undergo [54]. As the parent compound is broken down into its by-products, the emergence of micropollutants into the natural environment will not be completely prevented as part of the breakdown process (Table 13.2).

Table 13.2 Sources of micropollutants in municipal wastewater.

Source	Pollutants
1) Elimination by urination and defecation	• Pharmaceuticals • Ingredients in artificial sweeteners • Medications that are illegally obtained • Hormones
2) Material diffusion/emission/leaching onto city surfaces through rain runoff directly released into sewers	• Persistent organic pollutants • Pesticides and other biocidal substances • Inorganic compounds with high metal content • PARFUM: Polycyclic aromatic hydrocarbons

(Continued)

Table 13.2 (Continued)

Source	Pollutants
3) Transfer from dirty materials and surfaces	• Heavy metals • Fillers for plastics • Inflammation modifying agents • Water-repellent chemicals
4) Direct discard into sewers	• Personal care products • Surfactants • Corrosion inhibitors

13.7 Concentrations of Micropollutants in Aquatic Organisms

When determining how dangerous something is, a lot of attention should be paid to the path along which contaminants build up over time. As contaminants enter one level of the food chain, the way they get there can have a big influence on the way they move through the rest of the chain as well. It is a well-known fact that zebra mussels (Dreissena polymorpha) are one of the species of small freshwater mussels that have a significant effect on fish species when they are released into the water. Generally, it is easy to understand how a contaminant's lipophilicity is related to its level of bioconcentration since they are highly related to each other [55].

It is possible to find the different pathways through which micropollutants are transported by both the environment and the biological system. When contaminants are introduced into benthic systems, they have the potential to accumulate through two distinct pathways. The consumption of contaminated sediment can result in organisms acquiring contaminants from the sediment if the sediment is contaminated itself [56, 57]. Due to this fact, benthos plays one of the most important roles within the food chain because pollutants are both removed from sediment storage and added to the food chain, as they perform both functions [58, 59]. As a result, the significance of dietary accretion varies from organism to organism depending on some factors, such as the organism's position in the food chain, the rate at which it feeds and how it behaves [60], the rate at which it assimilates, the lipid content of the prey and the food [61, 62], and the source of contamination [63].

13.8 Methods for Micropollutant Removal

Due to the discovery of novel microcontaminants in contaminated water, conventional wastewater treatment plants that are already in operation are not capable of treating the water in a way that ensures compliance with environmental standards because of the presence of new and emerging microcontaminants. Every living organism on this planet has an impact on the quality of the water as a result of the release of these compounds into the environment. There are a variety of conventional technologies that are used in the treatment process of water that contains complex pollutants as well as emerging micropollutants such as pharmaceuticals, personal care products, and various other additives. In this section, we will discuss the most common treatment methods for removing micropollutants from drinking water and wastewater as well as their effects on the environment. Both types of water can contain micropollutants.

Adsorption: It has been estimated that approximately 94% of microcontaminants were removed from the water [64]. In all cases, regardless of the initial concentrations of compounds, there was an improvement in the percentage of micropollutants removed with increasing doses of powdered activated carbon (PAC). Although the compounds were present in different concentrations, this was true [64]. As a result of interactions between particles and contaminants, the effectiveness of the removal of trace pollutants is greatly determined by how effectively these interactions take place. The removal efficiency of activated carbon can be reduced if there is competition for the adsorption sites or if the pores of the activated carbon are clogged (by solid particles) [65, 66].

Ozonation: When ozonation is applied to drinking water treatment plants, it is usually done for fumigation and oxidation (e.g., to control odor, remove color, remove micropollutants, etc.), or a combination thereof [67]. As a result of ozonation, micropollutants can be oxidized in a variety of ways, either through direct ozone reactions or indirectly through the formation of oxidizing hydroxyl radicals through indirect reactions.

Membrane processes: Two processes are extremely effective at removing turbidity from water: microfiltration (MF) and ultrafiltration (UF). Because micropollutants have a molecular size of less than 100 nm, these methods are insufficient to remove them from the environment because of their small molecular size. As a result, micropollutants can either be reduced or eliminated, depending on whether they interact with natural organic matter (NOM) or are adsorbed onto the membrane polymers used in the filtration process. As a result of the hydrophobic membrane, there was a significant improvement in the removal efficiencies of ibuprofen and estradiol, achieving approximately 25 and 80% removal efficiency, respectively. It was found that the combination of the MF and RO processes to remove micropollutants from household wastewater so that the wastewater can be reused [68]. There was evidence that the MF treatment was capable of removing up to 50% of the di(2-Ethylhexyl) phthalate that had been present previously. As a result of integrating the MF and RO systems, a higher amount of micropollutants were removed as a result of the integrated system. Various micropollutants, except for ibuprofen and nonylphenol, had a removal efficiency of 65 to 90%, whereas the removal efficiencies of other micropollutants were low. There was also a study that combined the MF and RO systems and found that, except for mefenamic acid and caffeine, most of the micropollutants were removed with an efficiency greater than 95% [69].

Advanced oxidation processes: The removal of micropollutants has been attempted by using a variety of advanced oxidation processes (AOPs), including ultraviolet light (UV), ultraviolet light combined with chlorine (UV/Cl_2), and ultraviolet light combined with ozone (UV/O_3). The oxidation of micropollutants by an AOP is carried out by a radical species e.g., chlorine radicals (Cl), hydroxyl radicals (OH), and ozonide radical ions ($O3^-$). The rapid degradation of micropollutants is caused by the continuous radical attacks which occur from their parent compounds to their by-products during the release process. It has been reported that ozonation can remove the majority of micropollutants from water. The results of the study showed that all carbamazepine, diclofenac, indomethacin, and trimethoprim concentrations were reduced by more than 95% when 5 mg of ozone per liter of water was applied.UV-light-based processes (UV and UV/H_2O_2) have been demonstrated to remove 41 micropollutants when used in conjunction with ultraviolet light [70]. As far as oxidation is concerned, most of the time it does not result in complete mineralization, but in some cases, it results in a processed product that has a lower level of biological activity than the original compound in comparison to the processed product [71, 72]. The majority of organic compounds undergo a transformation that results in the production of hydrophilic by-products, which have molecular weights that are either comparable to or lower than those of the original compounds [73, 74]. In several studies, it has been suggested that UV/H_2O_2 could be responsible for an increase in controlled trihalomethane and haloacetic acid formation [75, 76]. Ultraviolet light (UV), hydrogen peroxide (H_2O_2), or combinations of these chemicals were studied in terms of their effects on the formation and speciation of haloacetamide, as well as the effects of free chlorine neutralizing additional H_2O_2 and providing residual disinfection [77]. According to the findings of [78], the UV/O_3 reaction was responsible for the degradation of 84% of the drug metoprolol. Recently, [79] demonstrated that the UV/Cl_2 process is highly effective for treating methoproprol. This is because the process is also responsible for generating OH radicals in addition to Cl radicals.

13.9 Mitigation of Aqueous Micropollutants

There is no doubt that the need for clean and risk-free water is steadily increasing, which means that there is a need for more effective strategies to reduce the amount of water that is polluted in the future. As a result of these strategies, we have to focus on reducing the number of hazardous chemicals that are used, which will in turn result in a reduction in the amount of pollution caused by these chemicals. The latter involves the cleaning up of contaminated sites as well as the treatment of wastewater and raw water to make them fit for consumption by humans. A large number of polluted areas are composed of diverse ecosystems, and these ecosystems are extremely complex to understand because they are composed of such a variety of ecosystems. Because there is insufficient knowledge about the system, the available understanding is often insufficient to optimize remediation techniques. However, the adsorption or retention capacity of these methods decreases as a result of interference with natural organic matter, depending on the amount.

In recent years, there has been a growing body of evidence that indicates that traditional methods such as pumps and treatment have proven to be relatively inefficient in the long run, because they require active treatment over a long period. These methods are not economically viable, primarily due to the large number of locations that have the potential to cause significant contamination of the water supply. As a result, it is important to consider strategies that are geared toward in-situ microbial or abiotic degradation as well as natural attenuation as a potential treatment option for long-term therapy.

13.10 Chemical Treatment of Wastewater Discharge

Currently, the chemical composition of surface wastewater is the combined result of urban, suburban, and rural water use activities of a vast majority of the world's population. Among the main sources of organic pollution in the environment, wastewater treatment plants have been identified as one of the primary sources. In addition to the physicochemical and biological processes that are used during the treatment of wastewater, a variety of chemical processes are also used, each of which may influence the effluent quality that is produced as a result. A good example of this would be the use of sodium hypochlorite as a disinfectant in the treatment of wastewater. This would be a method of oxidizing inorganic and organic components in the wastewater. It is due to this method that the final effluent that is discharged into the receiving water may contain by-products that have been produced during the process.

Another very important source of one of the most significant sources of aluminum in surface waters is the disposal of alum sludge from municipal sewage treatment facilities. As a result of this phenomenon, aluminum is present in surface waters. The by-product of coagulation and occultation processes, in which alum or aluminum sulfate is used for the removal of turbidity and/or color from raw water supplies, is alum sludge, which is formed as a result of processing raw water. It is also possible for pollutants to be present in drinking water if the potable water supply and distribution systems in place are not properly treated. As a result of ozonating drinking water that already contains bromide ions, a variety of organic compounds containing bromine are produced including bromoform and bromohydrins.

13.11 Conclusion

There is a need for an examination of the introduction of these micropollutants into the hydrological cycle from nonpoint sources, as well as discharge points. It is possible that drinking water can be derived directly from water that receives trace pollutants, thus making that water a possible source of pollution in the area. There is a consensus among economists that the ever-increasing demand for raw materials is a result of modern-day societies' desire for economic growth and a higher standard of living. It is important to note that among the different types of environments that can introduce contaminants into water, two distinct kinds of environments should be taken into consideration. Several phytohormones can be found in plants, including isoflavones and lignans. Because synthetic hormones can mimic the effects of natural hormones, doctors often recommend that these hormones regulate the functions of our bodies in a similar way to natural hormones. As a result of the release of these compounds into the environment by every living organism on this planet, every living organism has an impact on the quality of the water we drink. Conventional technologies are used to purify water that has been contaminated with pharmaceuticals, personal care products, and other synthetic chemicals.

Acknowledgment

It is with great appreciation that the authors would like to acknowledge the Department of Agronomy at Lovely Professional University for their constant support and encouragement throughout the research process.

Authors Contributions

In addition to contributing to the outline, the author is responsible for leading the draft and editing of the manuscript. It was S.R.D., P.D., and P.K. who conducted an extensive literature search and contributed to the writing of sections and the construction of figures and tables. In addition, they contributed to the construction of figures and tables. In addition to providing professional advice, P.K. helped revise the final version of the document and participated in its revision. In the end, all authors read and approved the final version of the manuscript.

Conflicts of Interest

None

References

1 Daughton, C.G. (2003). Cradle-to-cradle stewardship of drugs for minimizing their environmental disposition while promoting human health. I. rationale for and avenues toward a green pharmacy. *Environmental Health Perspectives* 111: 757e774.

2 Heberer, T. (2002). Tracking persistent pharmaceutical residues from municipal sewage to drinking water. *Journal of Hydrology* 266 (3–4): 175–189.

3 Lishman, L., Smyth, S.A., Sarafin, K. et al. (2006). Occurrence and reductions of pharmaceuticals and personal care products and estrogens by municipal wastewater treatment plants in Ontario, Canada. *Science of the Total Environment* 367 (2–3): 544–558.

4 Tavengwa, N.T. and Dalu, T. (2022). Introduction to emerging freshwater pollutants. In: *Emerging Freshwater Pollutants: Analysis, Fate and Regulations* (ed. T. Dalu and N.T. Tavengwa). Cambridge: Elsevier.

5 Galhardi, J.A., Luko-Sulato, K., Yabuki, L. et al. (2022). Rare earth elements and radionuclides. In: *Emerging Freshwater Pollutants: Analysis, Fate and Regulations* (ed. T. Dalu and N.T. Tavengwa). Cambridge: Elsevier.

6 Hashemi, S.H. and Kaykhaii, M. (2022). Azo dyes: sources, occurrence, toxicity, sampling, analysis and their removal methods. In: *Emerging Freshwater Pollutants: Analysis, Fate and Regulations* (ed. T. Dalu and N.T. Tavengwa). Cambridge: Elsevier.
7 Moodley, B., Kudu, I., and Pillay, V. (2022). Antiretrovirals in the environment. In: *Emerging Freshwater Pollutants: Analysis, Fate and Regulations* (ed. T. Dalu and N.T. Tavengwa). Cambridge: Elsevier.
8 Madikizela, L.M., Rimayi, C., Khulu, S. et al. (2022). Pharmaceuticals and personal care products. In: *Emerging Freshwater Pollutants: Analysis, Fate and Regulations* (ed. T. Dalu and N.T. Tavengwa). Cambridge: Elsevier.
9 Cristale, J. (2022). Flame retardants: sources, fate and occurrence in the aquatic environment. In: *Emerging Freshwater Pollutants: Analysis, Fate and Regulations* (ed. T. Dalu and N.T. Tavengwa). Cambridge: Elsevier.
10 Yardy, L., Al-Jaibachi, R., and Callaghan, A. (2022). Microplastics in freshwater ecosystems with special reference to tropical systems; detection, impact,and management. In: *Emerging Freshwater Pollutants: Analysis, Fate and Regulations* (ed. T. Dalu and N.T. Tavengwa). Cambridge: Elsevier.
11 Lapworth, D.J., Baran, N., Stuart, M.E., and Ward, R.S. (2012). Emerging organic contaminants in groundwater: a review of sources, fate and occurrence. *Environmental Pollution* 163: 287–303.
12 Pal, A., Gin, K.Y.H., Lin, A.Y.C., and Reinhard, M. (2010). Impacts of emerging organic contaminants on freshwater resources: review of recent occurrences, sources, fate and effects. *Science of the Total Environment* 408 (24): 6062–6069.
13 Environment Agency (1999). *Pesticides in the Aquatic Environment—1996.* Wallingford: Environment Agency.
14 Maund, S.J., Hamer, M.J., Warinton, J.S., and Kedwards, T.J. (1998). Aquatic ecotoxicology of the pyrethroid insecticide lambda-cyhalothrin: considerations for higher-tier aquatic risk assessment. *Pesticide Science* 54 (4): 408–417.
15 Oves, M. and Saghir Khan, M. (2016). Heavy metals: biological importance and detoxification strategies. *Journal of Bioremediation & Biodegradation* 07 (02).
16 Rainieri, S., Barranco, A., Primec, M., and Langerholc, T. (2017). Occurrence and toxicity of musks and UV filters in the marine environment. *Food and Chemical Toxicology* 104: 57–68.
17 Sanganyado, E., Lu, Z., Fu, Q. et al. (2017). Chiral pharmaceuticals: a review on their environmental occurrence and fate processes. *Water Research* 124: 527–542.
18 Gwenzi, W., Mangori, L., Danha, C. et al. (2018). Sources, behaviour, and environmental and human health risks of high-technology rare earth elements as emerging contaminants. *Science of the Total Environment* 636: 299–313.
19 Aus der Beek, T., Weber, F.A., Bergmann, A. et al. (2016). Pharmaceuticals in the environment—global occurrences and perspectives. *Environmental Toxicology and Chemistry* 35 (4): 823–835.
20 Dulio, V., van Bavel, B., Brorström-Lundén, E. et al. (2018). Emerging pollutants in the EU: 10 years of NORMAN in support of environmental policies and regulations. *Environmental Sciences Europe* 30 (1): 1–13.
21 Kandie, F.J., Krauss, M., Massei, R. et al. (2020). Multi-compartment chemical characterization and risk assessment of chemicals of emerging concern in freshwater systems of western Kenya. *Environmental Sciences Europe* 32 (1): 1–12.

22 Ngumba, E., Gachanja, A., Nyirenda, J. et al. (2020). Occurrence of antibiotics and antiretroviral drugs in source-separated urine, groundwater, surface water and wastewater in the peri-urban area of Chunga in Lusaka, Zambia. *Water SA* 46 (2): 278–284.

23 Ngumba, E., Gachanja, A., and Tuhkanen, T. (2016). Occurrence of selected antibiotics and antiretroviral drugs in Nairobi River Basin, Kenya. *Science of the Total Environment* 539: 206–213.

24 Madikizela, L.M., Ncube, S., and Chimuka, L. (2020). Analysis, occurrence and removal of pharmaceuticals in African water resources: a current status. *Journal of Environmental Management* 253: 109741.

25 Malkoske, T., Tang, Y., Xu, W. et al. (2016). A review of the environmental distribution, fate, and control of tetrabromobisphenol A released from sources. *Science of the Total Environment* 569: 1608–1617.

26 Melymuk, L., Robson, M., Csiszar, S.A. et al. (2014). From the city to the lake: loadings of PCBs, PBDEs, PAHs and PCMs from Toronto to Lake Ontario. *Environmental Science & Technology* 48 (7): 3732–3741.

27 Dris, R., Gasperi, J., Rocher, V. et al. (2015). Microplastic contamination in an urban area: a case study in greater Paris. *Environmental Chemistry* 12 (5): 592–599.

28 Wethington, D.M. and Hornbuckle, K.C. (2005). Milwaukee, WI, as a source of atmospheric PCBs to Lake Michigan. *Environmental Science & Technology* 39 (1): 57–63.

29 Birnbaum, L.S. and Staskal, D.F. (2004). Brominated flame retardants: cause for concern? *Environmental Health Perspectives* 112: 9e17.

30 Qu, W., Bi, X., Sheng, G. et al. (2007). Exposure to polybrominated diphenyl ethers among workers at an electronic waste dismantling region in Guangdong, China. *Environment International* 33: 1029e1034.

31 Vos, J.G., Dybing, E., Greim, H.A. et al. (2000). Health effects of endocrine-disrupting chemicals on wildlife, with special reference to the European situation. *Critical Reviews in Toxicology* 30: 71e133.

32 Moon, H.-B., Kannan, K., Lee, S.J., and Choi, M. (2007). Atmospheric deposition of polybrominated diphenyl ethers (PBDEs) in coastal areas in Korea. *Chemosphere* 66: 585e593.

33 de Boer, J., Wester, P.G., van der Horst, A., and Leonards, P.E.G. (2003). Polybrominated diphenyl ethers in influents, suspended particulate matter, sediments, sewage treatment plant and effluents and biota from the Netherlands. *Environmental Pollution* 122: 63e74.

34 Streets, S.S., Henderson, S.A., Stoner, A.D. et al. (2006). Partitioning and bioaccumulation of PBDEs and PCBs in Lake Michigan. *Environmental Science & Technology* 40: 7263e7269.

35 Birkett, J.W. and Lester, J.N. (2003). *Endocrine Disrupters in Wastewater and Sludge Treatment Processes*. London: Lewis Publishers.

36 Hornsby, A.G., Wauchope, R.D., and Herner, A.E. (1996). *Pesticide Properties in the Environment*. New York, NY: Springer-Verlag.

37 McKinlay, R., Plant, J.A., Bell, J.N.B., and Voulvoulis, N. (2008). Endocrine disrupting pesticides: implications for risk assessment. *Environment International* 34: 168e183.

38 Karen, D.J., Draughn, R., Fulton, M., and Ross, P. (1998). Bone strength and acetylcholinesterase inhibition as endpoints in chlorpyrifos toxicity to fundulus heteroclitus. *Pesticide Biochemistry and Physiology* 60: 167e175.

39 Leong, K.H., Benjamin Tan, L.L., and Mustafa, A.M. (2007). Contamination levels of selected organochlorine and organophosphate pesticides in the Selangor River, Malaysia between 2002 and 2003. *Chemosphere* 66: 1153e1159.

40 Focazio, M.J., Kolpin, D.W., Barnes, K.K. et al. (2008). A national reconnaissance for pharmaceuticals and other organic wastewater contaminants in the United States e II untreated drinking water sources. *Science of the Total Environment* 402: 201e216.

41 Kolpin, D.W., Furlong, E.T., Meyer, M.T. et al. (2002). Pharmaceuticals, hormones, and other organic wastewater contaminants in US streams, 1999– 2000: a national reconnaissance. *Environmental Science & Technology* 36 (6): 1202–1211.

42 Kolpin, D.W., Kalkhoff, S.J., Goolsby, D.A. et al. (1997). Occurrence of selected herbicides and herbicide degradation products in Iowa's groundwater, 1995. *Ground Water* 35: 679e688.

43 Gilliom, R.J., Barbash, J.E., Crawford, C.G. et al. (2006). The quality of our nation's waters-pesticides in the nations streams and ground water, 1992e2001. Circular 1291. U.S. Reston, VA: Geological Survey. 172.

44 Fytianos, K., Konstantinou, J., Kaidatzis, J., and Albanis, T. (1998). Multiresidue determination of insecticides and fungicides in fruits and vegetables by gas chromatography. *Bulletin of Environmental Contamination and Toxicology* 61: 684e689.

45 Padrón Sanz, C., Halko, R., Sosa Ferrera, Z., and Santana Rodríguez, J.J. (2004). Micellar extraction of organophosphorus pesticides and their determination by liquid chromatography. *Analytica Chimica Acta* 524: 265e270.

46 Watanabe, H. and Grismer, M.E. (2001). Diazinon transport through inter-row vegetative filter strips: micro-ecosystem modeling. *Journal of Hydrology* 247: 183e199.

47 Brady, J.A., Wallender, W.W., Werner, I. et al. (2006). Pesticide runoff from orchard floors in Davis, California, USA: a comparative analysis of diazinon and esfenvalerate. *Agriculture, Ecosystems & Environment* 115: 56e68.

48 Schlüsener, M.P. and Bester, K. (2006). Persistence of antibiotics such as macrolides, tiamulin and salinomycin in soil. *Environmental Pollution* 143: 565e571.

49 Bester, K. (2005). Fate of triclosan and triclosan-methyl in sewage treatment plants and surface waters. *Archives of Environmental Contamination and Toxicology* 49: 9e17.

50 Heidler, J., Sapkota, A., and Halden, R.U. (2006). Partitioning, persistence, and accumulation in digested sludge of the topical antiseptic triclocarban during wastewater treatment. *Environmental Science & Technology* 40: 3634e3639.

51 Richardson, S.D. (2003). Water analysis: emerging contaminants and current issues. *Analytical Chemistry* 75: 2831e2857.

52 Loos, R., Gawlik, B.M., Locoro, G. et al. (2009). EU-wide survey of polar organic persistent pollutants in European river waters. *Environmental Pollution* 157 (2): 561–568.

53 Zhou, J., Zhang, Z., Banks, E., Grover, D., & Jiang, J. (2009). Pharmaceutical residues in wastewater treatment works effluents and their impact on receiving river water. *The Journal of Hazardous Materials, 166(2-3)* 655–661.

54 Caliman, F.A. and Gavrilescu, M. (2009). Pharmaceuticals, personal care products and endocrine disrupting agents in the environment–a review. *CLEAN–Soil, Air, Water* 37 (4-5): 277–303.

55 Hawker, D.W. and Connell, D.W. (1988). Octanol-water partition coefficients of polychlorinated biphenyl congeners. *Environmental Science & Technology* 22 (4): 382–387.

56 Landrum, P.F. & Robbins, J.A. (1990). Bioavailability of sediment-associated contaminants: a review and simulation model. *Sediments: Chemistry and Toxicity of In-place Pollutants. In: Lewis, Chelsea, MI*, 237–263.

57 Lee Ii, H., Boese, B.L., Randall, R.C., and Pelletier, J. (1990). A method for determining gut uptake efficiencies of hydrophobic pollutants in a deposit-feeding clam. *Environmental Toxicology and Chemistry: An International Journal* 9 (2): 215–219.

58 Baudo, R. (1990). *Sediments: Chemistry and Toxicity of In-place Pollutants*. CRC Press.

59 Reynoldson, T.B. (1987). Interactions between sediment contaminants and benthic organisms. In: *Ecological Effects of in Situ Sediment Contaminants: proceedings of an international workshop held in Aberystwyth, Wales (1984)*. Pp. 53–66. Springer Netherlands.

60 Wilcock, R.J., Pridmore, R.D., Thrush, S.F. et al. (1993). Bioaccumulation and elimination of chlordane by selected intertidal benthic fauna. *Environmental Toxicology and Chemistry: An International Journal* 12 (4): 733–742.

61 Clark, K.E. and Mackay, D. (1991). Dietary uptake and biomagnification of four chlorinated hydrocarbons by guppies. *Environmental Toxicology and Chemistry: An International Journal* 10 (9): 1205–1217.

62 Harkey, G.A., Lydy, M.J., Kukkonen, J., and Landrum, P.F. (1994). Feeding selectivity and assimilation of PAH and PCB in Diporeia spp. *Environmental Toxicology and Chemistry: An International Journal* 13 (9): 1445–1455.

63 Fisher, S.W., Gossiaux, D.C., Bruner, K.A., & Landrum, P.F. (1993). Investigations of the toxicokinetics of hydrophobic contaminants in the zebra mussel (Dreissena polymorpha). In: *Zebra Mussels: Biology, Impacts, and Control*, Lewis Publishers, Boca Raton, FL. 1993, pp. 465–490, *8 fig, 4 tab, 46 ref. Ohio Sea Grant College Program Project.*

64 Nam, S.W., Choi, D.J., Kim, S.K. et al. (2014). Adsorption characteristics of selected hydrophilic and hydrophobic micropollutants in water using activated carbon. *Journal of Hazardous Materials* 270: 144–152.

65 Bolong, N., Ismail, A.F., Salim, M.R., and Matsuura, T. (2009). A review of the effects of emerging contaminants in wastewater and options for their removal. *Desalination* 239 (1–3): 229–246.

66 Snyder, S.A., Adham, S., Redding, A.M. et al. (2007). Role of membranes and activated carbon in the removal of endocrine disruptors and pharmaceuticals. *Desalination* 202 (1–3): 156–181.

67 Von Gunten, U. (2003). Ozonation of drinking water: part I. oxidation kinetics and product formation. *Water Research* 37 (7): 1443–1467.

68 Garcia, N., Moreno, J., Cartmell, E. et al. (2013). The application of microfiltration-reverse osmosis/nanofiltration to trace organics removal for municipal wastewater reuse. *Environmental Technology* 34 (24): 3183–3189.

69 Sui, Q., Huang, J., Deng, S. et al. (2010). Occurrence and removal of pharmaceuticals, caffeine and DEET in wastewater treatment plants of Beijing, China. *Water Research* 44 (2): 417–426.

70 Kim, J.W., Jang, H.S., Kim, J.G. et al. (2009). Occurrence of pharmaceutical and personal care products (PPCPs) in surface water from Mankyung River, South Korea. *Journal of Health Science* 55 (2): 249–258.

71 Huber, M.M., Ternes, T.A., and Von Gunten, U. (2004). Removal of estrogenic activity and formation of oxidation products during ozonation of 17α-ethinylestradiol. *Environmental Science & Technology* 38 (19): 5177–5186.

72 Sun, Q., Deng, S., Huang, J., and Yu, G. (2008). Relationship between oxidation products and estrogenic activity during ozonation of 4-nonylphenol. *Ozone: Science and Engineering* 30 (2): 120–126.

73 Altmann, J., Ruhl, A.S., Zietzschmann, F., and Jekel, M. (2014). Direct comparison of ozonation and adsorption onto powdered activated carbon for micropollutant removal in advanced wastewater treatment. *Water Research* 55: 185–193.

74 Wert, E.C., Rosario-Ortiz, F.L., Drury, D.D., and Snyder, S.A. (2007). Formation of oxidation byproducts from ozonation of wastewater. *Water Research* 41 (7): 1481–1490.

75 Dotson, A.D., Metz, D., and Linden, K.G. (2010). UV/H2O2 treatment of drinking water increases post-chlorination DBP formation. *Water Research* 44 (12): 3703–3713.

76 Metz, D.H., Meyer, M., Dotson, A. et al. (2011). The effect of UV/H2O2 treatment on disinfection by-product formation potential under simulated distribution system conditions. *Water Research* 45 (13): 3969–3980.

77 Chu, W., Gao, N., Yin, D. et al. (2014). Impact of UV/H2O2 pre-oxidation on the formation of haloacetamides and other nitrogenous disinfection byproducts during chlorination. *Environmental Science & Technology* 48 (20): 12190–12198.

78 Šojić, D., Despotović, V., Orčić, D. et al. (2012). Degradation of thiamethoxam and metoprolol by UV, O_3 and UV/O_3 hybrid processes: kinetics, degradation intermediates and toxicity. *Journal of Hydrology* 472: 314–327.

79 Nam, S.W., Yoon, Y., Choi, D.J., and Zoh, K.D. (2015). Degradation characteristics of metoprolol during UV/chlorination reaction and a factorial design optimization. *Journal of Hazardous Materials* 285: 453–463.

14

Physicochemical Properties and Stability of MOFs in Water Environments

Priya Saharan[1], Vinit Kumar[2], Indu Kaushal[3], Ashok Kumar Sharma[3], Narender Ranga[4], and Dharmender Kumar[5]

[1] *Centre of Excellence for Energy and Environmental Studies, DCRUST, Murthal*
[2] *Central Instrumentation Laboratory, DCRUST, Murthal*
[3] *Department of Chemistry, DCRUST, Murthal*
[4] *Department of Physics, DCRUST, Murthal*
[5] *Department of Biotechnology, DCRUST, Murthal*

14.1 Introduction

In recent decades, freshwater scarcity has become a global problem due to overexploitation, anthropogenic activities, and rapid population growth. In total, one-third of the world's population does not have access to drinkable water [1, 2]. Pollutants enter the water without treatment, causing serious water pollution and health risks. Therefore, water treatment, water saving, and recovery from wastewater or used water sources have become essential factors in a sustainable water management system. In recent decades, the fast-growing world population, urbanization, and industrialization cause great pressure on the consumption of fresh water, thus directly generating a massive amount of wastewater, and also unintentionally leading to serious pollution of water resources [2]. Refreshing contaminated water is the most sustainable way to tackle the above problems [3]. Numerous technologies have been established for wastewater remediation, such as adsorption, catalysis, coagulation, sedimentation, membrane filtration, and biological and advanced oxidation processes [1, 4].

Researchers are currently concentrating on improving existing technologies, developing new approaches, and providing eco-friendly, long-term solutions to water pollution problems. Over the last two decades, research has focused on a class of advanced porous materials referred to as metal-organic frameworks (MOFs). MOFs have been gaining tremendous attraction due to high surface area, permanent porosity, and customizable pores and composition. MOFs are exceptionally fit to perform a variety of functions. They can be utilized as either templates of an efficient selfless nature or precursors to accomplish extraordinary novel properties with high, unambiguous surface regions, a flexible range of natural ligands, different metal ions, and particular architectures.

Metal Organic Frameworks for Wastewater Contaminant Removal, First Edition. Edited by Arun Lal Srivastav, Lata Rani, Jyotsna Kaushal, and Tien Duc Pham.

The purpose of this chapter is to comprehensively discuss the most recently published research on the physicochemical properties of MOFs and the remediation of organic contaminants from wastewater, as well as future aspects in the area of aquatic environments.

14.2 Background and Future Scope of MOFs

The term MOF was coined in 1995, and the field gained traction in the late 1990s. Williams et al. released a paper in 1999 on HKUST-1, a MOF consisting of copper-based clusters and benzene tricarboxylate linkers. Yaghi et al. [5] followed up on their discoveries by investigating the synthesis of MOF 5, which is built from zinc-based clusters. The prominence of MOFs is determined by their importance in everyday life and their implications for the future of human technology. They are generally defined as a type of porous, network-structured material that can be made by self-assembly of organic ions and metal ligands to form a metal-ligand complex. They are also known as porous coordination polymers, which are made by the combination of metal ions and organic and inorganic ligands via coordination bonds. They have been rapidly developing as one of the major areas of coordination chemistry. The versatility by which the framework topology of the constituents can be altered has resulted in more than twenty thousand distinct MOFs being notified and studied in the last decade. The thermochemical stability of MOFs has presented them as suitable candidates for postsynthetic covalent Functionalization of organic and metal complexes. These capacities enable a significant improvement of gas storage in MOFs. They stimulated their exclusive investigation of organic reaction catalysis, small molecule activation, gas separation, and ion conduction. Methodologies are currently being explored to prepare nanocrystals and supercrystals of MOFs for amalgamation with different appliances [6].

Up until now, MOFs with eternal porosity are more versatile and numerous compared to porous materials of different classes. These improvements have made MOFs ideal carriers for fuel storage, carbon dioxide capture, catalysis applications, wastewater treatment, gas decontamination, light capture and energy conversion, separation technologies, drug delivery, etc. Information on MOFs has been increasing exponentially in recent years, but there are still significant gaps in our knowledge about their structure, stability, and other physicochemical characteristics. Detailed and systematic studies of the factors that lead to the disintegration of the crystal structure of some MOFs, the thermal stability, and the degradation mechanism are still lacking [7].

14.3 Techniques Used to Determine the Physicochemical Properties of MOFs

Before examining and identifying the ability of MOFs for different applications, a systematic description is desirable to give a precise conclusion about their structural and physicochemical properties. Various techniques have been reported to deduce the structure and properties of fabricated MOF materials. Methods that are used for this purpose include: powder X-ray diffraction (PXRD), Brunauer-Emmett-Teller

(BET) adsorption-desorption measurements for surface area analysis, thermogravimetric analysis (TGA), infrared spectroscopy (IR), and elemental and morphological analysis. In the next section, the basic principles of each characteristic technique and the kind of structure and morphology given by them are explored.

14.3.1 Powder X-Ray Diffraction (PXRD)

PXRD offers information about the structural configuration of MOFs. The basic constituting entities can be accomplished by relating the fabricated MOF's diffractogram to a previously published one in the literature, a simulated arrangement provided by single crystal X-ray and stored in a database, or by computational modeling. This method allows us to determine the crystalline nature, whether it is an amorphous or crystalline solid, and if it is crystalline, what type of crystal it is. Once the crystal structure of the MOF material is determined, crystallographic parameters (viz. unit cell size, lattice parameters, crystallite size, etc.) can be easily determined. Just like the fingerprint, every element or compound has its own distinct set of diffraction peaks that can be used to identify it. The crystallite dimension is obtained using the Scherrer equation:

$$D = \frac{K\,\lambda}{\beta \cos\theta} \tag{14.1}$$

K deduces shape factor, λ represents the X-ray wavelength, β is the Full Width at Half Maxima (FWHM), θ represents Bragg's angle in radians, and D is the mean size of the particle.

MOFs are often characterized using the XRD technique, allowing researchers to understand the crystallization of a material and verify the creation of a new material, hetero junction, or composites. In various water treatments, the MOF's stability was also procured by using the XRD method. After experimentation, the used MOFs were recovered from the media, washed, dried, and then re-examined using XRD. Any structural changes in comparison with the pristine MOF will predict its stability.

14.3.2 BET Surface Area Analyzer

Brunauer-Emmett-Teller (BET) theory clarifies gaseous molecules adsorption over solid structures, which is further used for total surface-area measurements. BET surface-area analyzer is usually applied to characterize MOFs because it permits the determination of the textural properties. This technique studies the adsorption of N_2 gas over the surface of solids. It is done at the boiling point of liquid N_2. The synthesized material has taken to higher temperatures under a vacuum to outgas. The outgassed material is retained in the sample tube under vacuum conditions. It is engrossed in a coolant bath of liquid N_2. The material allows deposition of the molecules of gas onto its surface during drying of sample holder. The resultant sample pressure is recorded and used to determine the surface area based on the quantity of gas adsorbed. The isothermal pattern depicts material pore distribution. Generally, MOFs are highly porous and have a large surface area (> 2000 m^2/g), which is depicted by a type 1 isotherm. The BET method calculates and compares the surface area of the MOF material.

14.3.3 Electron Microscopy and Elemental Analysis

Electron microscopic techniques include scanning electron microscopy (SEM) and transmission electron microscopy (TEM), which deliver information regarding the sample's surface topography, composition, grain size, particle size, crystallographic data, etc. Different software databases can be used to analyze the microscopic images. This characterization tool is very useful for the physicochemical analysis of MOFs and MOF-functionalized materials. These techniques, together with energy dispersive spectroscopy (EDS) or energy dispersive X-ray analysis (EDX or EDAX), permit examination of MOFs' elemental composition, both quantitative and qualitative. For example, Fe-MOFs were synthesized and characterized using the FESEM technique [8]. The obtained spindle-shaped hexagonal morphology having a length range of 0.9–1.0 μm and 0.5 μm diameter was observed for Fe-MOFs [9].

The elemental analysis technique is commonly performed to confirm the type of element inside the material. For its operation, samples are burned in the presence of excess oxygen. The analysis of volatile combustion products is utilized to calculate the composition of the material employing chromatography.

14.3.4 Thermogravimetric Analysis (TGA)

The quantitative changes observed in the material's mass as a temperature function is determined via TGA. It provides quantitative analysis associated with weight changes supplemented with the transition of the synthesized material and thermal degradation. Therefore, this technique can be utilized to ascertain the material's thermal stability.

14.3.5 Fourier-Transform Infrared (FT-IR)

The Fourier-transform infrared principle is grounded on the fact that particular frequencies are absorbed by a specific type of molecule allocated to a specific type of functional group. The IR spectra (4000–600 cm^{-1}) were executed on powdered samples. For instance, MOF-801 and calcium fumaric acid MOFs were prepared by the solvothermal process. The fabricated material was further used to remove fluoride from brick tea. The vibrational spectra indicated that the hydroxyl groups in the MOFs were substituted by fluoride, leading to its removal in an acidic medium [10].

14.4 Physicochemical Properties of MOFs and Their Effects on Various Applications

14.4.1 Porosity

MOF nanocrystals offer enhanced bio-availability, higher surface-area-to-volume ratios, higher control over MOF membrane production, and improved sorption kinetics compared to equivalent bulk phases. Due to its porous nature, they are becoming popular and have arisen as a superior agent for the material industry. The carbon chain length of the linker or the number of benzene rings in it determines pore size. In contrast, the inclusion of various substituents and functional groups in the linker permits further selectivity and unique chemical characteristics of the pores. For instance, Zhou and coworkers [11] illustrated a link installation approach in the presence of additional organic ligands having different chain length and functionality. The shape and pore size of the MOF cavities were altered on insertion of this ligand, purposely amid the defected sites of Zr6 clusters. The resulting MOF thus obtained had size-selective catalytic properties. Similarly, the addition of metal nodes controlled the pore size of MOFs owing to the difference in coordination bonding and radius of metal ions that can be presented by pore size change.

An isostructural MOF (SIFSIX-3) was successfully prepared and used for the adsorption of CO_2 gas [12]. The pore sizes varied between 3.84 Å, 3.80 Å, and 3.50 Å for SIFSIX-3-Zn, SIFSIX-3-Ni, and SIFSIX-3-Cu, respectively. The variation resulted in an increased uptake of CO_2 at lower concentrations and adsorption heats (from 45 to 54 kJ mol^{-1}). Wu et al. [13] have synthesized hierarchical zeolitic imidazolate frameworks (ZIF-8) in an aqueous system using CTAB surfactant and amino acid His in an organic solvent and used them to remove arsenic from wastewater. ZIF-8 is a member of the MOF family and is made up of porous crystalline materials containing open framework zeolite structures manufactured by using imidazolate-derived ligands and metal ions from transition element series'. The results illustrated that, due to the addition of CTAB and His, the porous nature of ZIF-8 nanospheres were enhanced drastically. The specific surface area and total specific pore volume rose as well (from 670 to 1167 m^2g^{-1} and 0.55 to 1.03 cm^3g^{-1}), which are much larger than the values obtained when regular ZIF-8 is produced in aqueous environments. The hierarchical pore structure thus demonstrated an excellent adsorption mechanism for arsenate ions.

14.4.2 Size and Morphology

MOFs are illustrious due to their structural diversity, different symmetry, and pore sizes and hence, by their properties. They exhibit different morphologies (Figure 14.1), making them a suitable candidate for various applications. The chemical reactivity and physical features of MOFs are highly governed by size and morphology. For instance, when MOF crystal size is reduced to the nano regime, they can possess better properties comparative to bulk counterparts. Fundamentally, the higher nucleation minimizes the size of the MOFs, which can be achieved via controlling nucleation rate and crystal size [14]. For the high yield of small MOFs' crystals, most of the precursors tend to be consumed prior to the crystal growth stage. That is attained by fast nucleation attained by the high energy given, such as microwave [15, 16] and sonochemical methods [17]. A change in solvent [7, 8] and the inclusion of modulators [9] have also been documented to improve MOF nucleation rates, resulting in narrower crystal size distributions. MOF development can also be slowed by interfering with the crystal formation stage, which limits seed aggregation. This method has been used successfully in various investigations where shorter time [10] and lower temperatures [11, 12] restrict crystal formation.

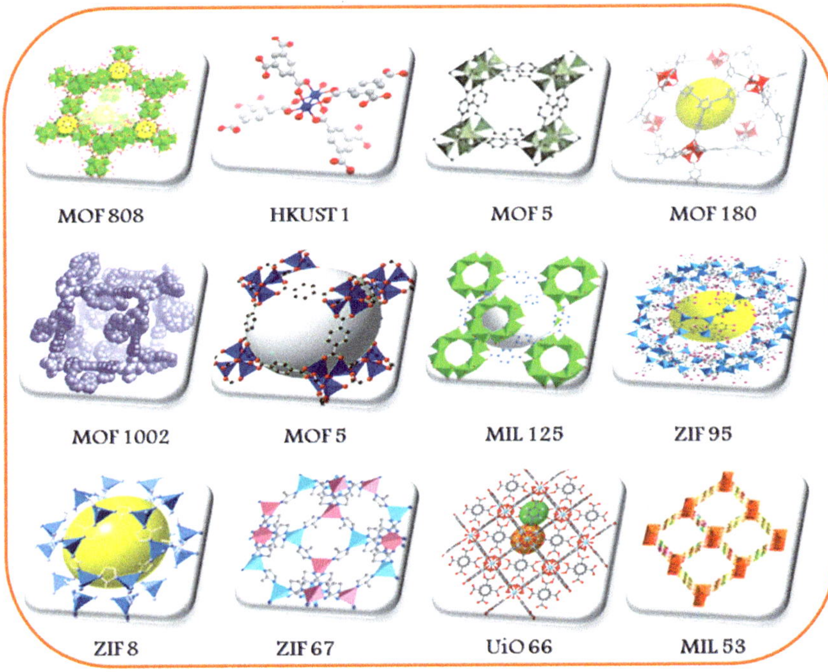

Figure 14.1 Different structures of metal-organic frameworks (MOFs).

Yin and his colleagues used a photoassisted multicomponent postsynthetic modification process in 2018 to link a zeolite imidazole framework-90 scaffold with a pyrimidinethione fragment. The synthesized framework were named zeolitic imidazolate frameworks ZIF90-THP (tetrahydropyran) and ZIF-90-THF (tetrahydrofuran). The BET surface area of ZIF90-THP and ZIF-90-THF at 77 K was 600 and 212 m^2/g, respectively. The synthesized materials were used to remove mercury (Hg[II]) with an adsorption capacity equal to 596 and 403 mg/g^{-1} using ZIF90-THP and ZIF-90-THF, respectively. The equilibrium was reached within ten minutes. The results showed that the surface areas and pore diameters of these MOFs make them a promising choice for rapid adsorption of Hg(II) ions because they allow ions to diffuse into functionalized sites.

Another type of hydrophillic MOF, MIL-101 (MIL: Material of Institute Lavoisier), can be fabricated utilizing metal precursors like Al, Fe, Cr, etc. They were abbreviated as (Cr) (MIL-101-Cr), aluminum (Al) (MIL-101-Al), and iron (Fe) (MIL-101-Fe). FESEM was used to examine the topography and particulate size of the synthesized MOF-Fe and amine-MOF-Fe, which revealed a hexagonal spindle-like shape with 0.9–1.0 m of length of and a diameter of 0.5 m. Functionalized amine-MOF-Fe showed a small structural alteration, with an increased length of 1.897 m and diameter of 0.451 m while retaining the hexagonal spindle-like structure. In a pseudo-second-order kinetics study on methylene blue dye, adsorption and adsorption capacity at the equilibrium was performed using methylene blue dye as the adsorbent. The results revealed that the amine-functionalized MOF removed more

MB at a lesser adsorbent dosage than the non-functionalized MOF, with MOF-Fe and amine-MOF-Fe reaching saturation points at 0.8 and 0.4 g/L of MOF loadings, respectively [10]. Liu et al. [18] synthesized three different types of ZIFs with different morphologies, viz. cubic, leaf-shaped, and dodecahedral ZIFs. It was observed that, within ten hours, the adsorption equilibrium for all ZIFs was attained, and about 95% of As(III) was eradicated from the water environment. Compared to dodecahedral ZIF, the leaf and cubic-shaped ZIFs displayed faster sorption at the initial stage, which caused desorption of the dyes; thus, they may be only appropriate for a rapid adsorption of As(III). In contrast, even while the concentration of residual As(III) increased somewhat after 24 hours for dodecahedral ZIFs, desorption was blocked when the adsorbent dose exceeded 0.4 g/L, indicating that dodecahedral ZIF is substantially more stable than the other two ZIFs.

Peng and coworkers described the synthesis of ZIF-8 MOFs using electro-spun PAN nanofiber membranes (ZIF-8–PAN) via an in-situ growth system, which comprised electro-spinning and hot pressing. The as-prepared sample's SEM scans indicated that thick ZIF-8 nanoparticles entirely coated the ZIF-8/PAN layer. Furthermore, the ZIF-8/PAN nanofiber's durability and tensile qualities were critical for the efficient capture of heavy-metal ions at high water flux and practical applications in complicated situations. These findings also showed that ZIF-8/PAN NF could endure a certain amount of water pressure while adapting to high-throughput heavy-metal ion filtration. The ZIF-8–PAN membrane displayed a high water flux of 12 000 $Lm^{-2}h^{-1}$ and enhanced filtration efficacy of 96.5% for Cu(II) removal. Furthermore, utilizing an amalgamation of adsorption and electrochemistry, the clearance rate of Cu(II) in four minutes was 34.1%. The highest capacity was 225.62 mg/g1. The experiments revealed that the better filtration performance of ZIF-8-PAN for Cu(II) ions was due to its large specific surface area, numerous reactive sites of the ZIF-8 particles, and lowered equivalent apparent pore size of the membrane due to hot pressing [19]. The removal of various water contaminants using different morphological structures of MOFs along with their adsorption efficacy/percentage have been illustrated in Table 14.1.

14.4.3 Chemical Reactivity

As mentioned earlier, MOFs are a class of materials highly used in many applications. Their structural versatility, tenability, and modularity is attributed to their high numbers of structures. The chemical reactivity of MOFs is an important feature that allows them to be considered for various applications. MOFs feature active sites present on the essential parts of the inorganic nodes or organic linkers. The pores act as trapping sites for the reactive guest molecules. MOF membranes possess high selectivity, permeability, and anti-fouling performance, making them suitable candidates for water applications [42]. Adsorption techniques are highly practiced sequestration methods of pollutants from water, being economical, convenient, and eco-friendly [43].

Compared to conventional adsorbents, MOFs offer structural diversity, adaptable pore properties, and greater thermal and mechanical stability, making them

Table 14.1 Removal of various water contaminants by MOFs based on different morphology.

MOFs	Shape/ Morphology	Contaminant	Adsorption capacity/ percentage	Ref
MIL-125	Decahedron	Hg (II)	946.0 mg/g	[20]
MOF-5	Rod-shaped	Pb (II)	412.7 mg/g	[21]
MOF-808-EDTA	Octahedral	22 metal ions	>99%	[22]
MOF-808-SO_4	Octahedral	Barium	131.1 mg/g	[23]
FJI-H9	Octahedral	Cd (II)	286 mg/g	[24]
thiol-laced MOFs	Octahedral	Hg	–	[25]
ZIF-8	Hexahedron	Pb^{2+} Cu^2	1119.80 mg/g 454.72 mg/g	[26]
ZIF-67	Wafer-like shape	Pb^{2+} Cu^2	1348.42 mg/g 617.51 mg/g	[26]
Co@ZIF-8	Spherical	Cu^2	1191.67 mg/g	[27]
Ni@ZIF-8	Layered	Cu^{2+}	1066.67 mg/g	[27]
ZIF-8 NPa	Rhombic dodecahedron	As(III) As(V)	9.49 mg/g 60.03 mg/g,	[28]
MIL-125–2-imino-4-thiobiuret	Decahedron	Hg(II)	946.0	[20]
ZIF-8/PAN nanofiber	Nanofiber	Cu^{2+}	96.5%	[19] P25
MOF-5	Not defined	Pb(II)	658.5	[21]
MOF-5/ZnO/C	Cubic	Methylene blue dye	86%	[29]
poly-o-anisidine/ ZnO/ ZIF-8	Rhombic	Malachite green	96%	[30]
ZIF-CNT	Rhombic dodecahedral	Malachite green	2034 mg/g	[31]
ZnO/MIL-101(Fe)	Spindle	Rhodamine B	97.1%	[9]
UiO-66-NH_2	Octahedral	Cr^{6+}	338.98 mg/g	[32]
MOF-235	Octahedral	Methylene blue	477 mg/g	[33]
ZIF-8@GO	Regular layered	Cu^{2+}	d 482.29 mg/g	[34]
Fe^3O^4@ZIF-8	Core-shell	Pb2+ Cu2+	719.42 301.33	[35, 36]
Fe-Mg MOF	Octahedral	Cd2+ Cu2+	191 mg/g 175 mg/g	[8]

Table 14.1 (Continued)

MOFs	Shape/ Morphology	Contaminant	Adsorption capacity/ percentage	Ref
Fe-MIL-100/ polydopamine	Polyhedral	Pb2+	99.8%	[37]
Ag-Fe/MOF	Rod and dot struture	Cd^{2+} Cu^{2+}	265 213 mg/g	[38]
ZIF-8	Flaky hexahedron	Pb^{2+}	1119.80	[26]
ZIF-67	Wafer-like shape	Pb^{2+}	1348.42	[26]
Fe_3O_4@SiO_2@ HKUST-1	Core-shell nanosphere	Hg^{2+}	264	[39]
MOF polymeric membranes	Membrane	Hg2+	98%	[40]
Cyclodextrin -MOF	Irregular	Cd2+	140.85	[41]

an ideal adsorption material. Various types of adsorption mechanisms have been employed for wastewater treatment. The different interactions that play a role in MOFs and adsorbates are electrostatic interaction, adsorption due to the formation of hydrogen bonding, Π-Π interactions, and acid-base interactions (Figure 14.2). In the electrostatic interaction mechanism, the dispersed MOFs with overall surface charge need to be attuned according to the solution pH to generate electric charge. The protonation and deprotonation can help control the MOFs' net charge. The electrostatic interactions does play a role afterward in the removal of the targeted adsorbates. Current trends illustrate that MOFs are highly impactful for extracting phosphate, arsenic, and fluoride [44].

In MOFs, their pores' morphological and chemical functionality can be designed to host inorganic or organometallic compounds in their frameworks and isolate guests via supramolecular forces. In a similar pattern to matrix isolation methods, the guest reactive molecule is leadingly stabilized through kinetic factors. There are a number of parameters controlling or directing the reactivity of MOFs. The selection of solvent, the temperature, or the gas atmosphere plays a vital role [7].

14.4.4 Chemical Stability

The chemical activity of MOFs is the sustainability and extent to which MOFs are resistant when exposed to various chemical environments such as acidic, basic, or aqueous environments. Generally, MOFs have low endurance, leading to structural degradation even when a moderate environment is present. The reason for lower

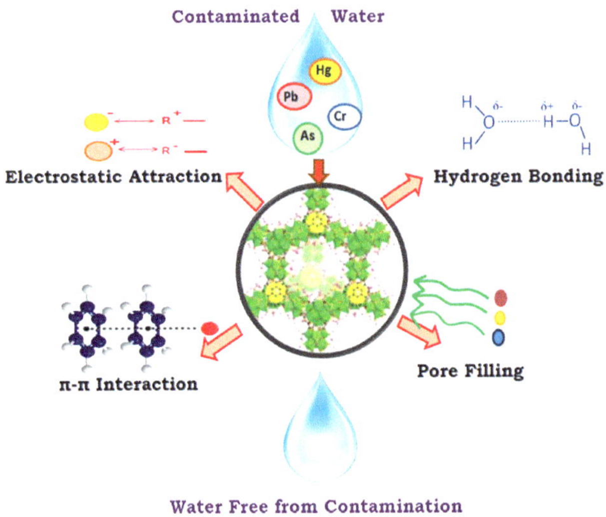

Figure 14.2 Different physisorption and chemisorption interactions of contaminants onto MOFs.

chemical stability is labile coordination bonds among MOF constituents [45]. Hence, the chemical stability is highly based on the intrinsic structures of MOFs, comprising charge density, basicity, configuration, ligand hydrophobic nature, etc. [46] Enormous structures of MOFs have been explored to date, but few have provided interesting features under non-inert atmospheres where they can be modified without disturbing a porous network. Alternate ways have been used to overcome the limited chemical stability of MOFs by transforming MOFs into metal compounds, carbonaceous materials, and their composites [47]. Metal ions with high valency and a high coordination number result in increased rigidity and less vulnerability to water molecules' attack. Wang et al. used high valence metal units including zirconium, iron, and chromium, as well as carboxylate-type bridging units, to create water-loving MOFs [43]. Low water stability is a big challenge and constraint on using MOFs for practical applications involving water [17]. In the same context, chemical stabilization is the most important criterion to form water-stable MOFs resistant to moist atmosphere [48].

In 2009, Low et al. [49] interrogated the impact of a moist environment on a sequence of MOFs via dual computational and experimental research. To some extent, this association was applicable to molecules such as phosphate, H_2S, SO_x, NO_x, NH_3, etc. These molecules cause friction with organic linkers and disrupt the ligand to cation bond. Chemical stability can be divided in subclasses such as moisture water stability, acidic or basic media stability, and harsh condition stability. The interplay between inorganic and organic moieties must be addressed while enhancing the stability. Another strategy to improve stability is to prevent or limit competing agents from entering the cation-ligand connection. The redox behavior, the geometry of coordination cation, the nuclearity and connectivity of the inorganic building-unit, the stiffness of the linker, the existence of open-metal sites, or defects are all important elements to consider when evaluating chemical stability [46]. Most of the time, assessing the chemical stability of a MOF is as simple as comparing the PXRD patterns of the sample before and after exposing the solid to a specific environment [50].

Actually, due to partial deformation of the MOF, linkers get released into the solution to some extent, leading to low pH. In addition, most of the times in the literature, insufficient examination of the solution is available, e.g., BET surface analysis, thermal gravimetric analysis, etc. [51].

14.4.5 Thermal Stability

A MOF compound's thermal stability may be referred to as the ability to resist irreversible structural deformation upon exposure to a high temperature. Thermal stability measures MOFs to sustain structural integrity under high temperatures or heat exposure. Materials with higher thermal stability are highly demanded and utilized in various industries. Poor thermal stability has been the main concern in the case of MOFs for industrial applications. The weak coordination bonds lead to lower thermal stability than zeolites and other porous inorganic materials. Thermal stability and hydrothermal stability are different aspects for studying MOF stability with temperature. The hydrothermal stability of MOFs indicates the metal's oxidation state, coordination number, and metal linker bond type (M–O or M–N). Depending on the kinetics of release required, selecting the appropriate inorganic sub-unit is critical for producing a MOF with suitable stability [6]. The weight loss in MOFs usually occurs because of changing chemical or physical aspects as temperature effects them [52]. Along with the increasing temperature, weight loss occurs. As the temperature goes beyond the decomposition limit, the discharge and combustion of the guest molecules and infringement of the metal-ligand bonds pursued by ignition of the organic spacer occur.

Usually, TGA is used for MOF compounds' thermal stability assessment. The loss found in the mass of the sample as a function of temperature in well-controlled conditions, generally under nitrogen gas, is measured using TGA. The thermal stability of MOFs is primarily determined by their structural formation and the strength of their chemical connections. The weight loss curve observed for the changes and a derivative of the weight loss curve can help to find the point where the weight released is most prominent. The MOF's curve pore volume can also be estimated by obtaining the solvated sample curve [6]. Wolfgang Kleist et al., after analyzing the thermal analysis of compound MIL-53 (Al), revealed that thermal stability relies on the proportion of linker molecules involved in the formation of MOFs [53].

Metal species are general constituents of MOFs. The thermal stability depends on the nature of a metal ion (as the oxidation state, ion radius, coordination number, and its interaction with a given linker through metal ligands) [54].

Depending on the donor heteroatoms, the linkers are mainly of two types: oxygenated and nitrogenated linkers affecting the thermal stability. Further, they are classified in two groups, i.e., aromatic and aliphatic, based on chemical structures. Based on the chemical structure point of view, the nature of linkers can govern the stability. Oxygenated linkers govern MOF structural designs, as nearly all the metal ions form coordination bonds with oxygen atoms with significantly varying strength. Pertaining to the nature of the molecule to which oxygen is attached, MOFs can be classified into various subgroups as carboxylate linkers sulfate or phosphonate. Because of their strong binding strength and chemical inertness, phosphonate and sulfate-based MOFs have superior thermal endurance than those with carboxylate and phenolate linkers that undergo decarboxylation and oxidation while heated. Apart from oxygenated linkers, nitrogen-containing ligands are capable of building MOF structures having divalent metal centers. In this scenario, ZIFs give convincing support. ZIF-8 is a member of the zeolitic imidazolate frameworks (ZIFs), a potential family of MOFs for water-phase adsorption with strong chemical and thermal stability and porosity comparable to zeolite [45]. Possibly single type of coordination bond exists in the ZIF family, which governs the formation of thermally stable divalent metal based MOFs. Additionally, MOFs containing basic ligands possess higher hydrothermal stabilities than acidic ligands [43]. For example, ZIF 8, constituting tetrahedral oriented imidazole anions, showed appreciable hydrothermal stability and was found to be stable in its original form, even under PBS buffer suspension at 37 ^{0}C for one week [6].

Thermal stability directly depends on the intrinsic structure, defects present, intrinsic linker network, etc. The POM-encapsulated HKUST-1 was shown to exhibit improved thermal stability. Reaching upto 483 K did not lead to complete decomposition of crystal structure [7]. Hydrophobicity can also help to upgrade the thermal stability of MOFs as depicted by many researchers [55].

Summarizing the above, a MOF structure with a short aromatic linker, metal ions with stable oxidation states, robust metal-ligand interaction, and a defect-free and densely packed molecular structure is a suitable material for industrial applications where thermal stability is a requisite. Hence, the utility of MOFs also need balanced features for wider applications than thermal stability. Therefore, the right balance between porosity, functionality, and stability shall remain a conciliation [51].

MOF stability and heat capacity have been examined employing TGA-DSC. Overall, little knowledge is available about the systematic thermal stability and heat capacity of MOFs. There have been no proper reports for thermal structure-property relationship. However, Mu and Walon employed a TGA coupled with DSC to investigate the thermal stability and heat capacity of typical MOF materials, including three corresponding ligands and five metals. They carried out TGA-DSC coupled TGA scan, and a DSC scan at an appropriate range of temperatures, scan rates, etc. Keeping constant pressure, the specific heat capacity (Cp) indicated the amount of energy required to heat one unit mass of a material by one Kelvin, which could be defined by the following Equation 14.2:

$$\mathrm{Cp(T)} = \frac{dH(T)}{\partial T} \cdot \frac{1}{m} \tag{14.2}$$

In the above equation, H is used for enthalpy and m denotes sample mass.

Individual Cp values at different temperatures can be determined using obtained DSC data using following Equation 14.3:

$$Cp(T) = \frac{DSC_{sample} - DSC_{baseline}}{DSC_{reference} - DSC_{baseline}} \bullet CP, \text{referenœ} \tag{14.3}$$

(Herein, Cp(T) denotes the sample's specific heat at temperature T; (DSC_{sample} – $DSC_{baseline}$ = h(s)) is the recorded DSC_{signal} of the sample and baseline correction; ($DSC_{reference}$ - $DSC_{baseline}$ = h(r)) is the DSC signal of sapphire corresponding to the baseline correction; and M for mass of sample). In this work, the thermal stability was claimed to be investigated systematically for the first time. The investigation outcomes demonstrated that the MOF's stability with temperature is very much dependent on the coordination atmosphere and corresponding surrounding ligands rather than the structural topology of the MOF [56].

14.4.6 Mechanical Stability

Viewing the past development phases of MOFs, a major concern has been the development of possibly larger porosity within the structures, which have resulted in weak intrinsic mechanical stability. Hence, studies and analyses of mechanical stability point of not much explored, and thus, insufficient literature is available addressing mechanical stability.

The main concern observed for mechanical stability is that some MOFs lose their crystallinity, as if a guest constituent is released from the structure. It is possibly due to capillary force-driven impact. To overcome such a problem, CO_2 can be used as a filler to replace the leaving guest molecules, thereby leading to negligible capillary force and none of the solvent surface tension, as well [57, 58]. Further, inducing hydrocarbon or fluorocarbon attachments with the metal site is another strategy that has been successfully practiced for Zr-MOFs with large channels. This way, hydrogen bonding is shortened between inorganic SBUs and the guest molecule.

Additionally, water cluster molecules introduced in pores also curtailed in size [59]. Although limited studies are available, some outcomes can be summarized. Mechanical loading capability gets elevated as the structure is denser with low porosities. So, to generalize, it can be depicted that inorganic subunits having high coordination numbers, reduced and rigid linkers, and reduced internal porosity leads to enhanced mechanical stability.

Summarizing that close-packed network, short and rigid linkers and suitable guest molecules as fillers are a few applied techniques to avoid the loss of porosity in MOFs, thereby enhancing the stability of MOFs [51]. Many researchers have worked on enhancing the mechanical stability of MOFs using various approaches. By incorporating water-soluble emulsifier, the thickening of MOF capsule walls has worked well to upgrade the mechanical stability of the MOF capsules [7].

14.5 Conclusion

MOFs are crystalline porous materials made up of metal ions or clusters, linked together by organic linkers. MOFs have attracted interest in various analytical applications due to their remarkable physicochemical features such as distinctive structures, high porosity with homogeneity and adjustability in the size of pores and cages, and good thermal and chemical durability. We conclude that using metal-organic frameworks in analytical chemistry provides an additional fascinating opportunity by expanding the analytical toolkit for trace metal analysis. MOFs have a high surface area, which leads to excellent extraction efficiency and enrichment factors. On the other hand, the instability of many MOFs in aqueous solutions is a considerable drawback. Chemical pre- or postsynthetic modification and surface functionalization are recent advancements in the creation of MOFs. These recent improvements in MOF production circumvent their well-known drawback of water instability, which limits their potential use to real-world sample analysis.

References

1 Saharan, P., Sharma, A.K., Kumar, V., and Kaushal, I. (2019). Multifunctional CNT supported metal doped MnO 2 composite for adsorptive removal of anionic dye and thiourea sensing. *Materials Chemistry and Physics* 221 (May 2018): 239–249.

2 Saharan, P., Kumar, V., Mittal, J. et al. (2021). Efficient ultrasonic assisted adsorption of organic pollutants employing bimetallic-carbon nanocomposites. *Separation Science and Technology* 56 (17): 2895–2908.

3 Saharan, P., Kumar, V., Sharma, A.K. et al. (2020). Scalable fabrication of chitosan-grafted silica bionanocomposite for the superb sequestration of anionic dye from aqueous solution. *Emergent Materials* 3 (6): 871–879.

4 Chaudhary, G.R., Saharan, P., Mehta, S.K. et al. (2013). Fast and efficient removal of hazardous congo red from its aqueous solution using γ-Fe2O3 nanoparticles. *Journal of Nanoengineering and Nanomanufacturing* 3 (2): 142–146.

5 Bosch, M., Zhang, M., and Zhou, H.-C. (2014). Increasing the stability of metal-organic frameworks. *Advanced Chemistry* 2014: 1–8.

6 Das, S., Kim, H., and Kim, K. (2021). Metal-organic frameworks history. *Inorganic Chemistry* 21 (05): 1–29.

7 Topologies, M.O.F., Stock, N., and Biswas, S. (2012). Synthesis of metal-organic frameworks (MOFs): routes to various. 933–969.

8 Abo El-Yazeed, W.S., Abou El-Reash, Y.G., Elatwy, L.A., and Ahmed, A.I. (2020). Facile fabrication of bimetallic Fe-Mg MOF for the synthesis of xanthenes and removal of heavy metal ions. *RSC Advances* 10 (16): 9693–9703.

9 Amdeha, E. and Mohamed, R.S. (2021). A green synthesized recyclable ZnO/MIL-101(Fe) for Rhodamine B dye removal via adsorption and photo-degradation under UV and visible light irradiation. *Environmental Technology (United Kingdom)* 42 (6): 842–859.

10 Ke, F., Peng, C., Zhang, T. et al. (2018). Fumarate-based metal-organic frameworks as a new platform for highly selective removal of fluoride from brick tea. *Scientific Reports* 8 (1): 1–11.

11 Yuan, S., Chen, Y.P., Qin, J.S. et al. (2016). Linker installation: engineering pore environment with precisely placed functionalities in zirconium MOFs. *Journal of the American Chemical Society* 138 (28): 8912–8919.

12 Lin, R.B., Xiang, S., Zhou, W., and Chen, B. (2020). Microporous metal-organic framework materials for gas separation. *Chem* 6 (2): 337–363.

13 Wu, Y.N., Zhou, M., Zhang, B. et al. (2014). Amino acid assisted templating synthesis of hierarchical zeolitic imidazolate framework-8 for efficient arsenate removal. *Nanoscale* 6 (2): 1105–1112.

14 Flügel, E.A., Ranft, A., Haase, F., and Lotsch, B.V. (2012). Synthetic routes toward MOF nanomorphologies. *Journal of Materials Chemistry* 22 (20): 10119–10133.

15 Ni, Z. and Masel, R.I. (2006). Rapid production of metal-organic frameworks via microwave-assisted solvothermal synthesis. *Journal of the American Chemical Society* 128 (38): 12394–12395.

16 Xiang, Z., Cao, D., Shao, X. et al. (2010). Facile preparation of high-capacity hydrogen storage metal-organic frameworks: a combination of microwave-assisted solvothermal synthesis and supercritical activation. *Chemical Engineering Science* 65 (10): 3140–3146.

17 Yin, W.H., Xiong, Y.Y., Wu, H.Q. et al. (2018). Functionalizing a metal-organic framework by a photoassisted multicomponent postsynthetic modification approach showing highly effective Hg(II) Removal. *Inorganic Chemistry* 57 (15): 8722–8725.

18 Liu, B., Jian, M., Liu, R. et al. (2015). Highly efficient removal of arsenic(III) from aqueous solution by zeolitic imidazolate frameworks with different morphology. *Colloids and Surfaces A: Physicochemical and Engineering Aspects* 481: 358–366.

19 Peng, L., Zhang, X., Sun, Y. et al. (2020). Heavy metal elimination based on metal organic framework highly loaded on flexible nanofibers. *Environmental Research* 188: 109742.

20 Adly, M.S., El-Dafrawy, S.M., Ibrahim, A.A. et al. (2021). Efficient removal of heavy metals from polluted water with high selectivity for Hg(ii) and Pb(ii) by a 2-imino-4-thiobiuret chemically modified MIL-125 metal-organic framework. *RSC Advances* 11 (23): 13940–13950.

21 Rivera, J.M., Rincón, S., Ben Youssef, C., and Zepeda, A. (2016). Highly efficient adsorption of aqueous Pb(II) with mesoporous metal-organic framework-5: an equilibrium and kinetic study. *Journal of Nanomaterials* 2016.

22 Peng, Y., Huang, H., Zhang, Y. et al. (2018). A versatile MOF-based trap for heavy metal ion capture and dispersion. *Nature Communications* 9 (1).

23 Peng, Y., Huang, H., Liu, D., and Zhong, C. (2016). Radioactive barium ion trap based on metal-organic framework for efficient and irreversible removal of barium from nuclear wastewater. *ACS Applied Materials & Interfaces* 8 (13): 8527–8535.

24 Xue, H., Chen, Q., Jiang, F. et al. (2016). A regenerative metal-organic framework for reversible uptake of Cd(II): from effective adsorption to: in situ detection. *Chemical Science* 7 (9): 5983–5988.

25 Yee, K.K., Reimer, N., Liu, J. et al. (2013). Effective mercury sorption by thiol-laced metal-organic frameworks: in strong acid and the vapor phase. *Journal of the American Chemical Society* 135 (21): 7795–7798.

26 Huang, Y., Zeng, X., Guo, L. et al. (2018). Heavy metal ion removal of wastewater by zeolite-imidazolate frameworks. *Separation and Purification Technology* 194 (September 2017): 462–469.

27 Shen, B., Wang, B., Zhu, L., and Jiang, L. (2020). Properties of cobalt-and nickel-doped zif-8 framework materials and their application in heavy-metal removal from wastewater. *Nanomaterials* 10 (9): 1–15.

28 Jian, M., Liu, B., Zhang, G. et al. (2015). Adsorptive removal of arsenic from aqueous solution by zeolitic imidazolate framework-8 (ZIF-8) nanoparticles. *Colloids and Surfaces A: Physicochemical and Engineering Aspects* 465: 67–76.

29 Hussain, M.Z., Schneemann, A., Fischer, R.A. et al. (2018). MOF derived porous ZnO/C nanocomposites for efficient dye photodegradation. *ACS Applied Energy Materials* 1 (9): 4695–4707.

30 Mohd, S., Wani, A.A., and Khan, A.M. (2022). ZnO/POA functionalized metal-organic framework ZIF-8 nanomaterial for dye removal. *A Chemical Engineer's* 3 (April): 100047.

31 Abdi, J., Vossoughi, M., Mahmoodi, N.M., and Alemzadeh, I. (2017). Synthesis of metal-organic framework hybrid nanocomposites based on GO and CNT with high adsorption capacity for dye removal. *Chemical Engineering Journal* 326: 1145–1158.

32 Chen, P., Wang, Y., Zhuang, X. et al. (2023). Selective removal of heavy metals by Zr-based MOFs in wastewater: new acid and amino functionalization strategy. *The Journal of Environmental Sciences (China)* 124: 268–280.

33 Haque, E., Jun, J.W., and Jhung, S.H. (2011). Adsorptive removal of methyl orange and methylene blue from aqueous solution with a metal-organic framework material, iron terephthalate (MOF-235). *The Journal of Hazardous Materials* 185 (1): 507–511.

34 Li, D. and Xu, F. (2021). Removal of Cu (II) from aqueous solutions using ZIF-8@ GO composites. *Journal of Solid State Chemistry* 302 (July): 122406.

35 Yadav, A., Bagotia, N., Sharma, A.K., and Kumar, S. (2021). Advances in decontamination of wastewater using biomass-basedcomposites: a critical review. *Science of the Total Environment* 784 (April): 147108.

36 Jiang, X., Su, S., Rao, J. et al. (2021). Magnetic metal-organic framework (Fe^3O^4@ ZIF-8) core-shell composite for the efficient removal of Pb(II) and Cu(II) from water. *The Journal of Environmental Chemical Engineering* 9 (5): 105959.

37 Sun, D.T., Peng, L., Reeder, W.S. et al. (2018). Rapid, selective heavy metal removal from water by a metal-organic framework/polydopamine composite. *ACS Central Science* 4 (3): 349–356.

38 Abo El-Yazeed, W.S., Abou El-Reash, Y.G., Elatwy, L.A., and Ahmed, A.I. (2020). Novel bimetallic Ag-Fe MOF for exceptional Cd and Cu removal and 3,4-dihydropyrimidinone synthesis. *Journal of the Taiwan Institute of Chemical Engineers* 114: 199–210.

39 Huang, L., He, M., Chen, B., and Hu, B. (2015). A designable magnetic MOF composite and facile coordination-based post-synthetic strategy for the enhanced removal of Hg2+ from water. *Journal of Materials Chemistry A* 3 (21): 11587–11595.

40 Gao, J., Li, Z., Deng, Z. et al. (2021). Rapid removal of mercury from water by novel mof/pp hybrid membrane. *Nanomaterials* 11 (10): 1–9.

41 Liu, C., Wang, P., Liu, X. et al. (2019). Ultrafast removal of cadmium(II) by green cyclodextrin metal-organic-framework-based nanoporous carbon: adsorption mechanism and application. *Chemistry: An Asian Journal* 14 (2): 261–268.

42 Ren, J., Langmi, H.W., North, B.C., and Mathe, M. (2015). Review on processing of metal-organic framework (MOF) materials towards system integration for hydrogen storage. *International Journal of Energy Research* 39 (5): 607–620.

43 Zhang, S., Wang, J., Zhang, Y. et al. (2021). Applications of water-stable metal-organic frameworks in the removal of water pollutants: a review ☆. *Environmental Pollution* 291 (May): 118076.

44 Fan, S., Qu, Y., Yao, L. et al. (2021). Journal of colloid and interface science MOF-derived cluster-shaped magnetic nanocomposite with hierarchical pores as an efficient and regenerative adsorbent for chlortetracycline removal. *Journal of Colloid and Interface Science* 586: 433–444.

45 Russo, V., Hmoudah, M., Broccoli, F. et al. (2020). Applications of metal organic frameworks in wastewater treatment: a review on adsorption and photodegradation. *Frontiers In Chemical Engineering* 2 (October): 1–13.

46 Ding, M., Cai, X., and Jiang, H.L. (2019). Improving MOF stability: approaches and applications. *Chemical Science* 10 (44): 10209–10230.

47 Wu, H.B. and Lou, X.W. (2017). Metal-organic frameworks and their derived materials for electrochemical energy storage and conversion: promises and challenges. *Science Advances* 3 (12): 1–17.

48 Advanced Science (2019). Liu – structural engineering of low-dimensional metal organic frameworks synthesis properties.pdf.

49 Yazaydin, A.Ö., Benin, A.I., Faheem, S.A. et al. (2009). Enhanced CO 2 adsorption in metal-organic frameworks via occupation of open-metal sites by coordinated water molecules. *Chemistry of Materials* 21 (8): 1425–1430.

50 Quintero-Álvarez, F.G., Rojas-Mayorga, C.K., Mendoza-Castillo, D.I. et al. (2022). Physicochemical modeling of the adsorption of pharmaceuticals on MIL-100-Fe and MIL-101-Fe MOFs. *Adsorption Science & Technology* 2022.

51 Mouchaham, G., Wang, S., and Serre, C. (2018). Metal-organic frameworks: applications in separations and catalysis, first edition. (i). 1–28.

52 Jang, S., Song, S., Lim, J.H. et al. (2020). Application of various metal-organic frameworks (Mofs) as catalysts for air and water pollution environmental remediation. *Catalysts* 10 (2).

53 Marx, S., Kleist, W., Huang, J. et al. (2010). Tuning functional sites and thermal stability of mixed-linker MOFs based on MIL-53(Al). *Dalton Transactions* 39 (16): 3795–3798.

54 Liu, D., Gu, W., Zhou, L. et al. (2022). Recent advances in MOF-derived carbon-based nanomaterials for environmental applications in adsorption and catalytic degradation. *Chemical Engineering Journal* 427 (May 2021): 131503.

55 Qiu, S., Wang, Y., Wan, J. et al. (2020). Applied surface science enhancing water stability of MIL-101 (Cr) by doping Ni (II). *Applied Surface Sciene* 525 (December 2019): 146511.

56 Mu, B. and Walton, K.S. (2011) Thermal analysis and heat capacity study of metal À organic frameworks.

57 Farha, O.K. and Hupp, J.T. (2010). Rational design, synthesis, purification, and activation of metal-organic framework materials. *Accounts of Chemical Research* 43 (8): 1166–1175.

58 Shalaby, M., Farag, S., Haddad, A. et al. (2020). Collagen polymer and magnetic collagen nanocomposite recycled from waste to reduce polluted water toxicity. *Polymers and Polymer Composites* 29: 1–13.

59 Deria, P., Chung, Y.G., Snurr, R.Q. et al. (2015). Water stabilization of Zr6-based metal-organic frameworks via solvent-assisted ligand incorporation. *Chemical Science* 6 (9): 5172–5176.

15

Metal-Organic Framework Adsorbents for Indutrial Heavy-Metal Wastewater Treatment

Gopal Sonkar

University of Delhi

15.1 Introduction

This chapter's purpose is to recognize the function of adsorbents under metal-organic frameworks (MOFs) for extracting metal contaminants from industrial wastewater. MOFs clearly have two classifications: organic (bio) and inorganic [1]. This chapter presents research related to bio adsorption. The bio adsorption method plays an important part in MOFS for eliminating metal contaminants from pollutant wastewater [2]. The growing demand for MOF techniques indicates that they are a desirable approach due to their low cost and efficiency [3]. The adsorbent materials called MOFs are made up of a well-organized arrangement of positively charged metal ions that are bound together by biological linker molecules [1]. The metals ions assemble themselves into bulges that connect the linkers' arms into an iterative cage [4]. MOFs have a very large internal surface area due to their hollow structure [5. 6]. MOFs provide a wide range of structural variety such as atomic-level physical uniformity, homogeneous pore assemblies, an abundance of variations and adaptability in network topology, programmable porosity, geometry, size, and chemical functionality in comparison to other adsorbent materials [7]. Scientists could design materials that selectively adsorb pollutants into specialized niches inside the MOF's structure by constructing it from a multiplicity of metal atoms and organic linkers [8]. Thus, there is a huge opportunity for the efficient integration and study of MOFs in a wide range of heavy metal contaminant (HMC) removal applications. MOFs are more versatile than any other known material class because they may be assembled randomly [9].

The 1989 Robson et al. study described a "framework comprised of three-dimensionally connected rod-like segments" as the first MOFs [5]. The field had begun to acquire attention in the late 1990s, and the term "metal-organic frameworks (MOFs)" was coined in 1995 [10]. Using copper-based clusters and benzene tricarboxylate linkers, Williams et al. published a study on a MOF called HKUST-1 in

Metal Organic Frameworks for Wastewater Contaminant Removal, First Edition. Edited by Arun Lal Srivastav, Lata Rani, Jyotsna Kaushal, and Tien Duc Pham.

1999 [11]. Later that same year, Yaghi et al. disclosed a MOF they called MOF-5; it was constructed from zinc-based clusters and benzene dicarboxylate linkers [1, 12]. It set new benchmarks due to its extreme porosity. The interior surface chemistry of MOFs can be modified, unlike that of other porous materials, which allows for modification of characteristics like hydrophilicity and acidity. However, the characteristic that distinguishes MOFs from other materials is their extremely high porosity (up to 90% free volume) and enormous internal surface areas (ISA) [11]. Some MOFs have been reported to have an ISA that is greater than 10,000 m^2/g^{-1}. Synthesis of MOFs typically takes place through either hydro- or solvothermal processes [1]. Recent years have seen an expansion of this practice into ecologically friendlier processes. In addition, flaws can be purposefully designed into a material in order to give more adsorption or catalytic sites. Nevertheless, functionalization of frameworks might bring about the most significant changes to features of MOFs. For instance, the active site of a catalyst could be located at the ligand, metal node, or as a guest encased inside pores [1].

The previously mentioned organic adsorbent experiemental studies were performed on MOFs. These experimental studies and literature reviews do not provide an assessment on bio-based adsorbents as a consolidated format under MOFs. The present research contributes an evaluation of bio adsorbents used under MOFs by combining the results of earlier experiments with the arrangement and evaluation of the validity of bio adsorbents for the removal of heavy metals.

15.2 The Applications of MOFs

Several applications are being developed utilizing cage-like structures in MOFs for use in a wide variety of industries. Some examples of these applications include HMC removal, gas separation and storage, liquid separation and purification catalysis, and detection and electrochemical energy storage. MOFs have been put to use as one-of-a-kind precursors in the production of organic functional materials that can be designed in ways that have never been seen before [1]. These structures include metal-based mixtures and their compounds. Organic materials are now receiving a lot of attention because of the wide variety of uses they may have [1].

15.3 Comparison Between MOF Adsorbents and Bio-Based Adsorbents

Chemical adsorbents like polymers, activated carbon, and activated charcoal are among the most often utilized adsorbents. Nevertheless, each has its limitations and drawbacks. Since a few decades ago, researchers have been looking for non-living organic waste that falls under natural alternative (bio) adsorbents [13–19]. The use of natural and environmentally friendly adsorbents is currently a top priority [3, 19–21]. The removal of HMC is a primary concern of a great number of researchers, and as a result, a great deal of investigation into various natural (bio) adsorbents has been carried out [22–25]. Many waste organic (bio) products have been reported to have adsorption capabilities, including waste tea and coffee [21, 26], activated carbon [27, 28], sawdust [20, 29, 30], peanut shells [31], rice husks [32], and orange peel [33]. Typically, cellulose in these organic bio products has strong heavy metal adsorption characteristics [3, 21]. The studies reportedly used watermelon rind to adsorb nickel and heavy copper metals [34]. Mishra and Patel (2009) investigated the adsorption of zinc and lead metals using affordable adsorbents [35]. The initial amounts of adsorbent, metal ion concentration, pH, and temperature were discovered to be factored into the adsorption process [8, 35]. Neem leaf powder was used to study chromium elimination from aqueous solutions, and it was discovered that pH significantly impacted the leaf powder's capacity for equilibrium adsorption [36]. Coffee husks were utilized by Waleska et al. (2008) as adsorbents for the purpose of removing copper, zinc, and cadmium from an aqueous solution. They found that the adsorption capability of each metal ion was at its highest at certain pH values [37].

Ideally, bio adsorption methods can be effective as a possible solution [38]. Bio adsorbents have the potential to be employed as an alternative conventional wastewater treatment method for the removal of metal contaminants [39]. Bio adsorbents procedures must be carried out to protect aquatic ecosystems and water resources at the same time [40]. The term 'bio adsorbent' refers to "biological matter including non-living biological matter such as orange peel, untreated coffee grounds, rice husk, and onion peel, including living biological matter like algae, moss, yeast, etc." [40–43]. There are various sources of bio adsorbent including both living and non-living. The bio adsorbent has become an increasingly popular organic adsorbent for adsorption purposes [19, 21, 44]. Organic adsorbents are a more environmentally friendly alternative beside using chemical adsorbents in adsorption methods. The organic adsorbent might be something like algae, which is a living organism [45–47], moss [48], yeast [7], or non-living organic products like orange peel [49], untreated coffee grounds and tea waste [8], potato peel waste, [50] etc. When organic adsorbents are utilized in an adsorption process, the process is known as the biosorption process [38].

The conventional methods used to remove metal contiminants are "biological treatment, ion exchange, electrochemical degradation, membrane-based separation technology, adsorption, chemical precipitation coagulation or flocculation, and filtering techniques" [4, 51–54] (Figure 15.1).

Several conventional methods (Figure 15.1) have been used for metal contaminant treatment. Among them, chemical precipitation is a frequent method in the wastewater treatment field. Electrochemical degradation and membrane-based separation technology are typically used to treat wastewater in this process. These methods have significant drawbacks such as excessive chemical usage and sludge production [55–57].

The membrane process and ion exchange have a high cost of operation and maintenance [40]. Chemical precipitation is by far the most common among adsorbents [62], but this adsorbent comes at a hefty price. Table 15.1 displays the primary merits and demerits of various conventional treatment options. This has led researchers to focus on discovering bio-based adsorbents that may be used at a lower cost. The adsorption method is straightforward to implement when extracting metal-ion contaminants.

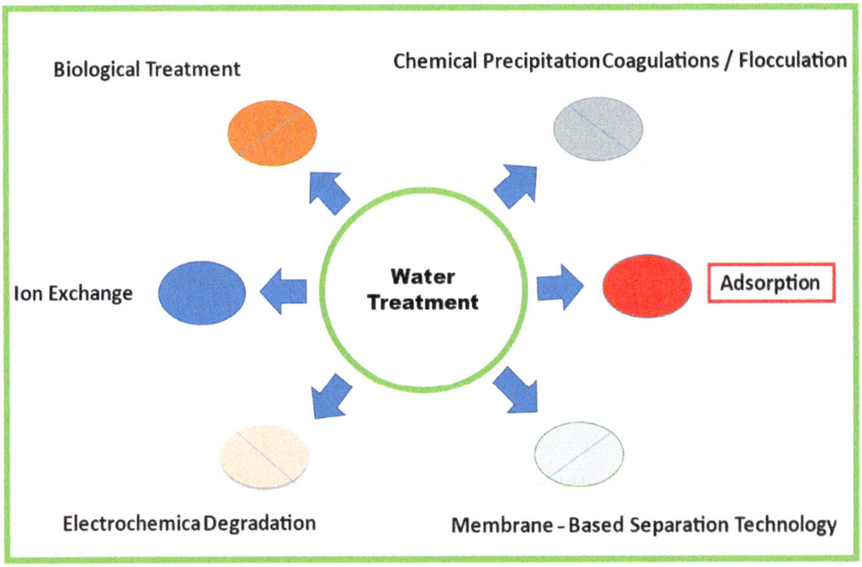

Figure 15.1 Conventional wastewater treatment methods.

Table 15.1 Merits and demerits of various conventional treatment methods.

Method	Merit	Demerit	References
Ion exchange	Limited space is required, low generation of sludge, recovery of metal is convenient	High operation and maintenance cost, fouling problem	[45, 58]
Coagulation / flocculation	Well settled sludge, bacterial inactivation potential	A large amount of sludge generation, pH and temperature dependent	[43]
Complexation	Low cost, simple operation	Toxic sludge is generated	[26, 59]
Chemical precipitation	Low cost, reliable	Toxic sludge not suitable for low concentration	[60]
Flotation	Suitable for fine particles, low retention time	Toxic sludge and metal concentration dependent	[48, 61]
Membrane process	Limited space is required, low waste	High operation and maintenance cost, fouling problem	[40]

15.4 Heavy Metal Contaminant Sources and Impacts

HMCs have become a really problematic environmental pollution issue [63]. HMCs are a direct consequence of the dramatic acceleration of industrialization in the modern age [64]. HMC entrance into aquatic systems gives the most cause for concern [16]. Extant HMC threats to living beings are due to their inherent chemical reaction [63]. There has been growing awareness related to this issue in the past few decades. The HMC discharge into water streams is caused by various industrial processes. HMCs are incapable of biodegradation so, as a result, they pose a considerable risk to all types of living beings [24]. Cadmium, nickel, copper, lead, zinc, chromium, and mercury are the most important HMCs that have toxicity in nature. HMCs have accumulated by waste emitted from industries such as tannery, steel, paint and pigments, metallurgy, mining, electroplating, petroleum refining, battery manufacturing, etc. (see Figure 15.2) [39, 64–66].

Figure 15.2 presents many industries that are producing various HMCs in the environment. HMCs are responsible for various adverse effects on living beings such as nausea, vomiting, pulmonary fibrosis, skin dermatitis, central nervous function disturbance, stimulus in lungs, changes in blood composition, and cancerous diseases and disorders [67]. HMCs can make their way into the food chain and harm the living being [63]. HMCs must be effectively eliminated from wastewater to shield the environment and living beings from potentially damaging effects. The growing concern around protecting the environment from HMCs has been encouraged by many countries through stricter laws. Worldwide scientists are constantly figuring out how to mitigate harmful consequences of HMCs [68].

Human civilization has utilized heavy metals for a long time. Fertilizers, sewage sludge, composts, and other waste items may include traces of these contaminants, and contaminants may also make their way into water and soil resources. HMCs themselves are not the problem, but issues arise when HMCs extend beyond safe

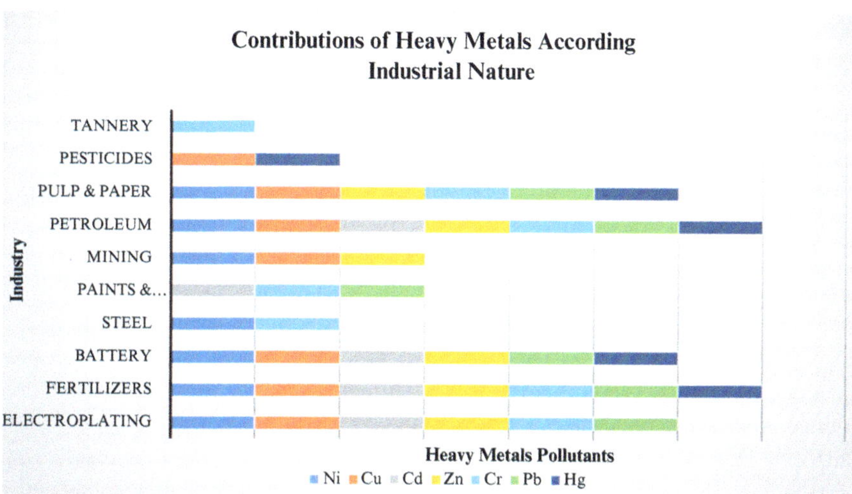

Figure 15.2 Metals contributions based on their industrial origin. *Sources:* "Y. C. Sharma, G. Prasad and D. C. Rupainwar, 'Removal of Ni (II) from aqueous solutions by sorption,' *International Journal of Environmental Studies*, no. 37, p. 183 191, 1991; G. Yan and T. Viraraghavan, 'Heavy metal removal in a biosorption column by immobilized M. Rouxii biomass,' *Bioresource Technology*, no. 78, p. 243 249., 2001; E. Remoudaki, A. Hatzikioseyian, K. Tsezos and M. Tsezos, 'The mechanism of metals precipitation by biologically generated alkalinity in biofilm reactors,' *Water Research*, no. 37, p. 3843 3854, 2003; G. Zhao, M. Li and H. Hu, 'Dissociation and removal of complex chromium ions containing in dye wastewaters,' *Separation and Purification Technology*, no. 43, p. 227 232, 2005."

levels in the ecosystem. The problem can be explained as any metal present beyond a specific limit that can bring harm to living beings.

Copper, nickel, lead, cadmium, mercury, chromium, silver, aluminum, selenium, caesium, cobalt, manganese, molybdenum, strontium, and uranium are some of the metals that are used in the environment in various ways to meet human needs [69]. Metals such as zinc, lead, copper, and cadmium are found naturally in ores and are utilized in a wide variety of productive processes and compounds like electroplating, alloys, plastics, paints and pigments, batteries, and many more.

The electroplating, battery, motor vehicle, petroleum, and steel sectors, as well as aircraft manufacturing are discharged into wastewater with nickel contamination. Steel, battery, and alloy production are three primary industries that utilize the most nickel metals [70]. Copper contaminants are elements that can be found in nature from various sources. It can exist either as a free element or joint with other elements to formulate different mixtures. The disintegration of rocks, soils, and industrial activities are major sources of copper contaminants. Industries like battery, petroleum, pesticide, pulp and paper, mining, electroplating, fertilizer, metal cleaning and plating, and metalworks are some of the various sectors that use copper and dispose of copper contaminants in their wastes and wastewater [71, 72]. Copper contaminants can have several adverse effects on people such as irritation of the central nervous system, gastrointestinal system, liver, and kidneys. Large amounts can possibly cause extensive damage to capillaries, the liver, and kidneys [42]. Even aquatic species living in their native water ecosystem are at risk. It has very harmful effects on humans and other animals and is therefore classified as a heavy metal. Sometimes, copper contaminants wreak havoc on the cardiovascular system, disrupt vitamin D metabolism in the kidneys, weaken bones, and produce life-threatening anemia [73].

The wastewaters produced by battery, petroleum, steel, plastic, alloy, and mining sectors are the most significant sources of cadmium. Also, cadmium is utilized in producing television screens, anticorrosion agents in cosmetics [72], and as a barrier to nuclear fission [73].

The primary sources of zinc contaminants include air pollution, agricultural runoff, pharmaceutical waste, wastewater from homes, and industrial waste. The zinc smelters, zinc mines, battery, petroleum, toy, alloy, detergent, coal combustion, and steel industries are also contributors [72]. Zinc has numerous negative consequences on both people and animals. Zinc fumes can lower the lung volume and induce other symptoms, such as chest pain, coughing, and nausea if inhaled. The corrosive effects, ulceration, blistering, and irreversible scarring that result from skin contact are all caused by this. Mouth, throat, and stomach damage may cause nausea, vomiting, and abdominal pain. It is also a leading cause of Alzheimer's [74].

Nickel is a transition metal that is found hard but ductile in nature, and it is a silvery-white color. Nickel has a high shine, and it is resilient to degradation. Nickel compounds can have a harmful effect on people and animals when present in sufficient quantities. The adverse effects of nickel may be broken down into several different types of body parts depending on the route of exposure. These include immunologic, neurological, reproductive, and carcinogenic consequences (inhalation, oral, or cutaneous). That is conditional on the length of time the substance was in the body, which can be either short-term (less than one hundred days) or long-term (one hundred days or more). The most common effect is an allergic skin reaction in humans. Beside this, it can affect the lungs and nervous systems and may produce tumotd and cancers [75].

15.5 Adsorption

Research is being conducted with the purpose of developing HMC-removal technologies that can be efficient and cost-effective [41, 76–78]. The removal of metals can now be done easily, efficiently, and cost-effectively through a process called bio adsorption. The bio adsorption can be a solid-liquid combination, gas-liquid combination, or even liquid-liquid combination [38]. The bio adsorbent's surface characteristics and pore number become the most essential components in establishing a stable equilibrium whenever adsorption occurs on a surface. The key benefits of the adsorption process are low capital investment and operational cost, broad pH ranges (2–9), broad temperature ranges (4–90°C), and the selective removal of metals at high concentrations. Many plants create bio waste that turns out to have helpful adsorption qualities. Most bio adsorbents are made up of cellulose-based compounds. Cellulose materials can effectively adsorb heavy metals. Mango and orange peels, tea, neem leaves, and sawdust are some examples of bio products that are effective adsorbents [13, 32, 33, 79].

The attractive flavor and health benefits of tea waste have led to widespread consumption of the beverage. Tea contains substances called xanthene, caffeine, and trace amounts of theobromine and theophylline that have the quality of stimulants. Tea waste has the qualities of an excellent adsorbent because it contains 30–40% polyphenols, amino acids, cellulose, hemicelluloses, lignin, and structural proteins [43, 80–83].

As a result, the adsorbent properties of waste tea could be put to use in the removal process of various HMCs. Heavy metals like cadmium and nickel have been subject to several experimental studies that have explored the adsorbent results of tea waste [84], including zinc, lead [3], and copper [59]. However, a significant amount of additional work is still necessary for tea waste.

15.5.1 The Adsorption Process

Generally, adsorption occurrs on the interface of the adsorbent and adsorbate that phase transitions between solid and liquid, gas and liquid, liquid and liquid, and liquid combinations [38]. A specific temperature and equilibrium are said to have been reached when the rate of reactions going forward and backward are equal. In order to determine adsorption equilibrium, it is crucial to take into account the adsorbent's surface properties and number of pores. This technology is straight forward to implement and offers many benefits including cheap cost, high efficiency, and ease of use [85]. The process may be carried out at temperatures and pH levels that are considered normal, and it does not result in the creation of sludge that needs to be disposed of in a separate manner. The used bio adsorbent can be recycled and put back into use in several different cycles. The bio adsorbent can also be used to recover metal absorbed by it. The adsorption process has an additional advantage over many other types of treatment procedures because of its excellent selectivity, efficiency, and capacity to deal with waste streams that contain low quantities of metal ions [44, 86, 87].

Organic (bio) adsorbents are a more environmentally friendly alternative to chemical adsorbents in the adsorption process. Living organisms, such as algae, can serve as the natural adsorbent [45–47]. Moss [48], yeast [7], and non-living organic material like orange peel [33], untreated coffee grounds [8], and potato peel utilized in an adsorption operation is referred to as the organic sorption process.

15.5.2 Adsorption Mechanisms

Adsorption mechanisms involve a variety of forms and are influenced by several different factors. Chemical adsorbents and organic (natural) adsorbents are the two main categories that can be used to describe adsorbents in a general sense. Physical adsorption is the name given to processes that take place when adsorption occurs as a consequence of attraction between the intermolecular forces of adsorbent molecules and the molecules of the metal. One of the most prevalent adsorption processes involves this mechanism. These forces are known as 'Van der Waals forces'. The adsorption processes for copper, nickel, cadmium, zinc, and lead are typically classified as being of the physical type [88]. However, Kuyucak and Volesky (1989) postulated that there might be a further instance of physical adsorption, often known as physisorption. According to this organic adsorbent hypothesis, fungi can biosorb cadmium, uranium, copper, zinc, and cobalt due to electrostatic contact between metal ions and cell walls.

Chemisorption is the ename given to processes that take place when particles of adsorbate and adsorbent interact chemically with each other. The new chemical bonds' formation and release of heat occur at the surface of the adsorbent. Cohesive forces that hold adsorbent and adsorbate molecules together are significantly enhanced in this particular form of adsorption. The process of chemisorption is almost always of an irreversible character. The processes can be going on at the same time in some circumstances. However, when a catalyst is incorporated into the process, chemisorption assumes greater significance.

15.5.3 Adsorption Parameters

Metal ion absorption equilibrium: Many experimental studies have explored bio absorption equilibrium. The initial metal ion concentration is changed while a constant amount of bio adsorbent is utilized in the experiment. Then, the experiment can calculate a percentage of the removal and adsorption capacity. This tendency can be credited to the increased accessibility of dynamic sites for the metal ions at the start of the experiment [60, 89].

Interaction Period: The length of interaction time controls the amounts of adsorption. In most cases, it was shown that as contact time rose, the percentage of elimination likewise increased until equilibrium was established [90–92]. The generally accepted adsorption magnitude of an adsorbent is measured during its state of equilibrium, which represents its maximum adsorption capacity [8, 47, 93].

pH's influential: The pH of a solution is a controlling parameter of the adsorption process in the most cases [94–96]. There is a correlation between the pH of a solution and the polarity of surface charge of an adsorbent used. In addition, metal ion precipitation is also affected by the solution's pH [97].

Dose of adsorbent: The amount of adsorbent quantity has a significant impact on removing metal ions from a solution. Many experiments have discovered that increasing the adsorbent quantity led to a higher rate of metal-ion removal [36, 98, 99]. A greater number of adsorbent surfaces became available when the adsorbent's surface area increased.

Effect of temperature: Temperature is particularly important among many factors involved in the process of adsorption. During various experimental studies, it was discovered that a rise in temperature led to a fall in the percentage of removal. The metal ion elimination percentage was increased in low temperatures. Low temperatures are optimal for the adsorption of substances. Heat is produced throughout the process, which creates an exothermic reaction [3, 4, 19, 21, 43].

15.5.4 Different Processes for Methods of Adsorption

15.5.4.1 Equilibrium Models

A normal equilibrium is created between deposited metal ions and the adsorbent at a specific temperature. Forward reaction rates become equal to backward reaction rates in a predetermined amount of time in the equilibrium model [3]. This takes place when the adsorption process is completed in its entirety. The adsorption isotherm at specific temperatures is defined as the link between the number of metal ions present in a solution and the concentration of ions that have adsorbed on the surface of the adsorbent. The metal uptake quantity (Qt) [7] at equilibrium is calculated by the following equation:

$$qt = \frac{(C_0 - C_e)V}{M} \tag{15.1}$$

Where "C_0 and C_e are initial equilibrium concentrations, V is the volume of the solution, and M is the mass of the adsorbent" [7].

Several models define adsorption isotherms, and each is based on either a physical image of adsorption or an empirical relationship with empirical parameters.

$$\%\text{Removal(R)} = \frac{(\text{Co} - \text{Ct})}{\text{Co}} \times 100 \tag{15.2}$$

15.5.4.2 The Langmuir Isotherm Model

The Langmuir equation has been utilized in a majority of situations, and the Langmuir isotherm model indicates "monolayer adsorption, homogeneous surface, and low contact forces between molcules" that have been adsorbed [100, 101].

$$q_e = \frac{Q_m K_L C_e}{(1 + K_L C_e)} \tag{15.3}$$

Or,

$$\frac{1}{q_e} = \frac{1}{Q_m K_L C_e} + \frac{1}{Q_m} \tag{15.4}$$

Where "Q_m and K_L can be found from the linear plot of $1/q_e$ vs $1/C_e$, the separation factor R_L is a dimensionless constant" well defined by following the relationship [100].

$$R_L = \frac{1}{(1K_{+L}C_0)} \tag{15.5}$$

The value of R_L indicates the "type of Langmuir isotherm to be favorable ($0<R_L<1$) or unfavorable ($R_L>1$) for adsorption" [28].

15.5.4.3 The Freundlich Isotherm Model

The Freundlich equation is as follows:

$$q_e = K_F C_e^{\frac{1}{n}} \tag{15.6}$$

The linear form of the Freundlich equation is

$$\log q_e = \log K_F + \frac{1}{n} \log C_e \tag{15.7}$$

If the value of the Freundlich constant n is from 1–10, it depicts favorable adsorption. A larger value of n indicates a strong interaction between metal and adsorbent molecules. The value of 1/n indicates the identical adsorption energies for all sites resulting from linear adsorption.

KF and n are the Freundlich constants. These constants relate to the adsorption capacity of the adsorbent, and n relates to the the intensity of the adsorption process. Unfortunately, the Freundlich model does not provide information about the monolayer adsorption capacity compared to the Langmuir model [4, 49].

15.5.4.4 The Dubinin-Radushkevich Isotherm Model

The heterogeneity of surface energies was modeled using this Dubinin-Radushkevich (D-R) model, which was employed for the estimate process. As a linear equation, the D-R isotherm formula looks like this:

$$\text{lnqe} = \ln Q_m \text{-} K_d \varepsilon^2 \tag{15.8}$$

Where

$$\varepsilon = RT\ln\left(1+\frac{1}{C_e}\right) \tag{15.9}$$

The D-R constants "Q_m and K_d were found from the linear plots of $\ln q_e$ vs 2." The mean free energy E (kJ/mol) of adsorption per molecule of the adsorbate can be calculated from the relationship [102].

$$E = \frac{1}{\sqrt{2K_d}} \tag{15.10}$$

The adsorption process precedes via chemisorption if the value of E lies between 8 and 16 kJ/mol^{-1}. If the value of $E < 8$ kJ/mol^{-1}, the nature of the adsorption process is an entirely physical type.

Further, Dubinin-Radushkevich's (D-R) isotherm model was applied to learn about the physical and chemical properties of the adsorption process [34].

15.6 A Specific Review on Tea-Waste Adsorption

The research looked at uranium adsorption by tea wastes in a laboratory setting. Results showed that the process was highly dependent on pH, with the maximal adsorption efficiency determined to be 29.4 mg/g at pH 4 and 298 K. Thermodynamic analysis confirmed that the reaction occurred spontaneously and endothermically [61]. Research into the efficacy of tea waste in removing ethylene gas was conducted. Researchers utilized micro-sized tea leaves (MSTL) and discovered that at 250 degrees Celsius, the reaction was spontaneous with an adsorbent's capacity of 6.93 umol/g [103]. Researchers loaded tea waste with the oxides of Al and Fe, creating a bio sorbent that is both cheap and effective in order to remove fluoride from drinking water. Fluoride removal was most effective between a pH range of 4.0 and 8.0 [6]. Tea wastes were used in batch and column adsorption trials to remove cationic and anionic dyes from textiles. The maximum efficacy in the removal of cationic and anionic dyes was tentative at pH 2 and 6 separately. Column experiments yielded a higher adsorption capacity for basic dyes than batch tests [104]. The examined efficacy of removing chromium metal ions using cooking tea waste and commercial activated carbon found that the tea waste was more effective at a pH of 2. Both adsorbents were revealed to be more operative in removing chromium metal ions as the pH value was raised [105]. The research used a surfactant-altered tea waste to take away basic red 12 dye in a feasible endothermic adsorption method [106]. Adsorption tests were conducted to eliminate copper and lead from tea leftovers, and 90% removal was achieved [21]. The research examined the efficacy of using tea fungus, a by-product of black tea fermentation, to adsorb arsenic from groundwater. With the prepared adsorbent, complete elimination was possible [107].

The initial results suggest that tea waste possesses favorable adsorption characteristics, and further investigation is needed to realize its potential fully. In addition, heavy metals can be removed with it when appropriate pre-treatments have been applied to increase the efficiency with which that task is accomplished.

15.7 Conclusions

This research adds an evaluation of bio adsorbents utilized within MOFs by combining the results of earlier experiments with the elution of the bio adsorbent's validity in the removal of heavy metals. The organic adsorbent showed significant improvement in the adsorption process effectiveness of various parameters. Bio adsorbents are shown to be natural adsorbents that are cheap, effective, and efficient in removing zinc, nickel, chromium, cadmium, copper, etc., HMC from aqueous solutions. Adsorbents are also found to be environmentally friendly. Furthermore, it has been found that pre-treatments of bio adsorbents significantly increase the effectiveness of the adsorption process. The pH of the solution affects the percentage of metal that is removed. Operational variables such as temperature, dose of adsorbent, initial metal ion concentration, and contact time all substantially impacted the adsorption process. Bio adsorbent technology was found to be quite useful since it makes good use of a material that would otherwise go to waste.

References

1 Berger, M. (2022). Nano werk. Nano-Society: Pushing the Boundaries of Technology, [Online]. Available: https://www.nanowerk.com/mof-metal-organic-framework.php.

2 Lv, S.W., Liu, J.M., Wang, Z.H. et al. (2019). Recent advances on porous organic frameworks for the adsorptive removal of hazardous materials *Environmental Sciences (China)* 80 (169–18).

3 Ahmaruzzaman, M. (2011). Industrial wastes as low-cost potential adsorbents for the treatment of wastewater laden with heavy metals. *Advances in Colloid and Interface Science* 166: 36–59.

4 Aksu, Z. (2002). Determination of equilibrium, kinetic and thermodynamic parameters of the batch biosorption of nickel ions on to Chlorella vulgaris. *Process Biochemistry* 38: 89–99.

5 Hoskins, B.F., Robson, R. (1989). Infinite polymeric frameworks consisting of three dimensionally linked rod-like segments. *Journal of the American Chemical Society* 111 (15).

6 Cai, H. M., Chen, G. J., Peng, C.Y. et al. Removal of floride from drinking water using tea waste loaded with Al/Fe oxides: anovel, safe and efficient biosorbent. *Removal of Floride from Drinking Water Applied Surface Science* 328: 34–44.

7 Can, C. and Jianlong, W. (2007). Removal of Pb2+, Ag+,Cs+ and Sr2+ from aqueous solution by brewery's waste biomass. *Journal of Hazardous Materials* 151: 65–70.

8 Azouaou, N., Sadaoui, Z., Djaafri, A. and Mokaddem, H. (2010). Adsorption of cadmium from aqueous solution onto untreated coffee grounds: equilibrium, kinetics and thermodynamics. *AzouaouN.,SadaouiZ.,Djaafri A and MokaddemH.,2010. Adsorption of Cadmium from aqJournal of Hazardous Materials* 184: 126–134.

9 Rasheed, T., Hassan, A.A., Bilal, M. et al. (2020). Metal organic frameworks based adsorbents: a review from removal perspective of various environmental contaminants from wastewate. *Chemosphere* 259: 127369.

10 Sud, D. and Kaur, G. (2021). A comprehensive review on synthetic approaches for metal-organic frameworks: from traditional solvothermal to greener protocols *Polyhedron* 193: 114897.

11 Ru, J., Wang, X., Wang, F. et al. (2021). UiO, series of metal-organic frameworks composites as advanced sorbents for the removal of heavy metal ions: synthesis, applications and adsorption mechanism. *Ecotoxicology and Environmental Safety* vol. 15, no. 15. 208: 111577.

12 David, J., Tranchemontagne, J. R., and Hunt, O. M. Y. (2008). Room temperature synthesis of metal-organic frameworks: MOF-5, MOF-74, MOF-177, MOF-199, and IRMOF-0 *Tetrahedron* 64 (36):8553–8557.

13 Nuria, M., Cesar, V., Ignasi, C. et al. (2010). Cadmium and lead removal from aqueous solution by grape stalk wastes: modeling of a fixed bed column. *Nuria. M, Cesar. V, Ignasi. C, Maria. M and Antonio. F., 2010. Cadmium and Lead RemovalJournal of Chemical Engineering Data* 55: 3548–3554.

14 Srinivasan, A. and Viraraghavan, T. (2010). Decolorization of dye wastewaters by biosorbents: a review *Journal of Environ. Manage* 55: 1915–1929.

15 Mustafa, Y., Gode, F., Pehlivan, E. et al. (2008). An economic removal of Cu 2+ and Cr3+ on the new adsorbents: pumiceand polyacrylonitrile/ pumice composite *Chemical Engineering Journal* 137: 453–461.

16 Wan, W. S. N. and Hanfiah, M. A. M. (2008). Removal of heavy metal ions from wastewater by chemically modified plant wastes as adsorbents: a review *Bioresource Technology* 99: 3935–3948.

17 Cetin, S. and Pehlivan, E. (2007). The use of fly ash as a low cost, environmentally friendly alternative to activated carbon for the removal of heavy metals from aqueous solutions *Journal of Colloids Surfactants* 298: 83–87.

18 Bayat, B. (2002). Comparative study of adsorption properties of Turkish fly ashes: II. The case of chromium (VI) and cadmium (II). *Journal of Hazardous Materials* 95: 276–290.

19 Amany, E. S., Ahmed, E. N., Azza, K., and Ola, A. (2007). Removal of toxic chromium from wastewater using green alga ulva lactuca and its activated carbon. *Journal of Hazardous Materials* 148: 216–228.

20 Ajay, K., Kadirvelu, M., Mishra, G. K. (2008). Adsorptive removal of heavy metals from aqueous solution by treated sawdust *Journal of Hazardous Materials* 50: 604–611.

21 Amarasinghe, B. M. W. P. K. and Willims, R. A. (2007). Tea waste as a low cost adsorbent for the removal of Cu and Pb from wastewater. *Chemical Engineering Journal* 132,: 299–309.

22 Atalay, E., Gode, F., and Sharma, Y. C. (2010). Removal of selected toxic metals by a modified adsorbent. *Practice Periodical of Hazardous Toxic and Radioactive Waste Management* 14: 132–138.

23 Qi, B. C. and Aldrich, C. (2008). Biosorption of heavy metals from aqueous solutions with tobacco dust. *Journal of Bioresource Technology* 99: 5595–5601.

24 Uysal, M. and Ar, I. (2007). Removal of Cr(VI) from industrial waste waters by adsorption: partI: determination of optimum conditions. *Journal of Hazardous Materials* 149: 482–491.

25 Nasernejad, B., Zadeh, T., Pour, B. B., Bygi,M. E., and Zamani, A. (2005). NasernejadB.,ZaComparison for biosorption modeling of heavy metals Cr(III),Cu(II),Zn(II) adsorption from wastewater by carrot residue. *Journal of Process Biochemistry* 40: 1319–1322.

26 Amir, H., Mahvi, D. N., Forugh, V., and Shahrokh, N. (2005). Teawaste as an adsorbent for heavy metal removal from industrial wastewaters. *American Journal of Applied Sciences* 2: 372–375.

27 Gulnaziyal, M. K. A. and NikM, S. (2008). Continuous adsorption of lead ions in a column packed with palm shell activated carbon. *Journal of Hazardous Materials* 155: 109–113.

28 Kadirvelu, K. and Namasivayam, C. (2003). Activated carbon from coconut coir pith as metal adsorbent: adsorption of Cd(II)from aqueous solution. *Advances in Environmental Research* 7: 471–478.

29 Bin Y., Zhang Y., Alka S., Shyam S.S and Kenneth L.D. (2000). The removal of heavy metal from aqueous solutions by sawdust adsorption-removal of copper. *Journal of Hazardous Materials* 80: 33–42.

30 Mehmet, E. A., Sukru, D., Celalettin, O., and Mustafa, K. (2007).Heavy metal adsorption by modified oak sawdust: thermodynamics and kinetics *Journal of Hazardous Materials* 4: 77–85.

31 Qin, L., Jianping, Z., Wenyi, Z. et al. (2007). Kinetic studies of adsorption of Pb(II),Cr(II) and Cu(II) from aqueous solution by sawdust and modified peanut husk *Journal of Hazardous Materials* 141: 163–167.

32 Kishore, K. K., Xiaoguang, M., Christodoulatos, C., and Veera, M. B. (2008). Biosorption mechanism of nine different heavy metals onto biomatrix from rice husk. *Journal of Hazardous Materials* 53: 1222–1234.

33 Ferda, G. and Selen, S. D. (2012). Adsorption study on orange peel: removal of Ni (II) ions from aqueous solution. *African Journal of Biotechnology* 11: 1250–1258.

34 Lakshmipathy, R. and Sarada, N. C. (2013). Application of watermelon rind as sorbent for removal of nickel and cobalt from aqueous solution. *International Journal of Mineral Processing* 122: 63–65.

35 Mishra, P. and Patel, R. K. (2009). Removal of lead and zinc ions from water by low cost adsorbents *Journal of Hazardous Materials* 168: 319–325.

36 Venkateswarlu, P., Venkata, R. M., Subha,R. D., and Venkateswara, R. M. (2007). Removal of chromium from an aqueous solution using Azadirachtaindica(neem) leaf powder as an adsorbent. *International Journal of Physical Sciences* 2: 188–195.

37 Waleska, E., Oliveira, A. S., Franca, L. S., and Sonia D. R. (2008). Untreated coffee husks as biosorbents for the removal of heavy metals from aqueous solutions. *Journal of Hazardous Materials* 152: 1073–1081.

38 Barakat, M. A. (2011). New trends in removing heavy metals from industrial wastewater. *Arabian Journal Chemistry* 4: 361–377.

39 Yan, G. and Viraraghavan, T. (2001). Heavy metal removal in a biosorption column by immobilized M. Rouxii biomass. *Bioresource Technology* 78: 243 249.

40 Aderhold, D., Williams, C. J., and Edyvean, R. G. J. (1996). Removal of heavy metal ions by seaweeds and their derivatives. *Bioresource Technology* 58: 1–6.

41 Zafar, M. N., Adeem, R.N., and Hanif, M. A. (2007). Biosorptions of nickel from protonated rice bran. *Journal of Hazardous Materials* 134: 478–485.

42 Ajmal, M., Khan, A. H., Ahmad, S., and Ahmad, A. (1998). Role of sawdust in the removal of copper (II) from industrial wastes. *Water Research* 32 (10): 3085–3091.

43 Shah, J., Jan, M. R., Haq, A., and Zeeshan, M. (2015). Equilibrium, kinetic and thermodynamic studies for sorption of Ni (II) from aqueous solution using formaldehyde treated waste tea leaves. *Shah J., Jan M. R., HaqA., Zeeshan M., 2015. Equilibrium, Kinetic and Thermodynamic Studies for sorptJournal of Saudi Chemical Society* 19: 301–310.

44 Ayhan, D. (2008). Heavy metal adsorption onto agro-based waste materials: a review. *Journal of Hazardous Materials* 157: 220–229.

45 Gupta, V. K. and Rastogi, A. (2007). Biosorption of lead from aqueous solutions by green algae spirogyra species: kinetic and equilibrium studies. *Journal of Hazardous Materials* 152: 407–414.

46 Dumitru, B. and Laura, B. (2012). Equilibrium and kinetics studies of heavy metal ions biosorption on green algae waste biomass. *Journal of Bioresource Technology* 103: 489–493.

47 Mohammmmad, M. M. R., Parisa, R., Atefeh, A., and Ali, R. K. (2011). Kinetics and equilibrium studies on biosorption of cadmium, lead, and nickel ions from aqueous solutions by intact and chemically modified brown algae. *Journal of Hazardous Materials* 185: 401–407.

48 Lee, C. K. and Low, K. S. (1989). Removal of copper from solution using moss. *Journal of Environmental Technology* 10: 395–404.

49 Ferda, G. and Selen, S. D. (2012). Adsorption study on orange peel: removal of Ni(II) ions from aqueous solution. *African Journal of Biotechnology* 11: 1250–1258.

50 Mohammed, A. A. and Devi, P. (2009). Kinetic and equilibrium studies for the biosorption of Cr(VI) from aqueous solutions by potato peel waste. *Journal of Chemical Engineering Research* 1: 51–62.

51 Meena, A. K., Mishra, G. K., Rai, P., Rajagopal,C., and Nagar, P. N. (2005). Removal of heavy metal ions from aqueous solutions using carbon aerogel as an adsorbent *Journal of Hazardous Materials* 122: 161–170.

52 Villaescusa, I., Fiol, N., Martinez, M. et al. (2004). Removal of copper and nickel ions from aqueous solutions by grape stalks wastes. *Water Research* 38: 992–1002.

53 Patmavathy, V., Vasudevan, P., and Dhingra, S. (2003). Biosorption of nickel (II) ions on Bake's yeast. *Process Biochemistry*38: 1389–1395.

54 Hasar, H. (2003). Adsorption of nickel (II) from aqueous solution onto activated carbon prepared from almond husk. *Journal of Hazardous Materials* B-97: 49–57.

55 Spearot, R. M. and Peck, J. (1984). Recovery process for complexedcopperbearing rinse waters. *Environmental Progress* 3: 124–129.

56 Brierley, J. A. and Goyak, G. M. (1986). A new wastewater treatment and metal recovery technology. *Fundamental and Applied Biohydromettallurgy* 291–303.

57 Peters, R. W., Young, K., and Bhattacharya, D. (1985). Evaluation of recent treatment techniques for removal of heavy metals from industrial wastewaters. *AICHE Symposium* 81: 1605–1707.

58 Tsezos, M. and Volesky,B. (1982). The mechanism of uranium biosorption by Rhizopusarrhizus. *Biotechnology Bioengineering* 24: 385–401.

59 Chen, X. C., Wang, Y. P., Lin, Q. et al. (2005). Biosorption of copper(II) and zinc(II) from aqueous solution by pseudomonas putida cz1.colloid and surface B. *Biointerfaces* 42: 101–107.

60 Han,R., Zhang, J., Han, P. et al. (2009). Study of equilibrium, kinetic and thermodynamic parameters about methylene blue adsorption onto natural zeolite. *Chemical Engineering Journal* 145: 496–504.

61 Li,X., LiF, J. Y., and Jiang, C. (2015). The uptake of uranium by tea waste investigated by batch, spectroscopic & modeling techniques. *Journal of Molecular Liquids* 209: 413–418.

62 Fornwalt, H. J. and Hutchins, R. A. (1966). Purifying liquids with activated carbon. *Chemical Engineering Journal* 73: 179–184.

63 Volesky, B. (1995). Biosorption of heavy metals. *Biotechnology Progress* 11: 235–250.

64 Zhao, G., Li, M., and Hu, H. (2005). Dissociation and removal of complex chromium ions containing in dye wastewaters. *Separation and Purification Technology* 43: 227–232.

65 Sharma, Y. C., Prasad, G., and Rupainwar,D. C. (1991). Removal of Ni (II) from aqueous solutions by sorption. *International Journal of Environmental Studies* 37: 183–191.

66 Remoudaki, E., Hatzikioseyian, A., Tsezos, K., and Tsezos, M. (2003). The mechanism of metals precipitation by biologically generated alkalinity in biofilm reactors. *Water Research* 37: 3843–3854.

67 Ahluwalia, S. S. and Goyal, D. (2005). Removal of heavy metals by waste tea leaves from aqueous solution. *Engineering Life Science* 5: 158–162.

68 Jiaping, P. (2012). *Decontamination of Heavy Metals: Processes, Mechanisms, and Applications*. New Delhi: CRC Press.Print.

69 Reena, S., Neetu, G., Anurag, M., and Rajiv, G. (2011). Heavy metals and living systems: an overview. *Indian Journal of Pharmacology* 43: 246–253.

70 Friberg, L. M., Piscaor, G. F., Nordberg, T., and Kjellstrom (1980). *Cadmium in the Environment*, Cleveland, OH: CRC Press.

71 Dean, J. C., Bosqui, F. L., and Lanovette, K. H. (1972). Removing heavy metal from waste water. *Environmental Science & Technology* 6: 518.

72 Garba, S. M. and Abubakar, M. A. (2018). Source and distribution of the heavy metals: pb,Cd,Cu,Zn,Fe,Crand Mn in soils of Bauchi Metropolis. *Nigeria American Journal of Engineering Research* 7 (2): 13–23.

73 Bernhoft, R. A. (2013). Cadmium toxicity and treatment. *The Scientific World Journal* 1–3.

74 Nriagu, J. (2007). Zinc toxicity in humans. *School of Public Health, University of Michigan* 1–6.

75 Das, K. K., Das, S. N., and Dhundasi, S. A. (2008). Nickel its adverse health effects & oxidative stress. *Indian Journal of Medical Research* 128: 412–425.

76 Xuan, Z., Tang, Y., Li, X. L. Y., and Luo, F. (2006). Study on equilibrium, kinetics and isotherm of biosorption of lead ions on to pretreated chemically modified orange peel. *Biochemical Engineering Journal* 31: 160–164.

77 Gupta, V. K., Jain, C. K., Ali, I. et al. (2003). Removal of cadmium and nickel from wastewater using bagasse fly ash a sugar industry waste. *Journal of Colloid Interface Science* 21 (328): 271.

78 Rio, M., Parwate, A. V., and Bhole, A. G. (2002). Rio M., parremoval of Cr6+ and Ni2+ from aqueous solution using bagasse and fly ash. *Waste Management* 830: 821.

79 Asma, S., Muhammed, I., and Akhtar, M. W. (2005). Removal and recovery of lead(II) from single and multimetal (Cd,Cu,Ni,Zn) solutions by crop milling waste (black gram husk). *Journal of Hazardous Materials* 117: 65–73.

80 Nadaroglu, H., Kalkan, E., Nazan, D. (2010). Removal of copper from aqueous solution using red mud. *Desalination* 251: 90–95.

81 Wasewar, K. L., Atif, M., Prasad, B., and Mishra, M. I. (2009.). Batch adsorption of zinc on tea factory waste. *Desalination* 244: 66–71.

82 Almajano, M. P., Carbó, R., Jiménez, J. A. L., and Gordon, M. H. (2008). Antioxidant and antimicrobial activities of tea in fusions. *Food Chemistry* 108: 55–63.

83 Cay, S. and Uyanik, A. A. (2004). Single and binary component adsorption of copper (II) and cadmium (II) from aqueous solutions using tea-industry waste. *Separation and Purification Technology* 273–280.

84 Ozdes, D., Celal, D., HasanBasri, S. et al. (2011). Adsorptive removal of Cd (II) and Pb (II) ions from aqueous solutions by using Turkish illiticclay. *Journal of Environmental Management* 92: 3082–3090.

85 Demirbas, A. (2008). Heavy metal adsorption onto agro based waste materials. *Journal of Hazardous Materials*157: 220–229.

86 Mohsen, A., and Hashem, M.A. (2007). Adsorption of lead ions from aqueous solution by okra waste. *Journal of Physical Sciences* 2 (7): 178–184.

87 Mckay, G. H. Y. S., Wase, D. J., and Forster, C. F. (2008). Study of the sorption of divalent metal ions on to peat. *Adsorption Science and Technology* 18: 637–650.

88 Fourest, E. and Roux, J. C. (1992). Heavy metal biosorption by fungal mycelia byproducts: mechanism and influence of pH. *Applied Microbiology Biotechnology* 37: 399–403.

89 Mausumi, M., Noronha, S. B., and Suraish, k. G. K. (2007). Kinetic modeling for the biosorption of copper by pretreated Aspergillusniger biomass. *Bioresource Technology* 98: 1780–1788.

90 Ghosh, U. and Goswami, S. (2005). Studies on adsorption behavior of Cr (IV) onto synthetic hydrous stannic oxide. *Water SA* 4 (31): 597–602.

91 Babu, B. V. and Gupta, S. (2006). Adsorption of chromium (VI) by a low cost adsorbent prepared from Tamarind seeds. *Chemcon* 1–6.

92 Kannan, N. and Veemraj, T. (2009). Removal of lead (II) ions by adsorption onto bamboo dust and commercial activated carbon-a comparative study. *E-journal of Chemistry* 2 (6): 247–256.

93 Maria, M., Nuria, M., Soraya, H., and Nuria, F. (2006). Removal of lead (II) and cadmium (II) from aqueous solutions using grape stalk waste. *Journal of Hazardous Materials* 184: 203–211.

94 Waranusantigul, P., Pokethitiyook, P., Kruatrachue,M., and Upatham, E. S. (2003). Kinetics of basic dye (methylene blue) biosorption by giant duckweed (Spirodelapolyrrhiza). *Journal of Environmental Pollution* 125: 385–392.

95 Babu, B. V. and Gupta, S. (2008). Adsorption of Cr(VI) using activated neem leaves. *Adsorption* 14: 85–92.

96 Hao, C., Guoliang, D., Jie, Z. (2010). Removal of copper (II) ions by a biosorbent-Cinnamomumcamphora. *Journal of Hazardous Materials* 177: 228–236.

97 Myroslav, S., Boguslaw, B., Artur, P. T., and Jacek, N. (2006). Study of the selection mechanism of heavy metal (Pb,Cu,Ni and Cd) adsorption on clinopttilolite. *Journal of Colloid and Interface Science* 304: 21–28.

98 Saifuddin, M. N. and Kumaran, P. (2005). Removal of heavy metal from industrial wastewater using chitosan coated oil palmshellcharcoal. *Electronic Journal of Biotechnology* 1: 43–53.

99 El-Said, A. G., Badaway, N. A., and Garamon, S.E. (2010). Adsorption of Cadmium (II) and Mercury (II) onto natural adsorbent rice husk ash (RHA) from aqueous solutions: study in single and binary system. *Journal of American Science* (12): 400–409.

100 KadirveluK, T. K. and Namasivayam, C. (2001). Removal of heavy metals from industrial wastewaters by adsorption onto activated carbon prepared from an agricultural solidwaste. *KadirveluK.,Thamaraiselvi K and Namasivayam C., 2001. Removal of Heavy Metals from Industrial wastewatJournal of Bioresource Technology* 76: 63–65.

101 Malko, C. E. and Yasar, N. (2007). Potential of tea factory waste for chromium(VI) removal from aqueous solutions: thermodynamic and kinetic studies. *Separation and Purification Technology* 54: 291–298.

102 Igwe, J. C. and Abia, A. A. (2003). Maize cob and husk as adsorbent for removal Cd,Pb and Zn ions from waste water. *Physical Science* 2: 83–94.

103 Tzeng, J. H., Weng,C. H., Huang, J. W. et al. (2015). A new non conventional and low cost biosorbent for ethylene removal. *International Biodeterioration and Biodegradation* 104: 67–73.

104 Khosla,E., Kaur, S., and Dave, P. N. (2013). Tea waste as adsorbent for ionic dye. *Desalination and Water Treatment* 51: 51.

105 Attia, H. G. (2012). A comparison between cooking tea waste and commercial activated carbon for removal of chromium from artificial waste water. *Journal of Engineering and Development* 16: 307–322.

106 Khosla, E., Kaur, S., and Dave, P. (2011). Surfactant modified tea waste as a novel adsorbent for the removal of basic dye. *Der Chemica Sinica* 2 (5): 87–102.

107 Murugesan, G. S., Satish, k. M., and Swaminathan, K. (2006). Arsenic removal from groundwater by pretreated waste tea fungal biomass. *Bioresource Technology* 97: 480–488.

16

Evaluation of MOF Applications for Groundwater Arsenic Mitigation of the Middle Ganga Plains of Bihar, India

Arun Kumar[1], Vivek Raj[2], Mohammad Ali[1], Abhinav[1], Mahesh Rachamalla[4], Dhruv Kumar[3], Arti Kumari[2], Rakesh Kumar[5], Prabhat Shankar[1], and Ashok Kumar Ghosh[1]

[1]*Mahavir Cancer Sansthan & Research Centre, Patna, Bihar, India*
[2]*Department of Biotechnology, Patna Women's College, Patna, Bihar, India*
[3]*School of Health Sciences & Technology UPES University,, Dehradun, Uttarakhand, India*
[4]*University of Saskatchewan, Canada,*
[5]*School of Pharmaceutical Sciences, Lovely Professional University, Phagwara, Punjab, India*

16.1 Arsenic Contamination in the Groundwater of Bihar

In recent decades, the contamination of groundwater by naturally occurring and man-made arsenic (As) has been highlighted as a major environmental disaster in many parts of the world, including Argentina, Bangladesh, Bolivia, Brazil, Chile, China, Cambodia, Ghana, Greece, Hungary, India, Japan, Korea, Mexico, Mongolia, Nepal, New Zealand, Poland, Taiwan, Vietnam, and the USA [1, 2, 3, 4, 5;]. Around 300 million individuals are thought to be impacted by arsenic poisoning, which poses health risks. Of these, 70 million live in India alone [6]. India is a sizable developing nation whose strong agricultural foundation is reliant on an irrigation system that uses both groundwater and surface water. Ganga-Meghna-Brahmaputra and Meghna are its two main river basins [7, 8]. The Ganga river basin in India is densely inhabited. States Uttrakhand and Uttar Pradesh make up the upper Gangetic plains, whereas Bihar makes up the middle Ganga plains, and West Bengal makes up the lower Ganga plains. India's states with the highest arsenic contamination levels are Assam, West Bengal, Bihar, and Uttar Pradesh [9, 10, 11, 12, 13, 14].

According to a number of studies, the Ganges delta's alluvial aquifers, which are made up of sediments transported from Bihar's mineralized regions rich in sulfides and other sulfides, are the main locations where groundwater is most likely to be contaminated with arsenic [15, 16]. The Brahmaputra alluvium and a sizable section of Indo-Gangetic alluvium, which extends further west, have higher arsenic contents in wells that are situated in late Quaternary and Holocene aquifers, according to recent studies [17,18].

Metal Organic Frameworks for Wastewater Contaminant Removal, First Edition. Edited by Arun Lal Srivastav, Lata Rani, Jyotsna Kaushal, and Tien Duc Pham.

The majority of the time, iron oxyhydroxides collected under typically oxidizing conditions present during the deposition of Holocene sediments that have been adsorbed to the surface. Arsenic is produced during the weathering of sulfide minerals. However, redox reactions in the sediments spark the reductive dissolution of iron oxides, which in turn causes biogeochemical reactions to transport a sizable quantity of arsenic to the aqueous phases [19, 20]. Sediments from rivers that were deposited during the late Quaternary or Holocene epoch about 12 thousand years ago are preserved in the Ganga-arsenic Brahmaputra basin's groundwater. Except for the thin alluvial sections in the Precambrian Highlands in the south, the Ganges River system's Quaternary deposits are more than 300 m thick [21] bordering the state of Jharkhand. According to a recent assessment by the Central Ground Water Authority and Directorate of Ground Water, Government of Bihar, the middle Ganga plain's renewable groundwater resource has a development level of 36% and is estimated to be 26.4 bcm [22]. These late Quaternary sediments are made up of clays, silts, and sands in their lithology. These sediments are rich in organic matter and have a feldspar, quartz, illite, and kaolinite mineral composition. The fine-grained shoreface facies [23, 24, 25, 26]. The various rivers that spring in the Himalayas in the north and northeast have spread a thick layer of younger alluvium that contains sand, silt, and clay. The majority of the arsenic-related environmental issues that have been identified so far are the result of mobilization in response to environmental factors. As a result, it has been determined that the presence of arsenic in the groundwater of the Gangetic Plains was due to mobilization of the metal during natural hydrogeological circumstances.

The method by which arsenic is mobilized in groundwater in the Ganges plain is still not completely understood. It is also necessary to determine why the Kosi and Gandak rivers' floodplains and the other important rivers that drain the Middle Ganges have less contamination than the main Ganges floodplain. For community-scale water supply, a deeper yield of low-arsenic aquifers within a subbasin must be established. In both the northern and southern Ganges plains, it is necessary to use assimilation isotopic techniques to clarify the groundwater recharge processes in contaminated shallow aquifers and deeper, low, arsenic-contaminated aquifers.

The groundwater resources in the state of Bihar are extremely abundant. The North and South Gangetic Plains are the two physical-geographical regions of Bihar that are separated by the Ganges River. Geologically, the state of Bihar is separated by younger Pleistocene and older Holocene alluvium, creating a two-tiered aquifer system. [27, 28]. A thick aquifer at nearly 30 m depth made of clay and sandy loam separates shallow aquifer systems at a 50 m depth from deeper aquifer systems at 120 to 300 m depth [29, 30]. While the deeper aquifer was confirmed to be clear of arsenic, the shallow aquifer was discovered to be contaminated with it [31, 32, 33]. According to a hydrogeochemical study, the state's shallow aquifers' hydraulic conductivity, irrigation water infiltration with fertilizers and pesticides, and recharge from rainfall infiltration are key hydrogeochemical processes that affect the groundwater chemistry of these aquifers [34]. Groundwater chemistry in deep aquifers is significantly influenced by leakage from low aquifers, ion exchange, and the weathering of rocks [35]. Arsenic (As), which precipitates with iron in oxidizing environments, was not found in the arsenic-free shallow wells that were excavated, which were typically 8–12 m deep [36, 37]. The shallow aquifer has a depth range of 3 to 90 meters, and the deep aquifer has a depth range of 21 to 314 meters. Thus, especially in rural regions, hand pumps are the most often used home water source in the state. Reports state that 91.4% of rural families get their drinking water from hand pumps. Arsenic pollution of groundwater has been recorded from 18 districts, posing a threat to more than 10 million people in the state. However, quality is a significant worry [38, 39, 40, 41].

After an occurrence involving an intense outbreak of diarrhea, a widespread transfer from microbiologically risky surface water to microbiologically safe groundwater led to millions of individuals in Bangladesh and India establishing their own manual tube wells. Changes in the groundwater's oxidation-reduction state and geochemical characteristics were the main causes of the significant arsenic pollution in these places. Due to excessive groundwater use for agriculture and consumption, arsenic has been leaking from minerals, resulting in this alteration [42, 43, 44, 45, 46, 47]. Arsenic poisoning, also known as arsenicosis, can result from consuming arsenic in food and water on a regular basis. Although West Bengal reported the first instance of arsenicosis in the early 1980s, it was not until 1995 that the widespread pollution was realized. A similar pattern was followed when Bangladesh's groundwater was discovered to be contaminated with arsenic much later. Since the discovery of arsenic pollution back in 2002, we have identified 12 districts in Bihar that are along the Ganga that are arsenic contaminated, and we have also identified patients with arsenical skin lesions in six districts. More are emerging with the ongoing surveys. We anticipate that districts located in the region where the Ganga and other tributaries coming from the Himalaya moved through time will be arsenic contaminated based on our most recent preliminary assessment from UP and Bihar. The Terai region of Nepal, which is bordered by the arsenic-contaminated states of UP and Bihar, may also be impacted.

Following the outbreak of the very first case of groundwater arsenic poisoning and arsenicosis in Bihar, a sample method was developed in 11 districts to investigate all the hand tube wells installed by the government, and a random sampling strategy was put in place to assess the content of arsenic in private hand pumps throughout the state [48]. It was discovered that all 11 districts and 50 community blocks were contaminated with arsenic, with raised arsenic levels typically being restricted to areas less than 10 km from the river Ganges [49].

The first instance of arsenic contamination of groundwater in Bihar was documented in 2002 in the village of Semaria Ojha Patti, in the Shahpur block of the Bhojpur district [50]. In 2007, Nixon et al. reported arsenic contamination of groundwater in 50 blocks in 11 districts. Recently, [51] estimated that there is a very high health hazard in arsenic-contaminated areas in the Maner block of Patna district. They discovered that drinking groundwater contaminated with arsenic carried a 192 risk-hazard ratio for developing cancer [52]. In addition to drinking water, Maner block in the Patna district has reported arsenic contamination of soil and food products [53, 54]. Communities in arsenic-affected areas are at risk for poor health due to exposure to arsenic in the air, water, soil, and food. The maximum value of 498 μg/L arsenic was discovered in Haldi Chhapra panchayat of Maner block, Patna, with an average of 142 μg/L arsenic found in this area. However, the Buxar District of Bihar reported the highest level of arsenic, which was 2182 μg/L. The residents of this region experience the normal effects of arsenic, including melanosis of the skin and hyperkeratosis on the soles of the feet and hands. Additionally, some displayed symptoms of various illnesses like cancer, diabetes, hypertension, appetite loss, tiredness, stomach discomfort, and hormone imbalance [55, 56, 57, 58, 59, 60, 61, 62, 63].

According to a study done in India, there is a direct link between exposure to arsenic and declines in mental abilities like vocabulary, numeracy, memory, and general cognition [64 Von Ehrenstein et al.,2007]. Dopamine levels drop as a result of arsenic's hormonal disruption and oxidative damage (Figure 16.1). Dopamine deficiency results in decreased brain development and neurobehavioral abnormalities such impulsivity, inattention, and a lack of concentration and attentiveness [65, 66, 67, 68].

Arsenic values over the WHO standards in Vaishali and over the BIS standards in Bhagalpur district were found in more than one new village in 2010–2011 and more than five new villages in 2010–2011, respectively [69, 70]. The greatest concentration was 21 μg/L in Vaishali, while it reached 599 μg/L in some of the communities of Bhagalpur [71, 72]. Arsenic levels greater than 50 μg/L were found in the Siwan community block in the Siwan district, with a maximum value of 150 μg/L [73, 74]. Drinking water in the Nautan community block of the West Champaran district can contain up to 397 μg/L of arsenic [76, 77]. Likewise, arsenic levels range from

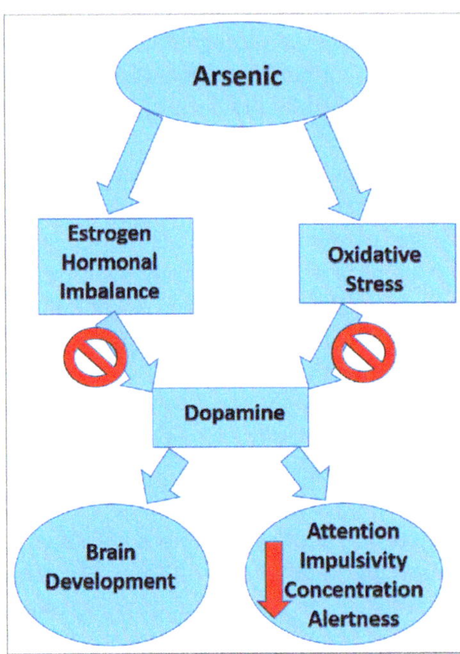

Figure 16.1 Schematic diagram showing arsenic causing hormonal imbalance. (*Source:* [75] Kumar et al., 2019).

<10 μg/L to 60 μg/L in new community blocks in the Samastipur district [78]. These remote districts of northern and southern Bihar, including Siwan, West Champaran, Darbhanga, Purnia, and Kishanganj demonstrate that the dispersion of arsenic is not confined to a 10 km radius around the river Ganges [79]. The groundwater arsenic-tainted community block is around 19 kilometers from Siwan's closest river, Ghaghara. Nautan, a village in West Champaran that is arsenic contaminated, is situated about 6 km from the Gandak River. The Ganges River, a perennial transboundary river that rises in the Tibetan Plateau, passes the Himalayas in Nepal, empties into the Sarda River near Brahmaghat and connects Bihar and Uttar Pradesh via the Siwan, has as its second-longest tributary the Ghagara River. This river basin has been identified as the most arsenic-affected region in Uttar Pradesh, a border state of Bihar [80]. The Ghaghara River basin might be responsible for Siwan's increased arsenic levels. The Gandak River, on the other hand, has its beginnings in Tibet, close to Nepal's border in the Himalayas. The river flows through West Champaran before entering Indian territory close to river Tribeni in Nepal, forming the border between Uttar Pradesh and Bihar. It then merges with the Ganges at Hajipur in Bihar. In recent research, the district of Muzaffarpur, which is situated along the Budhi Gandak River, was found to have arsenic concentrations over the BIS standard [81]. The origin of this river is West Champaran in Bihar, and it flows through Muzaffarpur and ultimately joins the river Ganges near the Khagaria district of Bihar [82].

Since the Ganges River is far from these locations, it is still too early to pinpoint their true sources of arsenic contamination. Therefore, a thorough investigation of arsenic-affected river basins and groundwater sources close to other rivers is required. For a better understanding of the genesis and distribution of arsenic in Bihar, hydrogeochemical analysis of shallow and deep aquifers and studies of surface soils, especially biogeochemical properties, are essential. Groundwater arsenic poisoning has thus far been looked into in 87 out of 532 community development blocks, or a total of 22 out of 38 districts. All of the examined sources in three community blocks had groundwater arsenic values below the testing method's detection limit, making them safe for consumption and other household works [83]. Twelve blocks of the district have arsenic values below the WHO guideline. Ten blocks had arsenic concentrations above the WHO standard but equal to or below the BIS standard [84, 85]. In this study, more than 50 blocks had arsenic concentrations above the BIS standard of 50 μg/L. Overall, the number of groundwater arsenic-contaminated blocks were too high. A total of 72 blocks were contaminated with arsenic levels in drinking water exceeding the WHO standard of 10 μg/L [86, 87, 88, 89, 90] (Figure 16.2).

It is tough to determine whether the whole studied area or only a part of the study area is affected with groundwater arsenic contamination. Previous research clearly indicates that groundwater arsenic contamination is most prevalent in rural areas, with more than 10 million people living in the arsenic-contaminated rural areas described (Table 16.1). Among the affected areas, more than 50% of the blocks in Khagaria were contaminated with arsenic. In Bhojpur, Munger, Patna, and Samastipur, more than 40% of the areas were contaminated with arsenic.

Figure 16.2 Arsenic map of Bihar. [*Source:* 55].

Table 16.1 Showing arsenic-affected regions.

Districts	Total no. of blocks	Arsenic-affected blocks above the WHO standard (%)	Total population[a]	Rural population[a]
Bhojpur	14	6(43%)	1 191 190	932 672
Bhagalpur	16	6(38%)	1 414 717	1 000 577
Begusarai	18	6(33%)	1 230 641	1 123 018
Buxar	11	4(36%)	608 512	525 344
Darbhanga	18	1(5%)	233 029	233 029
Khagaria	7	4(57%)	805 292	728 965
Kishanganj	7	2(29%)	391 645	277 937
Katihar	16	6(37%)	685 151	663 348
Lakhisarai	6	2(33%)	292 985	215 110
Munger	9	4(44%)	675 935	390 902
Patna	23	10(43%)	1 654 345	1 170 270
Purnia	14	2(14%)	494 058	296 847
Saran	20	4(20%)	790 229	516 140
Siwan	19	1(5%)	272 509	162 590
Samastipur	20	8(40%)	1 145 061	1 131 292
Vaishali	17	5(29%)	922 037	802 625
West Champaran	18	1(6%)	175 938	175 938
Total	253	72(33%)	12 983 274	10 346 604

Other districts showing as the least affected by arsenic pollution were Darbhanga, Siwan, and West Champaran. In the remaining districts, 14% to 38% of the territories were contaminated with arsenic. It should be noted that in Table 16.1, blocks contaminated with arsenic should not be considered as 100% contaminated. Community vulnerability is significantly shaped by the area's socioeconomic situation and biophysical state. The ability of a system to manage social, economic, and/or environmental effects is referred to as adaptive capacity. [91] reported details of these perceptions and arsenic susceptibility in Bihar. The most vulnerable districts in terms of socioeconomic and biophysical factors were Vaishali, Samastipur, Khagaria, and Katihar.

16.2 Status of Groundwater Arsenic Exposure in the Affected Population

Groundwater contaminated with arsenic is turning out to be a grave public health alarm in several parts of the world due to its unavoidable health risks. The health effects of arsenic completely depend upon the type and route of exposure from different arsenic compounds. It is a well-established fact that the inorganic form of arsenic is more toxic than the organic form of arsenic and out of it arsenite an inorganic arsenic [As^{III}] is than arsenate a less toxic inorganic arsenic [As^{V}]. The toxicity of arsenic also depends upon the form of availability. For instance, monomethylarsonic acid (MMA) and dimethylarsinic acid (DMA) as final arsenic metabolites are less toxic than inorganic arsenic [92, 93, 94]. On absorption, arsenic is stored in the liver, kidney, heart, and lungs. A lower amount of arsenic is observed in muscles and neuronal tissues [95]. The accumulation of arsenic in these tissues is associated with many disorders [96]. Acute arsenic poisoning is always related with some common symptoms like nausea, vomiting, abdominal pain, and severe diarrhea. The long-term exposure of arsenic ingestion through contaminated groundwater results in the accretion of arsenic in the vital organs and tissues, which ultimately gives rise to the inception of atherosclerosis, hypertension, ischemic heart diseases, diabetes, hepatotoxicity, nephrotoxicity, and cancer of the skin, bladder, liver, and lungs [97]. It has also been classified as a class I human carcinogen by several global agencies, including the International Agency for Research on Cancer (IARC1987).

The population that is at risk of injury from arsenic is exceptionally large, since groundwater is the main source of drinking water and makes up more than 80% of the drinking sources in rural Bihar. Prior to the 1980s, open well water was considered a reliable supply of drinking water [98]. Arsenic pollution in the Gangetic plains, however, has multiplied recently as a result of increased human activity and geogenic factors. Both natural and artificial forces are to blame for this. In the Indian state of Bihar, more than 10 million people are thought to drink water that has more arsenic than the ten micrograms per liter WHO/BIS permissible standard [99]. It was first stated that there was a problem with arsenic poisoning in groundwater in the hamlet of Simaria Ojha patti, which is situated in the district of Bhojpur in the state of Bihar [100]. The bulk of the community's citizens were compelled to leave their houses since the exposed population was severely harmed. In addition to other harmful health effects like skin manifestations, gastrointestinal disorders, neurological effects, hormone disruption, and infertility, drinking water contaminated with arsenic is the leading cause of cancers of the skin, lung, bladder, and kidneys [101]. These conditions together pose a threat to the health of people all over the world. Arsenic may enter the body by drinking, where it passes through the digestive tract before entering the bloodstream. From there, it moves on to the vital bodily organs, where it interferes with their metabolic processes and ultimately leads to illness in those organs [102].

Various epidemiological studies have been conducted worldwide to understand the actual status of arsenic exposure in the arsenic-affected population [103]. The first footprint of arsenic contamination of groundwater was reported in 2002 in Ojha Patti village, the Shahpur block of the Bhojpur district in Bihar, where tube wells replaced dug wells about 20 years ago. The research article was published in 2003, entitled "Arsenic groundwater contamination in middle Ganga plain, Bihar, India: A future danger". The research work was carried out by School of Environmental Studies, Jadavpur University, Kolkata. They analyzed 206 tube wells for arsenic content, and they found 95% of the tested tube wells were arsenic contaminated above 50 ppb. Adults (390) and children (160) were included in this study for medical examination; 13% of the adults and 6.3% of the children had typical skin lesions indicating clear symptoms of arsenicosis. Traces of arsenic were also detected in urine, nails, and hair samples, which showed a positive correlation with elevated groundwater arsenic contamination. Arsenic-typical neuropathy was diagnosed in 63% of the adults. The overall health status of the people residing in this village was very poor [104].

In the last two decades there has been a sudden upsurge in the population affected by arsenic poisoning. A cross sectional study of 2000 cancer patients from the different geographical areas of Bihar were included to examine the increased incidences of cancer with prolonged arsenic exposure through various sources, including arsenic-contaminated groundwater and arsenic-contaminated food. The study exposed high blood arsenic levels in the cancer patients, which strongly correlated the relationship of arsenic with cancer, especially the carcinoma type which is more susceptible [105].

A study was conducted in the arsenic-exposed districts of Bihar, India, to evaluate the risk factors associated with arsenic-contaminated food over arsenic-contaminated drinking water. In this study, an increase in lifetime cancer risk was also predicted. The presence of inorganic arsenic in the food samples were much more elevated than the suggested levels. The highest concentration of arsenic was observed in cooked rice along with the drinking water. The lifetime cancer risk factors in the population consuming contaminated food were as high as those observed in the case of drinking water. The result of the study clearly indicated the increased lifetime cancer risk was nearly equal in both sources of exposure [106]. Epidemiological studies were performed after frequent reporting of different types of cancer like breast, renal, skin, and thyroid, including the common symptoms of arsenicosis from the Saran district of Bihar [107].

Economically weaker sections were more sufferer due to arsenic poisoning. Arsenic-contaminated drinking water creates an unwanted health burden on economically weaker sections of society. A study was conducted in two districts of the state Bihar to analyse the impact of arsenic contamination and its health burden. The incidence rate of arsenicosis was found to be higher among females than males and children. The annual wage loss, cost of treatment, and cost of illness created unwanted economic pressure [108, 109, 110].

The Chapar village of the Samastipur district, Bihar, was also targeted to understand arsenic-related health hazards. The health status of the exposed population was assessed and correlated with groundwater arsenic contamination and its presence in hair and urine samples. General health parameters related to arsenic poisoning were also included in this study. A huge probability of cancer in the exposed population of the Chapar village was predicted. The carcinogenic risk in the exposed population consuming arsenic-contaminated water was more than 99% for children, 95% for females, and 97% for male subjects [111]. A recent study was conducted in the state of Bihar to understand the rapid urban shift of arsenic poisoning together with the rural population. This study was conducted in the state capital of Bihar Patna [112, 113, 114,115, 116,117].

According to Globocan, there were more than 15 million new cancer diagnoses, approximately 9.5 million deaths, and nearly 43 million relapses (within five years of survival) in 2018. There were 3.9 million new cases of cancer in India in 2016 according to the National Cancer Registry Program of the Indian Council of Medical Research (ICMR). Uttar Pradesh had the most cancer cases overall. This was followed by Maharashtra and Bihar. Tamil Nadu registered the highest incidence of cancer in southern India, compared to Karnataka, Andhra Pradesh, Telangana, and Kerala cases. The alarming increase in cancer cases has been a major challenge for the authorities in the state of Bihar. Despite extensive investigation, the exact cause of cancer cases in this area remains unclear [118].

Apart from exposure through drinking water in Bihar, India, a recent study investigated exposure through food. The levels of arsenic detected were very high and contributed to the high risk of cancer in the region. The inhabitants of Bihar, India, who were tested for inorganic arsenic (iAs) exposure found that their overall iAs exposure from food was virtually comparable to that from water, as assessed by measuring iAs intake through the consumption of rice, wheat, and potato. It appeared that a sizable proportion of the drinking waters consumed by the population under study may have had arsenic concentrations that were lower than the WHO guideline value of 10 g/L, as 77% of the households had an arsenic concentration in their drinking water that was lower than the guideline value, and 37% of the households were using some form of improved water for drinking. Contrary to earlier research, it was shown that when the arsenic content in drinking water was above 10 g/L, food contributed more than water by 6%. When iAs levels in drinking water were < 10 g/L, the median proportion of food to overall iAs exposure was 36%. This demonstrated that higher total As concentrations in food may occur from the presence of high As concentrations in water. In addition to the increased As on heating, naturally high total As levels in grains and potato may have also contributed to the food's greater total As concentration. Therefore, for a sizable fraction of the people studied, iAs exposure through food was as essential as, if not more important than, iAs exposure by drinking water. Because of the relatively high concentration of As in cooked rice, ingestion of this food type is a major contributor to total exposure to iAs. Although it is urgently necessary to supply arsenic-free irrigation water for the growing of crops and vegetables in arsenic-endemic areas, India also has to pay more attention to food safety rules. Changing cooking practices at home, such as using low-arsenic water, boiling rice in extra water, and kneading wheat flour in low-arsenic water, may help reduce arsenic intake through food [119, 120, 121, 122, 123, 124,125, 126,127,128,129,130].

16.2.1 Mitigation Status in the Arsenic-Exposed Area of Bihar

The safe supply of water is essential to a country's progress and a basic human right worldwide. Arsenic contamination in drinking water is predominantly a public health problem. Therefore, understanding the public health dimension of the problem is an essential prerequisite to a successful arsenic mitigation program. Mitigation measures have included tube-well testing and replacing; usage of deeper wells; surface water preservation and treatment; use of sanitary dug wells, river sand, and pond sand filters; rainwater collection and storage; household-scale and large-scale arsenic filtrations; and rural pipeline water supply installation. Apart from that, shallow tube-well installation has been discouraged and efforts have been made to increase people's awareness toward groundwater arsenic contamination and its health consequences (Milton et al., 2012). In an arsenic hot spot of south-eastern Bangladesh, arsenic removal filters (ARFs), surface water filter membranes, pond sand filters (PSFs), rainwater harvesters (RWHs), and As-safe tube wells were distributed among the severely exposed population to compare the different mitigation approaches and their sustainability. Arsenic-safe tube wells were found to be a widely accepted option, mainly because of their easy operation and the availability of water, good water quality, and negligible maintenance (Hossain et al., 2015). One of the ready solutions of arsenic elimination is reduction at source and providing an alternate source of drinking water. The complete reduction process can be done in two ways: (i) alternative arsenic-free water source and (ii) treatment of arsenic-contaminated water. Alternative sources include groundwater deep tube wells, dug wells, infiltration galleys, surface waters, and rainwater harvesting. Some treatments have been used to mitigate arsenic. Solar oxidation of arsenic kept in transparent bottles efficiently reduced the arsenic concentration by increased oxidation from As(III+) to As(V+). Another effective way to remove arsenic is through coprecipitation and adsorption processes and, subsequently, reduction of As(III+) to As(V+) by using coagulants [131].

In our recent intervention in the arsenic-exposed area of Bihar, we have installed five arsenic filter units, which regularly carry out monitoring of the water quality. This ensures trust in the exposed population that they are getting arsenic-free water [132, 133] (Figure 16.3).

16.2.2 Application of MOFs in Arsenic Removal from Groundwater

MOFs have been appealing a inordinate amount of attention in the past decade. It is, by definition, a class of crystalline porous materials comprising metal nodes (i.e., metalions or clusters) and organic linkers linked via coordination bonds [109 Furukawa et al., 2013, 134, 135]. It is a new class of hybrid porous material manufactured from organic linkers and inorganic metal (or metal-containing clusters). The frequently used metals for the building of MOFs include Ag(I), Cu(II), Mg(II), Zn(II), Co(II), Fe(III), Al(III), Ln(III), Zr(IV), and so on, which can adopt various coordination geometries, such as trigonal bipyramidal, pyramidal, square, tetrahedral, and octahedral. However, the hydrothermal permanency of MOFs remains a challenge, as most

Figure 16.3 Mitigation unit installed in the Buxar district of Bihar. [*Source:* 67]

MOFs are sensitive to water. Recently, some water-stable MOFs have been developed and used for heavy-metal ion decontamination [136, 137]. MOFs are considered to be a class of materials holding excellent features like chemical constancy, high porosity, multiple functional groups, and large superficial areas. These properties can be effectively channelized to make MOFs excellent adsorbents for the removal of arsenic from contaminated water and make it drinkable [138,139]. The first study on water-stable MOFs for the removal of aquatic arsenic was done by Wang C et al. in 2015. In their study, water-stable zirconium MOFs (UiO-66) were synthesized and applied as adsorbents to remove aquatic arsenic contamination. The as synthesized UiO-66 adsorbent functions excellently across a broad pH range of 1 to 10 and achieves a remarkable arsenate uptake capacity of 303 mg/g at the optimal pH, i.e., pH = 2. Several other metals have also been used in the past in the form of MOFs to eliminate groundwater arsenic. Iron (Fe) is one of them, and several studies reflect that such a type of study advocates for the development of technologies like MOFs as a future mitigation approach for arsenic and other heavy metals.

16.2.3 Conclusion

The exposure of groundwater arsenic poisoning causes a lot of health hazards in bigger rural populations in the middle Ganga plains of Bihar. Exposed populations are exhibiting various health consequences including typical symptoms of arsenicosis and even the disease of cancer. Therefore, there is a need for mitigation intervention at a large scale, and the state government needs to make urgent health policies for the exposed population.

References

1 Murcott, S. (2012). *Arsenic Contamination in the World: An International Sourcebook 2012*. IWA Publishing.

2 Naujokas, M.F., Anderson, B., Ahsan, H. et al. (2013). The broad scope of health effects from chronic arsenic exposure: update on a worldwide public health problem. *Environmental Health Perspectives* 121: 295–302.

3 WHO (2001). *Arsenic and Arsenic Compounds (Environmental Health Criteria 224)*, 2e. Geneva: World Health Organization International Programme on Chemical Safety.

4 WHO (World Health Organization) (2011). *Arsenic in Drinking-Water. Background Document for Development of WHO Guidelines for Drinking-Water Quality*. Geneva: World Health Organization.

5 World Health Organization (2018). Available online at: http://www.who.int/news-room/fact-sheets/detail/arsenic (accessed September 20, 2019).

6 Hassan, M.M. (2018). *Arsenic in Groundwater: Poisoning and Risk Assessment*, 1e. Boca Raton: Taylor & Francis. CRC title. ISBN 9781439839270.

7 Acharyya, S.K., Lahiri, S., Raymahashay, B.C., and Bhowmik, A. (2000) Arsenic toxicity of groundwater in parts of the Bengal basin in India and Bangladesh: the role of Quaternary stratigraphy and Holocene.

8 Chakraborti, D., Sengupta, M.K., Rahman, M.M. et al. (2004). Groundwater arsenic contamination and its health effects in the Ganga-Meghna-Brahmaputra plain. *Journal of Environmental Monitoring* 6 (6): 74N–83N.

9 Chakraborti, D., Rahman, M.M., Das, B., et al. (2013). Groundwater arsenic contamination in Ganga–Meghna–Brahmaputra plain, its health effects and an approach for mitigation. *Environmental Earth Sciences* **70**: 1993-2008.

10 Ahamed, S., Sengupta, M.K., Mukherjee, A. et al. (2006). Arsenic groundwater contamination and its health effects in the state of Uttar Pradesh (UP) in upper and middle Ganga plain, India: a severe danger. *Science of the Total Environment* 370 (2–3): 310–322.

11 Katiyar, S. and Singh, D. (2014). Prevalence of arsenic exposure in population of Ballia district from drinking water and its correlation with blood arsenic level. *Journal of Environmental Biology* 35 (3): 589–594.

12 Yunus, F.M., Khan, S., Chowdhury, P. et al. (2016). A review of groundwater arsenic contamination in bangladesh: the millennium development goal era and beyond. Naidu R, Rahman MM, eds *International Journal of Environmental Research and Public Health* 13 (2): 215.

13 Bhattacharya, P., Welch, A.H., Stollenwerk, K.G. et al. (2007). Arsenic in the environment: biology and chemistry. *Science of the Total Environment* 379 (2–3): 109–120.

14 Das, B., Rahman, M.M., Nayak, B. et al. (2009). Groundwater arsenic contamination, its health effects and approach for mitigation in West Bengal, India and Bangladesh. *Water Quality, Exposure and Health* 1 (1): 5–21.

15 Amaya, A. (2002). Arsenic in groundwater of alluvial aquifers in Nepal, extent of contamination, genesis and aspects of remediation in the districts of Nawalparasi and Kathmandu. *TRITA-LWR Master Thesis, Department Land and Water Resource Engineering*, Stockholm.
16 Smedley, P.L. and Kinniburgh, D.G. (2002). A review of the source, behaviour and distribution of arsenic in natural waters. *Applied Geochemistry* 17 (5): 517–568.
17 Saha, D., Upadhyay, S., Dhar, Y.R., and Singh, R. (2007). The aquifer system and evaluation of its hydraulic parameters in parts of South Ganga Plain, Bihar. *Journal-Geological Society of India* 69 (5): 1031.
18 Brouns, M., Janssen, M., and Wong, A. (2013). Dealing with arsenic in rural Bihar, India Evaluating the successes and failures of mitigation projects and providing a long-term mitigation strategy. Department of Technology Dynamics and Sustainable Development. Delft, the Netherlands: Delft University of Technology.
19 Chakraborti, D., Biswas, B.K., Basu, G.K. et al. (1999). Possible arsenic contamination free groundwater source in Bangladesh. *Journal of Surface Science and Technology* 15 (3/4): 180–188.
20 Datta, D.K., Gupta, L.P., and Subramanian, V. (1999). Distribution of C, N and P in the sediments of the Ganges–Brahmaputra–Meghna river system in the Bengal basin. *Organic Geochemistry* 30 (1): 75–82.
21 Nickson, R., McArthur, J., Burgess, W. et al. (1998). Arsenic poisoning of Bangladesh groundwater. *Nature* 395 (6700): 338-338.
22 Saha, D. (2009). Arsenic groundwater contamination in parts of middle Ganga plain, Bihar. *Current Science* 97 (6): 753–755.
23 Saha, D., Dwivedi, S., and Sahu, S. (2009). Arsenic in ground water in parts of middle ganga plain in Bihar-an appraisal. *Editorial Board* 24 (2&3).
24 Saha, D., Sahu, S., and Chandra, P.C. (2011). Arsenic-safe alternate aquifers and their hydraulic characteristics in contaminated areas of Middle Ganga Plain, Eastern India. *Environmental Monitoring and Assessment* 175 (1): 331–348.
25 Ghosh, A., Singh, S.K., Bose, N., and Chowdhary, S. (2007 August). Arsenic contaminated aquifers: a study of the Ganga levee zone in Bihar, India. In: *Symposium on Arsenic: The Geography of a Global Problem*, London: Royal Geographical Society.
26 Ghosh, J., Das, J., Manna, P., and Sil, P.C. (2009). Taurine prevents arsenic-induced cardiac oxidative stress and apoptotic damage: role of NF-κB, p38 and JNK MAPK pathway. *Toxicology and Applied Pharmacology* 240 (1): 73–87.
27 Goodbred, S.L., Jr (2003). Response of the Ganges dispersal system to climate change: a source-to-sink view since the last interstade. *Sedimentary Geology* 162 (1–2): 83–104.
28 Hossain, M.S., Ahmed, F., Abdullah, A.T.M. et al. (2015). Public health risk assessment of heavy metal uptake by vegetables grown at a waste-water-irrigated site in Dhaka, Bangladesh. *Journal of health & pollution* 5 (9): 78–85. doi: 10.5696/2156-9614-5-9.78.
29 Opar, A., Pfaff, A., Seddique, A.A. et al. (2007). Responses of 6500 households to arsenic mitigation in Araihazar, Bangladesh. *Health & Place* 13 (1): 164–172.
30 Phan, K., Sthiannopkao, S., Kim, K.W. et al. (2010). Health risk assessment of inorganic arsenic intake of Cambodia residents through groundwater drinking pathway. *Water Research* 44 (19): 5777–5788.

31 Sastry, K.V. and Gupta, P.K. (1978). Histopathological and enzymological studies on the effects of chronic lead nitrate intoxication in the digestive system of a freshwater teleost, Channa punctatus. *Environmental Research* 17 (3): 472–479.

32 Singh, S.K. (2011). *Arsenic Contamination in Water, Soil, and Food Materials in Bihar*. Germany: LAP LAMBERT Academic Publishing.

33 Smedley, P. (2005). Arsenic occurrence in groundwater in South and East Asia.

34 Kumar, A., Rahman, M.S., Iqubal, M.A. et al. (2016). Ground water Arsenic contamination: a local survey in India. *International Journal of Preventive Medicine* 7: 100. doi: 10.4103/2008-7802.188085.

35 Chakraborti, D., Mukherjee, S.C., Pati, S. et al. (2003). Arsenic groundwater contamination in Middle Ganga Plain, Bihar, India: a future danger? *Environmental Health Perspectives* 111 (9): 1194–1201. doi: 10.1289/ehp.5966.

36 Singh, S.K. and Ghosh, A.K. (2012). Health risk assessment due to groundwater arsenic contamination: children are at high risk. *Human and Ecological Risk Assessment: An International Journal* 18 (4): 751–766.

37 Prakash, C., Soni, M., and Kumar, V. (2016). Mitochondrial oxidative stress and dysfunction in arsenic neurotoxicity: a review. *Journal of Applied Toxicology* 36 (2): 179–188.

38 Abhinav, S., Navin, S., Verma, S.K. et al. (2016, June). Groundwater and blood samples assessment for arsenic toxicity in rural population of Darbhanga district of Bihar, India. In *Arsenic research and global sustainability-proceedings of the 6th international congress on arsenic in the environment*, 419–420. London: CRC Press, Taylor & Francis.

39 Haque, R., Mazumder, D.N., Samanta, S. et al. (2003). Arsenic in drinking water and skin lesions: dose-response data from West Bengal, India. Epidemiology (Cambridge, Mass.) 14 (2): 174–182. doi: 10.1097/ 01.EDE.0000040361.55051.54.

40 Kumar, A., Rahman, M.S., Kumar, R. et al. (2019a). Arsenic contamination in groundwater causing impaired memory and intelligence in school children of Simri village of Buxar district of Bihar. *Journal of Mental Health and Human Behaviour* 24: 132–138. doi: 10.4103/jmhhb.jmhhb_31_18.

41 Kumar, A., Ali, M., Kumar, R. et al. (2020a). High arsenic concentration in blood samples of people of village GyaspurMahaji, Patna, Bihar drinking arsenic contaminated water. *Springer Nature J Expos Health* 12: 131–140. doi: 10.1007/ s12403-018-00294-5.

42 Kumar, A., Ali, M., Kumar, R. et al. (2021a). Arsenic exposure in Indo Gangetic plains of Bihar causing increased cancer risk. *Scientific Reports* 11 (1): 2376. doi: 10.1038/s41598-021-81579-9.

43 Kumar, A. and Ghosh, A.K. (2021b). Assessment of arsenic contamination in groundwater and affected population of Bihar. In: *Arsenic Toxicity: Challenges and Solutions* (ed. N. Kumar). [ISBN 978-981-33-6067-9; ISBN 978-981-33-6068-6 (eBook)]. doi: 10.1007/978-981-33-6068-6_7.

44 Kumar, A., Kumar, R., Rahman, M.S. et al. (2021c). Assessment of arsenic exposure in the population of Sabalpur village of Saran District of Bihar with mitigation approach. *Environmental Science and Pollution Research International* 28 (32): 43923–43934. doi: 10.1007/s11356-021-13521-5. Advance online publication. doi: 10.1007/s11356-021-13521-5.

45 Kumar, A., Rahman, M.S., Ali, M. et al. (2021d). Assessment of arsenic exposure and its mitigation intervention in severely exposed population of Buxar district of Bihar, India. *Toxicology and Environmental Health Sciences* doi: 10.1007/s13530-021-00086-6.

46 von Ehrenstein, O.S., Poddar, S., Yuan, Y. et al. (2007). Children's intellectual function in relation to arsenic exposure. *Epidemiology (Cambridge, Mass.)* 18 (1): 44–51. doi: 10.1097/01.ede.0000248900.65613.a9.

47 Rodrıguez, V.M., Jiménez-Capdeville, M.E., and Giordano, M. (2003). The effects of arsenic exposure on the nervous system. *Toxicology Letters* 145 (1): 1–18.

48 Singh, A.P., Goel, R.K., and Kaur, T. (2011). Mechanisms pertaining to arsenic toxicity. *Toxicology International* 18 (2): 87.

49 Singh, A.M.R.I.T.A. and Choudhary, S.K. (2010). Arsenic in groundwater in five villages under Nathnagar block of Bhagalpur District, Bihar. *Ecoscan* 4 (2–3): 213–216.

50 Bhatia, S., Balamurugan, G., and Baranwal, A. (2014). High arsenic contamination in drinking water hand-pumps in Khap Tola, West Champaran, Bihar, India. *Frontiers in Environmental Science* 2: 49.

51 Singh, S.P. (2014). Spatial relationship of various parameters in drinking water in Siwan town of Bihar (India) with special emphasis on Arsenic contamination in groundwater. *Journal Chemical Science Review and Letters* 2 (7): 588–595.

52 Gupta, V., Singh, J., Kumari, P. et al. (2014). Arsenic in hand pump water of different blocks of Samastipur district (Bihar), India. *International Journal of Basic and Applied Science Research* 1 (1): 1–5.

53 Yasunori, Y., Kenichi, I., Akihiko, K. et al. (2012). Arsenic polluted groundwater and its countermeasures in the middle basin of the Ganges, Uttar Pradesh State, India. *Journal of Environmental Protection* 2012.

54 Mumtazuddin, S., Azad, A.K., Bharti, P., and Ranjan, R. (2012). Physico-chemical analysis of groundwater of the BudhiGandak belt in Muzaffarpur district, India. *International Research Journal of Environment Sciences* 1 (1): 7–11.

55 Nickson, R., Sengupta, C., Mitra, P. et al. (2007). Current knowledge on the distribution of arsenic in groundwater in five states of India. *Journal of Environmental Science and Health, Part A* 42 (12): 1707–1718.

56 Kumar, A., Rahman, M.S., Ali, M. et al. (2022). Assessment of disease burden in the arsenic exposed population of Chapar village of Samastipur district, Bihar, India, and related mitigation initiative. *Environmental Science and Pollution Research* 29: 27443–27459. doi: 10.1007/s11356-021-18207-6.

57 Singh, S.K. and Vedwan, N. (2015). Mapping composite vulnerability to groundwater arsenic contamination: an analytical framework and a case study in India. *Natural Hazards* 75 (2): 1883–1908.

58 Domingo, J.L. (1995). Prevention by chelating agents of metal-induced developmental toxicity. *Reproductive Toxicology (Elmsford, N.Y.)* 9 (2): 105–113. doi: 10.1016/0890-6238(94)00060-3.

59 Petrick, J.S., Ayala-Fierro, F., Cullen, W.R. et al. (2000). *Monomethylarsonous acid (MMAIII) is more toxic than arsenite in chang human hepatocytes*, 163 (2),203–207. doi:10.1006/taap.1999.8872

60 Shankar, S. and Shanker, U. undefined Shikha (2014). Arsenic contamination of groundwater: a review of sources, prevalence, health risks, and strategies for mitigation. *The Scientific World Journal* 2014: Article ID 304524. 18. doi: 10.1155/2014/304524.

61 Klassen, C.D. (1996). Heavy metals and heavy metal antagonists. In: *Goodman and Gilman's the Pharmacological Basis of Therapeutics* (ed. J.G. Hardman, A.G. Gilman, and L.E. Limbird), 1649–1672. New York: McGraw-Hill.

62 Ratnaike, R.N. (2003). Acute and chronic arsenic toxicity. *Postgraduate Medical Journal* 79 (933): 391–396. doi: 10.1136/pmj.79.933.391.

63 Singh, S.K., Ghosh, A.K., Kumar, A. et al. (2014). Groundwater arsenic contamination and associated health risks in Bihar, India. *International Journal of Environmental Research* 8 (1).

64 IARC (2012). Arsenic, metals, fibres, and dusts IARC monographs on the evaluation of carcinogenic risks to humans. *Iarc Monographs* 100: (Arsenic,metals, fibres, and dusts).

65 Rodrigues, J.L., Batista, B.L., Nunes, J.A. et al. (2008). Evaluation of the use of human hair for biomonitoring the deficiency of essential and exposure to toxic elements. *Science of the Total Environment* 405 (1–3): doi: 10.1016/j.scitotenv.2008.06.002.

66 Argos, M., Ahsan, H., and Graziano, J.H. (2012). Arsenic and human health: epidemiologic progress and public health implications. *Reviews on Environmental Health* 27 (4): 191–195. doi: 10.1515/reveh-2012-0021.

67 Chakraborti, D., Mukherjee, S.C., Pati, S. et al. (2003). Arsenic groundwater contamination in Middle Ganga Plain, Bihar, India: a future danger? *Environmental Health Perspectives* 111 (9): 1194–1201.

68 Kumar, A., Ali, M., Kumar, R. et al. (2021). Arsenic exposure in Indo Gangetic plains of Bihar causing increased cancer risk. *Scientific Reports* 11 (1): 1–16. doi: 10.1038/s41598-021-81579-9.

69 Mondal, D., Rahman, M.M., Suman, S. et al. (2021). Arsenic exposure from food exceeds that from drinking water in endemic area of Bihar, India. *The Science of the Total Environment* 754: 142082. doi: 10.1016/j.scitotenv.2020.142082.

70 Thakur, B.K. and Gupta, V. (2019). Valuing health damages due to groundwater arsenic contamination in Bihar, India. *Economics & Human Biology* 35: 123-1321570-677X. doi: 10.1016/j.ehb.2019.06.005.

71 Richards, L.A., Kumari, R., Parashar, N. et al. (2022). Environmental tracers and groundwater residence time indicators reveal controls of arsenic accumulation rates beneath a rapidly developing urban area in Patna, India. *Journal of Contaminant Hydrology* 249: 104043. ISSN 0169-7722. doi: 10.1016/j.jconhyd.2022.104043.

72 Sharma, D.C. (2016). Cancer data in India show new patterns. *The Lancet. Oncology* 17 (7): doi: 10.1016/S1470-2045(16)30202-9.

73 Milton, A. H., Hore, S. K., Hossain, M. Z., & Rahman, M. (2012). Bangladesh arsenic mitigation programs: lessons from the past. Emerging health threats journal, 5, 10.3402/ehtj.v5i0.7269. https://doi.org/10.3402/ehtj.v5i0.7269.

74 Mondal, D., Banerjee, M., Kundu, M. et al. (2010). Comparison of drinking water, raw rice and cooking of rice as arsenic exposure routes in three contrasting areas of West Bengal, India. *Environmental Geochemistry and Health* 32 (6): doi: 10.1007/s10653-010-9319-5.

75 Mondal, D. and Polya, D.A. (2008). Rice is a major exposure route for arsenic in Chakdaha block, Nadia district, West Bengal, India: a probabilistic risk assessment. *Applied Geochemistry* 23 (11): doi: 10.1016/j.apgeochem.2008.06.025.

76 Mwale, T., Rahman, M.M., and Mondal, D. (2018). Risk and benefit of different cooking methods on essential elements and arsenic in rice. *International Journal of Environmental Research and Public Health* 15 (6): doi: 10.3390/ijerph15061056.

77 Raab, A., Baskaran, C., Feldmann, J., and Meharg, A.A. (2009). Cooking rice in a high water to rice ratio reduces inorganic arsenic content. *Journal of Environmental Monitoring* 11 (1): doi: 10.1039/b816906c.

78 Sanyal, T., Bhattacharjee, P., Paul, S., and Bhattacharjee, P. (2020). Recent advances in arsenic research: significance of differential susceptibility and sustainable strategies for mitigation. *Frontiers in Public Health* 8: 464. doi: 10.3389/fpubh.2020.00464.

79 Rowsell, J.L.C. and Yaghi, O.M. (2004). Metal-organic frameworks: a new class of porous materials. *Microporous and Mesoporous Materials* 73 (1–2): 3–14.

80 Zhou, H.C., Long, J.R., and Yaghi, O.M. (2012). Introduction to metal-organic frameworks. *Chemical Reviews* 112 (2): 673–674.

81 Wang, C. ,Liu, X., Chen, J. P. et al. (2015). Superior removal of arsenic from water with zirconium metal-organic framework UiO-66. *Scientific Reports* 5: 16613. doi: 10.1038/srep16613.

82 Xiang, S., He, Y., Zhang, Z. et al. (2012). Microporous metal-organic framework with potential for carbon dioxide capture at ambient conditions. *Nature Communications* 3: 954 doi: 10.1038/ncomms1956.

83 Singh, S.K. (2015). Groundwater arsenic contamination in the Middle-Gangetic Plain, Bihar (India): the danger arrived. *Ernational Research Journal of Environmental Sciences* 4 (2): 70–76.

84 Rajasekaran, S., Devi, K.R.S., Pinheiro, D. et al. (2022). Metal organic frameworks to remove arsenic adsorption from wastewater. In: *Inorganic-Organic Composites for Water and Wastewater Treatment. Environmental Footprints and Eco-design of Products and Processes* (ed. E. Lichtfouse, S.S. Muthu, and A. Khadir). Singapore: Springer. doi: 10.1007/978-981-16-5928-7_1.

85 Mukherji, A., Rawat, S., and Shah, T. (2013). Major insights from India's minor irrigation censuses: 1986-87 to 2006-07. *Economic and Political Weekly*, 115–124.

86 Government of India (2012). Census of India 2011: Houses, household amenities and assets. Latrine facility.

87 Kumar, M., Rahman, M.M., Ramanathan, A.L., and Naidu, R. (2016). Arsenic and other elements in drinking water and dietary components from the middle Gangetic plain of Bihar, India: health risk index. *Science of the Total Environment* 539: 125–134.

88 Thakur, M., Rachamalla, M., Niyogi, S. et al. (2021). Molecular mechanism of arsenic-induced neurotoxicity including neuronal dysfunctions. *International Journal of Molecular Sciences* 22 (18): doi: 10.3390/ijms221810077.

89 Chakraborti, D., Rahman, M.M., Ahamed, S. et al. (2016). Arsenic groundwater contamination and its health effects in Patna district (capital of Bihar) in the middle Ganga plain, India. *Chemosphere* 152: 520–529. doi: 10.1016/J.CHEMOSPHERE.2016.02.119.

90 Kumar, A. and Ghosh, A.K. (2021). Assessment of arsenic contamination in groundwater and affected population of Bihar. In Arsenic Toxicity: Challenges and Solutions. doi: 10.1007/978-981-33-6068-6_7.

91 Kumar, A., Kumar, R., Ali, M. et al. (2016 June). Groundwater arsenic poisoning in Buxar District, Bihar, India: health hazards. In: *Arsenic Research and Global Sustainability: Proceedings of the Sixth International Congress on Arsenic in the Environment (As2016), June 19- 23,2016, Stockholm, Sweden*, 378. CRC Press.

92 Kumar, A., Ali, M., Rahman, S.M. et al. (2015). Ground water arsenic poisoning in "Tilak Rai Ka Hatta" village of Buxardistrict, Bihar, India causing severe health hazards and hormonal imbalance. *Journal of Environmental & Analytical Toxicology* 5 (4): 1–7.

93 International Agency for Research on Cancer (IARC) (1987). *IARC Monograph on Arsenic and Arsenic Compounds*. 7. Lyon, France: IARC. 100.

94 Mukherjee, A., Fryar, A.E., Eastridge, E.M. et al. (2018). Controls on high and low groundwater arsenic on the opposite banks of the lower reaches of River Ganges, Bengal basin, India. *Science of the Total Environment* 645: doi: 10.1016/j.scitotenv.2018.06.376.

95 Kumar, A., Kumar, R., Ali, M. et al. (2016). Groundwater arsenic poisoning in Buxar district, Bihar, India: health Hazards. *Arsenic Research and Global Sustainability – Proceedings of the 6th International Congress on Arsenic in the Environment, AS 2016* doi: 10.1201/b20466-178.

96 Verma, S. and Sinha, A. (2022). Appraisal of groundwater arsenic on opposite banks of River Ganges, West Bengal, India, and quantification of cancer risk using Monte Carlo simulations. *Environmental Science and Pollution Research* doi: 10.1007/s11356-021-17902-8.

97 Rodriguez-Lado, L., Sun, G., Berg, M. et al. (2013). Groundwater arsenic contamination throughout China. *Science (New York, N.Y.)* 341 (6148): 866–868. doi: 10.1126/science.1237484.

98 Durmusoglu, E., Taspinar, F., and Karademir, A. (2010). Health risk assessment of BTEX emissions in the landfill environment. *Journal of Hazardous Materials* 176 (1–3): doi: 10.1016/j.jhazmat.2009.11.117.

99 Meena, R.S. and Jha, R. (2022). Flood inundation modeling using coupled 1D–2D HEC-RAS model in lower Kosi River Basin, India with limited data. In: *Recent Trends in River Corridor Management: Select Proceedings of RCRM 2021*, 177–188. Singapore: Springer Nature Singapore.

100 Obiri, S., Dodoo, D.K., Okai-Sam, F., and Essumang, D.K. (2006). Cancer health risk assessment of exposure to arsenic by workers of AngloGold Ashanti-Obuasi gold mine. *Bulletin of Environmental Contamination and Toxicology* 76 (2): doi: 10.1007/s00128-006-0907-0.

101 Shaji, E., Santosh, M., Sarath, K.V. et al. (2021). Arsenic contamination of groundwater: a global synopsis with focus on the Indian Peninsula. *GeoscienceFrontiers* 12 (3): 101079. doi: 10.1016/J.GSF.2020.08.015.

102 Tarafdar, A. and Sinha, A. (2017). Cancer risk assessment of polycyclic aromatic hydrocarbons in the soils and sediments of India: a meta-analysis. *Environmental Management* 60 (4): doi: 10.1007/s00267-017-0920-6.

103 Kumar, A., Rahman, M.S., Iqubal, M.A. et al. (2016). Ground water arsenic contamination: a local survey in India. *International Journal of Preventive Medicine* 7.

104 Chakraborti, D., Ahamed, S., Rahman, M.M. et al. (2004). Risk of arsenic contamination in groundwater: response from Chakraborti et al. *Environmental Health Perspectives* 112 (1): A20–A20.

105 sea-level fluctuation *Environmental Geology* 39: 1127–1137.

106 Kumar, A., Ali, M., Kumar, R. et al. (2021). Arsenic exposure in Indo Gangetic plains of Bihar causing increased cancer risk. *Scientific Reports* 11: 2376. doi: 10.1038/s41598-021-81579-9.

107 Kumar, A., Kumar, R., Rahman, M.S. et al. (2021c). Assessment of arsenic exposure in the population of Sabalpur village of Saran District of Bihar with mitigation approach. *Environmental Science and Pollution Research* 28: 43923–43934. doi: 10.1007/s11356-021-13521-5.

108 Furukawa, H., Gándara, F., Zhang, Y.B. et al. (2014). Water adsorption in porous metal-organic frameworks and related materials. *Journal of the American Chemical Society* 136 (11): 4369–4381.

Index

Metal Organic Frameworks for Wastewater Contaminant Removal, First Edition. Edited by Arun Lal Srivastav, Lata Rani, Jyotsna Kaushal, and Tien Duc Pham.

f

o

p

t